WORKS ISSUED BY
THE HAKLUYT SOCIETY

———

THE DISCOVERY OF RIVER GAMBRA
BY RICHARD JOBSON
(1623)

THIRD SERIES
NO. 2

HAKLUYT SOCIETY

Council and Officers 1998–1999

PRESIDENT
Mrs Sarah Tyacke CB

VICE PRESIDENTS

Lt Cdr A. C. F. David
Professor P. E. H. Hair
Professor John B. Hattendorf
Professor D. B. Quinn HON. FBA
Sir Harold Smedley KCMG MBE
M. F. Strachan CBE FRSE
Professor Glyndwr Williams

COUNCIL (with date of election)

Peter Barber (1995)
Professor R. C. Bridges (1998)
Dr Andrew S. Cook (1997)
Stephen Easton (co-opted)
Dr Felipe Fernández-Armesto (1998)
R. K. Headland (1998)
Francis C. Herbert (1996)
Bruce Hunter (1997)
Professor Wendy James (1995)
Jeffrey G. Kerr (1998)
James McDermott (1996)
Rear-Admiral R. O. Morris CB (1996)
Royal Geographical Society
  (Dr J. H. Hemming CMG)
A. N. Ryan (1998)
Dr John Smedley (1996)

TRUSTEES

Sir Geoffrey Ellerton CMG MBE
†H. H. L. Smith
G. H. Webb CMG OBE
Professor Glyndwr Williams

HONORARY TREASURER
David Darbyshire FCA

HONORARY SECRETARY
Anthony Payne
c/o Bernard Quaritch Ltd, 5–8 Lower John Street, Golden Square, London W1R 4AU

HONORARY SERIES EDITORS
Dr W. F. Ryan
Warburg Institute, University of London, Woburn Square, London WC1H 0AB

Professor Robin Law
Department of History, University of Stirling, Stirling FK9 4LA

ADMINISTRATIVE ASSISTANT
Mrs Fiona Easton
(to whom queries and application for membership may be made)
Telephone: 01986 788359   Fax: 01986 788181   E-mail: haksoc@paston.co.uk

Postal address only:
Hakluyt Society, c/o The Map Library, The British Library, 96 Euston Road,
London NW1 2DB

Website: www.hakluyt.com

Registered Charity No. 313168   VAT No. GB 233 4481 77

INTERNATIONAL REPRESENTATIVES

Australia: Ms Maura O'Connor, Curator of Maps, National Library of Australia, Canberra, ACT 2601

Canada: Dr Joyce Lorimer, Department of History, Wilfred Laurier University, Waterloo, Ontario, N2L 3C5

Germany: Thomas Tack, Ziegelbergstr. 21, D-63739 Aschaffenburg

Japan: Dr Derek Massarella, Faculty of Economics, Chuo University, Higashinakano 742–1, Hachioji-shi, Tokyo 192–03

New Zealand: J. E. Traue, Department of Librarianship, Victoria University of Wellington, PO Box 600, Wellington

Portugal: Dr Manuel Ramos, Av. Elias Garcia 187, 3Dt, 1050 Lisbon

Russia: Professor Alexei V. Postnikov, Institute of the History of Science and Technology, Russian Academy of Sciences, 1/5 Staropanskii per., Moscow 103012

South Africa: Dr F. R. Bradlow, 28/29 Porter House, Belmont Road, Rondebosch, Cape 7700

USA: Dr Norman Fiering, The John Carter Brown Library, Box 1894, Providence, Rhode Island 02912 *and* Professor Norman Thrower, Department of Geography, UCLA, 405 Hilgard Avenue, Los Angeles, California 90024–1698

Western Europe: Paul Putz, 54 rue Albert 1, 1117 Luxembourg, Luxembourg

River Gambia: the middle river, near Basse, evening, July 1979

THE DISCOVERY
OF RIVER GAMBRA
(1623)

BY RICHARD JOBSON

Edited, with additional material, by

DAVID P. GAMBLE
AND
P. E. H. HAIR

THE HAKLUYT SOCIETY
LONDON
1999

Published by the Hakluyt Society
c/o The Map Library
British Library, 96 Euston Road,
London NW1 2DB

SERIES EDITORS
W. F. RYAN
ROBIN LAW

© The Hakluyt Society 1999

ISBN 0 904180 64 6
ISSN 0072 9396

British Library Cataloguing-in-Publication Data
A catalogue record for this book is
available from the British Library

Typeset by Waveney Typesetters, Wymondham, Norfolk
Printed in Great Britain at
the University Press, Cambridge

CONTENTS

List of maps and illustrations	viii
Preface	ix

Introduction
The Sources	1
The 'Discovery' of River Gambia	4
The English and River Gambia	9
English trading voyages to Senegambia 1587–1621	9
The 1618 and 1619 English voyages to River Gambia	16
The English in River Gambia 1620–1621	24
Richard Jobson	38
The African Setting: The River	46
The African Setting: Polities and Peoples	55

Part I: Jobson's texts

The Golden Trade; or, A Discovery of the River Gambra (1623)	75
Jobson's 'Large Journall' (1620–1621)	185
Jobson's Petition (? 1626)	198
Appendix A. Jobson's Itinerary	207
Appendix B. The middle river base	223
Appendix C. Text of the Gerbier map, in translation	231
Appendix D. Account of the Upper River by Governor MacDonnell, 1848	234

Part II: Other early sources on River Gambia

Cadamosto (1455 and 1456 voyages)	244
Usodimare (1455 voyage)	256
Diogo Gomes (1455/1456 and (?)1458 voyages)	258
Duarte Pacheco Pereira (c.1508)	265
Valentim Fernandes (c.1508)	267
João de Barros (1552)	271
Francisco de Andrade (1582)	273
André Álvares de Almada (c.1594)	274
André Donelha (1625)	284
Francisco de Lemos Coelho (1669/1684)	291

Bibliography	313
Maps of River Gambia, 1468–1980	323
List of Marginal Notes (Sidenotes)	325
Index	331

LIST OF MAPS AND ILLUSTRATIONS

MAPS

Jobson's River Gambia 1620–1621	xiv/xv
Route Survey ['Upper Gambia' section], 1881	193
The 'Middle River'	222

COLOUR PLATES

River Gambia: the 'middle river', near Basse, 1979	iv
A Mandinka village: Kwinela in Kiang, 1966	xvi

ILLUSTRATION AND PLATES

'Draught of a Pholey Town and Plantation about'	102

between pages 311/312

I. Unloading salt carried from River Saalum to River Gambia, 1950
II. Landing place at Kau-ur, 1947
III. Swamp fishing, Wolof Panchang, 1947
IV. Boy studying, Fatoto, 1953
V. Teaching House, Kumbija, 1953
VI. Mandinka wife, recently married, Kerewaan, 1947
VII. Shed for newly-circumcised boys, Kundam, 1953
VIII. Newly-circumcised boys wearing traditional dress, Bunting, 1951
IX. Fula cattle with herdsman's hut, Saare Mansajang, 1953
X. Blacksmith at work, Njau, 1947
XI. Man playing *bolombato*, 1953
XII. Fiddler, 1953
XIII. Underarm drum, Njau, 1953

LINE DRAWINGS

Man fishing with a barbed spear	96
The game board	105
Traditional spears now used as ceremonial staffs by village heads	109
Woman pounding grain with mortar and pestle	117
Mandinka xylophone	152
Pipe smoking	161
Using the ridge hoe. The ridge hoe.	164
Basket hive	169

PREFACE

This is primarily an edition of three writings by Richard Jobson (fl. c.1620), as follows:

1. The Golden Trade: / OR, / A discovery of the River *Gambra*, and / the Golden Trade of the *Aethiopians*. / ALSO, / *The Commerce with a great blacke Merchant, called Buckor Sano, and his report of the* / *houses covered with Gold, and other strange* / *observations for the good of our* / *owne countrey;* / Set downe as they were collected in travelling part of / the yeares, 1620. and 1621. / by *Richard Jobson*, Gentleman. / London, / Printed by *Nicholas Okes*, and are to be sold by / Nicholas Bourne, *dwelling at the entrance* / of the Royall Exchange, 1623.

   STC 14623

2. 'A true Relation of Master Richard Jobsons Voyage, employed by Sir William Saint John, Knight, and others; for the Discoverie of Gambra, in the Sion, a ship of two hundred tuns, Admirall; and the Saint John fiftie, Vice-Admirall. In which they passed nine hundred and sixtie miles up the River into the Continent. Extracted out of his large Journall', in Samuel Purchas, *Purchas his Pilgrimes* … (4 vols, London, 1625), Part I, Book 7, Chapter 1, pp. 921–6.

3. 'The discovery of the Cuntry of Kinge Solomon his rich trade and trafique within Twenty daies saile of England …', undated manuscript, signed Richard Jobson, British Library, Royal MS L8 A LVIII, 295, 5 ff.

In Part II of the edition, extracts relating to River Gambia from other early sources are given in translation. The annotation of Jobson's material includes cross-references to material in Part II, but the latter is only minimally annotated and cross-referenced. The Introduction comments briefly on the relevance of Jobson's material to the history and literature of the English out-thrust, as well as to the history of the African region, but in the main attempts to clarify the details of a set of obscure episodes, the 1618–1621 English voyages to River Gambia and the consequent Anglo-African encounter.

* * *

Modern writings in English have almost without exception referred to River Gambia as 'the Gambia (River)', or, less frequently, 'Gambia River', but have normally been limited to discussing the territory of the British colony of Gambia, and its successor independent state within the same boundaries. Since the post-1965 state has taken as its name 'The Gambia', and since the length of River Gambia described in Jobson's book extends much further inland than the territory of The Gambia, stretching into the modern state of Senegal, to avoid confusion the river is hereafter always termed 'River Gambia' – thus partly echoing Jobson's own term, 'the River Gambra'. It should be further noted that titles of past writings normally use 'Gambian' to refer only to the limited territory and therefore not to the whole region of the river.

Within the limits of the colony and state, the terms 'lower' and 'upper' were, and are, employed to denote, administratively, a Lower and an Upper Province or Division, while modern geographers have at times referred to supplementary or different divisions within the territory. In this edition these previous distinctions are not employed, the river instead being divided into three parts: the lower river (from the mouth up to Kasang), the middle river (between Kasang and Baarakunda), and the upper river (beyond Baarakunda, hence entirely in modern Senegal).

Finally, the term 'district' used in this edition has its general sense and does not refer to the administrative Districts of The Gambia. But for the purpose of consulting latterday maps and texts, it may be helpful to indicate the present-day divisions: Lower River Province/Division (Jarra, Kiang, MacCarthy Island, North Bank, and Western Districts), Upper River Province/Division (Sandu, Basse/Fuladu East, Wuli, and Kantora Districts).

Consistency in the spelling of place names has been attempted, but with difficulty. Added to changes over the centuries (perhaps at times reflecting actual pronunciation changes), and differences between English and French orthography (e.g. Dialacoto/Jalakoto), there have been recent revisions, adopted for various reasons. Thus Salum/Saluum has become Saalum. In important instances the variant spellings are shown when a place name is first introduced. However, few places on River Gambia have deliberately adopted an entirely new name in recent decades, one exception being the capital of The Gambia, Banjul, formerly Bathurst. Hence most place names appear to have a long history – although a few denote settlements which seem to have shifted their location somewhat, for instance, from one side of the river to the other.

All dates relating to Jobson's voyage (and to the later English voyages of Vermuyden, Hodges and Stibbs) are Old Style. This means that Jobson's dates for travel on River Gambia have to be advanced by ten days (and the dates of the post-1700 voyages by eleven days) to match the statements of average seasonality of the local climate, and hence the river, given in modern sources. Since, however, the river seasons fluctuate from year to year, often by at least ten days, the difference has been ignored.

* * *

The major document of this edition is Jobson's book of 1623. The transcription presented here follows standard practice in that the printed letters v/V and i/I are given their alternative forms of u/U and j/J, when so required by modern spelling conventions. Occasional inversions of n and u in common terms are corrected silently. Otherwise the original text is reproduced exactly, including the italicization. Although idiosyncratic, the punctuation of the text has been retained. Its generous commas are useful, but larger breaks are shown erratically by colons or semi-colons, with little distinction between the two, and these regularly serve to mark the end of a sentence (even the end of a paragraph), the full stop being seldom employed.

The text includes frequent side notes, which not only in some instances provide additional information but in general guide the reader to specific passages, since there are very few sectional headings. The side notes have therefore been retained.

The text contains spellings which exceed the normal inconsistency of the period, and the syntax is not always that which gives easy reading and comprehension. Comparison of the spelling and syntax of Jobson's manuscript Petition with those of the printed book is not very revealing. Some apparent idiosyncrasies of spelling are repeated, some not, but given the freedom in spelling in the period, no firm conclusion can be drawn. Similarly, the syntax has an inconclusive proportion of resemblances. On balance, it is impossible to say whether – assuming that the Petition was actually composed and written solely by Jobson himself – any changes between the styles of spelling and syntax of the manuscript and those of the book were made by anyone other than the author of the book, although the printer might have been suspected to have intervened. As for the Extracts, it is assumed that comparison is pointless because they almost certainly contain rewriting by Purchas while summarizing.

One thing is, however, clear. The text of the book contains many misprints, perhaps because Jobson's handwriting was difficult even for a contemporary to read; and patently the text was poorly proof-read, if at all. Spelling and syntax of the time being what they were, it is possible that at certain other points of the text apparent obscurities are the result of undetectable misprints. The modern reader will not find much difficulty with most of the contemporary spelling, but in some instances editorial aid has been supplied, by the insertion within square brackets of either additional letters within or at the end of a word, or the word again in modern spelling. Where it seems certain that the printer misread a word, the correct word is similarly added in square brackets, the only regular instance being where the printer set 'which' for 'with'. (Jobson's petition shows that in his hand the abbreviated forms of the two words were extremely alike, making it difficult to distinguish one from the other. But the error also shows that the printer did not read the words of the text contextually.) In a few instances doubt remains as to the correct emendation, in which case a query mark appears within the square brackets. Very rarely [*sic*] is inserted, but this is done, for instance, when a word is doubled. Apart from these editorial corrections, obsolete or rare terms have had added a modern or meaningful equivalent, in square brackets and introduced by *sc*.

The printer misnumbered many pages. Page 134 was given twice and 135 omitted. Pages 137, 140 and 141 were given as 145, 148 and 149. The final fifteen pages, which should be 144–158, were given as 152–166. The explanation of these errors is not obvious. Double numbers have been supplied in this edition, e.g. 144/152, or, for emphasis, 144=152.

The annotation of the text of Jobson's book and of Purchas's extracts from Jobson's 'Journall' is intended to assist the reader of these, and is therefore limited. More extensive and detailed annotation, for scholarly reference, will be found in the Introduction and the Itinerary. Inevitably there is occasional repetition of information.

* * *

The first editor, David P. Gamble, Professor Emeritus of Anthropology at San Francisco State University, has studied the peoples of the Gambia region for half a century. Since carrying out field work between 1946 and 1958, he has regularly visited the region, and has published extensively on its bibliography. The second editor, P. E. H. Hair, Emeritus Professor of Modern History, University of Liverpool, having over the years been generously supplied with information relevant to the River Gambia section of the early Portuguese texts on Guinea he was translating and editing, suggested a joint edition of the first English writing on the river, Jobson's 1623 text. The material in the present edition relating to the river and its peoples has been largely contributed by the first editor, who also prepared an initial typed copy of Jobson's book; the material on the English aspects of Jobson's voyage has been the work of the second editor, who has also prepared Part II and put together the final text of the edition. The first editor supplied the line drawings and photographs. Although the editors have worked together solely by correspondence, the edition represents the combined approaches of an anthropologist and an historian, with a handful of passages where the reservations of one or the other have produced a cautious compromise.

* * *

We acknowledge the friendly assistance of various libraries and archives in the USA and Britain, and of scholars who have answered our queries. In particular, we name: Adrian Ailes (Public Record Office, London); Dr John Appleby (University of Liverpool); Peter Barber (British Library); Professor Louis Brenner (School of Oriental and African Studies, University of London); Dr Stephan Bühnen (Bremen); Susan Carpendale; Dr Sean Cunningham (Public Record Office); Professor Matthew Hill (Department of Anthropology, University of Waterloo, Canada); Aidan M. Ireland (National Archives, Dublin); Sidia Jata (Banjul); Ian Maxted (Exeter Public Library); Linda Salmon (San Francisco); Adele M. Sock (Banjul). And we thank the anonymous relevant staff of the Bancroft Library, University of California, Berkeley; Brisbane Public Library; library of the Institut Fondamental d'Afrique Noire, Dakar; Gambian National Archives, Banjul. Dr Peter Weil read an early draft of the section on the peoples of River Gambia and offered helpful comments. For permissions to publish material, we thank the Centro de Estudos de História e Cartografia Antiga, Lisbon, and the British Library. Finally, we thank the Hakluyt Society and its editors, Will Ryan, Sarah Tyacke and Robin Law, and its designer, Stephen Easton, for encouraging us to proceed and for bringing the edition to publication.

1.1.1999

D.P.G.
P.E.H.H.

KEY
to abbreviations in the notes

GT + number = page number of Jobson's book (find on pp. 75–184 in this edition)
PJ + number = page number in Purchas's extracts (find on pp. 185–97 in this edition)
Purchas, *Pilgrimes*, 1/9/13, 1568 = Samuel Purchas, *Purchas his Pilgrimes*, Part 1, Book 9, Chapter 13, p. 1568; and other references to this work (except to the section signalled as *PJ*) are in this form.
  Note: in the Glasgow reprint the above reference is instead given as II.ix.1568. = Volume II, [Part 1], Book 9, [Chapter 13], p. 1568; and other references therein are in this form.

APC = *Acts of the Privy Council* – in the Bibliography, see English/British State Papers
CSP = *Calendar of State Papers* – in the Bibliography, see English/British State Papers
DNB = *Dictionary of National Biography*
HCA = High Court of Admiralty
OED = *Oxford English Dictionary*
PRO = Public Record Office, Kew, London

DPG = David P. Gamble
PEHH = P. E. H. Hair

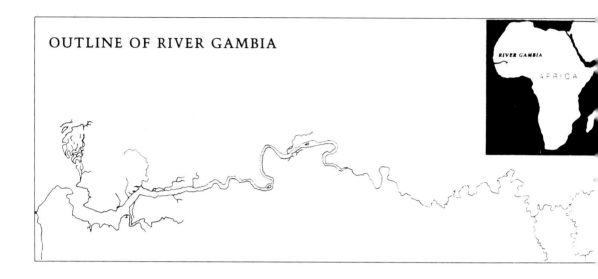

Jobson's River Gambia 1620-1621

Key: Localities and districts mentioned by Jobson in **bold**. Equivalent names in modern spelling in *italics*. Major localities and districts not mentioned by Jobson or not existent at that date, but mentioned in annotation, in modern spelling in italics. District names in large type.

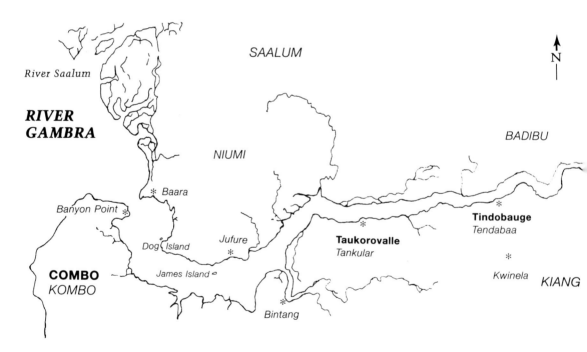

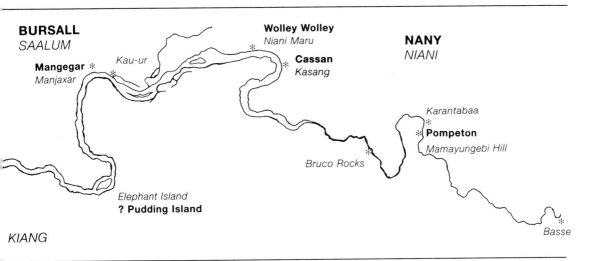

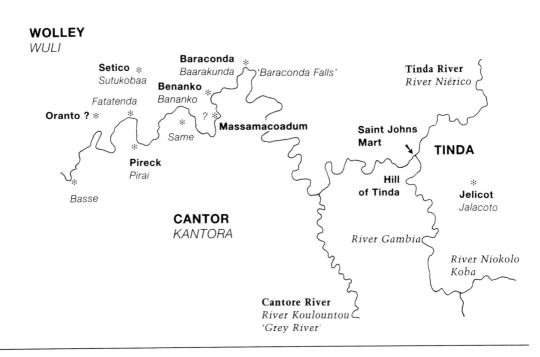

A Mandinka village: Kwinela in Kiang, September 1966

INTRODUCTION

The Sources

In 1623 a book with the following title appeared.

> The Golden Trade: OR, A discovery of the River *Gambra*, and the Golden Trade of the *Aethiopians*. Also, *The Commerce with a great blacke Merchant, called* Buckor Sano, *and his report of* the houses covered with Gold, and other strange *observations for the good of our owne countrey*; Set downe as they were collected in travelling, part of the yeares, 1620, and 1621. By *Richard Jobson*, Gentleman.[1]

In a text of some 55,000 words, Richard Jobson, a man known only from his writings, thus discussed a 1620–21 voyage up River Gambia in West Africa. Two years after this publication, in 1625, Samuel Purchas included in his vast work, *Purchas His Pilgrimes*, a text of some 5,000 words, entitled 'A true Relation of Master Richard Jobsons Voyage ... Extracted out of his large Journall'.[2] Not content with this, Purchas also included, much later in *Pilgrimes*, another text of some 10,000 words, 'Larger Observations of Master Richard Jobson ... then in his Journall is contayned, gathered out of his larger Notes'.[3] In a marginal note, Purchas remarked that, because of 'faults' in his previous inclusion, 'as also for better and fuller intelligence, this is added'. This second inclusion is, however, merely a brief potted version of the text of Jobson's published book, although it was most probably composed from Jobson's manuscript before the book actually appeared.[4] It deserves no further attention.[5] So far, Jobson's original writings are his book and the edited extracts

---

[1] The work was entered as 'The Discovery of the golden trade of Aethiopia written by Captaine Jobson' at Stationers' Hall on 18 June 1623 (E. Arber, ed., *A Transcript of the Register of the Company of Stationers of London 1554–1640*, 5 vols, London, 1875–94, IV, 1877, p. 61). For details of later reprint editions, see the Bibliography. References to Jobson's book are hereafter given in the form *GT* + page number.

[2] Samuel Purchas, *Purchas his Pilgrimes ...*, 4 vols, London, 1625, [reprinted 20 vols, Glasgow, 1905, the original Book and page references shown in the margin], Part I, Book 7, Chapter 1, pp. 921–6. References to this section containing the extracts from the Journall are hereafter given in the abbreviated form *PJ* + page number.

[3] Purchas, *Pilgrimes*, Part I, Book 9, Chapter 13, pp. 1567–76 [hereafter 1/9/13, 1567–76].

[4] Purchas's *Pilgrimes*, published in January 1625, went to the printer in stages between 1621 and 1624. While Books 3–4 of Part I are thought to have gone to press by the end of 1621, the subsequent Books of Part I – including both the Jobson pieces in Books 7 and 9 –, together with Part II, are believed to have gone during 1622, or at least, with even later material, before November 1623 (P. A. Neville-Sington in L. E. Pennington, ed., *The Purchas Handbook*, Hakluyt Society, London, 1997, pp. 522–4). It would seem very likely, therefore, that Purchas had the Jobson pieces in print, although not published, before Jobson's book appeared.

[5] Although this summary of Jobson's book appeared after the book was published, it was most likely in print earlier than the book (see the previous note). This would explain why Purchas fails to mention the title or indeed the existence of Jobson's book, and would indicate that he prepared his summary from a manuscript copy, presumably one supplied by Jobson. Moreover, when, in a marginal note later in his own work Purchas cites Jobson's account, he does so in terms of his own summary. Thus, mentioning a hanging bird

from his Journall, or voyage log, as published by Purchas. But we also have an undated manuscript petition of some 4,000 words, written by Jobson and addressed to the crown.[1] In the manuscript Jobson refers to a previous petition to the crown, 'two years since', but this document is not now extant, and it was, perhaps, only an earlier version of the known petition. As the latter is addressed to King Charles, it cannot be earlier than March 1625, and it is probably of 1626 or even 1627.

Purchas and Jobson knew each other, to an uncertain extent. In the introduction to his published account Jobson remarks that he has 'received this caveat from that worthy gentleman, Mr Samuel Purchas', the caveat being that formal private reports of an overseas venture should not preclude a full public account intended for 'such as take pleasure in reading of other men's adventures'. Jobson had therefore written his account, 'being likewise incouraged by him, after he had seene and read my journall, breefly relating each dayes particulars.'[2] These remarks do not prove that the two men were in verbal contact since Jobson might only have meant that he had received inspiration from Purchas's published writings and from corresponding with him. But in a marginal note to his summary of Jobson's book, Purchas introduces a point with 'He told me ...', indicating a conversation between the two.[3] And in another marginal note, when elsewhere in *Pilgrimes* discussing elephants in Angola, Purchas notes that 'M. Jobson gave me one of the tayles ...'.[4] At least some of these contacts must have taken place between 1621 when Jobson returned to England and 1623 when his book appeared.[5]

(2/5/2, 965), he cites 'Jobson Tom. 1. l[iv]. 9.' (i.e. pt I, bk 9), representing his own summary (the passage being on p. 1576 of Chapter 13). Another marginal note (2/5/3, 978) limits itself to saying that 'Jobson mentioned the like [ant hills] neere Gambia' and gives no reference, but the information can be found in the summary (p. 1570). These notes occur in parts of Purchas's work which were most probably in print before Jobson's book was published. It is, however, curious that Purchas referred to Jobson's text only this small number of times, and no more often, given that many other items in Jobson's account matched those in later accounts in *Pilgrimes*. The actual summary is a brisk and fairly careful abbreviation, with few inaccuracies, although it becomes increasingly selective as it advances through Jobson's text. The style is not the normal one of either Jobson or Purchas, being much simpler, but the employment of an occasional conceit (e.g. 'for defence against Mankinde-beasts, [rather] then beastly unkinde Men'), suggests strongly that the author was not a hack but Purchas himself. A few added marginal notes represent the only material not in Jobson's book (contra Pennington, *Purchas Handbook*, p. 415), and patently these were the work of Purchas himself and are knowledgable and helpful. In the final edition of his other great work, *Purchas his Pilgrimage ...*, published in 1626 (the previous edition being in 1617), Purchas noted Jobson in a list of 'The names of manuscripts, travellers and other authors, the most of which are published in our Bookes of Voyages [i.e. *Pilgrimes*] ...', but did not include Jobson in the list of 'Authors mentioned in this work'. Although the sections on Africa are extensive, Jobson is not named in the index or in the marginal references, and none of the text appears to derive from Jobson, the fairly detailed account of River Gambia being drawn from Cadamosto, Barros and Jarric (*Pilgrimage*, pp. 710 ff.). Nevertheless, at least twice Purchas invites the reader to gain more information on specific Guinea topics by consulting Jobson 'in my Voyages', pp. 623, 625.

[1] British Library, Royal MS L8 A LVIII, 'p. 275', 5 ff [hereafter cited as Petition]. For two manuscript maps of River Gambia with a possible bearing on Jobson's voyage, see the Bibliography.

[2] *GT* 1–2. It might be deduced that Jobson had written a business report to his employer, the Guinea Company, as his predecessor Thompson certainly had done in 1619 (*GT* 6, 'hopefull letters'), but the records of the company are largely not extant.

[3] Purchas, *Pilgrimes*, 1/9/13, 1575.

[4] Ibid., 1/7/3, 1000.

[5] At these dates Purchas was living and working in London, so most likely contacts were made there, where Jobson most probably was also living. Presumably there were at least two contacts, once when

INTRODUCTION

In the first chapter of *Pilgrimes*, Purchas considers, at great length, the significance of the voyages of King Solomon's 'navy' and the likely locations of 'Ophir', remarking that 'some I have knowne which place Ophir neere Gambra'. He continues: 'Of this minde was Captaine Jobson, which travelled up that River, nine hundred and sixtie miles, and [the reader can] hear such golden reports of the In-land Countreys, as this Worke will from him deliver to you.'[1] There is no further mention of Gambia or Jobson later in the chapter, Purchas having a different view about the identity of Ophir.[2] The reference to Jobson looks like a late insertion in a chapter composed most probably in 1624.[3] Jobson's argument that Ophir was located in the River Gambia district appears, not in his printed texts, but at length in his petition.[4] Purchas must therefore have ascertained Jobson's view by 1624, either in discussion, or by seeing a copy or draft of the earlier petition (supposing that this resembled the extant later petition in respect of the Ophir reference). This is of interest inasmuch as it raises the possibility that Jobson was composing a petition by 1624. Although there is little common ground between what the two men write about Ophir, it is possible that Jobson inserted the Ophir argument into his petition(s) after Purchas drew his attention to the issue.[5]

Presumably Purchas obtained Jobson's 'Journall' from its author not only to read but in order to publish material from it. The Journall is not extant, nor are any other accounts of the 1620–1621 voyage. Nor are there any accounts of its immediate predecessors, the English voyages to River Gambia which set off in 1618 and 1619. For the history of all three voyages, we are dependent, to a very small extent on contemporary documentation, and almost overwhelmingly on the two printed sources: Jobson's published account and the material 'Extracted' from the Journall by Purchas.

The two sources, although composed by the same individual, vary in nature and only loosely fit together. The Journall is a chronological account of travel up, and then down, River Gambia, between November 1620 and June 1621. If Jobson's

---

Purchas saw (and was given or copied) Jobson's Journall, and at a later date, when he saw the manuscript of the book.

[1] Ibid., 1/1/1, 1–48, passage cited on 28. It being assumed in this period that the earliest historical record was in the Old Testament, the sending of a 'navy' to Ophir by Solomon (I Kings 9: 26–8, 10: 11), was regarded as the beginning of long-distance marine enterprise. Purchas was one of many contemporary scholars who learnedly discussed the venture and the identity of Ophir, and he drew on a 70-sheet manuscript on the subject written by Dr John Dee in the 1570s, now lost: for more, see H. C. Porter in Pennington, *Purchas Handbook*, pp. 186–8.

[2] Purchas had already committed himself in earlier editions of *Pilgrimage* to the view that ancient Ophir, if not Sumatra, was Sofala (Purchas, *Pilgrimage*, 1617, pp. 557, 696, 859–60; *Pilgrimage*, 1626, p. 756).

[3] However, according to Purchas himself, Book I was 'one of the last printed' (Purchas, *Pilgrimes*, 'To the Reader'), presumably going to the press in 1624. It is generally supposed that this remark referred specifically to chapter 1 of Book I, because this has a separate pagination and presumably went to the press later than the rest of Book I, the composition of the sermon-like material on Ophir causing the delay. Thus, it is likely that, whereas the Jobson material, in the second pagination of Book I, was in print before 1624 (p. 1, n. 5), this reference to Jobson and Ophir was only written by Purchas in 1624.

[4] That a discussion on or even a reference to Ophir was included in Jobson's Journall, but was cut by Purchas in the published version, seems very unlikely, given the nature of the Journall.

[5] This could have been either in discussion or as a result of Jobson reading *Pilgrimage*, in the earlier editions which appeared before he set out for River Gambia. But the latter seems ruled out by the total lack of reference to Ophir in Jobson's 1623 book.

statement above ('each dayes particulars') is to be taken literally, the Journall had daily entries and was therefore somewhat of the nature of a ship's log. However, Purchas published no entries when the ships were at sea (that is, between 5 October and 14 November 1620, and after an unstated date a few days later than 9 June 1621 when they left River Gambia).[1] More pertinently, representing the seven months or about 210 days the English were in the river, Purchas supplies entries that are dated, or can be dated, for only some fifty days. While there are statements which individually may cover several of the missing days, it would seem likely that Purchas sharply abbreviated what he had received.[2] Presenting the material in narrative form, he probably also summarized certain individual day entries, and he may have run some of them together under a single date. Since he did not indicate the extent of his abbreviating or specify omissions, it is now impossible to know how much other material the Journall contained, while the 'faults' Purchas admitted to may have included some miscopying or misordering.

The other text is very different. Composing a book for public consumption, Jobson abandoned chronology and presented an analytical account, dealing in separate sections, for instance, with climate, ethnography and natural history. Allowing for his strange style – but no doubt not as strange to the reader of his own day – Jobson provided an interesting, sometimes lively, and generally novel account of a region not previously known to English readers. However, the book contains only scattered and elusive references to the exact history of the voyage – although, oddly, it provides the fullest known references to the earlier English voyages to the river. The sequence of the 1620–21 voyage has therefore to be based mainly on Purchas's extracts from the Journall, and this being so there are points of uncertainty. An Itinerary of the voyage is given below (Appendix A). While this details the context of Jobson's account, it is not necessary to digest the chronology and topography expounded therein before reading the main texts, which largely stand on their own. Annotation of these texts has therefore been normally confined to other matters.

The 'Discovery' of River Gambia

The Portuguese were the first Europeans to reach River Gambia. This was around 1450, a few years after they attained River Senegal and Cape Verde, at the western extremity of Guinea. By 1460 the river had been penetrated for several hundred

---

[1] Similarly Purchas omitted most details of the outward journey and all details of the homeward journey of the 1621 Algiers expedition (Purchas, *Pilgrimes*, 1/6/5, 882, 886, as noted in Pennington, *Purchas Handbook*, pp. 132, 222).

[2] This corrects an earlier opinion that Purchas published the Journall 'almost in full' (P. E. H. Hair, 'Material on Africa (other than the Mediterranean and Red Sea lands) and on the Atlantic islands, in the publications of Samuel Purchas 1613–1626', *History in Africa*, 13, 1986, p. 126; also, the article slightly abbreviated, in Pennington, *Purchas Handbook*, p. 202).

INTRODUCTION

miles, and in the 1460s and 1480s, brief accounts of early explorations of the river were composed.[1] From the start it was realized that the river provided access to an interior region which either had gold itself or was traversed by African gold merchants. The river runs a generally westward course to the sea, of some 700 miles, although its meandering middle section, and the NW course of its upper reaches, mean that, in the interior, the main stream is never more than 250 or so miles from the sea. The lower half of the course is navigable by smaller ocean vessels.[2] Diogo Gomes, a Portuguese agent in the 1450s, travelling in an ocean vessel of the period, penetrated most of the lower half and reached 'Cantor', a centre for the gold trade. By the early sixteenth century, African-European trade was established along a considerable stretch of the river, about 200 miles. On the European side this involved official and unofficial Portuguese commerce. However, once the Portuguese had reached the eastern half of the Guinea coast and had inserted themselves into the long-standing and rich gold trade of the area they came to call Mina – the later 'Gold Coast' – the importance of Gambian gold exports gradually declined.[3] By 1600 the Portuguese export trade of River Gambia tended to be in humbler products and to be conducted by a small number of Afro-Portuguese, persons of mixed ancestry, either operating from the Cape Verde Islands, with mainland bases on the 'Little Coast' of Senegal (the stretch south of Cape Verde and almost up to River Saalum), or actually resident in the river.[4] Nevertheless these Afro-Portuguese were generally successful in keeping out from the riverain trade the next wave of Europeans, the French; and, as Jobson's book evidences, the first English attempts to trade in the river were met with spasmodic Portuguese hostility. Only after 1650 did the English become the dominant European nation trading in River Gambia.

---

[1] The explorers were Cadamosto and Gomes: see D. Peres, ed., *Viagens de Luís de Cadamosto e de Pedro de Sintra*, Lisbon, 1948, pp. 72–4; T. G. Leporace, *Le navigazione atlantiche del veneziano Alvise da Mosto*, Venice, 1966, pp. 112–14; G. R. Crone, trans. and ed., *The Voyages of Cadamosto and other Documents on Western Africa in the Second Half of the Fifteenth Century*, Hakluyt Society, London, 1937, pp. 75–7; Frédérique Verrier, trans. and ed., *Voyages en Afrique noire d'Alvise Ca' da Mosto (1455 et 1456)*, Paris, 1994, pp. 100–104; T. Monod, R. Mauny, and G. Duval, *De la première découverte de la Guinée[:] récit par Diogo Gomes (fin XVe siècle)*, Bissau, 1959, pp. 34–45. See the extracts in Part II below. For the name 'Gambia', see *GT*, p. 84, n. 3.

[2] For a detailed description of the river, see the section below, 'The River'. In summary on its navigation, the river is tidal for a considerable distance, but the water level is seasonally affected, at points and times to an extreme degree, making navigation in some upper stretches impossible in the dry season and dangerous at the height of the rainy season. A modern view is that, under best conditions, ocean vessels up to 13 feet draught can penetrate 150 miles (to Kuntaur) and not exceeding 6½ feet 288 miles (to Fatoto); launches and canoes perhaps 320 miles (passing from the state of The Gambia into the state of Senegal): R. J. Harrison Church, *West Africa*, London, 1957, p. 217; *Africa Pilot* Vol. I, 12th edn, London, 1967, p. 294 (slightly different figures).

[3] World circumstances helped the decline. The immense accession of gold to western Europe from Spanish America during the 16th century reduced the value of marginal gold trades such as the Gambian one. African merchants and producers probably tended to switch back to the older trans-Saharan export route to Islamic North Africa and the Middle East.

[4] From the mid-16th century onwards, Portuguese sources continually complained about the decline in Portuguese trade with Guinea caused by the competition of other European nations. For instance, a trader, complaining apparently with reference to River Gambia, wrote in 1606 that 'the quantity of wax in these parts can be seen in the customs books, before other nations came here' (Fernão Guerreiro, ed., *Relaçam anual das cousas que fizeram os padres da Companhia de Jesus nas partes da India Oriental, & em algũas outras ...*, 5 parts, Evora/Lisbon, 1603–11, reprinted, ed. A. Viegas, 3 vols, 1930–42, Coimbra, part 1607, liv. 4, cap. 9, f. 157, reprint, II, p. 211; English version of Guinea sections, in P. E. H. Hair,

English interest in the river and its trade came about indirectly. In 1591 a leading polity in the far interior of West Africa was attacked and conquered by an army which had crossed the Sahara Desert from Morocco. The spoil sent back to Morocco included large quantities of gold, and this, having been noted and reported by English merchants in Morocco, was announced in print by Hakluyt in 1599.[1] As English marine power increased, it occurred to certain London merchants and promoters that the gold trade of the West African interior, inaccessible from the north because of Islamic resistance, might be tapped from the south by means of the Guinea rivers – a notion that had long before occurred to and excited the Portuguese. River Gambia was the most obvious target.[2] Comparatively easy to reach from England, the river was flanked on the west by the 'Little Coast' of Senegal, a district with which the English had developed a trade, mainly in hides, since the 1580s, although probably never penetrating into the neighbouring river. English attempts to sail up River Gambia to the gold-trade areas began in 1618, and ended (for the time being) in a voyage of 1620–21 which successfully surveyed the district but gained (it appears) only a tiny amount of gold, the voyage described by Richard Jobson in his book.

The English, late European arrivals anywhere in Guinea, were very late arrivals in River Gambia. The 'discovery' announced in Jobson's title was a discovery for the English but very old news to the Portuguese.[3] It is doubtful if any of the English,

---

trans. and ed., *Jesuit Documents on the Guinea of Cape Verde and the Cape Verde Islands 1585–1617*, Department of History, University of Liverpool, 1989, item 35).

[1] Richard Hakluyt, *Principal Navigations* ..., 3 vols, London, 1598–1600, II/2, pp. 192–3. A report about West African gold arriving in Morocco was sent in 1594 to Anthony Dassell, 'merchant of London' who from 1588 had participated in an English trading venture to Senegal (see the following section). Hakluyt also published a Spanish report of 1591 about the trans-Saharan gold trade, written from Arguin on the Saharan coast and recommending that it be tapped from there (ibid., II/2, p. 188). A reference to this gold trade also appeared in a minor work published in 1609 (R. C., *A True Historical Discourse of Muley Hamet's rising to the three Kingdoms of Moruecos, Fes and Sus*, London, 1609; as reprinted in Purchas, *Pilgrimes*, 1/6/2, 872–3). A marginal note in a work of 1600 suggested that English merchants might participate in the interior trade of West Africa by supplying a much desired import – 'Great scarcity of salt in Tombuto [Timbuktu], which commodity might be supplied by our English Merchants to their unspeakable gaine' (John Pory, *A geographical historie of Africa*, London, 1600, as reprinted in Robert Brown, ed., *The History and Description of Africa* ..., 3 vols, Hakluyt Society, London, 1896, III, p. 824 – the reference copied in Purchas, *Pilgrimes*, 1/6/1, 828, as noted by Carol Urness in Pennington, *Purchas Handbook*, p. 130).

[2] Possibly the peace between England and the Spanish-Portuguese state in 1604, and the Anglo-Spanish rapprochement in the mid and late 1610s, encouraged the belief that intrusion into the Portuguese commercial monopoly in River Gambia would be tolerated, or at least not sharply resisted. In 1617 Raleigh was allowed to go to River Orinoco to test Spanish tolerance.

[3] In proclaiming 'Discovery', Jobson was, however, following the language of the patent authorizing the company which employed him. This claimed that the Guinea Company's monopoly had been justifiably gained, because 'divers of our loving subjects by their long travel and industry and at their great charges and expenses discovered and found out a trade into certain places of Africa hereafter mentioned' (C. T. Carr, ed., *Select Charters of Trading Companies A.D. 1530–1707*, London, 1913, p. 99). But in a petition to Parliament of 1624 (with an echo in the 1631 charter of the successor Guinea Company), the claim of first discovery was challenged. It was argued that it was false that 'some of the patentees had been the first discoverers of that trade ... [since] it is apparent that these parts had first been traded to near fifty years since by Sir John Hawkins deceased and some by divers other merchants' (*Journal of the House of Commons*, London, n.d., I, pp. 710, 794). The argument was strained. It turned on the contemporary meaning of 'discovery', and on the identification of the 'places' to which the 1618 charter was referring. Whereas the 1555 charter of the Muscovy Company referred to the 'discoverie of Regions ... unknowen' (Richard Hakluyt, *Principall Navigations* ..., London, 1589, pp. 259, 304), and such geographical discovery was sometimes distinguished from trade ('as well for discoverie as Trade': *Acts of the Privy Council* [hereafter *APC*] *1628*

INTRODUCTION

including Jobson, penetrated any locality where Portuguese individuals had not explored earlier, perhaps much earlier; and it is certain that earlier the Portuguese had traded regularly in all but the uppermost localities.[1] Yet Jobson's book is considerably more than a notice of one small part of the English delayed but potent overseas out-thrust, although it is certainly that, being the earliest substantial

July–1629 April, item 440), nevertheless the term was also used in a wider sense, to indicate economic as well as geographical discovery, for instance, in the 1576/7 charter of the 'Fellowship of English Merchants for Discovery of New Trades' (Carr, *Select Charters*, 28 – the title still in use in 1614: PRO, High Court of Admiralty [hereafter HCA] 13.42, 12.11.1614). The proposed trade to River Gambia, and the projected subsequent gold trade, were both new English trades yet to be 'discovered'. However, the 1618 patent was unduly vague about the geographical direction of its new trade. The trade that living persons, including some of the Guinea Company Adventurers, had initiated and practised was the Senegal trade from the 1580s and more especially the recent trade of John Davies and others to Sierra Leone and Sherbro. The patent did not specify River Gambia or either of these other areas, but instead, when defining 'places', gave the greatly expanded claim of 'Gynney and Binney', thus covering the whole Guinea coast – Senegal, River Gambia, Sierra Leone, Gold Coast, and the Niger Delta (River Benin). ('Guinea and Benin' as a description of the whole coast was not novel, having been used earlier, e.g. in 1611 in a suit involving John Davies, one of the 1618 Adventurers: PRO, HCA 24/75, no. 33). As the 1624 critics could have learned from reading Hakluyt, the earliest regular English voyages to Guinea, seventy years earlier, had visited and traded at Gold Coast and Benin, so that in those places there could indeed be no new 'discovery' of either sort. Yet, in the 1620s there was no English interest in Gold Coast (let alone far-away Benin), where the 1618 company was indeed never to operate, whereas a trade in Sierra Leone was being developed. The critics therefore ignored the earliest voyages to Gold Coast when making their case, and instead highlighted John Hawkins who never visited Gold Coast but did visit Sierra Leone. (They may have thought this line particularly telling since Hawkins's son was a member of the 1618 company.) The reference to John Hawkins therefore challenged the supposed claim that John Davies and other Adventurers had 'discovered' the trade of Sierra Leone. (Similarly, after Davies' death, his successor Humphrey Slaney, in 1628 was to claim, against the Guinea Company's monopoly, that he had 'planted and set people in the partes of the countries that were first discovered by themselves where never others were before' (*APC 1628 July–1629 April*) – the geographical references being probably to Sherbro, unvisited by Hawkins.) In sum, the 1618 Guinea charter had much justification in its claim of past and intended 'discoverie', but only insomuch as this related to Senegambian trade; while the 1624 critics had a modicum of justification in pointing out that even the merchant pioneers among the Company members were not the 'first discoverers' of trade further east. But Jobson's claim for his own and the 1618 company's 'discovery' of an opportunity for English intervention in the gold trade of River Gambia was fully justified.

However, a claim to novel discovery, in either sense, was not essential for a charter. 'According to Elizabethan practice, it was recognized that either re-discovery, or the effective prosecution of a branch of foreign trade, was a sufficient ground for exceptional privileges. This was covered by the clause in early grants, which stated that certain places had "not been commonly frequented" by English merchants' (Scott, *Joint-Stock Companies*, I, p. 179).

[1] Whereas the Portuguese regularly sailed to 'Cantor', Jobson voyaged further to 'Tinda' (see Appendix A, Itinerary, p. 217, n. 4). Jobson's predecessor, Thompson, had also reached Tinda but left no record; and Portuguese references to voyages beyond Cantor are vague ('twenty leagues', 'one hundred leagues') and significantly supply no toponyms. Thus, Jobson's references to Tinda appear to be the earliest extant record of that toponym. However, the Portuguese at Cantor had almost certainly made inquiries about the interior gold trade and most probably knew of Tinda, even if they had not visited it; and Thompson probably learned about Tinda from Portuguese contacts on the river. The Portuguese had certainly penetrated the interior much further. In 1487 and 1488 the Portuguese crown made rewards to two men who had been sent to 'Mandimansa', the Mandinka potentate probably residing on River Niger in the far interior, and a little later trade to Cantor is documented (*Portugaliae monumenta africana* vol. 1, eds Luís de Albuquerque, Maria Emília Madeira Santos, *et al.*, Lisbon, 1993, pp. 199, 214; vol. 2, Lisbon, 1995, pp. 55, 63, 179–80). Again, in 1534 an envoy was sent to Mandimansa 'on affairs of the trading-place of Cantor', as reported in print by Barros (João de Barros, *Ásia* ... [1552–1613], ed. H. Cidade and M. Murias, 4 vols, Lisbon, 1945, déc. 1, liv. 2, cap. 12). It is virtually certain that these agents travelled up River Gambia, but whether they sailed beyond Cantor and perhaps as far as Tinda, or otherwise went overland from Cantor, is not known. They appear to have been Europeans, not Africans or Afro-Portuguese. Most probably Jobson

account of any part of Black Africa by an Englishman – as well as almost the earliest substantial account of the same, in print, in English.[1] Perhaps unexpectedly, it also represents the earliest detailed account of the River Gambia region, in print.[2] Although a series of Portuguese accounts of western Guinea included extensive descriptions of the Gambia region, including three by Cape Verde Islands traders composed in the 1590s, 1620s and 1660s, none of these accounts was published until centuries later.[3] Moreover, if at an earlier date any Portuguese did explore the upper parts of the river reached by Jobson, none left a record. Jobson therefore provided an intellectual, if not geographical, discovery of River Gambia, not only for readers in Britain but for scholars across western Europe.[4] Moreover, while the Portuguese sources form a useful supplement to, and a check on, Jobson's account, the Englishman at times supplies fuller ethnographic and natural history details.

---

was unaware of the reference in Barros, for although the work was published in 1552 there was no English translation, and there seems to be no reference to the information on the contacts with Mandimansa in any pre-1620 English publication, not even in Hakluyt, as there were to be none in Purchas. Writing in the late 1660s, a Portuguese trader on the river lent support to Jobson as an explorer, by stating that 8 leagues beyond Baarakunda was 'the furthest point reached by whites that is known about' (Part II below: Lemos Coelho, 1669, 29). However since this writer was unaware of the English venture a generation earlier which had certainly gone further than he indicated, he may also have been unaware of even earlier Portuguese explorations.

[1] Eden in 1555, Willes in 1577, and Hakluyt in 1589 and 1598–1600 had published accounts of English voyages to Guinea which included limited material on the localities visited, and the account in Hakluyt of the Lancaster voyage of 1591–2 to the East Indies included limited material on the Cape of Good Hope and Zanzibar (Hakluyt, *Principal Navigations*, II/2, pp. 103–5). Purchas in his 1617 edition of *Pilgrimage* included scattered extracts from the account of Angola by Andrew Battell, and also, in relation to the Cape of Good Hope, from various journals of English voyages to the East (Hair, 'Material on Africa', pp. 129, 132). As regards foreign material in English, translations of Pigafetta on the Congo and Leo Africanus on the northern parts of West Africa had appeared in 1597 and 1600 (P. E. H. Hair, 'Guinea', in D. B. Quinn, ed., *The Hakluyt Handbook*, 2 vols, Hakluyt Society, London, 1974, pp.197–8); the 1617 *Pilgrimage* included extracts from a translation of de Marees on Gold Coast, and for translations of Dutch works which included limited material on Black Africa, see P. E. H. Hair, 'Dutch Voyage Accounts in English Translation 1580–1625: a checklist', *Itinerario*, 14, 1990, pp. 95–106.

[2] Reference to River Gambia in Barros 1552 was limited to a dozen sentences (see Part II below: Barros).

[3] Descriptions of River Gambia, not least of its navigation and trade, were produced by André Álvares de Almada, *c.* 1594, André Donelha, 1625, and Francisco de Lemos Coelho, 1669, enlarged 1684, each a trader from the Cape Verde Islands who had had experience of River Gambia commerce over several decades before writing his account. Although parts of the account by Álvares de Almada were summarized in a printed Jesuit collection of 1605, and thereafter, by borrowing, in other printed works, the River Gambia material was abbreviated to a few sentences. See Part II below for André Álvares de Almada, *Tratado breve dos Rios de Guiné*, ed. L. Silveira, Lisbon, 1946, chs 5–6 (in English, *An Interim Edition of Almada's "Brief Treatise on the Rivers of Guinea"*, trans. and ed. P. E. H. Hair, Liverpool, 1984, chaps. 5–6); André Donelha, *Descrição da Serra Leoa e dos Rios de Guiné do Cabo Verde (1625) / An Account of Sierra Leone and the Rivers of the Guinea of Cape Verde (1625)*, ed. A. Teixeira da Mota and P. E. H. Hair, Lisbon, 1977, chs 10–12; and Francisco de Lemos Coelho, *Duas descrições seiscentistas da Guiné* [1669, enlarged version 1684], ed. D. Peres, Lisbon, 1953, ch. 2 (in English, *Description of the Coast of Guinea (1684)*, trans. and ed. P. E. H. Hair, Department of History, University of Liverpool, 1985, ch. 2).

[4] For the general 'intellectual discovery' of Guinea on the part of the Portuguese, see P. E. H. Hair, 'Discovery and Discoveries: The Portuguese in Guinea 1444–1650', *Bulletin of Hispanic Studies*, 69, 1992, pp. 11–28, reprinted in P. E. H. Hair, *Africa Encountered: European Contacts and Evidence 1450–1700*, Aldershot, 1997, item I.

INTRODUCTION

The English and River Gambia

English trading voyages to Senegambia 1587–1621

After a late and then desultory interest in trade with West Africa, in the 1580s the English began regular trading with the nearest part, the coast of Senegal, particularly with its 'Little Coast' to the east. During the sixteenth century the marine export trade in this area had largely passed from the Portuguese to the French, the trade being mainly in hides and to a lesser extent in gum arabic, wax and ivory.[1] Between 1600 and 1620, although Portuguese (and Afro-Portuguese) agents remained the commercial intermediaries on land, and although Portuguese vessels occasionally tried to intervene,[2] it was mainly French, Dutch and English ships which vied for the trade – and not infreqently fought each other. But the Portuguese retained for much longer a virtual monopoly of the export trade from the neighbouring River Gambia.[3] It is doubtful whether the Dutch ever entered the river before the 1640s.[4] The French made a few unprofitable sorties into the river in the

---

[1] Almada, *Tratado breve*, ch. 2/17–18 (English version, ch. 2/7–8); Hakluyt, *Principal Navigations*, II/2, pp. 188–92, 'The voyage of Richard Raynolds and Thomas Dassel to the rivers of Senega and Gambra adjoyning upon Guinea, 1591, with a discourse of the treasons of certain of Don Antonio his servants and followers', esp. 189; Nize Isabel de Moraes, 'Le Commerce des peaux à la Petite Côte au XVIIe siècle (Sénégal)', *Notes africaines*, 134, 1972, pp. 37–45, 111–26. A number of Portuguese, some of them Jewish refugees, remained as residents on the coast, but appear to have traded more with French and English shippers than with Portuguese.

[2] For a 1601 attack on a Dutch vessel on the Little Coast by three Portuguese vessels 'heading for River Gambia', and a 1606 Dutch counter-assault on a Portuguese vessel on the same coast, see Nize Isabel de Moraes, ed., *À la découverte de la Petite Côte au XVIIe siècle (Sénégal et Gambie)*, 4 vols, Dakar, 1993–5, I, pp. 62–3. (Note that this work, despite its title, does not extend its comprehensive survey of documentation on Senegal to that on Gambia.)

[3] Some intrusion into the river by non-Portuguese vessels is indicated by Jobson's reference to the Africans in the estuary being reluctant to contact ships because they have been seized and carried off 'many times, by severall nations' (*GT* 27). Writing in 1625 but discussing River Gambia which he visited in the 1580s, a Portuguese writer complained, not only that the commodities of the river were carried to Senegal 'to the French, English and other nations', but that these foreigners 'even come up the river and ... draw immense profit' (Part II below: Donelha, f. 27^v). But the last remark most probably only applied to the French after 1610 and the English in 1620–21 (see p. 10, nn. 1–3, p. 11, n. 1), and the profits were almost certainly not 'great'. A Jesuit writing in Sierra Leone in probably the early 1610s said that the 'great [Portuguese] trade' on River Gambia had had the effect that 'at times pirates come here', probably a reference to French irruptions (Manuel Álvares, *Ethiopia Minor and a Geographical Account of the Province of Sierra Leone (c. 1615)*, trans. P. E. H. Hair, Department of History, University of Liverpool, 1990, f. 9).

[4] The Dutch traded in Senegal from about 1600, and by 1610 were regarded in Portuguese sources as the chief rival in districts to the east (e.g. Guerreiro, *Relaçam anual*, f. 157, English version, p. 211). A lengthy discussion of Dutch trade on the Little Coast of Senegal and between there and Cacheu includes a mention of River Gambia in relation to the Portuguese trade in kola, but otherwise ignores the river (Dierick Ruiters, *Toortse der Zeevaerten*, Vlissinghen, 1623, reprinted, ed. S. P. L. Naber, Linschoten-Vereeniging, 's Gravenhage, 1913, pp. 273–80; cf. K. Ratelband, ed., *Reizen naar West-Afrika van Pieter van den Broecke 1605–1614*, 's Gravenhage, 1950, pp. 13–18). Another Dutch account refers to the commodities coming from River Gambia, including ambergris, a large lump of which was found in the river in 1606, but adds that this commodity was purchased on the Little Coast, implying that it was brought there by the Portuguese, who were then said 'to dominate River Gambia, where they have many bases where they alone trade' – thus leaving it unclear whether the Dutch entered the river (Moraes, *Petite Côte*, I, pp. 137–8). However, the Gerbier map, which probably represents pre-1620 information, contains one strange toponym, Wingnosberge, possibly a corruption of a Dutch term.

sixteenth century;[1] and in 1612 followed these with an unsuccessful attempt to establish a base there.[2] In 1616, however, a French ship was trading in the river for three months, although the crew of one of its boats was massacred, perhaps by the Portuguese.[3]

The start of English interest in Senegambia coincided with the presence in England of the Portuguese pretender, Dom António, and in 1588 an arrangement was made whereby an English group of merchants was given a charter of monopoly from the English crown, and a licence from the impecunious pretender, to trade in Senegambia.[4] Although this worked badly, with growing friction between the English and the Portuguese, thereafter English voyages to Senegal became increasingly regular, first, in the later 1590s, under further charters from the English crown, later, up to 1618, as private concerns.[5] These voyages are poorly

---

[1] For 16th-century sorties, Visconde de Santarém, *Quadro elementar das relações políticas e diplomáticas de Portugal t. III*, Paris, 1854, p. 511 (Dieppe vessel attacked in the river, 1570); Hakluyt, *Principal Navigations*, 'Voyage of Raynolds and Dassel', II/2, p. 189 ('The Frenchmen never use to go into the river of Gambra: which is a river of secret trade and riches concealed by the Portugals. For long since one Frenchman entered the river with a small barke which was betrayed, surprised, and taken by two gallies of the Portugals'); Part II below: Almada, 6/10 ('For 70 leagues up-stream ... the inhabitants have ... canoes, such large ones that they have attacked French launches'); Part II below: Donelha, f. 29 (enemy boats up to Cantor *c*.1608); Part II below: Lemos Coelho, 53 (at a distant time, perhaps in this period, a French ship sailed up to Baarakunda, but was then destroyed).

[2] For the attempted base, about which very little is known, see Moraes, *Petite Côte*, I: 48, note 55. In an editorial note to the printed version of the 'Memoires du voyage aux Indes Orientales du General Beaulieu' (in Melchisedek Thévenot, ed., *Relation de divers voyages*, 4 vols, Paris, 1661–4), II, p. 128), based on the manuscript *mémoires* and information from relatives, the following was stated. 'Auguste de Beaulieu estoit à Rouen, son premier voyage fut en la riviere de Gambie à la coste d'Affrique, où il alla en 1612, avec le Chevalier de Briqueville de Normandie, pour s'y fortifier et y establir une Colonie, mais ils y perdirent presque tous leurs gens de maladie, pour y estre arrivez dans l'arrier saison; ce contre-temps rendit leur armement inutil qui d'aileurs estoit considerable, Beaulieu commandoit une Patache'. Moraes also cites Masseville, *Histoire sommaire de Normandie*, Paris, 1668, pp. 75–6, which adds little, and no contemporary documentation appears to have been traced. The disastrous sickness blamed on arrival in the wrong season foreshadows the English experience a few years later.

[3] Moraes, *Petite Côte*, I, pp. 170–88 – for the massacre, see p. 12, n. 2. The Gerbier map (see Appendix C) may represent information collected by a Frenchman, and may relate to French activities in the river before 1618. In 1626 it was stated that – 'Il y a quelques Francoys quy ont traffiqués dans la riviere de Gambye' (ibid., II, p. 228).

[4] Hakluyt, *Principal Navigations*, II/2, pp. 123–6; Mario Alberto Nunes Costa, 'D. António e o trato Inglês da Guiné', *Boletim cultural da Guiné Portuguesa*, 8, 1953, pp. 683–797, on pp. 705–20. Both the charter and the licence refer to trade, not only between Rivers Senegal and Gambia but also within the rivers. The charter stated that the English merchants had been 'perswaded and earnestly mooved by certaine Portingals resident within our Dominions', and when in 1593 the merchants renounced the Portuguese licence, they claimed that their trade with Senegal was begun 'uppon the perswation and motion of Dom Anthonye kinge of Portugall and other Portyngalles resident within this realme' (Nunes Costa, 'O Trato Inglês', p. 794).

[5] The only authoritative account of English trading in western Guinea between the 1580s and 1620s is to be found in three articles by the late John W. Blake, 'English Trade with the Portuguese Empire in West Africa, 1581–1629', *Quarto congresso do Mundo Português*, VI/1, Lisbon, 1940, pp. 314–33; 'The English Guinea Company, 1618–1660', *Proceedings of the Belfast Natural History and Philosophical Society*, III/1, 1945/6, pp. 14–27; 'The Farm of the Guinea Trade, 1631', in H. A. Cronne, T. W. Moody and D. B. Quinn, eds., *Essays in British and Irish History in Honour of James Eadie Todd*, London, 1949, pp. 85–105. Based on detailed research in the High Court of Admiralty archives [hereafter HCA], these articles fill out earlier references in W. R. Scott, *The Constitution and Finances of English, Scottish and Irish Joint-Stock Companies to 1720*, 3 vols, Cambridge, 1910–12, esp. II, pp. 12–14; and have only been supplemented by a fuller study of the period 1588–95: Nunes Costa, 'O trato Inglês'. The English voyages began in 1587, as vaguely stated in the 1588

INTRODUCTION

recorded.[1] However, a single published account of trade on the Little Coast of Senegal, particularly in relation to a voyage of 1591–1592, appeared in Hakluyt's collection of 1598–1600 and must have been known to the promoters of the 1618–1620 voyages to River Gambia, and to Jobson. Although the account does not indicate any English penetration of River Gambia, it provided English readers with the following information: 'Gambra river: The commodities are rice, waxe, hides, elephants teeth, and golde'. It further noted that 'resident by permission of the Negros ... in the townes of Cantor and Cassan are many Spaniards and Portugals'.[2] It is not inconceivable that English vessels in the Senegal trade entered River Gambia before 1618, but no English source provides proof.[3]

One episode in the sequence of early English voyages to Senegal, made public by

---

charter (Hakluyt, *Principal Navigations*, II/2, p. 123), but confirmed in a report of 11 March 1588 from a Spanish agent in England (J. M. Gray, *History of the Gambia*, London, 1940, p. 18, relying on *Calendar of State Papers* [hereafter *CSP*] *Spanish 1587–1603*, ed. M. A. S. Hume, London, 1899, item 237). For comment on the 1588 charter, see Scott, op. cit., II, pp. 10–11. The 1598 extension of the original ten-year charter referred to the same region (between and within Rivers Senegal and Gambia) and not instead only to River Gambia, as stated by Gray (op. cit., p. 19, a misreading of *CSP Domestic 1598–1601*, ed. M. A. E. Green, London, 1869, p. 16). Blake correctly argued that between 1598 and 1618 English voyages to Senegal were regular and probably increasing, although the 'much evidence' for this he noted (Blake, 'English Trade', p. 326), being fragmentary and not easily accessible, remains to be analysed – see the next note.

[1] The sources for the 1590s are almost entirely in HCA documents and English and Portuguese state papers: see the articles by Blake and Nunes Costa (p. 10, n. 5 above), the latter publishing a large number of documents in Portuguese and English, many of them among the personal papers of D. António only latterly recovered by the Portuguese state archives from an archive in Belgium (where the prince died). A Portuguese source of the mid 1590s went so far as to claim that the export trade of the Little Coast of Senegal was 'carried on more by the English than the French, because the English being stronger, drove the French out of the trade': Almada, *Tratado breve*, chap. 2/8. Certain English voyages between 1600 and 1618 are noted in HCA documents and patently many others occurred which did not involve episodes that led to HCA documentation. For instance, the London Port Books record four ships sailing to 'Guinea' in the twelve months between Christmas 1615 and Christmas 1616 (information supplied by Dr John Appleby), yet none of these sailings appears to be noted in HCA documents. Study of the English in Senegal in the early 17th century has yet to be made. Perhaps this is because, in 1957, it was discouragingly remarked that, although material for studying the period of Anglo-African trade up to the 1630s existed, 'it remains doubtful if the results would be commensurate with the effort' (K.G. Davies, *The Royal African Company*, London, 1957, p. 38, n. 1).

[2] Hakluyt, *Principal Navigations*, II/2, pp. 188–92, passage cited 189, 191. Cantor was mis-printed 'Canton'. The English liked to associate Spaniards and Portuguese but it is unlikely that there were any of the former in River Gambia. Although the information was no doubt obtained from the Afro-Portuguese in Senegal, this appears to be the earliest show of interest in River Gambia on the part of the English. Hakluyt's references are probably the earliest mentions of Cassan/Kasang and 'Cantor' in English printed works.

[3] A series of statements claiming English involvement in River Gambia before 1618 will now be examined. **(a)** In 1650 it was alleged that an English factory had been set up in River Gambia 'about 35 years since' (i.e. *c*. 1615) under the first Guinea Company (in fact, established only in 1618), and that for fifteen years thereafter no other factory was built (*CSP Colonial 1574–1660*, ed. W. N. Sainsbury, London, 1860, p. 339 = PRO, SP Colonial XI.15, as cited in Scott, *Joint-Stock Companies*, II, p. 12). Blake has shown that the later part of the claim is false (Blake, 'English Trade', p. 329); and the first part is too. Nothing resembling a permanent post was established during the 1618–21 voyages, and nothing in Jobson's account confirms any recent English activity in the river. **(b)** The title of the Hakluyt account (p. 9, n. 1) reads 'The voyage ... to the rivers Senega and Gambra ... 1592', but the title was inserted editorially and is inaccurate, since the text makes no reference to either river having been entered. Perhaps the error arose because both the charter and the licence specifically permitted trade within the river (p. 10, n. 4). **(c)** A 1588 report of a Spanish agent stated that a Portuguese pilot took 'some Englishmen in two ships to River Gambia, near Cape Verde, and they recently came back with much ivory and hides' (*CSP Spanish*

Hakluyt, surely influenced Jobson's subsequent activities in River Gambia. As Hakluyt printed it – 'In our second voyage and second yeere [1588] there were by vile trecherous meanes of the Portugals and the king of the Negros consent in Porto d'Ally and Joala about forty Englishmen cruelly slaine and captived, and most or all of their goods confiscated: wherof there returned onely two, which were the marchants.'[1] Jobson's suspicious and hostile attitude to the Portuguese in Senegambia was, however, fuelled not only by an event in Senegal of thirty years earlier, but by an event on River Gambia occurring only twenty months or so before he arrived in the river, as explained below.[2]

1587–1603, item 237, dated 11.3.1588 but possibly 1588/9). This is plausible evidence. But since all the English voyages that immediately followed went only to Senegal, the agent may have been mistaken. **(d)** A Spanish source on Senegal of apparently *c.* 1600 incidentally referred to trade on 'Rio de Jambra' (i.e. Gambra/Gambia), trade seemingly deriving from the Grand-Fulo in the interior and being conducted by the English as well as by the Portuguese, French, and Flemings (A. Teixeira da Mota, 'Un Document nouveau pour l'histoire des Peuls au Sénégal pendant les XVème et XVIème siècles', *Boletim cultural da Guiné Portuguesa*, 24, 1969, pp. 781–860, also repaginated (1–86) in série separatas LVI, Agrupamento de estudos de cartografia antiga (Lisbon, 1969, pp. 49, 55–6). Equally unevidenced in English sources but more securely mentioned by foreign sources was an English post on River Senegal, said to have been abandoned *c.* 1610 (P. E. H. Hair, Adam Jones and Robin Law, eds, *Barbot on Guinea: The Writings of Jean Barbot on West Africa 1678–1712*, Hakluyt Society, London, 1992, pp. 57–8, n. 16). In the period before 1618, English activity in River Senegal is more likely to have been the case than the same in River Gambia (since Portuguese resistance was weaker in the former); moreover, the references to the Grand-Fulo and to the Flemings/Dutch fit River Senegal better than River Gambia. This suggests that the Spanish source confused the two rivers. **(e)** In 1625 a Portuguese with much earlier experience of the river claimed that 'the French, English and other nations … even come up the river' (see p. 10, n. 5 above). But, as regards the English, it was most probably the 1618–21 voyages that the writer had in mind. **(f)** In 1617 a French agent was sent to meet an international fleet of pirate ships cruising around 'the seas of Africa, Spain and the Indies, and various other places', and said to 'winter in Africa or the Guinea coast', the fleet being led by some English captains, allegedly Roman Catholic escapees. The fleet was later contacted off Ireland and then took refuge in Morocco. According to a second-hand report, when the agent 'went to fetch the ships, on reaching Gambia on the Guinea coast, the soldiers and sailors of his ship mutinied and put him on land' (*CSP Venetian 1619–1621*, ed. A. B. Hinde, London, 1910, items 340, 395, 424, 523). The final reference, dubious and vague, has been interpreted to mean that the pirate fleet had 'an ideal base' in River Gambia (Gray, *History of the Gambia*, p. 20). There is no supporting evidence, and the English voyagers in 1618–21 appear to have gained no knowledge of any pirate base in the river, or to have met or heard of any pirate ships there.

All in all, doubt remains as to whether the English ever traded in River Gambia before 1618, at least after the possible exception of 1587.

[1] Hakluyt, *Principal Navigations*, II/2, p. 189. There is no other English evidence relating to this episode. But in 1592 a Portuguese officer of D. António referred to a 1587 English voyage of three ships to 'the coast of Cape Verde' (i.e. Senegal), where, allegedly because the ships failed to carry an Antonine agent, most of the English were killed or captured, apparently by the local Portuguese (Nunes Costa, 'O trato Inglês', p. 757). This is surely the incident mentioned in the English account, misdated by the Portuguese. If the second English voyage sailed, like later voyages, in September–November, the attack may in fact have occurred in early 1589. It is perhaps unlikely to have been the same as a Guinea episode reported in September 1588, in limited and barely accurate detail, in respect of 'a Taunton ship seized by Spaniards there inhabiting', *APC 1588*, London, 1897, 30.9.1588. Taunton merchants were not associated with the 1588 charter, although they were with a 1592 charter to trade in the Sierra Leone district (Hakluyt, *Principal Navigations*, II/2, p. 193).

[2] It is also just possible that Jobson heard about an episode of 1616. A number of French and English crewmen, set adrift when a French privateering vessel was attacked by a Dutch ship, reached land between River Gambia and Cape Roxo and made their way to the river. They were well treated by a local ruler, a *Farran* (possibly that of Quiam/Kiam/Kiang: Part II below, Lemos Coelho, 18), and eventually contacted a French ship trading in the river (p. 10, n. 3). A few were taken by canoe to 'Cape Verde' to be picked up by shipping there, including at least one Englishman. A handful joined some of the ship's crew in a boat which continued trading, but all in the boat were killed – with the exception of a 'little English boy', but including one Englishman – allegedly by Africans. However the news was

INTRODUCTION

In the 1610s the English extended their trading in Guinea from Senegal to Sierra Leone, some hundreds of miles to the east, the new trade mainly in timber, particularly dyewood. The leading merchant was John Davies.[1] It has been argued that Davies was the major influence behind the formation of a group of promoters, described as 'Gentlemen and Merchants' – but in numbers consisting mostly of the former – which in 1618 obtained a royal charter for a monopoly of trade in 'Ginney and Binny' (that is, western and eastern Guinea).[2] In the same year the group initiated a series of voyages to River Gambia. This was in keeping with the design of the Company seal, which showed 'a coast and river of the said part of Africa with a ship

---

conveyed by Portuguese who may well have been themselves responsible. Details of the episode, which in several aspects foreshadowed the English experience three years later, can be extracted from a series of documents in Moraes, *Petite Côte*, I, pp. 170–88. If the English survivor(s) picked up at Cape Verde reached England, the story may have reached Jobson.

[1] To Blake's account of Davies's pioneering activities in Guinea may be added the following: in 1614 a ship of the East India Company met at River Sess (in modern Liberia) an 'Indian' [*sc.* African], of importance locally, who was known as 'John Davis' and spoke good English, having spent 'two years in England with Mr Davis' (F. C. Danvers and William Foster, eds, *Letters Received by the East India Company*, 6 vols, London, 1896–1902, II, p. 329). One of Davies' competitors was Humphrey Slaney. In 1618, the year the Guinea Company was established, a will was proven in respect of an individual who had died on the *Hearts Desire*, a ship 'bound for Guinea for Mr Humphrey Slaney' (probably to trade at Sierra Leone): E. Stokes, ed., *Index of Wills proved in the Prerogative Court of Canterbury, vol. V, 1605–19*, London, 1912, p. 70. Not one of the founding members of the Company, Slaney must have had some agreed association with it, since in 1620 he was, conjointly with the Governor of the Company, hiring from its owners the *Sion* for the third voyage to River Gambia: PRO, HCA 3.95 (6.2.1621). In 1628 he, with others, petitioned the Privy Council in complaint against the Company's monopoly; this resulted in an 'Accommodation' (which at one stage he accused Crispe, his partner in a subsequent venture, of blocking), after which he apparently became a member – or perhaps (given the obscurity of the Company's institutional procedures) a full or executive member (*APC 1628 July–1629 April*, items 440, 547, 559, 570; *CSP Domestic 1629–1631*, London, 1860, ed. J. Bruce, pp. 155–6).

[2] Blake, 'Guinea Company', pp. 17–19. The charter of the 'Company of Adventurers trading to Gynney and Bynney' is printed in Carr, *Select Charters*, pp. 99–106, and includes a list of 36 members of the Company, of whom 16 were knights, 15 esquires, and five merchants. (Blake counts 37, by mistakenly assuming that the first names on the list, 'Sir Robert Rich, Knt, Lord Rich', represented two individuals, whereas 'Lord Rich' was a courtesy title for Robert Rich, who succeeded his father as Earl of Warwick in 1619 – an earlier source counted correctly (Edward D. Collins, 'The Royal African Company', Ph.D. thesis, Yale, 1899, p. 11).) By the mid 1620s there were 32 members (PRO, SP14/124, no. 115). The initial membership is analysed in T. K. Rabb, *Enterprise & Empire: Merchant and Gentry Investment in the Expansion of England*, Cambridge, Mass., 1967, *passim*. Although Rabb's listing (pp. 232–410) contains, in the case of the 'Africa Company', many errors and misidentifications (for instance, he counts 38 initial members by including Slaney who joined later), his general conclusions hold. Many members were MPs; two-thirds had, or were to have, interests in other overseas ventures; yet with a few exceptions, these interests were limited. As Blake points out, there were some family connections. The first Governor, Sir William St John of Heighleigh, Glamorgan, was possibly a distant relative of Sir Oliver St John (later Viscount Grandison), Lord-Deputy of Ireland, but was certainly married to Oliver's niece; and through her was brother-in-law to the Deputy Governor, Sir Allen Apsley, and also to the elder step-brother of the Duke of Buckingham, Lord High Admiral and Court favourite (C. D. Squibb, *Wiltshire Visitation Pedigrees 1623*, Harleian Society, London, 1954, pp. 168–9). But it may be doubted whether this added up to much influence at Court. Certain of the Gentlemen seem to have been mere investors, others mere figureheads; for a probable instance of the latter, Sir Richard Hawkins, the elderly son of John Hawkins. If one or two were lively entrepreneurs, the remainder were somewhat undistinguished in this aspect. However, it is undeniable that the Adventurers included men with a professional as well as an economic interest in voyages and overseas enterprise. Half a dozen were, or had been, naval commanders, including the current 'Vice-Admiral of England' and the two who became the first and third Governors of the Company. (Four of the Gentlemen Adventurers, Sir Robert Mansell, Sir William St John, Sir Thomas Button, and Thomas Love, were to command vessels in the 1621 naval expedition to Algiers: Purchas, *Pilgrimes*, 1/6/5, 881–2). A dozen had a past, or developing, active

at anchor and another with many little shallops with oars issuing out and going up the river upon discovery'.[1] Three voyages to the river, begun in 1618, 1619, and 1620 – the last chronicled by Jobson – produced only heavy losses, in capital,[2] and in men,[3] while a final voyage, beginning in late 1624 or early 1625, went no further than Dover.[4] Opposition to the Company's charter came from London clothiers, who complained that the monopoly increased the cost of two commodities used in

interest in plantations in Ireland and America. The third Governor, Sir Thomas Button, another Glamorgan gentleman as well as a naval man, had been a distinguished explorer of Hudson's Bay (portrait in the 1905 edition of Purchas, *Pilgrimes*, VI: opp. p. 136). Nevertheless, Blake considers, probably justly, that the establishment of the Company was mainly a device of John Davies, the first-named of five 'Merchants' in the patent, to enable him to gain a monopoly of trading at Sierra Leone, which he had sought in vain in 1614 (*APC 1613–1614*, London, 1921, p. 633). To that extent, the voyages to River Gambia in search of gold were a useful diversion since they gained Davies the support of influential gentlemen. However, while many of the original Gentlemen Adventurers retired after the failure of the Gambian enterprise, a number continued to lend active support to the Company; and Davies himself appears to have managed many of the Company's affairs up to his death in 1626, admittedly coincidental with a shift of its interests from River Gambia to Sierra Leone.

[1] From the charter: see previous note. Regrettably, no references to the voyages to and from River Gambia, to the ships or to Richard Jobson, have yet been traced in the London Port Books, PRO, E.190, partly because there are gaps in the sequence of the books covering the relevant dates, and partly because of the nature, form and present condition of the volumes. But the possibility that one or more references exist in the books, in some form somewhere, cannot be utterly excluded.

[2] On the first disastrous voyage the Company allegedly lost £1,856 (figures rounded), perhaps not including the loss on the ship itself (even if it was only hired and was perhaps insured); on the second, the sum borrowed for the voyage was £1,988 and the return (in hides) £80, a loss of £1,908; on the third, with two ships, cargo and related charges costing £1,920 and freight and wages £1,300, against a return cargo worth £1,386, the Company lost £1,835 (figures from PRO SP 14/124, no.115, and printed in Scott, *Joint-Stock Companies*, II, p. 12). Blake thus stated the total capital loss as £5,600 8s 4d, Blake, 'Guinea Company', p. 18. Jobson appears to suggest that the third voyage broke even because unsold trade goods were returned to England, but admits to ignorance as to how the accounts were calculated (*GT* 164/156–165/157) – as indeed we too are ignorant.

[3] By sickness in the river, the third voyage, with two ships of 200 and 50–60 tons respectively and probably some 70–90 sailors and traders, probably lost about 25–35. (For estimates of manning levels, see p. 26, n. 1.) The second voyage was of a ship of 50–60 tons, with a crew of probably 20–25. According to Jobson, by sickness in the river 'many of them lost their lives ... and ... few returned' (*GT* 164/156), suggesting a loss of 10–15. The first voyage, of a ship of 120 tons, therefore with sailors and traders numbering probably 40–60, had an uncertain number murdered at Kasang, an uncertain number dying of sickness in the river, and an uncertain number escaping overland and returning to England, with eight men remaining in the river and seven of them being rescued by Jobson. The loss is uncertain but may have been 20–30. The total loss was therefore, on these very rough estimates, probably between 55 and 80 men out of 130–175 man/voyages, between a third and a half. The diseases endemic in River Gambia to which Europeans were particularly and immediately susceptible because of lack of acquired immunity had inflicted sickness and high mortality on crews of visiting European ships from the time of the earliest recorded encounter: cf. Part II below: Cadamosto, [24]; Gomes, f. 278.

[4] Blake's account of this episode (Blake, 'Farm of the Guinea trade', pp. 94–5), combines information from an 1628 Chancery law suit (PRO, Chancery Proceedings, C2, Charles I, S44/26), with statements by Jobson in his undated petition to King Charles (see below), in which he complains that the aborted voyage was in 'a Spanish bottome; bought here in the River of Thames, which proved rotten'. The law suit refers to an unnamed bark of 100–120 tons, bought from Edmund Scarborough in 1624, by Sir William St John, Governor of the Guinea Company and certain other members of the Company, with part payment and a bond covering the remainder, and then in November 1624 leased to the Company at £80 a month for ten months – patently for a voyage to River Gambia. Further expense for re-rigging the ship was incurred before it sailed, presumably in late 1624 or early 1625, but the vessel was so unseaworthy that it had to put in to Dover, where it was sold. The law suit related to the bond and the debts, and in 1625 St John was replaced as Governor (and at a later date was for a time imprisoned over the debts issue).

INTRODUCTION

cloth manufacture, dyewood and gum arabic, as well as from other merchant interests. The Company, attacked in the Parliamentary debates on monopolies in the mid 1620s, and racked by financial difficulties, became relatively inactive.[1] Jobson in vain presented a petition, first to James I and then to the new king, Charles I, seeking royal intervention to bring about further voyages to River Gambia.[2]

---

[1] For the diminished role of the Company, see Blake, 'Farm of the Guinea trade', which convincingly argues that part of the problem was the novelty of the financial arrangements, and that, despite all the set-backs, an active minority of members of the 1618 Company continued to take an interest in trade with Guinea up to the time of the reformed and more successful Guinea Company of 1631. The 1618 Company limited its collective activity but licenced trading by individual members, as its charter permitted, causing quarrels and lawsuits over trading regions between Davies and his rival and successor in commercial influence, Humphrey Slaney. Meanwhile, in 1622 the London clothiers and the Eastland merchants complained about the Guinea monopoly, leading in 1624 to Privy Council and Parliamentary discussions (*APC 1623–1625*, London, 1933, p. 247; *Journal of the House of Commons*, I, pp. 710, 793–4) – as noted by Jobson in his petition. In 1626–7 further complaints about the monopoly were made (*APC 1626 June–Dec.*, London, 1938, p. 385; *APC 1627 Jan.–Aug.*, London, 1938, items 400, 436). Nevertheless, the Company did not become completely moribund and it did more than merely 'privateering and exacting licences' (Gray, *History of the Gambia*, p. 28). Its Council met and discussed finances; voyages, especially to Sierra Leone, continued under its name and general auspices; in 1625 it was asked to send home exotic African objects such as an elephant's head (*CSP Colonial 1574–1660*, p. 75); in 1626 its new Governor, George Digby, sought through Sir John Coke, Secretary of State, the loan of two crown vessels (*Twelfth Report of the Historical Manuscripts Commission*, Appendix 1, London, 1890, p. 275); and in 1628 it proposed a further Company voyage to River Gambia (see the following note).

[2] Did Jobson's petitions to the crown have any result? In 1627, on the grounds that 'the King intended to adventure with the Guinea Company', a prize ship in crown possession, the *St Anne*, was passed to the Company – for many references to the fitting out of this ship, see *CSP Domestic 1625–1626*, ed. J. Bruce, London, 1858. However, in May 1627 and again in December of the same year, the Governor of the Guinea Company (now Sir Thomas Button) and other Adventurers, were issued with 12-months letters of marque for the *St Anne* and two other vessels, which probably thereafter concentrated on privateering rather than trading (ibid., items 439, 576; *CSP Domestic 1628–1629*, ed. J. Bruce, London, 1859, p. 297; *CSP Colonial 1574–1660*, p. 82; HCA 13.46 [19.6.1627, 25.6.1627]). (Combination of the two activities was not, however, new in Guinea voyages: the early voyages to Senegal took prizes when they could: Nunes Costa, 'O Trato Inglês', pp. 720–22). The disposal of prize ships by the crown, by loan to individuals or companies, was commonplace at the time. Probably therefore the royal desire 'to adventure' referred more to adventure in taking prizes than to adventure in trading, since a third of the prize money gained by loaned ships went to the crown (*APC 1626 June–December*, Council decision of 25.10.1626; Joyce Lorimer, ed., *English and Irish Settlement on the River Amazon 1550–1646*, Hakluyt Society, London, 1989, p. 287, reference to the loan of the *St Anne* to the Guinea Company as a precedent for the loan of a ship to the Guiana Company). For the importance of prizes and their income in 1626–7, see Roger Lockyer, *Buckingham*, London, 1981, p. 303 (and for references to the position of the navy in this decade, *passim*). That the king had taken shares in the Company, indicated by the handing over of the *St Anne* (Scott, *Joint-Stock Companies*, II, p. 13), is unlikely and unevidenced. In fact, Jobson's first petition of 1624 or 1625 faced an unmentioned difficulty: Sir William St John, Governor of the Guinea Company from 1618 to 1625, after 1623 was in disgrace with the crown. Dismissed from his post as a vice-admiral over alleged financial irregularities, during investigations up to 1628 he suffered spells of imprisonment (many references in the State papers, e.g., *CSP Domestic 1619–1623*, ed. M. A. E. Green, London, 1858, pp. 511, 561; *CSP Domestic 1628–1629*, p. 219). Nevertheless, in July 1626 his successor as Governor asked the crown for the loan of two ships (see previous note), and it was probably this approach, rather than Jobson's petitions, perhaps never read by either monarch, which eventually led to the crown offering the *St Anne*. Yet Jobson sought not merely renewed Company activity but further voyages to River Gambia. In 1626 the Governor requested the two ships 'for prevention of danger [presumably to form convoys against foreign privateers] and to prosecute a hopeful discoverie' – the latter clause suggesting a disposition to pursue the Gambia adventure. And in 1628, when the Company reached an Accommodation with Slaney (see p. 13, n. 1), he was not allowed to exclude River Gambia from his range of future operations, as he seems to have wished, but had to agree, in principle at least, to send ships there, this indicating that the Company, if not Slaney, was still interested in the river: *APC 1628 July–1629 April*, items 440, 570. Thus, Jobson's activities, his book and his petitions, may well have

During the later 1620s, individual members of the Guinea Company – singly or in partnerships, on licence or privately – organized voyages to Guinea and did continue and did expand English trading in Senegal and Sierra Leone, especially in the latter's Sherbro district. In this district, however, the Company appears to have established some control over licenced trading by its members, by appointing resident factors from at least 1627.[1] In 1628 the Company planned a further voyage to River Gambia but gained insufficient financial support to proceed.[2] Otherwise, it appears, no attempt to follow up the English penetration of River Gambia was made in the later 1620s – indeed, it is doubtful whether any significant English activity in the river was attempted before the 1640s.[3] Yet it is worth noting that when the 1618 Guinea Company was reconstituted in 1631, by concentrating its subsequent activities on a new area, the 'Gold Coast', it resumed what had been the prime aim of the 1618 Company, intervention in the West African gold trade – Jobson's 'Golden Trade'.[4]

The 1618 and 1619 English voyages to River Gambia

Most of what we know about the English voyages to River Gambia of 1618 and 1619 is from Jobson, the references in his book being occasionally confirmed by a small number of documents. The 'Company of Adventurers of London trading to

---

been a factor in persuading the Company in 1626 as well as in 1628 (see n. 2 below) to consider another voyage to River Gambia.

[1] In 1628, when the Company's stock in Guinea was bought out by Nicholas Crispe and associates (Blake, 'Farm of the Guinea Trade', pp. 100–101: R. Porter, 'The Crispe Family and the African Trade in the Seventeenth Century', *Journal of African History*, 9, 1968, pp. 59–60), documents refer to a chief factor, apparently at Sherbro, and to 'all other Englishe ffactors of the Ginny Companie ... [and] and all the Ginny Companies goods'. One man in 1627 'as a ffactor was sente to Ginny and Binny by that Company and was placed in a ffactory in the ryver of Sherbarrow in Binny, where the examinante boughte for the use of that Company one hundred and ten tonns or therabouts of red wood and some Elephants teeth'. Another man was 'ymploied to gett and provide goods for the joynte Company of Ginny and Binny', and another employee referred to 'goods sente thither [i.e. from England] by the joynte stocke of the whole Company interreste in the yeare 1627' (HCA 13.48 [14.7.1629]). Although the Company's name was freely used, and no doubt misused, by the individual traders, this appears to be conclusive evidence that the Company retained an active interest in trade on the coast. But no firm evidence has been traced to indicate that the Company continued to organize voyages, and it appears to have used resident factors to buy and store commodities on the coast and then either sell them on the spot to licenced traders or load them on the ships of these traders. At this date a 'factory' probably meant no more than the presence of a resident agent or 'factor', with his residence used as a storehouse.

[2] Blake, 'Farm of the Guinea Trade', p. 98.

[3] In the course of contradictory evidence as to whether or not in 1622–3 any of Slaney's ships had traded in Sierra Leone, contrary to his licence, it was stated that Slaney had sent out, apparently in September 1622, a pinnace, the *Helpe* (Joshua Lowe), intended for River Gambia; or alternatively, had sent out a ship, the *Harts Desire*, and a pinnace, the *Nightingale*, intended for 'Sherbrowe and Gambra in the West Indyes [*sic*]' (HCA 13.44 [11.2.1622/3, 14.2.1622/3]). Since the object seems to have been to return with 'brasil', that is, red-wood, it may be doubted whether any of these vessels entered River Gambia, since it did not normally provide red-wood – the false evidence being given in order to conceal trading in red-wood at Sierra Leone. No voyages to River Gambia in the 1630s have been traced, but occasional voyages occurred in the 1640s (HCA 13.60, 4.5.1646, pinnace seized 1643; 13.72, 17 and 20.11.1648, slave ship).

[4] The pioneering English trading voyages to Guinea in the 1550s and 1560s had sought mainly gold and hence visited 'Gold Coast'; but the later voyages had instead traded for less valuable commodities, pepper (at Benin in the 1580s and 1590s), hides (in Senegal from the 1580s), and dyewood (at Sierra Leone from the 1610s).

INTRODUCTION

Gynney and Binney' was incorporated by letters patent in November 1618.¹ But already, according to Jobson, its first ship had sailed.

> In the yeare 1618, in the month of September, they set forth a ship called the *Catherine*, burthen 120. tun,² and in her imployd on[e] *George Thompson*, a man about fifty yeares of age, who had lived many years a Marchant in *Barbary*,³ the carcazon [*sc.* cargo] of goods he carried with him amounted unto 1856£. 19s. 2d.⁴ having his instructions from the Governour and Company to enter in the River of *Gambia*, and with such shallops, as hee had, and were thought convenient for him, to follow his trade, and to discover up the River, leaving the shippe in a secured Harborough: All which in his part being carefully performed, in his absence, through the overmuch trust of our English hearts, and faire familiarity wee use to all nations, with whom we are in amity, the shippe was betrayde, and every man left in her, his throat cut, by a few poore dejected *Portingals* and *Molatos* to whom they gave free recourse aboord ...⁵

The ship, the *Catherine* (or *Katherine*), was seized and not heard of again. The attack occurred at Cassan (Kasang), 'the shippe then riding before the Towne', but the

---

¹ In the patent the company was incorporated 'in the name of *Governor and Company of Adventurers of London trading into the parts of Africa*'. But in the course of the patent the parts of Africa were defined as 'all and singular regions countries dominions territories continents coasts and places now or at any time heretofore called or known by the name or names of Gynney and Bynney', Carr, *Select Charters*, pp. 100, 103. Hence the company was usually known by the title in the text, or more simply, as either the Guinea Company or the Africa Company, titles employed by most later writers.

² That the ship left before November might be confirmed by a statement of 25 October 1618 if this related to the River Gambia voyage: 'Captain Brockett gone to Senega, to procure for the Guinea company a better trade than that of the East Indies' (*CSP Domestic 1611–1618*, ed. M. A. E. Green, London, 1858, p. 587). However, the master of the *Catherine* was Samuel Lambert (PRO, SP 14/124, no. 115), and no further reference to 'Captain Brockett' has been traced (unless he was the Edward Brockett 'gentleman' who died 'at sea' in 1620: R. H. Ernest Hill, ed., *Index of Wills Proved in the Prerogative Court of Canterbury, vol. VI, 1620–9*, London, 1912, p. 43).

³ In 1599 George Thompson (or Tomson) was a factor in Morocco, together with Jasper Thompson, perhaps a brother, working for Richard Thompson, a merchant with wide interests (see *DNB*), perhaps his/their father (T. S. Willan, *Studies in Elizabethan Foreign Trade*, Manchester, 1959, p. 303). (It does not seem to be known whether these Thompsons were related to the Thompsons, including a George Thompson, who in the 1640s were active traders to America.) In 1600 George Thompson sent to Robert Cecil, the English Secretary of State, a report on happenings in Morocco, written at Saffi, the major port for European vessels (*CSP Domestic 1598–1601*, p. 461). At one point Jobson refers to him as 'Captain Thompson', that is, as leader of the up-river expedition, if not as commander of the whole voyage ('Master Tomson' *PJ* 922, 924; 'Captaine Thompson' *GT* 7). Since the object of the expedition was to contact those in the auriferous districts who presently dealt with traders from Morocco, Thompson was most likely chosen because he had some command of Arabic, and if it was thought that he would actually meet Moroccan merchants, because of his experience and contacts when in Morocco. Moreover, it is likely that his Moroccan experience meant that Thompson had also some command of either Spanish or Portuguese, or both, languages known to be useful in Senegambia. However, his successor, Jobson, appears to have had no Arabic and it is doubtful whether he had any Portuguese – possibly Thompson had signalled that alternative qualifications would be appropriate for an assistant.

⁴ The sum given by Jobson as the value of the cargo was stated by the Company as its loss on the voyage (PRO, SP 14/124, no. 115, adequately calendared in *CSP Colonial 1574–1660*, p. 27). See p. 14, n. 2 above.

⁵ *GT* 5. A marginal note in Purchas's summary of Jobson's book reads as follows: 'Hector Nunez etc. which under colour of trade waited their time to kill the English and take their ship, Tomson and others being in their trade in the Countrey, others on shore, and divers sick: after much love and pretended kindness' (Purchas, *Pilgrimes*, 1/9/13, 1569). The reference to 'others on shore, and divers sick' is original, but it is not clear whether this information derived from Jobson or was merely a supposition by Purchas.

'bloody murther' was condemned – so Jobson claimed – by the local Africans, purportedly with the result that the Portuguese removed themselves.¹ How many Englishmen were killed is nowhere stated, but unless most were already sick, it is likely that the number overwhelmed by a 'fewe' Afro-Portuguese, even in a surprise attack, was small.² Assuming that the master of the *Catherine* stayed with his ship, he was presumably among those killed. For this disaster, Jobson blamed the 'want of care and Judgement of those Sea-men and Merchants who had the managing, by over-much trust of supposed friends'.³

Thompson and the rest of those aboard had gone up-river, apparently in shallops. Exactly how far the whole party went is not clearly stated, but

> *Thompson* upon intelligence, being gotten farre uppe into the River, and finding the inhabitants to use him curteously, with the Kings allowance of the Country, seated himselfe uppon the land ...⁴

---

¹ *PJ* 922; *GT* 30–31, 44. Jobson states that the leading Portuguese, those 'that were of the best and most ablest estates', took care 'to quit their dwellings, and to seeke out else-where, leaving none but a few poore snakes', apparently residents at points on the river other than Kasang. As for the Kasang Africans, 'some of them [were] spectators' of the massacre. But Kasang may not have been as friendly to the English as Jobson made it out to be. (For a 1616 episode when a party from a French ship was massacred in the river, allegedly by Africans but perhaps incited by the Afro-Portuguese, see p. 12, n. 2 above). When the *Saint John* went up-river, no pilot and interpreter could be hired at Kasang, purportedly, as was explained to Jobson, because 'the Portuguese had hired men of that Towne to kill us as we went up' (*PJ* 922). This suggests that the influence of the Afro-Portuguese was still strong at Kasang. However, the history of more than a century of Portuguese relations with Kasang included at least one incidence of violence. At an unstated date but probably around 1580, ten Portuguese and ten townspeople were one day killed in an affray (see Part II below: Almada, 5/6).

² It being unclear how many men Thompson was able to take up-river in his shallops, it is uncertain how many of the English were left aboard the *Catherine* at Kasang. The date of the attack is nowhere stated, nor are the dates of the subsequent adventures of the survivors before Jobson rescued a few of them in December 1620. The *Catherine* probably reached River Gambia before November 1618 and Kasang in November or December. The attack therefore probably occurred in early 1619. If so, this hardly allows enough time for the general sickness and high mortality which affected the crews of later vessels that stayed longer in the river, although no doubt, as Purchas states (p. 17, n. 5 above) 'divers' were already sick. The leader of the Afro-Portuguese, allegedly one 'Hector Nunez', i.e. Heitor Nunes, had his boat and goods seized in reprisal by the English at River Saalum in 1620, the local Afro-Portuguese 'knowing and cursing his villanie' (*PJ* 921). A Heitor Nunes was agent at Seville for trade from the Cape Verde Islands at an uncertain date in the period 1590–1610 (Maria Emília Madeira Santos, ed., *História geral de Cabo Verde, volume II*, Lisbon/Praia, 1995, p. 111), but may not have been the same man. Since some of the Afro-Portuguese traded on the Little Coast with English and other non-Portuguese vessels, to the distress of the Portuguese authorities, their disavowal of Nunes may not have been totally insincere. If, as Purchas states, some of the crew were ashore when the attack took place, but were nevertheless killed, it might be inferred that the local Africans were not as guiltless as Jobson was to claim.

³ *GT* 163/155–164/156. Anglo-Portuguese relations in Guinea were complex and varied, particularly during the wars and armed truces between England and the joint monarchy of Spain and Portugal. Broadly, whereas official relations were hostile, traders on the coast often sought, through opportunist co-existence, a *modus vivendi*, and this was sometimes insisted on by local African polities. But attacks by English vessels on Portuguese ships from Guinea as they approached Europe were common, and privateers found it difficult to resist extending their activities to Guinea. Thus, on the Little Coast of Senegal, in 1607 a Portuguese ship was seized by an English privateer, to the distress of the regular English traders in port, who were imprisoned by the local African ruler until restitution was made, it being the case that 'free trade was permitted there unto all that come thither in peaceable maner for trade': PRO, HCA 13.38 (1.7.1607). Similarly, the Afro-Portuguese of Senegambia veered between peaceful trading with the English, and acts of hostility, such as the massacres of 1588 and 1619, the latter because of either official pressure or momentary economic nervousness.

⁴ *GT* 5–6.

INTRODUCTION

It seems likely that this was where Jobson later found the remaining English, at Oranto or at a neighbouring riverside port, somewhere on the middle river, and many scores of miles above Kasang. Thompson appears to have learned about the disaster when he sent a party down-river to the ship. The English, an unstated number of them, were then instructed by Thompson – or otherwise they simply decided for themselves – to abandon the expedition.[1] Jobson tells the story twice.

> ... and thorough the kindnesse of the inhabitants, neere those parts where the shippe was lost, some of the English who came downe from *Thompson*, where [were] safely conveighed many dayes travaile over land, untill they found meanes, to meete with shipping to transport them home, with their woefull tidings ...[2]

Elsewhere Jobson says that the party of Englishmen came down-river, not to 'neere' Kasang but actually to the town.

> And further, when some of our people, who were above in the River, not knowing of this evil accident, and were upon occasions returning to the shippe,[3] whom they found so miserably lost, and carried away, the people of the Towne, especially some principall, and most powerfull men, tooke such compassion upon them, that they fed them, and lodged them, with a great deale of loving care, and that for no small time, until they had devised and concluded amongst themselves, what course to take; and having resolved, to take a tedious journey by Land, in seeking to crosse the country to the North-ward, untill they came to *Cape de Verde*, where they were sure to meete with shipping, they not onely fitted them, with such necessaries as they could, but also sent of their owne people, as guides with them: and being in that manner commended from one King to another, were lovingly entertained, lodged, and fed, and with new guides still conveyed, never leaving them, untill their desire was satisfied, and they safely arrived, where they found convenient shipping, and still the commendations that went alongst with them, from one blacke King to another, was, in regard their shippe was betraied, and taken away by the *Portingall*, whereby they found such compassion, that in some places they had horses to ride on, and in other places were entreated to rest, and recreate themselves, longer then they were willing.[4]

---

[1] The English who instead stayed with Thompson up-river were described by Jobson as 'those people of ours ... willing to stay behind' (*GT* 128), which may indicate that Thompson gave the others the choice of staying or attempting to make their own way back to England. However, it is possible that they were sent down-river before Thompson knew of the loss of the *Catherine*, his intention being that they should join the ship which would then carry them to England with his report.

[2] *GT* 6. In 1616 some survivors from a French marine disaster had been conveyed by canoe from River Gambia to 'Cape Verde', similarly to meet European shipping and be transported home (see p. 12, n. 2 above).

[3] The wording 'on occasions' may mean that the English did not return in a single party to Kasang, but in several groups, implying that they remained in contact with Thompson up-river, and perhaps that he instructed them to attempt a journey overland.

[4] *GT* 31–2. Jobson is of course anxious to stress that Senegambian Africans are friendly and helpful, only the Afro-Portuguese being hostile, although a later reference shows that this was not always the case. This was almost the earliest reported instance of Englishmen travelling any distance overland in Senegambia. In the 1590s, Richard Rainolds, trading on the Little Coast of Senegal, was sent three horses in order to travel to the residence of a friendly African ruler one league and a half inland, but he decided not to go (Hakluyt, *Principal Navigations*, II/2, p. 191). However, in 1616 a party of survivors from a marine disaster, mainly French but including a few Englishmen, after landing at an uncertain distance east of River Gambia travelled overland to the river (apparently being passed through tolerant African polities): see p. 12, n. 2 above.

This party of Englishmen reached 'Cape de Verde', that is, the Little Coast of Senegal, whose ports were regularly visited by English ships, after covering overland a distance of some 150 miles.[1] Meanwhile, Thompson remained up-river, accompanied by seven other Englishmen. This was apparently in the Oranto district, if not at Oranto itself, since the English were under the protection of a local governor, the *Ferambra*, as well as that of local ruler, both of whom Jobson later met.[2] According to an African informant, 'Thompson did use to lye at this Ferambras house, who had shewed himselfe a faithfull friend, in time of neede unto him, and his company', to the extent that 'when the Portingals had got the King of Nany to send horsmen to kill Thompson and his small company, hee did preserve them, and put himselfe & Country in armes for their defence'[3] Purchas's extracts add that Ferambra 'conveyed them over the River to his brother … and saved their goods'.[4] Why the ruler of Niani, a polity whose territory seems to have stretched down-river to just beyond Kasang and up-river to have bordered with Oranto, should have been so much influenced by the local Afro-Portuguese is not explained.[5]

---

[1] Unfortunately nothing is known about the circumstances of their rescue or how they were received in England. But clearly they reported the disaster to the Company, and since a relief vessel was then sent out, apparently during the summer of 1619, they must have reached England fairly expeditiously, probably arriving between April and June, and thus less than six months since the disaster at Kasang. Counting backwards, if before setting out overland they deliberated 'for no small time' at Kasang, this suggests that they reached there not long after the attack on the ship, and that in turn this occurred not long after the ship arrived there and once the up-river party had left.

[2] *GT* 141.

[3] *GT* 100.

[4] *PJ* 923.

[5] Since Niani is not mentioned elsewhere, the relationship between the unfriendly King of Nany and the friendly King of Cassan is not explained, but the latter, as a lesser ruler, was probably subject to the former (who in turn was subject to the *buur Saalum*). According to a Portuguese who wrote in 1625 but who had visited Kasang in the 1570s, the ruler then was 'Sandeguil, lord of this town, whom the *tangomaos* [Afro-Portuguese adventurers] call duke, since he is the person [who commands] after the king' (Part II below: Donelha, f. 25ᵛ; and cf. Almada, 2/9). A map of 1602 noted the *Ducade de Cassan* 'dukedom of Kasang' (Donelha, *Descrição*, fig. 3). 'Sandeguil' appears to be a version of the Mandinka term *santigi*, the title of a particular sort of local ruler or governor (with both secular and spiritual powers), and if the title still applied in 1620, the 'king of Cassan' may have ruled only the town and a limited neighbourhood. As Jobson later states, one man was deposed as 'king of Cassan' in favour of another by the overlord, the 'King of Bursall', that is, the *buur Saalum*, the ruler of Saalum (*PJ* 925; *GT* 59–60). This ruler had some degree of control over part of the north bank of the river and was perhaps particularly interested in the port of Wolley Wolley, close to Kasang and seemingly under Kasang rule, which apparently was the site of a depot for the salt trade from River Saalum. There is no suggestion, however, that the deposition of the king had anything to do with the operations on the river of the Europeans, either Portuguese or English. The polity of Niani, although neighbouring, and probably with the same overlord, thus appears to have had no direct control over Kasang. The ruler of Niani may at this date have lived at a 'traditional' royal site, Pallan/Palang, some 3 miles from Kasang (or at a later-evidenced royal site, at Kataba, some 6 miles from Kasang); and if the Afro-Portuguese actually left Kasang after the attack on the *Sion*, they may simply have moved to a place under his protection, which might explain their influence with him. Be that as it may, the English who travelled overland to the coast and were 'commended from one King to another', after leaving Kasang may not have been able to avoid passing through part of Niani, but, if so, were apparently not molested. This may have been because only a few hours' journey was required before they reached safety in territory more directly under the *buur Saalum*. Alternatively, the Afro-Portuguese may have seen their exodus as a relief, but later instigated African action against those English who had not left the river.

INTRODUCTION

Having received news of the disaster to the 1618 expedition, the Company rapidly prepared and sent out a relief ship, the *Saint John*, a pinnace of 50 tons, under Thomas Coxe, with a cargo, the whole operation costing some £2,000.[1] Again, we rely on Jobson's account.

> Whereupon the noble Adventurers, with all expedition set forth a Pinnace of fifty tunnes, called the *S. John*, and in her a new supply of goods, and direction to *Thompson*, either for his repaire, withall his Company home, or as he did affect his trade, or had hope of his discovery, to make use of those goods, and abide there: He utterly refused to come away, and therefore sent away the *S. John*, who for that they came in an unseasonable time, which then experience made them understand, and thorough some other abuses, which more conveniently else where I shall set downe, which [with] losse of many of her men returned, and [brought] as little comfort of gaine to the Adventurers, onely hopefull letters from *Thompson*, inviting them to a new supply, and by the next season to send unto him a shippe and pinnace, with some especiall commodity hee made mention of, confidently affirming, they should no wayes doubt of a hopefull discovery, where the Moores of *Barbary* traded, and a valewable returne for their losses sustained, promising in the meane time, which [with] such company as he had left with him, being in all onely eight persons, in his small boate to search up the River …[2]

Jobson supplies no dates but makes much of the 'unseasonable time' of the year when the *Saint John* was in the river, and by this he almost certainly means the months of the rainy season, June/July to September/October.[3] It is not stated where the pinnace contacted Thompson, but it was presumably up-river, either at Thompson's base or, perhaps less likely, at Kasang, so the pinnace must have been in the river for at least 4–5 weeks. Whereas, to explain the sickness and deaths of Englishmen, 'our Seamen directly charge the unholsomnesse of the ayre, to be the sole cause', a circumstance which would have inhibited trading in River Gambia at all times, Jobson instead argued, more conveniently, that the danger was seasonal, the problem being that the first rains brought poison with them or washed it out of the earth, thus polluting the river water. It followed that, 'for men to water [*sc.* obtain drinking water], in those pestilent times, and in the open Rivers, as the *Saint Johns* men in their first voyage did, I say it was a desperate attempt, and might have beene the confusion of them all, as indeed there were but few of them escaped'.[4] This suggests that the pinnace was in the river during the early part of the rains, perhaps in

---

[1] Two members of the Company, Thomas Love and Abraham Williams, put up between them £1,988 6s 0d, as stated in the later account of the Company's losses, but whether the total outlay was more or less is not clear (PRO, SP 14.124, no. 115). The pinnace was perhaps named in honour of the Governor of the Company, Sir William St John.

[2] *GT* 6–7. There is no suggestion in Jobson's statement that Thompson had any intention of handing over command of the expedition when the 1620 voyage arrived, and therefore it would seem that Jobson was sent out, not to replace Thompson, but to assist him. Admittedly, we do not have the Company's instructions to Jobson, but it would seem unlikely that they involved the recall of Thompson. Without the instructions, we cannot judge the degree of assistance Jobson was intended to supply.

[3] *GT* 9, 11, 13, 125–6.

[4] *GT* 126–8. Jobson repeats in the 'Conclusion' to his book that blame 'may some-wayes be laid uppon the Sea-men, whose understanding should have avoyded unseasonable times, and especially Discretion should have led them to have shunned watering in the very height of unseasonablenesse, but it may be excused for want of experience, insomuch as there had never beene any triall made, so high in the River before to any effect, to discover the unholsomnesse, with the operation thereof, whereby so many of them

July–August 1619. It did very little trading, returning to England with only £80 worth of hides.[1]

While the Company was – rather bravely – preparing to undertake what Thompson's 'hopefull letters' had demanded, Thompson himself was active. His men survived at his base; 'remayning there almost three yeeres, there was not one of them dyed, but [eventually] returned all into their owne countrey, being eight of them in number'.[2] Thompson himself, having learned, perhaps from caravans carrying salt from the sea coast to the interior and passing his base, both going and returning, that a leading merchant lived further up-river, in the Tinda region, decided to seek him out.[3] He proceeded -

> in a payre of Oares, takeing onely two of his owne Company with him, the rest people of the Countery,[4] which [with] whom hee past up the River, and got to *Tinda*,[5] a place hee aymed at; in hope to have had conference with a blacke Merchant, called *Buckor Sano*, ... [but] fayling of [meeting] him, for that hee was then in his travailes within the land, hee stayd not many houres above ...[6]

Thompson thus preceded Jobson in the whole extent of the latter's 'discovery' up-river. Jobson supplies no details of the journey, one that took Thompson some 200 miles up-river beyond his base, other than that he 'travelled some way by land'.[7] At or in 'Tinda', Thompson made inquiries about the interior gold trade, and he sent a marabout (a literate Muslim) to 'Jaye, a towne nine dayes travell higher in the countrey'.[8] Perhaps Jobson exaggerated the brevity of Thompson's visit to Tinda.

---

lost their lives ... and those few that returned [testified to the unseasonableness]' (*GT* 164/156). A further reference to 'the losse of so many lives and the dangerous sicknesse of others' (*GT* 39), probably refers principally to the 1619 experience. In the quotation in the text above, Jobson refers to 'some other abuses' which damaged the voyage of the *Saint John*, but nowhere specifies what he had in mind.

[1] PRO, SP 14.124, no.115.

[2] *GT* 6, 128. The wording of the former reference, 'being in all onely eight persons', makes it uncertain whether the total of eight included or excluded Thompson himself; but a later reference to a group of five men (*GT* 32), if in fact these were survivors and they numbered five at a time when Thompson and two other survivors were up-river, would indicate that the former alternative was the case.

[3] *GT* 84–5.

[4] The *OED* meaning of 'a payre of Oares', repeated in *GT* 84, a boat rowed by two men, seems inappropriate here, for such a long journey and for a party of a large number of men, probably most of them 'people of the Country', that is, hired Africans, who would surely be expected to be more than mostly passengers. Nevertheless *OED* does not lend support to 'a payre of Oares' instead meaning two row-boats. Perhaps the Africans took turns in rowing.

[5] A region of the interior beyond the point where the up-stream course of River Gambia, after penetrating inland broadly eastwards, turns more or less due south: in various writings Tynda, Tinda, Tanda, Tenda. Although Jobson at one point refers to Tinda as a 'Towne' (*GT* 83), he never himself visited any town in Tinda, and no town of that name is evidenced elsewhere, the major town of the district evidenced in recent centuries being Bady/Badeyo. For the king of Tinda, see the Itinerary, Appendix A, p. 218, n. 1.

[6] *GT* 6–7.

[7] *GT* 85. Thompson may have gone overland around the difficult river passages, as Jobson later did at **Baa**rakunda, as well as travelling from the river into Tinda district, which Jobson did not attempt.

[8] *GT* 101. 'Jaye' is perhaps Jenne on River Niger, although this, being some 500 miles away, would seem more than merely nine days' journey, even if mostly by river. Alternatively, it might be a poor attempt at Dhiaka, in the auriferous district of Bambuk, only some 100 miles away. Jobson suggested that 'Jaye' represented 'Gago' (*GT* 13, 102 – the text has 'Fay', a misprint for 'Jay'), that is, Gao, also on River Niger, but some 800 miles away.

INTRODUCTION

Eventually Thompson set out again for his base down-river, 'but found his expectation so satisfied, in that he had hard of the Moores of *Barbary*, and was come so neere where they frequented, that hee talkt of nothing, but how to settle habitations, and fortefie the River to defende themselves, and keepe out other nations'. Moreover,

> such an extasie of joy possest him, as it is and hath beene aleadged against him, that growing more peremptory then he was wont, and seeming to governe with more contempt, by a quarrell falling out amongst them one of his Company slew him, to the utter losse of what he had attaind unto, who in regard of emulation in striving to keepe others hee affected not [*sc.* others he had no affection for] in ignorance, committed nothing to paper, so as all his endeavours and labours were lost with him.[1]

The extraordinary murder of Thompson occurred in March 1620.[2] It is not clear whether this was during the journey down-river from Tinda, or after arrival at the base.[3] The news did not reach England before the next expedition set out, in September 1620.[4] Jobson learned of it when the ships reached Tendabaa, some 70 miles up-river, on 25 November, from a Portuguese who stated that 'Master Tomson was killed by one of his Company, and that the rest were in health'.[5] Eventually Jobson met a man called Brewer and an unnamed man, the two Englishmen who had accompanied Thompson on his fatal journey.[6]

Thompson was the pioneer of English up-river exploration of River Gambia, and his efforts were not entirely wasted. It was presumably from the body of survivors that Jobson learned about Buckor Sano, whom he was eventually to meet; and

---

[1] *GT* 7, 85. All the information about Thompson's final journey Jobson must have obtained from one or both of Thompson's companions, and Jobson travelled up-river with Matthew Broad, who may have been one of the two, or if not, could have learned about it from them (*PJ* 922, 924). The companions perhaps had reason for claiming that Thompson was mentally disturbed. No doubt Thompson, with his Moroccan experience, was excited at the possibility of establishing contact with merchants from Morocco, but it is possible that he was suffering from a fever and was slightly delirious – or that he had been drinking. The suggestion that River Gambia might be fortified to keep out 'other nations', if ever made, certainly verged on the absurd at the time, although in fact forty years later the English establishment of James Fort, in the lower river, went some way towards doing just this. However, what Thompson suggested may have been misunderstood and transmitted to Jobson wrongly. Since the English would compete with the 'Arabecks' of Barbary for the gold trade, the fortification may have been to protect the former against the latter. Jobson himself later claimed – surely unconvincingly – that the English could advance further towards the sources of the gold by using their guns, although the potential enemy was not in this case the Arabecks but the savage peoples in between (*GT* 91). But he also suggested that future English traders should operate from boats in the river 'to defend our selves, from the rage of the Barbary Moore' (*GT* 14). Purchas discreetly omitted mention of Thompson's murder.

[2] *GT* 7.

[3] Hence it is not possible to work out in what month Thompson visited Tinda, and indeed whether this was in late 1619 or in January–March 1620.

[4] This is surprising. The news must have circulated within the Afro-Portuguese community in Senegambia, and it might have been picked up on the Senegal coast and brought to England by a visiting English vessel.

[5] Since Jobson was able to interview the survivors, it can be ruled out that the murder was committed by other hands, Portuguese or African. No action against the unnamed murderer is recorded, either by Jobson while on River Gambia or by the Company after his return to England. Possibly the survivors collectively testified to guilt and a plea of necessity.

[6] *PJ* 922. Jobson did not name Brewer in his book.

Jobson must have gained from Brewer, and from the second companion of Thompson's Tinda journey, information about the up-river course – not least if one or both also accompanied Jobson on his own journey up-stream.[1]

The English in River Gambia 1620–21

Jobson's two texts tell how the English ascended River Gambia in 1620–21 and returned down-river. The topographical details are presented below in the Itinerary, and later sections of this Introduction discuss the nature of the river and the various peoples encountered. We begin the present section with a brief summary of the journey. Two ships sailed from England in late October 1620. Entering River Gambia in mid November, the larger vessel halted at a point some 70 miles up-river, but later moved up-river to the major centre of **Kasang**. Meanwhile, the smaller vessel, on which Jobson travelled, had sailed beyond Kasang to a point some 200 miles upriver. There, in late December 1620, the survivors of the 1618 voyage were contacted and trade was undertaken at a middle river base. But Jobson, with a few Englishmen and some Africans, travelling in a small boat, led his party a further 150 miles up-river. At the final point on the upper river, at the beginning of February 1621, he contacted an African trader of **Tinda**, Buckor Sano, made inquiries about the interior gold trade, and was rewarded with certain privileges by an African ruler. Leaving after nine days, the English rejoined the smaller vessel, en route observing a circumcision ceremony. Jobson, however, soon sailed again up-stream, a short distance, to visit the trading and Islamic centre of **Sutuko**. On his return, the smaller vessel sailed downstream, and for a month in March–April, lay with the larger vessel at Kasang. Both ships left the river in early June 1621.

In detail. The ships sailed from Dartmouth on 26 October 1620. The master of the *Sion*, a ship of 200 tons, Richard (?) Molhughes, died in River Gambia in early March 1621, after the ship had been lying off for five months, first at Tendabaa, then higher up-river at Kasang. Jobson said this mariner was 'known for an excellent Artsman, but in the government of himself, so farre from knowledge, that [from leaving England] … he was never twenty dayes sober'.[2] The master of the *Saint John*, a pinnace of 50 or 60 tons, Hugh Musgrove, survived (and continued to work for the Company for at least a decade), although this vessel lay off at the middle river base for some two months, and like the *Sion*, was in the river for altogether six months.[3] The 'principall Factor' of the expedition was Henry Lowe, a man who 'had spent many yeares, and made many voyages upon that continent [*sc.* Africa]'.[4] At one point, Jobson mentions 'the Merchaunts' at the middle river base,

---

[1] Another of the survivors, Broad, certainly accompanied Jobson, but we do not know if Broad accompanied Thompson, or if Brewer, who did accompany Thompson, travelled with Jobson. Possibly the apparent enthusiasm of Buckor Sano to meet Jobson – he arrived rapidly and with a large retinue – was because Thompson had made a good impression when he visited Tinda, partly by promising regular English voyages. Otherwise it is difficult to understand why the merchant was so interested in the arrival of a few strangers in a small boat.

[2] *PJ* 926; *GT* 40; PRO, SP 14/124, no.115.

[3] PRO, SP 14/124 no. 115; HCA 13/48, 14.7.1629.

[4] Since Thompson had been a merchant in Morocco, it is possible that Lowe's African experience was also in Morocco. But if the experience of 'many voyages' was in relation to Guinea, the voyages could only have been to Senegal and/or Sierra Leone.

but how many traders were under Lowe is not stated. No doubt some were left in the *Sion*, and operated down-river at and from Kasang, and when the *Saint John* returned to Kasang at least one 'factor' was there.¹ While Jobson went up-river in the smaller shallop, Lowe in the other shallop traded from the base, probably moving down-river, but most of his dozen crew became sick, and before or after the shallop reached the ship (which ship is not clear) all but two died.² We are not told whether Lowe survived. A 'Chirurgion' from one of the ships served at the base, and although he eventually died, being allegedly another drunkard, he was still alive in mid February 1621.³ Among those stung by an electric fish at Kasang was 'the Cooke' of 'the shippe', that is, of the *Sion*, an unnamed individual who thus appears to have survived at least five months in the river.⁴

As the *Saint John* went up-river in December 1620, three men were left at Mangegar, Humfrey Davis, John Blithe, and 'one Nicholas', but when the ships returned in March 1621 only Davis was still alive.⁵ Three survivors of the 1618 expedition, who had remained at a place known to the English as Oranto (an uncertain location) and were now rescued by Jobson are named: Matthew Broad, Henry Bridges, and Brewer. As mentioned above, Brewer was one of the two Englishmen who had accompanied Thompson up-river; and at least one of the three, Broad, later accompanied Jobson up-river.⁶ Among the English, only Lowe, the three men left at Mangegar, and the three survivors had their names recorded in the extracts from the Journall, while Jobson's book supplies no English names, other than that of Thompson, perhaps deliberately.

Since the 1620 party seems to have known at least approximately where the 1618 survivors were to be found, it is possible that the 1620 voyage brought out from England one or more of those seamen who had been up-river previously, being either 1618 survivors who had made their own way home, or participants in the 1619 voyage.⁷ When the *Saint John* had to be towed up-river, the two shallops were manned by 25 Englishmen, presumably selected from among the total number of sailors and traders then aboard the pinnace.⁸ Some of these had most likely been transferred from the *Sion*, although its subsequent movement up-river from Tendabaa to Kasang suggests that its crew was not greatly reduced. A normal complement for an ocean-going ship the size of the *Saint John* was probably 20–25. Given

---

¹ *PJ* 922; *GT* 40, 60, 101. In an unnamed country 'where we had much and often trade', the English 'for the most part, kept a Factor lying' (*GT* 58). But presumably this man was picked up on the return journey and may have been the same man as the one aboard at Kasang.

² *GT* 40. Lowe's movements are very inadequately described by Jobson, perhaps deliberately. For an interpretation of scattered references, see Itinerary, p. 214, n. 4.

³ *PJ* 925; *GT* 113.

⁴ *GT* 24. Since Jobson appears to have been an eye-witness and since 'the shippe' (as distinct from the pinnace) was only encountered by him at Kasang on the down-river journey, this must have been in April–May 1621. Of course, the original cook may have died.

⁵ *PJ* 922. 'Davis' may have been a relative of and agent for John Davies, the leading merchant in the Company.

⁶ *PJ* 922, 924.

⁷ A less likely alternative possibility is that Thompson had sent home detailed instructions, with conceivably a map.

⁸ *PJ* 922.

that the *Sion* was a much larger vessel, it is likely that the total number of Englishmen – seamen and traders – in River Gambia during the expedition was some 80–100.[1] It is also likely that more than a third of them failed to survive the expedition.

On returning to the *Sion* in March 1621, Jobson found not only the master dead but 'many others, and not above foure able men in the Company'.[2] The master's mate, Robert Elving, was one of those who died in the river.[3] Sickness in the *Saint John* recalled Jobson from Sutuko. Not only the surgeon but 'sundry officers that were of his societie, with their lives paid for their riotous order'. By the time the ships left the river, all their carpenters were dead.[4] Other deaths are noted above. Jobson, however, boasted that he had taken nine men and himself up-river and returned to the middle river base after six weeks without any sickness among them. Later, travelling for a fortnight to and from Sutuko with perhaps a partly different crew, 'two or three fell sicke' but none died. As for Jobson himself, he claimed that

---

[1] In the 1560s the manning/ton ratio of English ships travelling to Guinea was 1:3+: P. E. H. Hair and J. D. Alsop, eds, *English Seamen and Traders in Guinea 1553–1565: the New Evidence of their Wills*, Lewiston/Lampeter, 1992, pp. 107–8. But in the early 17th century ships may have been more fully manned. On the Algiers expedition of 1620, nine merchantmen of 180–300 tons were manned in a fairly consistent ratio, one man to 2.3–2.8 tons, although the smallest ship, one of 100 tons, had a ratio of 1:2.0 (Purchas, *Pilgrimes*, 1/6/5, 881–2). While this was a military expedition, it is unlikely that an overseas exploratory expedition would have a very different manning ratio. Again, 21 ships sailing to Virginia in 1621 had a ratio of one man to just over 2 tons, but these ships were probably carrying a number of settlers, so the seaman ratio was probably somewhat higher (ibid., 1783). Other references in the same source suggest a normal ratio of between 1:2.5 and 1:3, although very small vessels sometimes managed with tiny crews (ibid., 2/3/13, 556; 2/3/14, 567; 2/325, 574; 22/4/1, 699; 2/4/3, 711; 2/6/16, 1267; 2/8/12, 1654). The *Sion* of 200 tons therefore probably carried a crew of 50–60. As well as seamen, the ships carried a small number of traders.

[2] *PJ* 925. The high mortality of the seamen is probably to be attributed in large part to mosquitoes reaching the ships and conveying malaria to the seamen from the local African population, all of whom almost certainly carried the parasite responsible. Moreover, when African men and women were aboard they could be bitten, and any mosquito surviving on board for the parasite's incubation period of twelve days would then pass on the disease to the whites. In addition, some men may have slept ashore. Finally, the seamen ashore during the day, when net fishing, for instance, would be in close contact with Africans, although the mosquitoes would not in daylight be so active. A Portuguese writer thought Kasang 'very healthy' (Part II below: Lemos Coelho, 29). But such judgements are relative, and no doubt the Afro-Portuguese had acquired a degree of immunity in respect of the prevalent diseases.

[3] M. Fitch, ed., *Index to Testamentary Records in the Commissary Court of London 1571–1625 vol. III*, London, 1985, p. 138, reference to the *Sion* and River Gambia. According to the index, although the relevant register (no. 24) contains many entries relating to seamen dying between 1620 and 1625 on named ships other than the *Sion* or *Saint John*, it has only a handful of entries relating to seamen dying in 1620–21 on unnamed ships, and therefore our expectation that the wills of the many other members of the crews of the *Sion* and *Saint John* who died in River Gambia would be found in entries adjoining that of Elving seems to have been erroneous. None of the known names of individuals on the three Guinea company voyages of 1618–21, none of the names of the three ships, other than the single entry relating to the *Sion*, and no other references to River Gambia, occur in the Index above, or in the following: Stokes, *Index to Wills 1605–19*; Hill, *Index to Wills 1620–9*. It is likely, however, that concealed within the large number of entries in these indexes relating to seamen dying 'beyond the seas' on unnamed ships are entries relating to individuals who died on the River Gambia voyages. In any case, previous experience with seaman's wills indicates that any information about the actual voyages the missing wills may contain is unlikely to be commensurate with the labour of tracing and investigating them (see Hair and Alsop, *English Seamen*).

[4] *PJ* 925, 926; *GT* 40.

INTRODUCTION

he was 'never one quarter of an houre sick'.[1] He was, of course, anxious to claim that, with the right procedures, visiting the river was not necessarily fatal to Englishmen.[2]

The exact role of Jobson in the 1620–21 voyage is far from clear, partly because we know nothing of his earlier career and little of what followed. Since Thompson's death was not known when he left England, presumably Jobson was intended to be an assistant in up–river exploration, and if so, his capacity to take on a more decisive role does him credit. In his petition Jobson was to describe himself humbly as 'a poore seaman',[3] but nothing in the extant material shows him doing more than navigating a shallop up an admittedly difficult river. He travelled from England aboard the smaller vessel, and although in 1621 he sailed down-river from Kasang in the larger vessel, it cannot be assumed that he replaced the dead master or master's mate, or that he returned to England in this ship. At Sutuko, when he was accompanied by two Englishmen, his African guide introduced him as 'the Captaine and Commander of our people'.[4] He certainly represents himself as the leader of the group of three Englishmen on this occasion, and if his account is trustworthy, as the leader of the up-river expedition from the middle river base. The marabout may have considered Jobson to have been the leader of the section of the English who had travelled up-river on the *Saint John*, but may have been unaware of the existence of a larger vessel down-river, which since the marabout returned from up-river to his home situated far short of Kasang he never actually saw. Purchas calls Jobson normally 'Master Richard Jobson' or 'M. Jobson', but on one occasion 'Capt. Jobson', however he may have been following the passage quoted above, or have simply used the term as a general honorific.[5] Perhaps more significantly, the entry in the Stationers' Register refers to the author as 'Captaine Jobson'. This may have been a marine title, a military title, or again, merely a general honorific for one in command.[6] Whatever his standing as 'captain', Jobson makes it clear that he was subordinate to the chief factor, Lowe, who selected the best men for his own shallop.[7] Possibly Jobson downplays Lowe's role and activities. As probably one of the few leading men to survive, Jobson was able to tell the story his own way; possibly he had stepped into dead men's shoes, and no doubt he tended to over-rate his enterprise and capacity. Nevertheless, it is difficult not to believe that he was, throughout

---

[1] *GT* 40. Similarly, in 1724, although aboard the ship lying off for five weeks at Cuttejar Stibbs found 'most of the Sailors sick, and one dead', he boasted that, of those who went up-river, 'we lost neither one Man, nor was there one hardly that were sick; on the contrary, those that were in a weak Condition on our setting out, grew afterwards very healthful, fat and strong' (Francis Moore, *Travels into the Inland Parts of Africa*, London, 1738, pp. 296–7).

[2] This was, of course, contrary to the experience of the Portuguese, but Jobson was unaware of the Portuguese sources which noted the unhealthiness of the river for whites (Part II below: Fernandes, f. 105; Almada, 6/19).

[3] It may be relevant that at one point Jobson refers to the 'Star-board side' of the river (*PJ* 923).

[4] *GT* 64. Similarly, at one point Jobson referred to 'Captaine Thompson', *GT* 7.

[5] Purchas, *Pilgrimes*, 1/1/1, 28. Probably on the strength of the reference in the text, a later voyager, Hodges, spoke of 'Captain Jobson' (Thora G. Stone, 'The Journey of Cornelius Hodges in Senegambia 1689–90', *English Historical Review*, 39, 1924, p. 90).

[6] See p. 1, n. 1 above. 'Captain Thompson' (p. 17, n. 3) was a 'factor' or trading agent, and as far as we know, neither a marine nor a military officer.

[7] *GT* 40.

the expedition, a man in some authority, with a measure of expert knowledge, possibly both in seamanship and trading, who had been entrusted with a particular task, assisting the up-river 'discovery'. A plausible hypothesis is that he represented one of the leading 'Gentlemen' members of the Company, an individual whom he had previously served in some capacity, albeit, seemingly, not in Guinea. No doubt he was somewhat overshadowed in the early stages by those who had had such experience, and who were therefore, most likely, the agents of the leading merchant, John Davies. His inexperience, however, encouraged him to write an account of River Gambia of striking freshness, as well as of intelligence and empathy.

Jobson's first mention of a need for African assistants is when he notes that, at Kasang, he was unable to recruit a 'Blackman to goe with us to be our Pilot and Linguist'.[1] He fails to mention any local pilot on the lower river and the ships may have done without one, although it is not inconceivable that they picked up one near the mouth, or even earlier. However, Jobson does give the impression that, voyaging as far as the middle river base, the English did without any African assistance.[2] At Batto Jobson hired a young marabout, Fodee Careere, whom Jobson calls his 'Alcaide' or personal representative. Jobson speaks with respect about this man, with whom he formed 'an affectionate league', and from whom he gained much information about Islam and the local Islamic community. 'I did not only aske and require his advice, but in most things allowed and followed it'.[3] Jobson also was joined by a run-away youth, Sanguli, who had spent the last two or three years in the company of Thompson and his English party, and who as a result had a good command of English and at least at times acted as interpreter.[4] Jobson attended the youth's circumcision ceremony and says nothing to his detriment (other than that, in adolescent fashion, he blatantly disregarded his mother's plea to leave the English and return home).[5] At Baarakunda, Jobson hired two more Africans, chiefly to supply physical force to assist in manoeuvring the boat up-stream, but they remain shadowy figures. Jobson's account has far more references to the two Africans, Fodee Careere and Sanguli, than to any one of the English, who are almost all unnamed and noted only in generality.

Apart from the set-pieces describing his meetings with kings, with the merchant Buckor Sano, and with a dying chief marabout, also his attendance at a circumcision ceremony, Jobson supplies occasional glimpses of the behaviour of the English on the river. They fished from the ship, or at Kasang used nets from the shore; later,

---

[1] *PJ* 922.

[2] Exceptionally, according to a casual and unexplained reference in the Journall (but not mentioned in the *GT*), as the party approaches the Oranto district a message is sent to the survivors by a 'slave' (*PJ* 922). For an attempt to explain the presence of this individual in the English party, see *PJ*, p. 187, n. 6.

[3] *PJ* 922; *GT* 63, 76, 109. He called him 'the first of the Country, who ever I entertained and continued with me'. 'Fodee' (Mandinka *fode*) is a title of respect for a learned Muslim, and 'Careere' perhaps represents an Arabic personal name, Khalīl.

[4] *PJ* 922.

[5] *PJ* 922; *GT* 83, 109.

up-river, they fished from the boat with 'hooks'.¹ For food, as well as the fish, fowl and animals they killed, they bought 'beeves', goats, hens, milk, butter, oranges – and down-river 'aboundance of Bonanos' – and they sampled palm wine and the local beer.² At times they were given foodstuffs, such as 'a brace of partridges', and once Jobson was persuaded to try elephant meat.³ The English ate the local staple, rice, and 'even to us, it is a very good and able sustenance'.⁴ When up-river, the party sent the canoes to the shore for wood, which must mean that they did some cooking aboard the shallop.⁵ Jobson had strong views about not eating and partaking of 'hot drinks', that is, alcohol, during the hottest part of the day. But his party had liquor in the early morning and as a sundowner – 'the course we used, [being] a small cup before meate', and again, 'another after'.⁶ The ship and the boats carried *aqua vitae*, that is, brandy, and also sack, both to trade to Africans (or donate to kings), and to sustain the English; but Jobson complained that insufficient liquor had been provided for good health, possibly because of his notion that drinking river water was dangerous.⁷ Jobson carried spices and perfumes, *spica romana* and orras, perhaps for personal use rather than sale, and a cordial, *rosa solis*.⁸

The main commodity imported seems to have been iron, but this proved of little interest up-river (where iron came from the interior). Instead, salt was the commodity mainly in demand, Jobson wishing he had more than the 40 bushels he carried,⁹ and blaming Lowe, and perhaps by implication the organizers of the voyage, for the wrong provision of goods.¹⁰ Thompson had recommended an 'especiall commodity' for import but Jobson does not specify what it was – it may well have been salt.¹¹ To marabouts Jobson sold small quantities of paper, and to the chief

---

¹ *GT* 23, 24, 96. Fishing with nets was impossible near the river banks because of the fallen tree trunks (Moore, *Inland Parts*, p. 268).
² *GT* 36, 39, 43, 82–3, 96, 117, 131–2.
³ *GT* 111, 149/141.
⁴ *GT* 125.
⁵ *GT* 83.
⁶ *GT* 40, 42, 82, 87–8. Some of the drink was in bottles – 'two or three bottles of our best liquor' (*GT* 100).
⁷ *GT* 40, 42.
⁸ *PJ* 924; *GT* 71, 76, 97.
⁹ *PJ* 922, 923; *GT* 13, 16, 88, 90, 95, and note 165. The imported salt was not mined salt but 'bay salt' (that is, salt evaporated from salt water), as was the local salt, but was better, because the latter, presumably because less refined, was a 'course and durty kinde' (*GT* 80, 88). A Portuguese, however, thought the local salt as good as the salt of Portugal, only darker (Part II below: Lemos Coelho, 2).
¹⁰ Up-river Jobson feared that 'our provision, which was small, might faile us, and poorely (God knowes) we were provided of those materials, that would have helpt to maintaine that principall', *GT* 16. This probably refers, first, to the provision of food, and secondly, to the shortage of gunpowder and shot that limited the shooting of animals and birds for food.
¹¹ *GT* 6. Although Jobson gives the impression that his salt came from England, it is conceivable that some at least was Guinea-coast salt purchased at the salt depots down-river, notably Wolley Wolley and Kasang, or even when at River Saalum. The accounts of River Gambia not known to Jobson contain many important references to salt, its production in River Saalum (and also in the salt-water lower reaches of River Gambia – it is curious that Jobson never mentions the salt water of the river), the depots in River Gambia, and the trade to the interior by canoe and caravan: see Part II below: Gomes, f. 279; Almada, 6/11; Lemos Coelho, 2, 3, 16, 18, 55, 56, 61. Two centuries later, Park still noted that 'in the interior country the greatest of all luxuries is salt' (Mungo Park, *Travels in the Interior Districts of Africa*, London, 1799, ch. 21).

marabout's wives he gave pewter rings and to other women beads.¹ To Buckor Sano he gave 'a string of Christall, and a double string of Currall', and a companion gave a silver chain.² As on previous voyages, the goods purchased were, down-river, hides, and up-river, ivory, cotton, and cloths.³ Jobson also mentions 'some gold', but makes so little of this that the quantity must have been trifling.⁴

The ships carried light cannon, Jobson at one point measuring the width of the river by the range of his ship's 'Ordnance', but there is no evidence that they were ever fired in the river. Jobson and some other individuals had 'peeces', including fowling-guns and 'five Musketts'. But Jobson had also a 'stonebow or pelletbow', and there are references to 'our guilt swords' and to killing peacocks with 'our speeres'.⁵ However, Jobson grumbled that his party was short of both guns and powder, or at one point short of 'a good Peece', hence, to scare away hippopotamus at night, candles, fixed on boards, were floated down-stream.⁶ In his Journall Jobson comments on the 'immoderate heat in the Suns height', but in his book makes surprisingly few comments on the effect of the heat on the English, although he does state that his boat-party rowed only during the cooler part of the day, even if the tide was then favorable, and that when the men had to enter the water 'naked', to haul the boat, they found it 'a good refreshing', and 'comfortable'.⁷ We are not told what clothes the English wore, other than that Jobson had a 'gown' aboard which Buckor Sano apparently appropriated.⁸ (The interpreter, Sanguli, wore 'shirt, breeches, and a cap of stript stuffe'.)⁹ The shallop carried 'a small Globe, and our Compasse', presumably in Jobson's care.¹⁰ Jobson collected souvenirs to bring back to England, cola nuts, feathers from a crane, and a paper with writing in Arabic script; while he notes that Guinea fowl were brought to England, perhaps by earlier travellers if not in 1621, and that his party purchased a monkey, although it is not clear whether the animal survived the down-river and homeward voyage.¹¹ In order not be confused with the local musicians, who were despised, the English took care to 'neglect and especially in their [*sc.* the Africans'] hearing to play upon any Lute or Instrument which some of us for our private exercise did

---

¹ *GT* 65, 78, 94, 112.

² *PJ* 924.

³ *GT* 89. Jobson fails to mention wax, a product of River Gambia available in large quantities. In 1606 a Portuguese trader claimed in the past to have bought 200 quintals of wax in the river in one year (Guerreiro, *Relaçam anual*, part 1607, liv. 4, cap. 9, f. 158, reprint, II, p. 211; English version, item 35). However, although some wax was available higher up (see Part II below: Donelha, ff. 27–8), the trade in wax was principally near the mouth of the river where Jobson does not record that the English traded in 1620–21. The trade there in wax and honey is discussed in Stephan Bühnen, 'Geschichte der Bainunk und Kasanga', Ph.D. thesis, Justus-Liebig University, Giessen, 1994, pp. 233–6.

⁴ *PJ* 923; *GT* 14.

⁵ *PJ* 922–5; *GT* 32, 84, 90, 156; Petition, f. 4.

⁶ *PJ* 923; *GT* 21.

⁷ *PJ* 923; *GT* 15, 18, 21, 137.

⁸ *GT* 88.

⁹ *GT* 113.

¹⁰ *PJ* 924. The globe was presumably carried to impress and educate Africans. On the river the compass can only have been of use to to obtain the general trend of the course, which Jobson may have roughly charted. Cf. Appendix D, p. 235.

¹¹ *GT* 134, 143–4, 151/143–152/144, 155/147.

INTRODUCTION

carry with us'. We can presume that, going up-river and away from local settlements, the boat party was provided with music. At the furthest point up-river, 'our manner was to set our watch with a Psalme' (probably a metric psalm) – perhaps this was a practice throughout the journey up-river and even on the ship throughout the voyage – and the psalm may have had an instrumental backing.[1] It is unclear whether all the English slept in the boat or went ashore at night, but Jobson had his 'bed' aboard.[2]

At an onshore base, the English kept a 'small dogge', which hid under a bed when felines prowled outside.[3] There, 'through daily recourse' with the local Africans, they were visited by children, who would 'many times steale from their homes, and come and hang about us', but whose parents were distressed when the English offered them wine or 'sweet things' (the former contrary to Islamic discipline, the latter perhaps only to a domestic dietetic regime).[4] One passage in Jobson's book has been widely quoted by historians and other writers. Offered three women slaves, he replied that – 'Wee were a people, who did not deale in any such commodities ...'.[5] The reply has been hailed as lies and hypocrisy, the detractors pointing to the Hawkins slaving voyages. These occurred sixty years earlier and Jobson's statement was in fact generally correct for his day. Jobson did not buy slaves, the English voyages to River Gambia did not, the voyages of the 1618 Guinea Company did not, and the many English voyages to Guinea between the 1580s and the 1630s almost without exception did not.[6] Moreover, it is likely that the episode has been wrongly construed, for while Jobson undoubtedly replied to the offer in terms of slaving ethics – or at least affected to have done so when he wrote his book (the phrase is not in the Journall extracts) – it was surely made in quite other terms. The African merchant offering 'certaine young blacke women' and having heard Jobson's reply, 'seemed to marvell much at it', slaves being regularly sold down-river 'to white men, who earnestly desired them, especially such young women'. As a courtesy, and one not uncommon in Afro-European relations, the party of English

---

[1] PJ 924; GT 108. In 1724 Stibbs took aboard 'a Balafeu (which is a Country Musician) to chear up the Men, and recreate them in an Evening' (Moore, *Inland Parts*, p. 253). But the party included a large proportion of local Africans, and it was probably to cheer them, rather than the English, that the player on the local xylophone (*baloo/balafon* – see GT, p. 152, n. 1) was hired. However, in the 1590s a hostile source claimed that the English seduced the Portuguese of Senegambia by 'holding banquets for them, to the sound of music from violins and other instruments': Almada, *Tratado breve*, chap. 2/8. And for English psalm-singing in Guinea at an earlier date, see Hair, *Africa Encountered*, item II, p. 204.

[2] GT 87.

[3] GT 138.

[4] GT 75. Or perhaps there was some dark suspicion of aliens poisoning?

[5] GT 88–9.

[6] While research remains to be completed on the history of English trade in Guinea in the period, received statements about alleged instances of slave trading are inaccurate, e.g., J. Holland Rose, A. P. Newton and E. A. Benians, eds. *Cambridge History of the British Empire, Vol. I*, Cambridge, 1929, pp. 139, 155. Thus, the alleged slave cargo of one 1629 ship, the *Benediction* (as cited in Porter, 'The Crispe family', p. 60; Moraes, *Petite Côte*, p. 224), is contra-indicated by the port of lading (River Senegal), the intended port of unloading (London), and the known trade of the owner (Slaney), and is not confirmed by the documented list of goods aboard: PRO, HCA 13.49 (25.1.1630, 114.9.1631). There is, however, evidence that the English very occasionally acquired slaves from captured foreign vessels. Since the ships involved did not sail to America and patently did not convey the slaves to England, how they disposed of them is puzzling, but perhaps they sold them to other foreign vessels, or to the coastal Portuguese, in Senegambia or conceivably at S. Tomé.

males was being offered the women for sexual services, and Jobson's refusal, perhaps based on a lack of understanding, must indeed have puzzled the African.¹ As he pointed out, young women were much desired by white males down-river. Jobson is silent about the sexual behaviour of the English in the river. He himself was patently attracted by the physical appearance of the Fula women, who a generation later, when similarly bringing provisions to vessels in the river, were described by a Portuguese as 'extremely beautiful and very neat, but not at all chaste'. Of the wives of the local musicians, Jobson reported discreetly that 'if there be anie licentious libertie, it is unto these women, whose outward carriage is such wee may well conceit it'.² Probably longer-visiting Portuguese were more likely to recruit black bedmates than Englishmen briefly passing by in crowded boats.³ But a number of the English had spent over two years at Oranto and it is implausible that they had remained celibate throughout that time. Indeed, it is conceivable that when Buckor Sano spoke of white men down-river desiring young black women, he had in mind not only the Portuguese but the English refugees at Oranto, whom he had most likely heard of through commercial channels.⁴ Nor is it likely that the sailors and traders aboard the two vessels lying for months down-river in 1620–21 lacked occasional female company.⁵

It is difficult to be sure how much of the information about the river and its peoples supplied by Jobson represented what he himself had observed and noted. He contended that his account was 'written from mee either as an eye witness, or what I have received from the Country people'.⁶ Although 'Country people' normally

---

¹ For white visitors being offered members of their African host's family as 'comfort women', see Part II below: Fernandes, f. 106v; also Lemos Coelho, *Descrições*, ch. 9, pp. 38–9 (at Sierra Leone, with restrictions). Cadamosto was presented with a twelve-year old girl 'for the service of my chamber', by a Muslim ruler in Senegal, who was later said to lawfully sleep with the 'young negro girls in attendance upon each of his wives' (Crone, *Cadamosto*, chs xx, xxii).

² *GT* 34, 37, 107; Part II below: Lemos Coelho, 44.

³ Jobson does not make clear the extent to which, during the boat journeys, the English stepped ashore. They had, however, 'a small Cano, that was ready at all times to put ashoare: and when wee came to an anker, to fetch wood, or any other provisions, or likewise to carry us ashoare, and bring others to us' (*GT* 83). He does indicate that at 'divers times' they walked alongside the river, sometimes to shoot at animals (*PJ* 923; *GT* 119), and probably they went ashore to defecate. Since the boats must have been crowded, it is possible that some men slept ashore. A century later, those travelling up-river in canoes went ashore 'to dress our Victuals, and to avoid the scorching Heats of the Sun', also 'to refresh ourselves', and Stibbs' lieutenants wrote a letter 'under a green Shade by the River-side' (Moore, *Inland Parts*, pp. 280, 281, 283). Nevertheless, Stibbs 'chose always to anchor in deep Water, and the middle of the River, if possible, for fear of Accidents' (ibid., p. 284) – it is not clear whether he feared sudden changes in water levels or attacks from the local Africans.

⁴ He may have actually met them. 'Thompson ... hearing of divers Caravans, that past in the country, and went down to the King of *Bursals* dominions for salt, had learned that the principallest man ... was Buckor Sano ... who maintained and kept 300. Asses following that tedious travell. ... [He] was travelled higher into the Country, in the sale ... of his salt Commodity' (*GT* 84). Clearly Thompson had not met Buckor Sano before he set out for Tinda, but after Thompson's death the merchant may have travelled downriver to procure salt and conceivably have encountered the English survivors at Oranto – or at least he may have passed nearby and heard about them. Probably the local friend of the English, Ferambra, supplied women for the survivors, the women to report back to him on the general conduct of their bedmates.

⁵ A Portuguese source reported that, at Kasang, 'the black females are the most beautiful on the whole river ... Even if a white gets into trouble with one of these women – and may God deliver him from this – not even then will the blacks rob him' (Part II below: Lemos Coelho, 29).

⁶ *GT* 8.

INTRODUCTION

refers to the local Africans, here it may have been intended to include the local Afro-Portuguese, from whom he almost certainly also received information.[1] (The problem of translation is considered below.) A source of information not signalled by Jobson but almost certainly important was the discourse of the English survivors of the 1618–1619 expedition, whose knowledge of local matters after a residence of over two years on the river must have been considerable, at least in certain aspects. Nevertheless, there are many statements in the account which seem to arise directly from Jobson's personal experience and which give some impression of his curiosity and effectiveness as an observer. Among the glimpses of the writer and his immediate comrades, apparently Jobson's testimony 'as an eye-witness',[2] are ones of them confronting baboons, shooting elephant, concealing themselves in the tops of anthills while hunting, admiring weaverbirds' nests, and being surprised by an electric eel; also of meeting a fly-blown cowman, watching Islamic ritual ablutions, testing and admiring the leaf response of a mimosa plant, handling a baby after its rescue from being carried away by an 'ounce', and finally, by invitation, feeling the scarification on a woman's back and visually assessing a male circumcision.[3] Jobson also heard strange noises, such as the calling of crocodiles at dawn 'much resembling the sound of a deepe Well', the 'yalping' of a jackall, and general African music ('a noyse and din', 'the horrible din').[4]

Nevertheless, much of what Jobson reports came not from his own direct experience but from information passed to him by others. Some came from the Englishmen who were rescued, and some possibly came directly from Sanguli, the youth who spoke English. But most came from individuals who did not speak English. Unfortunately Jobson eschews the significant problem of communication and gives very little guidance as to how information was obtained across language barriers. Since the English irruption into River Gambia was novel and very recent, we can be reasonably certain that few if any of the inhabitants of the river had acquired even a smattering of English.[5] Jobson does not detail the process of translation which therefore must often have occurred. On several occasions the English met and conversed with Afro-Portuguese. At Tendabaa a Portuguese 'told us' about Thompson's death, the 'us' being the undivided English party, therefore a member other than Jobson may have translated. At Mangegar, the name of a Portuguese was obtained, but no speech is recorded. At Pompeton, 'above which dwels no Portuguese in this

---

[1] Indicated for instance by Jobson's use of Portuguese rather than Mandinka terms for occupations, e.g. 'Ferraro or Smith' (*GT* 120), Portuguese *ferreiro* 'blacksmith'. Such terms may have reached Jobson in conversations with Afro-Portuguese, or they may have been supplied by an interpreter who thought that these terms, being used by the Afro-Portuguese, were common European terms and therefore appropriate to give to an Englishman.

[2] *GT* 8.

[3] *GT* 23–4, 35, 55, 69–70, 114, 135, 138, 153/145, 158/150.

[4] *PJ* 923; *GT* 46, 111.

[5] A possible exception might be any of the Portuguese traders who moved between the river and the Little Coast of Senegal, which English shipping had been visiting for several decades. On that coast, during the later 16th century a number of African traders and notables had spoken French, but a Dutchman claimed that after 1600 some spoke English instead (Ratelband, *Reizen*, p. 13); however, it is unlikely that any of these coastal African traders and notables were encountered by the English in River Gambia.

River', Jobson and others on the *Saint John* were entertained by a Portuguese, who explained a particular phenomenon at length, but again Jobson need not have been the translator.[1] At Kasang, an African, 'in regard he had some small knowledge of the Portingall tongue, had great recourse amongst us', so those aboard the *Sion* must have included individuals who understood Portuguese.[2] Possibly therefore there was some acquaintance with Portuguese among those who went further up-river in the *Saint John*. These episodes, as recorded, unfortunately leave it uncertain whether Jobson himself had any Portuguese. Yet it is almost certain that some of his information was obtained through that language, as the recording of certain Portuguese terms suggests, and the transmission of such information must have been either directly, through one of the Afro-Portuguese contacted, or indirectly, through an African or Englishman who had been in verbal contact with an Afro-Portuguese. It is likely that, although the Afro-Portuguese on the river had some fluency in Mandinka, reciprocally some Mandinka were fluent in Portuguese. It is therefore conceivable that one of Jobson's African assistants had a measure of Portuguese, although Jobson says nothing to confirm this. A further possibility is that Jobson's shallop had aboard a 1618 survivor who knew some Portuguese. While it is unlikely that the survivors had had any length of contact with any Afro-Portuguese on the river, they may have included men who had been selected for the 1618 voyage precisely because they already had Portuguese, it being known by the organizers that this would be required on the river. All in all, it remains unproven that Jobson himself had any Portuguese.

Many of the Africans with whom Jobson reported verbal discourse are unlikely to have spoken, under the circumstances, in other than Mandinka. Some of the rescued English may have acquired a degree of command of that language and translated for Jobson, who is unlikely in a few months to have acquired more than a sprinkling of courtesy terms. But at one point Jobson speaks of hiring 'Blacke-men, as I had occasion to use them, as interpreters, likewise to send abroade, and to helpe to row'.[3] The two men hired at Baarakunda, apparently Mandinka-speaking, certainly both rowed and acted as messengers, and in the latter capacity may indeed also have interpreted with non-Mandinka Africans higher up the river. But the reference to 'interpreters' is perhaps more likely to have had specific reference to earlier 'Blacke-men', and these were undoubtedly Jobson's assistants, the marabout Fodee Careere, and the youth Sanguli. The latter was described by Jobson as follows; 'hee was come to speake our language, very handsomely, and him I used many times as an Interpreter'.[4] On one occasion when Sanguli was silenced, Jobson lost the information spoken in Mandinka by Buckor Sano. In his book, Jobson refers to the episode as follows: 'by reason of a disaster that fell in the way betwixt mee and my chiefe interpreter, I was hindred from understanding divers particulars'. In the extracts from his log, the incident is blamed on others present who 'would not suffer

[1] *PJ* 922; *GT* 118.
[2] *GT* 24.
[3] *GT* 18.
[4] *GT* 83.

my blacke boy to let me know what he [Buckor Sano] spoke'.[1] Unless Purchas miscopied the log at this point, it would seem that, on this occasion at least, Sanguli was the chief interpreter. The subsidiary interpreter need not have been Fodee Careere but may have been one of the English. Yet Sanguli, being uncircumcised and therefore considered a child, would not have been an acceptable person to deal directly with Mandinka men, especially notables. Assuming that questions were normally addressed to Mandinka-speakers by the senior assistant, it is possible that Sanguli generally served to interpret between the marabout and Jobson, and vice-versa. However, much information reached Jobson from Fodee Careere, and Jobson portrays lengthy discussions with his assistant as if they were directly man-to-man. Were they really conducted only through an intermediary? On this awkward basis, could it be that the two men managed to achieve not only a businesslike but an affectionate relationship? This would seem surprising. Could their conversations instead have been more direct?

There is nothing in Jobson's account to suggest that the marabout had had any connection with the English before Jobson arrived at his town, yet it cannot be ruled out that he had worked with Thompson and his men and had thereby acquired some English. This is, admittedly, as conjectural as the marabout's knowledge of Portuguese. There is, however, one episode which may clarify matters. With the marabout present, Sanguli is withdrawn to participate in the circumcision ceremony, yet Jobson and encountered Mandinka are represented as continuing to exchange verbal discourse.[2] The only possible explanations would seem to be, either that one of Jobson's companions on this occasion was a rescued Englishman with some acquired capacity in Mandinka who now obliged by translating, or else that Jobson and Fodee Careere managed to communicate directly, either in their respective commands of Portuguese, or perhaps much less likely, in English. Be that as it may, it seems that Jobson obtained those parts of his information not based on actual observation largely through translations of answers to inquiries, perhaps at times through double translations. Information passed through the instant translation of a third party is likely to contain misunderstandings and errors, and some caution is therefore necessary in accepting *au pied de la lettre* the statements allegedly made by Fodee Careere, or, for that matter, by other encountered Africans when the information was distilled through an interpreter.

Although we only have the account of one man to go by – and Jobson no doubt had his own axe to grind – it is clear that relations among the English were not always happy. The extraordinary murder of Thompson by one of the English, an event made more sinister by Jobson's reluctance to go into details and name the murderer, was indeed exceptional. But clashes among the English appear to have been not uncommon. The key word is 'emulation', meaning at that period, the desire to excel to the extent of producing jealousy of rivals. Thompson's 'emulation' took the form of keeping in ignorance of his discoveries regarding the gold trade,

---

[1] *GT* 93; *PJ* 924. The episode is discussed in *GT*, p. 143, n. 1.
[2] *GT* 112–14.

those of his companions whom 'hee affected not'.¹ Jobson does not use the term elsewhere in his book, but it appears in his Journall. Lowe, the chief factor, took the larger boat, but, as Jobson slyly notes, had less success in keeping his men healthy and alive. At the middle river base, 'Lowes emulation hindred us with delayes, both now and after'.² When Jobson is trading at the highest point and finds that salt is the major commodity required, he refers to carrying 'but forty bushels at first'; and earlier, in relation to the middle river base, he comments that 'the neglect of bringing Salt thorough ignorance or emulation was a hinderance'. Although Lowe was at the base with Jobson, the criticism, apparently of not bringing enough salt up to the base, perhaps means that the chief factor had left a quantity of salt down-river with the *Sion* or with the three traders stationed at Mangegar.³ But it was not only with Lowe that Jobson had difficulties. Up-river, when obtaining valuable information from Buckor Sano, 'more I might have knowne, had not the emulation of my companie hindred'. It is not Lowe who is here indicted and the mysterious reference to the interpreter, Sanguli, being silenced, is heightened by the alternative version in the book, Jobson losing information because of 'a disaster that fell in the way betwixt mee and my chiefe interpreter' – the episode is marginated as 'An unhappy accident'.⁴ It is not easy even to conjecture the nature of the 'disaster'. It is just possible to suppose that 'my companie' referred to all present, and that it was the marabout who silenced the youth, perhaps objecting to the merchant disclosing to the English too many trade secrets about the gold trade. But, more plausibly, 'my companie' may have referred to the other English. It is difficult to see where 'emulation' came in, but conceivably at least one of them may have been fearful that something was about to be disclosed relevant to the murder of Thompson. Jobson had already had a problem with his companions in the boat. When the messengers were sent away to seek Buckor Sano and did not return on the expected day or the next day, 'our men began to grumble, and my especiall consort to speake out, there was no reason wee should hazard our selves by staying any longer, in regard it was fallen out, as we were told below, that they [the local people at Tinda] were a bloody, and dangerous people, and therefore those people we sent up were murdered; and if we stayed our turnes would be next'.⁵ Jobson gives no clue as to the identity of 'my speciall consort', but perhaps he was the rescued Englishman, Broad, who may have been up-river before. Given that Thompson had fallen out with his companions

---

¹ *GT* 7. It is, admittedly, difficult to see why the undisputed leader of the party should have suspiciously regarded his comrades as rivals to keep in the dark. It is perhaps more likely that he simply saw no need to pass on to probably less well educated individuals the information he had obtained. Further more, *pace* Jobson, it seems unlikely that the comrades had absolutely no idea about the contacts Thompson had made and the information gathered. In fact, they told Jobson about his sending a marabout to Jaye.

² *GT* 40; *PJ* 922. Jobson refers to 'the principall Factor' but does not name Lowe in his book and the charge of 'emulation' is only in the Journall, which may indicate that Lowe survived and was alive in 1623 when the book appeared. The accusation that Lowe caused high mortality by being stingy with the ration of liquor aboard his boat, resulting in the men stealing and drinking too much, is a very skewed one.

³ *PJ* 922. Although Jobson does not express this, he may have thought that the English should have bought local salt when at Kasang.

⁴ *PJ* 924; *GT* 93.

⁵ *GT* 85.

INTRODUCTION

and then been murdered, and that at least one of those companions was a member of the boat party, Jobson was no doubt extremely relieved when, the following day, the messengers returned with Buckor Sano and his entourage.

In less directly confrontational terms, as noted above Jobson accused the master of the *Sion* of being a drunkard, and also the surgeon and other officers of the *Saint John* who indulged in 'riotous order'. The sailors in Lowe's shallop behaved with 'great greedinesse' by stealing liquor from the stock aboard.[1] The three English voyages had suffered very high mortality, but Jobson, calling for further voyages to the river, was desperate to argue that it was not inevitable mortality caused by the climate. He therefore blamed the 'Sea-men' who had led the second voyage for not avoiding 'unseasonable times' and in particular for permitting 'watering' in the river in those times, although allowing that this might have been through inexperience. As for the disaster on the first voyage, that was caused, according to Jobson, by 'the want of Care and Judgement of those Sea-men and Merchants who had the managing', in their over-trusting the Portuguese.[2] Wrong provisioning in England for the third voyage was not merely due to inexperience or poor judgement, but also to 'the neglect of some ill wishing persons, who deceive the trust the worthy Adventurers apposed upon them'.[3] Replying perhaps to the 1622 petition to the Privy Council seeking restraints on the Company, Jobson argued 'against all Maligners and secret opposers'.[4] Finally, in respect of his complaints against associates, after his return to England Jobson was involved in the fiasco of the 1624 or 1625 voyage, and in his petition to the crown he blamed this on the 'differences betweene the Gentlemen and the Merchaunts'. These had led to the merchants by 'secret practice' denying the Company suitable ships.[5] However, when an unsuitable ship sailed, the ship's

---

[1] *GT* 40.
[2] *GT* 164/156. In obscure language, the final paragraph of the 'Conclusion' seems to be saying that, since short-term profit has been denied, 'they seeke to discourage you, yet by all publique and secret meanes can be devised, they both have and doe still addresse themselves, to proceed and goe on by the same adventure' (*GT* 165/157). 'You' being the Gentlemen, 'they' are presumably the Merchants, and probably the latter, especially Davies, were vetoing a further voyage to River Gambia – where Jobson optimistically claimed a prospect of long-term profit – and were instead seeking licences to trade elsewhere in Guinea where the conditions were already proved to be more favourable and involved less lay-out of venture capital.
[3] *GT* 21–2.
[4] *GT* 162/154.
[5] The 18th-century anonymous editor of Astley's *Voyages* was sure that 'Jobson's design in publishing this work was, in great measure, to expose the malice and underhand dealings of the merchants (against whom he seems to be much irritated) as well as to set forth the laudable endeavours of the Company'. He summed up the position percipiently but incautiously as follows. 'This opposition was set on foot by certain Merchants of the new Company, who envious perhaps that Gentlemen and others, not bred Merchants, should share with them in the trade, or any but their own servants should be employed in it (both which are suggested in the author's obscure way) laboured all in their power to overthrow the undertaking. The method they took was to join one Low (Lowe) in commission with Jobson, and get others of their creatures into the ship, who carried every thing their own way, and obstructed the likeliest measures. By this means they were not sufficiently provided with goods, powder, and other materials, and by artful insinuations, that the people on the river were barbarous, grounded on the killing of Captain Thompson, a sufficient cargo was not sent. These men, at the same time that they gave all the opposition they could in secret, and exclaimed against the voyage by their emissaries, pretended in public to be very forward in promoting it. This their hypocrisy and craft Jobson endeavours to expose': [Thomas Astley], *A New General Collection of Voyages* ..., 4 vols, London, 1745, I, p. 176. Since it is unlikely that the editor had any original information about events more than a century earlier, the passage is partly interpretation and partly speculation, and as such cannot

company was 'devided', because 'divers' of the those aboard were Roman Catholics, shipped by their co-religionists among the Adventurers.[1] Both the merchants and the gentry therefore were responsible, and Jobson appealed over their heads to the crown, in the interests of 'the honor, wealth and preferment of our owne nation'.[2]

Richard Jobson

Nothing certain is known about the life of Richard Jobson other than what is contained in his own writings.[3] The name Jobson remains uncommon nationally, but in the early seventeenth century appears to have occurred most frequently in Yorkshire and in London. In 1621 a Richard Jobson held a Crown grant of tithe corn in a Yorkshire manor. In his petition, our Richard Jobson described himself as 'a poore seaman'. A Richard Jobson in 1604 held a minor post in the Admiralty at Portsmouth, and in 1608 a 'Jobsun' was an officer of the Lord Admiral of Ireland.[4] Whether any of these was the same man as the writer cannot be established. His account provides one clue about his earlier life. His fairly frequent references to Ireland and his knowledge of native customs there strongly suggest that he had lived and worked in that island.[5] In 1607–10 the Plantation of Ulster took place. A Francis Jobson, who had gone to Ireland in 1579, and after 1585 had been appointed 'measurer' (surveyor), to map and draw into plots forfeited lands in Munster, produced in 1589 a map of 'The Provence of Munster'. He was then employed in mapping Ulster and his name appears on many maps.[6] When Richard Jobson went up-river, he carried a small globe and a compass, and he describes keeping an accurate account of the river, 'all which way is perfectly known, and from part to part

---

be accepted as sound evidence for the specific charges against the merchants. It is to be noted that the editor confused the killing of Thompson, which was by one of his own men, not by the 'barbarous people on the river', with the massacre of the crew of the *Catherine* by the Afro-Portuguese.

[1] Petition, f. 2. The reference to Roman Catholic Adventurers was presumably aimed at George Digby, the second Governor 1625–6, and his relatives. Apart from causing division aboard ship by *odium theologicum*, probably one objection to Roman Catholics was they might act sympathetically towards, or even take the side of, the Portuguese on the coast.

[2] As he put it in *GT* 2.

[3] *DNB*; Robert Munter and Clyde L. Grose, *Englishmen Abroad, being an Account of their Travels in the Seventeenth Century*, Lewiston, 1986.

[4] *CSP Domestic 1619–1623*, p. 328; Petition, f. 4; *CSP Domestic 1603–1610*, ed. M. A. E. Green, London, 1857, p. 169; *Calendar of State Papers, Ireland, 1606–8*, eds, C. W. Russell and J. P. Prendergast, London, 1874, p. 550. No Richard Jobson has yet been traced in a variety of early 17th-century genealogical indexes, and our Richard Jobson seems to have published nothing before or after his 1623 work.

[5] He compares Fulbe women favourably to Irish women, and mentions 'kernes', the term for light armed Irish soldiers (*GT* 37). The cries and lamentations at an African burial (*GT* 71), are compared to those of the Irish. A griot (or musician) is likened to an Irish rimer (*GT* 105), and Gambian weapons are compared to those of the Irish (*GT* 45). The 'manner of taking away their wives' in the Gambia is 'in some sort used in Ireland at this day' (*GT* 56).

[6] R. Dunlop, 'Sixteenth-century maps of Ireland', *English Historical Review*, 20, 1905, pp. 314–5; *CSP Ireland 1598 Jan.–1599 Mar.*, London, 1895, pp. 445–6. Of his experiences in Ulster Francis Jobson wrote: '[it] ... is inhabited with a most savage and rebellious people, from whose cruelty at that time God only, by his Divine power, delivered me, being every hour in danger to lose my head.' In 1598 he wrote a tract recommending the despatch of 11,000 chosen men to eight different localities in Ulster.

INTRODUCTION

observed, and every reach in order by me set downe, and carefully kept', this being to 'order and give directions, what boates or vessels are most apt and proper to follow the discovery withall, as well for speedier passage, as also for the most advantage, to a more profitable returne.'[1] No connection between the two Jobsons can be evidenced, but given a degree of common interest in surveying they may have been related.

Jobson appears to have been engaged by the Guinea Company only at the time of the 1620 voyage, his account giving no indication that he had accompanied the earlier voyages or indeed ever been in Guinea before 1620. He was, however, knowledgeable about the earlier voyages, suggesting that he had seen the Company's records or had been briefed by one of the members. Apart from its conventional dedication, his book mentions by name none of the members. But in both its dedication and its 'Conclusion' it addresses only 'the noble gentlemen Adventurers', pointedly excluding the merchant members. After his return in 1621, Jobson published his book in 1623, sailed on the aborted voyage of 1624 or 1625, and wrote his first petition probably in early 1625. The extant petition to King Charles, probably of 1626 or 1627 and possibly a mere repeat of the earlier one, grumbled about the 'differences betweene the Gentlemen and the Marchaunts' which had helped to halt further ventures, but Jobson was still in the employ of the Company. Since he spoke bitterly of 'the secret practice of the Marchaunts', it seems unlikely that he was connected with this group of members. Probably, instead, he had a patron among the Gentleman, and if so he had most likely been a servant of his patron and accordingly been recommended to the Company. Since the Governor of the company in 1620 was a former naval commander and the members included the Lord High Admiral and other naval officers, it is conceivable that Jobson was employed through Admiralty connections. Appealing to the 'noble and worthy Gentlemen', Jobson specifically referred to those who in the present peaceful times had a 'cessation from those sea affaires, they were wont to be busied in'.[2] However, his less uncertain Irish connections may have been a more decisive factor. Many of the Gentlemen had had, or still had, interests in Ireland, often in the form of official positions. It is uncertain whether the Governor of the Company, Sir William St John, was related, or at least closely related, to Sir Oliver St John who in 1620 was Lord Deputy of Ireland; but the Oliver St John, esquire, who was another founding member of the Company may have been (at the time there was more than one man with this name) the son and heir of Sir Oliver. The Deputy Governor, Sir Allen Apsley, had been victualler of Munster as well as to the navy, and the third individual named in Jobson's dedication, Sir Thomas Button, was currently admiral in charge of the Irish coasts.[3]

Jobson's book tells us little about his educational background. His style of writing can present multiple clauses in enormously long sentences, and although this was partly the influence of current style his excesses seem a personal foible.[4] The style

---
[1] *GT* 155. See p. 30, n. 10 above.
[2] *GT* 4.
[3] For all the individuals with Irish connections, see *DNB*.
[4] This awkwardness in style rules out that it was ghosted for Jobson.

accentuates a tendency to ambiguity and obscurity, but these do not seem to have been deliberate, and the vocabulary and general expression indicate a well-schooled writer. The material is almost wholly original, and the impression is given that he had read very little of what was available on Guinea before going there. He mentions no sources, although at one point there are echoes of a story put into circulation by Cadamosto and often borrowed by secondary writings; and it seems likely that he knew of the reference in Hakluyt to the 1588 massacre of English trading in Senegal, and that he borrowed the variant 'Gambra' from this work. It must be said, however, that little on River Gambia was available in English, or indeed in any European language. Jobson includes a few Latin terms and refers to Quintus Curtius and the exploration of Jugurtha, suggesting a modest grammar school education.[1] It is possible, although very far from certain, that he had some Portuguese, although this would have been unlikely to have been acquired if his earlier career was entirely in England or Ireland. The freshness of his description of Islamic practices in the Gambia region appears to derive largely from his ignorance about the religion. He does not grasp that the books of the marabouts are written in Arabic, but thinks the script 'like Hebrew'.[2] And his naive belief that the knowledge of Old Testament names by the Muslims indicated a willingness to be persuaded to convert to Christianity suggests that he had no knowledge of Islamic history.

Jobson's book expressed the conventional piety of his day, with many interjections of 'please/bless/praise God', most of them seemingly routine but a few perhaps expressing a deeper religious feeling.[3] Hoping that the book might be to 'the advancement of God's glory' was mere pious rhetoric, even if that preceded a claim of the national good.[4] As a Protestant Englishman, he cited the Bible, although only a few times, surprisingly, deploying some interpretations that latterly seem curious but may not have seemed so at the time.[5] Rowing up-river, the boat party kept the Sabbath, although 'only two houres in the evening', and, at least when their destination was reached, watch was kept with a psalm.[6] On Christmas day 1620, 'at a dinner … we had varieties of meate'.[7] Despite his presumed earlier career in Ireland, Jobson's attitude to Roman Catholics was not notably fierce. His objection to

---

[1] The reference in *GT* 2–3 to the Latin sources, Quintus Curtius and Sallust (*The Jugurthine War*), being standard ones, may well have come from a secondary source in English (but they appear to be not in either of Purchas's compilations).

[2] *GT* 68. Purchas has a more knowledgable margin note; 'It seemes they speake some Arabike words: in which their Law is written' (Purchas, *Pilgrimes*, 1/9/13, 1572).

[3] *GT* iii, 2, 10, 15, 16, 33, 41, 80, 82, etc. The extracts from the Journall totally lack these, and it seems unlikely that Purchas, a clergyman, when extracting removed pious asides.

[4] *GT* 2.

[5] *GT* 53, 76, 123, 156/148.

[6] *GT* 83, 109. It is not clear exactly what Jobson means when he says 'our whole time spent … amounted to but foure score and eight houres, in which time, our Sabboth day was observed, onely two houres in the evening'. The journey was stated to be of 'eleven dayes' (although from the 'evening' of the 15th to the 'morning' of the 26th is actually somewhat short of ten full days, while elsewhere Jobson called it '12 dayes': *GT* 12). Eleven days, at eight hours rowing per day, makes Jobson's 88 hours. Yet if Jobson is saying that on the one Sunday (20 January) only two hours were rowed the calculation no longer holds. The other interpretation, that Sunday was only observed for two hours in the evening, seems unlikely, however.

[7] *GT* 149/141.

INTRODUCTION

English Roman Catholics on the abortive fourth voyage limited itself to stating that their presence caused division.[1] The Afro-Portuguese on River Gambia he despised, and noted that they 'have amongst them, neither Church, nor Frier, nor any other religious order', yet he allowed that they retained, 'with a kind of affectionate zeale, the name of Christians'. Unlike Jesuit critics of the *lançados* ('run-aways'), he did not accuse them of adopting local religious 'superstition', but only of uniting with black women and, when they died, leaving their children under African subjection.[2] His discovery that Islam was monotheistic – 'they do worship the same as we do, the true and only God, to whom they pray, and on his name they call' – and did 'acknowledge' a historical Jesus, encouraged him to assess the religion fairly positively, unlike many Christian contemporaries who believed that it involved the worship of Mohammed.[3] His judgement on the non-Muslim Africans was more commonplace. He noted a number of ritual practices, but especially – like all other writers on Senegambia in earlier times – the wearing of magical amulets which 'encrease their superstition'.[4] However, while not offering any form of justification of polytheistic animism, he failed to condemn it loudly, either because he thought that overall conversion to Islam was only a matter of time, or more likely because he wished to give a not too unfavourable picture of the majority of the Africans on River Gambia.

In the extracts from the Journal, the 'fearefull spirit', Hore, was termed only a 'Spirit', but in the book becomes 'their divell'.[5] This led Jobson into an area of intellectual difficulty, one equally experienced by the contemporary Jesuit missionaries in Guinea. Since seventeenth-century Christians believed in regular and patently-detectable diabolic intervention in human affairs, African paranormal phenomena could not be easily dismissed as both diabolic and unreal. Jobson accepted that the seemingly miraculous knowledge of the future arrival of the English was conveyed, via African diabolatry, to an up-river Portuguese, but consoled himself that 'the devil was ignorant' about the details.[6] His expressions of divine providence move from difficulty to confusion as he ties himself up in theological knots. Because 'these unbeleeving people' (he appears to be thinking mainly of the Muslims) do not accept God's Son, they 'shall feele and feare his omnipotent power', this being offered as an explanation of the 'contagious clymates' under which, in his view, the Africans suffered. Moreover, whereas they were punished for 'the hawty aspirings of sinnefull man', the English homeland was spared climatically.[7] (In the same vein, although God had imposed hard labour on mankind, for 'in the sweate of thy browe shalt thou eate thy bread', nevertheless He had given the English, but not the

---

[1] Petition, f. 2ᵛ.
[2] *GT* 28, 30.
[3] *GT* 67, 73. They only 'acknowledge Mahomet'.
[4] *GT* 50.
[5] *PJ* 925; *GT* 115.
[6] *GT* 118–9. In some parts of Black Africa such news could be conveyed by African drum language, but among the Mandinka drums are not employed for communication other than for simple signalling. (The many references to Gambian talking drums in Alex Haley's *Roots* are therefore erroneous and false.) Despite what Jobson alleges, given the twists in the river and the fairly slow progress of the English vessel, a message must have been conveyed overland and passed on before the English arrived.
[7] *GT* 129–30.

Africans, knowledge of how to use the beasts of the field to lessen their labour.)¹ Another view of God's providence was little more than tired mental reflex (and represented poor observation): ordained to operate only at night were lions and 'those beasts which are most tyrannous to others, and boldest against man, as ashamed of their bloody actions'.² Finally, in his 'Conclusion' Jobson turned another unhappy intellectual somersault. Noting the many deaths on the second voyage, he argued that it was convenient – 'wherein the power of God was manifested' – that some died but not all, those lost making the point of the danger, and those returning being able to explain that the deaths were unnecessary, being occasioned by drinking poisoned water. This is Pangloss with a vengeance.

It is well to recognize that Jobson was, in many aspects, a man of his time, lest his degree of fairly tolerant and kindly judgement be unduly assessed as that of a man before his time, an apostle of liberal relativism. Nevertheless, whatever the number of his strained interpretations, he was capable of careful observation and sensible deductions. He gave evidence that fear of Hore produced the actual reaction of an initiate that local people claimed to result from the power of the spirit.³ But his conclusion was that Hore was a hoax, 'some illusion' and – in very twentieth-century terms – invented 'to forme and keepe in obedience those younger sort'.⁴ He had doubts about the existence of the unicorn.⁵ His inquiries about Islam were intelligent. If he was wrong in referring to 'priests' and the 'high Priest', he correctly spelled out the fundamental objection of orthodox Islam to the divinity of Jesus, although careful to precede it with a personal statement of orthodox Christian belief.⁶ His relating the alleged large size of the penis among the (cursed) male descendants of Ham to an Old Testament text and female disorders was ingenious, if dubious and essentially derogating.⁷ But male circumcision,⁸ which as an un-English ritual he could have been expected to ascribe wholly to specific social and

---

¹ *GT* 123–5.
² *GT* 136–145/137.
³ *GT* 117.
⁴ *GT* 117–8.
⁵ *GT* 154/146.
⁶ *GT* 73.
⁷ *GT* 52–3.
⁸ *GT* 115. It is just possible, although Jobson does not express the notion, that there was some connection in his mind between the need for male circumcision and the size of the male organ. In the extracts from the Journall Jobson notes that 'the uncircumcized Females he [Hore] regards not' (*PJ* 925). This is Jobson's only reference to female circumcision, and may well be merely the result of Purchas misinterpreting while summarizing, since the comparable statement in the book reads – 'some of their uncircumcized sons are instantly taken away, females he meddles not with'. Whether the existence of female circumcision among the Mandinka was known to Jobson is therefore unclear. (For a specific reference to clitoridectomy, see Part II below: Lemos Coelho, 17.) Since today the circumcision ceremony for girls is normally held in a different year from that for boys, if he did know about female circumcision and the female ceremony it is likely that he could have gained the knowledge only casually and indirectly, perhaps from the English survivors, who would have become acquainted with the operation through their contacts with females. If he really wrote that Hore disregarded the 'uncircumcized Females', the statement is not transparently correct. The girls' circumcision ceremony differs from the boys' ceremony in that Hore does not attend, and perhaps Jobson had been told this and that this is what he meant. Yet, while girls, unlike boys, would not be 'eaten' by Hore, females in general would normally be only distant spectators of male ceremonies – as Jobson noted, at the male circumcision they stood 'a little distance off',– or totally absent, as if fearful of the power of the male

ethnic motives, as indicated in the Old Testament in the case of the Hebrews, or to specific historical conservatism as the Africans themselves were content to claim, he instead approvingly concluded was solely for sound medical reasons – 'of meere necessitie, as a morrall law for the preservation of their lives and healthes'.[1]

In more general terms, Jobson was often open-minded and sensitive. He believed that the marabouts were 'wonderfully desirous to confer with me about our Religion', and although he found Islamic ablutions odd, particularly in their attention to the genitalia, he spoke warmly of several leading marabouts, tolerating, with understanding, even being ritually spat on by one of them. He approved of the self-discipline and abstinence of the marabouts, and of their literate culture – their interest in 'bookes of great volumes'.[2] That, behind their courtesy and civility, there must have lain darker thoughts about the infidel English he perhaps only dimly perceived, when on occasions they were elusive about interior trade.

Jobson felt for the Fulbe who were oppressed by the Mandinka, and even more for African womenkind ('no woman can be under more servitude').[3] Here, however, when he worried because the women ate and drank separately, and did heavy domestic labour, he was blindly disregarding, unrealistically and unempathically, deeply embedded cultural markers, as perhaps most evident when he puzzled over why, in public, Africans displayed no 'familiar dalliance, insomuch as I thinke, there is hardly any Englishman can say, he ever saw the Blackeman kisse a woman', and why co-wives did not 'brawle, or scold' each other.[4] As discussed earlier, Jobson has been widely remembered for his claim, when offered three women slaves, that the English 'did not deale in any such commodities'.[5] Jobson expressed this, not as a personal opinion on slave trading, but as a general English opinion on the subject at the time. His intention was to distinguish the English from the Portuguese, and it is a moot point how general the opinion in fact was among the English, or at least among the educated English. Moreover, there were practical reasons why the boat party could not take aboard the women Jobson was offered.[6] Jobson has few other references to slaves in Africa and does not comment on their status or condition, but mentioning the slaves in Spanish America, 'in their Mines, or in other servile uses', he perhaps implies that their treatment was bad. However, he speaks of the slaves being sold to the Portuguese by 'the country people, when they commit offences', and promises to discuss this further – but forgot to do so.[7] He

---

spirits. So while Hore disregarded females, females did not totally disregard Hore. But in this there would be no distinction between circumcised and uncircumcised females.

[1] Mungo Park, however, expressed a similar sentiment, circumcision being 'considered not so much in the light of a religious ceremony, as a matter of convenience and utility' (Park, *Travels*, ch. 20). Possibly these writers, as Christians, wished to dissociate themselves from the ritual and religious meaning of the Jewish practice.

[2] *GT* 62–80.

[3] *GT* 35, 53–4.

[4] *GT* 54–5. On the final point, Jobson was in fact probing a delicate matter, for while African co-wives share a kitchen, etc, quarrels over their respective children are far from unknown, leading many modern Africans to prefer household monogamy.

[5] *GT* 89.

[6] Partly, perhaps, because these were young women, for whose services the men aboard might quarrel among themselves.

[7] *GT* 28–9.

thus does not display the sympathy for slaves that he displayed for the Fulbe and African women.

Jobson nowhere uses the term 'African/s', but instead refers to 'black/s', not however extending the term to the Fula who are 'tawny' (as is the Barbary Moor).[1] Many times Jobson refers to the 'Barbary Moor', but only occasionally does 'More/s' replace 'black/s', and then normally only when it refers to Muslims.[2] When being more specific he often distinguishes 'the men' from 'the women', or else he uses status and occupational terms, and thus he is not much given to broad generalizations about all Gambians (or all Africans) and avoids some of the conventional stereotyping. Several 'Kings' are despised, mainly because of their heavy drinking, but it is an African merchant who supplies him with the hardest judgement: 'the Kings of our Country ... eate and drinke, and lye still at home amongst their women'.[3] However, Jobson has 'much familiar conference' with the King of Cassan, the King of Jelicot was obliging, and the 'Ferambra', a local governor, 'shewed himself a faithfull friend, in time of need', to the English.[4] The marabouts ('Marybuckes') are regularly distinguished and usually treated sympathetically. But a marabout showed up divisions among the Mandinka, controverting Jobson when the Englishman described the party from Tinda, led by Buckor Sano – no marabout or strict Muslim since he drank alcohol – as 'good people, and very loving to us'.[5] The non-Mandinka visitors from the south side of the river were described as 'a more savage kinde of people', who wore animal skins rather than cotton clothes, and consequently Jobson's Mandinka companions did 'in scorne jest and deride' them.[6] Aspects of Mandinka dress Jobson thought odd, but he was more intrigued by the body-markings of the women, which they proudly showed him, although he added that they concealed their genitalia.[7] Jobson's description of the Gambian craftsmen is essentially neutral in tone, although he admires the workmanship in leather, hardly to be improved upon in England, and he records the very twentieth-century complaint that primary products such as hides and ivory, are exported to benefit foreign skills deliberately kept secret from Africans.[8] His emotional involvement is, however, considerable when he considers the 'laborious travell [*sc.* travail]' of the entire population in agriculture, particularly in tilling the ground without animal assistance.[9] Less justified was his feeling for 'the misery of the people'

---

[1] *GT* 33, 37, 92. The term 'Negro/es' is twice employed but not directly of 'blacks', the Afro-Portuguese disliking being called 'Negro' and local cloths being called 'Negros' cloths' (*GT* 30, 89).

[2] In the book, the only use of 'Moor' for 'black' is in relation to children 'sporting' at a point in the river which has been blessed by a marabout to keep away crocodiles, and perhaps the implication that these were Muslims led Jobson to use the term (*GT* 24). In the extracts from the Journall (and assuming that the term was not introduced by Purchas), the Africans travelling up-river with Jobson thrice become 'Moor/s' – not all were marabouts but probably all were Muslims (*PJ* 923).

[3] *GT* 92. But cf. earlier – 'eate, drinke and sleep, and keepe company with their women, and then in this manner consume their time untill Time consumes them' (*GT* 59).

[4] *GT* 59, 93,

[5] *GT* 101.

[6] *GT* 94.

[7] *GT* 50, 55–6.

[8] *GT* 131.

[9] *GT* 123–4.

INTRODUCTION

because the rainy season was, in the English appreciation of it, 'violent and fearefull'.[1] Jobson's curiosity about Gambian behaviour extended to that of the animal kingdom, and he patently enjoyed his observation of the exotic flora and fauna and even his close contact with elephants, hippopotamuses and crocodiles. If he gives a slight shudder when considering the last, he is cheered by what he considers the African admiration of the closeness of his contact to all (which was probably instead a head-shaking over the inexperience and folly of the whites).

What, indeed, did the Africans think about Jobson and his English comrades? Given that these were almost the first non-Portuguese to penetrate to the upper courses of the river, they were most likely barely distinguished from the Portuguese, being seen as occasional infidel aliens. Although they brought welcome trade, there can have been little confidence that their visits would be other than spasmodic – if they ever returned. While Jobson made courtesy visits and gifts to kings, and while he attempted to establish mutually beneficial trading relations with merchants such as Buckor Sano, and also, by showing interest in 'book' matters, flattered the self-esteem of the marabouts, he was – inevitably – not always circumspect. Angry with the man who had operated on his young African assistant, Jobson refused to give him the customary present; and 'four ancient men', showing 'great displeasure', stopped him from entering the bush of the newly-circumcised.[2] He and his 'consorts' had to be prevented from reckless spying on the voice of Hore.[3] At the Whiteman's Town, a friendly 'ancient Marybucke' intervened to settle 'falling out betweene the people and us', although the confrontation may not have been Jobson's fault.[4] He attempted to challenge customary attitudes regarding women, by trying to have them eat with the men, (seemingly) providing them with alcoholic drinks, and by dancing with them.[5] Yet, all in all, Jobson was tolerated as – correctly – a passing and unimportant visitor.

Jobson, like Thompson, had promised Africans that the English would return. Thompson's promise was fulfilled, Jobson's was not. It is most likely that the English voyages made no lasting impression on the affairs of River Gambia, and that Jobson was soon forgotten by his contacts. Yet there are grounds for supposing that the experience of 1618–21 and the reading of Jobson's book contributed significantly

---

[1] *GT* 125. Such malediction of the weather of Guinea was commonplace among European visitors in earlier times. Probably their fear of storms was due to the risk to life the storms threatened at sea, an aspect which did not affect Africans, who therefore seem not to have shared the fright of the visitors, despite European claims to the contrary.

[2] *GT* 114. Given that the English were themselves uncircumcised, as no doubt had become known to the Africans, their attempt to penetrate the secrets of the circumcision rite was perhaps additionally resented, since their condition might damage the spiritual efficacy of the rite.

[3] *GT* 118. In this aspect, he risked a severe reaction. That this was not forthcoming may have been because he was accompanying Muslim clerics, whose attitude to such animist practices was at best ambivalent, many regarding such things as contrary to Islamic beliefs, and hence the manifestations man-made or imaginary – as Jobson considered them.

[4] *GT* 69.

[5] *GT* 54, 61, 112. However, his dancing with women took place at the time of circumcision ceremonies, when reversals of customary behaviour are today tolerated and regarded as part of the fun. But he was disturbed that women watched male circumcision: *GT* 113.

to the eventual English decision to return to the river.[1] This happened in the 1650s and 1660s, although the later experience was to confirm the experience of the earlier in two unhappy aspects, that the climate was fatal to Englishmen and that access to the interior gold trade was a more or less unattainable vision.

The African Setting: The River

River Gambia reaches the West African coast after flowing some 700 miles, generally westward. Below its upper course it has no large tributaries, and its considerable but very varying flow depends essentially on the seasonal rainfall of the interior. During and for some time after the interior wet season of April/May to September, the water level in the upper and middle courses can rise up to thirty feet above the dry season level, with overflow often swamping neighbouring low land. However, despite frequent fierce currents, the higher flow helps navigation over the sandbars and rocks which impede passage at certain points in the meandering middle course and at many points in the narrower upper course (beyond Baarakunda). Another help to navigation is provided by the strong tides, which operate, with seasonal variation, up to 150–200 miles up-river. The winds tend to blow down-stream in the dry season, up-stream during the rains; but in either season, in stretches where the course meanders, and the available passage is too narrow for tacking, the prevailing winds are often a hindrance rather than a help. However, the wind usually drops at night, at times allowing progress to be made against the prevailing wind on an overnight tide, at least in those stretches where the course is clear. In earlier times, when wind and tide failed to assist, ships were regularly towed by rowed boats.

The tides bring salt water up-stream, producing the mangrove swamps which line the river along most of its lower and part of its middle course. In other parts, the banks have high cliffs, and stretches of the navigable course have woodland beyond the swamps or cliffs, with the trees actually impeding passage at points in the middle and upper courses where the main channel moves close to the bank. Landing places are therefore restricted in number, and where villages have been established on dryer land, up small streams navigable for at least canoes and launches, the entrances to the streams are often difficult to detect from the river. Nevertheless the traffic on the river, both in ferrying across and along it, and in conveying goods up and down stream, has always been such that a series of regular landing places has for centuries been in use, as witness the frequency of toponyms containing (often as a suffix) the Mandinka term *tenda* 'landing/crossing place'. Other sources mention the lively canoe traffic; very curiously Jobson does not.[2]

---

[1] In 1651 a new set of patentees attempted to prospect for gold beyond Baarakunda Falls, and in 1652 Prince Rupert, visiting River Gambia to attack Commonwealth vessels, appears to have been persuaded that the river afforded vast prospects, leading him after the Restoration to be a leading member of a new company which instantly set under way an exploration of the river (Gray, *History of the Gambia*, pp. 31–2, 35, 38, 52–3). Remarks in the papers of those involved in 1651–2 suggest that they had knowledge of Jobson's book (and perhaps also of the Gerbier text), and it is conceivable that Rupert had seen the petitions which Jobson had sent to his uncle and cousin.

[2] See Part II below: Cadamosto, [17], [22]; Gomes, f. 279; Donelha, f. 27^v.

INTRODUCTION

Such landing places did not always have habitations nearby but instead served a settlement further back from the river. The Portuguese called the landing places *portos*, 'ports' or 'harbours', but the facilities were of course normally rudimentary. (In recent centuries the British called the landing places 'wharfs', probably because larger vessels were now employed and the building of wharfs into the river allowed them to be loaded in deeper water, although many places given the description actually had no wharf, vessels being still handled from the banks or beaches.) Shifting of sandbars, and no doubt other changing circumstances, have led over the centuries to certain landing places being abandoned and some new ones established. Trails or 'roads' from the interior today reach the river at ferry crossings or general landing places, and there is some evidence that the roads shown on the maps produced since the 1880s were traditional highways, probably in use in Jobson's time.

Navigation on water is affected not only by physical conditions but by the form of vessel. Whereas the physical conditions of River Gambia have probably changed very little, the vessels using the river have in the last century and a half changed greatly. But before that, between the fifteenth and the eighteenth centuries, the changes were inconsiderable. Non-machine-powered navigation up River Gambia, such as that of the English in 1620–21, was described by several early Portuguese sources, which usefully detail the course of the river. But it was evidenced even more relevantly by Cornelius Hodges[1] and Bartholomew Stibbs[2] who described up-river

---

[1] Stone, 'The Journey of Cornelius Hodges', pp. 89–95. Hodges travelled in a sloop drawing 2½ feet of water, heavily laden with 1,000 iron bars, and three 'boats', the party consisting of thirty-four men. To pass the Baarakunda Falls he may have unloaded his cargo, portaged it, and reloaded it. He claimed to have taken the sloop an estimated 78 miles beyond Baarakunda and then, in only thirteen days, to have continued in 'a small sciffe', with three men, for a further estimated 148 miles, which would have taken him well beyond Jobson's final point. However, he gives the distance from James Island to Baarakunda as 'more than 558 miles', a two-thirds increase on the correct distance, so his other distances may also be exaggerated. His account is thin and vague but his list of toponyms, allegedly 'past the confines of Cantore', a polity thought at this period to extend at least as far as the Grey River, includes two rivers on the north side, the 'Nyolaein', probably a misreading of Nyolacin or even Nyelacon, and thus representing River Niériko, a tributary of River Gambia, and 'Nyolo Ba', probably representing River Niokolo Koba, a later tributary. Although both rivers had seemingly been reached earlier by Vermuyden (see p. 48, n. 1 below), in each case Hodge's account appears to supply the earliest record of the local names. Hodges also reported three identifiable ethnic names of the further interior (Badjady = Pajady, Cunnyage = Koniaji, Batchore = Bassari), a general feature name (Futa, as in Futa Jalon), and a string of unidentified names (Bibo, Tallo, Lumbo), the localities apparently further in the interior. While it is difficult to believe that he could have penetrated a considerable distance in so few days, given that he travelled as late in the dry season as April, so that perhaps he merely collected the later toponyms, nevertheless there is a case for believing that he travelled at least as far as Vermuyden. It is notable that after these voyages of 1661 and 1689–90, there is no further record of European travel on the final part of the upper course of River Gambia before the mid-19th century, and no evidence of European knowledge of the area.

[2] 'A voyage from James Fort up the River Gambia by command of the Royal African Comp. of England' (a report and letters mainly from Bartholomew Stibbs, 'Conductor' of the 'expedition'), in Moore, *Inland Parts*, pp. 245–97. Stibbs travelled in a ship up to Cuttejar, further progress being by a sloop, of unstated tonnage, ten individuals aboard, with five canoes of varying size and draught (3'2"–4'9") carrying forty-eight men and three 'Women Slaves for Cooks'. The sloop halted at Baarakunda, the canoes, previously at times carried aboard (282), proceeded further up-river. Stibbs does not provide a clear topographical indication of the points passed beyond Baarakunda, some being named but all the further ones unidentified from later sources. However, he claimed that from the final halt 'to Tinda … by Land is but a small Day's Journey', yet he estimated the distance travelled from Baarakunda as only

journeys in 1689 and 1723–4, in vessels little dissimilar to those on which Jobson travelled seventy years and one hundred years earlier, Stibbs providing a detailed log of progress up-stream almost as far as Jobson ventured. A third post-Jobson, up-river voyage, by Vermuyden in 1661, was the first to penetrate further than Jobson – being surpassed in turn by Hodges – but unfortunately lacks navigational details.[1] Hodges travelled in later March, Stibbs in January–February, each later than Jobson's up-stream voyage. Certainly, in Stibbs' case, the later dates seem to have worsened the difficult conditions of the upper river (beyond Baarakunda), although the large seasonal variations in water levels are of course regularly modified by small annual variations in the period and amount of rainfall.[2]

Jobson provides limited detail about the modes of navigation of his party. The pinnace was towed up-river 'in calmes' by two shallops (large rowed boats) and 'the

---

59 miles or, elsewhere, 25 leagues or 75 miles (285, 294). This contrasts with the 130 miles beyond Baarakunda travelled by Jobson and suggests that, unless Stibbs under-estimated his distance, Jobson went much further, which given the difficulties faced by Stibbs at a later date in the dry season, is plausible. At the start Stibbs thought that his journey was 'too late in the Year' (249).

[1] A 1661–2 journey up-river, in search of gold deposits, by Colonel John Vermuyden, 'engineer to Prince Rupert' (and perhaps a relative of the famous engineer, Cornelius Vermuyden), produced a manuscript map and an account (Gray, *History of the Gambia*, pp. 72–4, 186). Vermuyden took 'three barges and a wherry' up to Elephant Island, reporting from there in March 1661/1662 (Eliot Warburton, *Memoirs of Prince Rupert*, 3 vols, London, 1849, III, pp. 538–40). The account of a journey further upstream, in a boat of seven tons, with seven other Englishmen and four Africans, an account anonymous but presumably by Vermuyden, and apparently written thirty years after the event, was only published in 1738 (Moore, *Inland Parts*, Appendix III – a MS text dated 1693, ST 28, Henry E. Huntington Library, San Marino, California, is identical; Robin Hallett, *The Penetration of Africa*, London, 1966, pp. 82–4). The account claims travel and finding of gold far beyond Baarakunda, but is topographically inexplicit and supplies no local toponyms, only invented names; moreover the claim to have found gold is usually thought a dubious one. Nevertheless, and despite an odd similarity between certain points of the description of the journey and corresponding ones in Jobson's narrative, the account appears to be genuine and to evidence, albeit vaguely, a lengthy journey up-river in December 1661 to April 1662, starting from below Sutuko. In 1724 Stibbs cited a 1661 Vermuyden 'Journal', now seemingly lost, describing a voyage beyond Baarakunda, past 'Matlock Tar' and 'York River', the latter locality estimated by the Journal to lie 51 miles up-river, and he also mentioned a map (Moore, *Inland Parts*, pp. 254–5, 263, 277, 285, 288, 294). The extant map matches Stibbs's references in certain respects (for instance, in employing the toponym 'Maiden's Breasts' for 'Arse Hill') and therefore tends to confirm what the account claims – 'never any Boat, nor any Christian, have been so high in that River' (BL, King. Top. CXVII, 96, two copies; the reproduction in R. A. Skelton, *Geographical Magazine*, 1956, p. 153, and *Explorers' Maps*, London, 1970, pp. 279, 293, fig. 183, is curiously only of the section up to Kasang, and therefore not of the exploring section). The invented toponyms of the upper river shown on the map can be related to the Vermuyden voyage, inasmuch as 'York River' refers to James, Duke of York, principal of the Royal Adventurers, formed in 1660, after whom 'James Fort', in the lower river, was named in 1661; 'Pembroke River' to the Earl of Pembroke, a backer of the expedition; and 'Stoakes River' to Stoake(s) an officer on it. 'Matlock Tar' [?Tor], is not shown on the map but the name, with its reminiscences of Derbyshire, points to the miners who accompanied Vermuyden when he explored for the Adventurers. 'York River', emerging on the south bank, is probably Grey River, and 'Pembroke River', emerging later on the north bank, probably River Niériko, and if these are so, then 'Stoakes River', also emerging on the north bank, is presumably River Niokolo Koba. The map concludes at unidentified 'Falls', apparently representing the conclusion of the Vermuyden voyage, which therefore would seem to have penetrated at least 500 miles up-river, at least 40 miles beyond Jobson.

[2] Variations in rainfall may also have been affected over the centuries by climatic cycles. While the experience of the last three centuries appears to have been fairly consistent, it is conceivable that in earlier centuries the seasonal water levels were less extreme and that this permitted the Portuguese, and perhaps Jobson, to travel more easily to the upper reaches.

INTRODUCTION

Boat', presumably the ship's boat.[1] Higher up (beyond Baarakunda), a shallop met new problems. 'Having now the stream against us [that is, they were beyond the tidal reach], we durst not [row] for feare of Rockes in the night, nor could for immoderate heat in the Suns height proceed, but were forced to chuse our houres in the morning till nine, and after three in the afternoone.'[2] Returning down-river, 'the wind and streame served, but we durst not sayle for the sholde, nor row by night.'[3] Up-stream, the shallop had to avoid sunken tree trunks and to be hauled over sand shoals and rock ledges, until the water level fell so low that no further progress could be made.[4] The picture is closely confirmed but much illuminated by the fuller details in Stibbs' account of the 1723–4 voyage. His ship regularly sailed 'in the teeth' of a prevailing wind, and both his ship and his sloop used the floodtide ('taking all Opportunities of the Tide, both by Day and night'), or had to be towed, the sloop at one stage specifically by two canoes. Normally two periods were sailed daily, morning and evening, either period occasionally occupying part of the night ('anchor'd … it being so dark we could not see our Way'), adding up to 6–10 hours each 24. The afternoon was often used for going ashore. Up-stream, the canoes used met the same problems as Jobson's shallop, tree trunks, rocks, and frequent shoals.[5]

To describe the course in detail. The entrance to River Gambia, wide and with few dangers, leads to a broad 'estuary' where the Europeans established their main bases, notably at the island which the English, two generations after Jobson's day, christened James Island. Navigation up-river some 300 miles was undoubtedly undertaken for centuries before the European arrival, the pioneers probably the African traders whose canoes carried salt from the salines beyond the mouth of River Gambia, in River Saalum, to the middle reaches, from where some was carried overland to the salt-starved interior peoples. The Portuguese do not seem to have impeded this trade, even although they could themselves bring cheap and probably better salt from the Cape Verde Islands;[6] while Jobson deplored that his English party had not loaded more of this very desirable commodity. The middle river, with its twists and turns and other impediments, proved difficult for European vessels, particularly at points where only a narrow and at times shallow passage lay through rocks blocking the stream – by fond local tradition one of these narrows had been formed by a passing African army whose soldiers threw in stones to provide a ford. But the upper course, beyond Baarakunda, although forced by the Portuguese, by Jobson, and by the later English, seems to have proved too much for the canoe traders, and although there must have been ferry crossings,[7] the use of

---

[1] *PJ* 922.
[2] *PJ* 923; cf. *GT* 15–16.
[3] *PJ* 925.
[4] *PJ* 923; *GT* 12.
[5] Moore, *Inland Parts*, pp. 252–86.
[6] One Portuguese source, however, speaks of 'all the [Portuguese] settlers on the river owning ships' buying salt at River Saalum and therefore competing with the Africans in the River Gambia trade (see Part II below: Lemos Coelho, 2).
[7] Stibbs spoke of seeing bark logs, or floats made of bamboo canes (cf. 'two or three bundles of Palmeta leaves bound up together' (*GT* 16), used for crossing the river, and in the dry season at many points beyond Baarakunda the river could be forded (Moore, *Inland Parts*, pp. 269, 282, 284–5).

the river for trading up and down seems to have been beyond the capacity of the local peoples, or at least not worth the trouble involved. The first Portuguese to penetrate the middle river reached 'Cantor' in the 1450s; Jobson in 1621 went further, to the fringes of the district or region termed Tinda, certainly further up-river than Portuguese traders had regularly gone, although perhaps not further than any Portuguese (or Afro-Portuguese) had ever gone.

While earlier and later visitors to River Gambia had extensive dealings with localities and inhabitants of the lower river, Jobson's account almost ignores this area.[1] He does not name the first regular stopping place in the river, known to Europeans as Barra, where dues to the King of Niumi (Nyoomi) were often collected;[2] nor does he name the island near which his ship paused. Nor does he mention the important creeks on the south side along and up which the Portuguese had much trade, or the unimportant creeks on the north side named on the Gerbier map. He limits himself to commenting that this first stretch of the river, actually some 50 miles, 'doth spread it selfe, into so many rivers, bayes and creekes', too 'intricate' for him to bother to discuss.[3] On the return journey, however, the two English ships lay off the south side of the lower estuary, their depleted crews landing at an unstated locality in the kingdom of Kombo to prepare the ships for the journey home. Jobson notes, justly, the mosquitoes of this low-lying, swampy area – an area including or neighbouring the presentday urban locality significantly known as 'Half-Die'.

On their way up-river, the English ships missed the first apparently-intended calling place, Tankular, a settlement still to be found on the south side where higher ground runs down to the river, opposite a wall of mangroves on the north bank. The first business reported was transacted some 70 miles up-river at Tindobauge (Tendabaa = Mandinka *tenda* 'landing/crossing place', *baa* 'big'), where dues were paid to a local king, where the larger ship was left, and where a Portuguese resident must have startled Jobson by informing him about Thompson's murder by one of his own men. Here the river is still almost a mile and a half wide.

Jobson records that, from this point onwards, boats towed the remaining smaller ship 'up in calmes'. This was probably a less frequent occurrence than the wording suggests, since it is likely that progress was normally determined as much by the tide as by the wind, the ship anchoring during the ebb and riding on the flood. However Stibbs records that his ship, once past Kasang, had to be towed on several occasions, particularly through difficult passages around islands.[4] Proceeding up-river, Jobson's ship passed, without Jobson mentioning them, a number of landing places noted in Portuguese accounts, many of them remaining in operation in later centuries. Perhaps the English, all the way up the river, deliberately limited possible contacts with Portuguese who might be hostile. Jobson also failed to mention two localities marked on almost all early maps, the significantly-termed 'Devil's Turn', and Elephant Island – although the latter is probably Jobson's 'Pudding Island'. In

---

[1] Whether the English bought provisions and paid customs in the estuary is discussed in the Itinerary, p. 209, n. 3.
[2] But see *GT* 10, and p. 84, n. 4.
[3] *GT* 10. See Itinerary (Appendix A), p. 210, n. 1.
[4] Moore, *Inland Parts*, pp. 253, 256–8, 264–5, 269.

fact, Jobson supplies no description of this part of the river. The first port of call noted after Tendabaa is one 64 miles further up-river, Mangegar – a trading place eventually superseded by nearby Kau-ur.[1] This area was visited by traders from the Wolof and Wolof-dominated polities to the NW, in order to make contact with Mandinka and Jaxanke traders, partly to exchange salt for up-river goods; and in 1620 three Englishmen were left behind at Mangegar, presumably to trade.[2]

The next 30 miles up-river requires vessels to pass a series of low islands built up from silt, but beyond these is Niani Maru, a port of importance in the eighteenth and nineteenth centuries. Here, or nearby, seems to have been Jobson's 'Wolley Wolley', described by him as 'bigger than any [town] we had yet seene' – the name has disappeared but the settlement appears to have been incorporated into Niani Maru, or merely renamed.[3] The mangrove border to the river now begins to thin out, and it eventually disappears. A little further on from Wolley Wolley was Cassan (Kasang), the major calling place on the middle river for European vessels between the sixteenth and the nineteenth centuries, and hence marked prominently on many maps – yet latterly abandoned and the site overgrown, it being replaced by Kuntaur, a short distance up-stream. At Kasang, Jobson spent more time than anywhere else on the river, and Kasang was one of the two African towns which he described in any detail, the other being Sutuko. To reach Kasang from Tendabaa took Jobson seven or eight sailing days, and Stibbs seven sailing days, with some halts. The distance being almost 100 miles, the average speed of each was therefore 12–15 miles per day – Stibbs, however, recorded that he ran more slowly approaching Kasang because of the islands in the channel.[4]

After Kasang, the meandering of the river becomes extreme and hence navigation for wind-powered vessels very difficult, with the river swinging in all directions.[5] Jobson helpfully pointed out that by 'north side' and 'south side' he referred to the banks of the river which generally lay in those directions, even although at particular points 'north' and 'south' actually lay to the west or east – or even, in fact, at times to the south and north.[6] From Kasang the course turns first to the south, then makes a loop to the north. On this loop, on the west side, where hills overlook the river, is 'Monkey Court', which is probably where Jobson first saw his baboons. Further around the loop is, or rather, used to be, Pisania, the point from which Mungo Park set out on his 'travels in the interior', the place now abandoned but marked by a monument honouring the explorer. Unlike Jobson, Park chose to avoid the windings of the river by striking out eastwards overland, on a more direct

[1] Kau-ur, which already existed in Jobson's day (Part II below: Donelha, f. 24), in the 18th century became a major port, and today takes ocean-going vessels for the export of groundnuts.
[2] *PJ* 922.
[3] *PJ* 922. Writing in 1669 but based on experience from the 1640s, Lemos Coelho referred to 'Nanhimargo' but not to Wolley Wolley (see Part II below: Lemos Coeho, 1669, 13).
[4] Stibbs conducted business with various river locations en route. He noted his sailing hours on most days, Jobson does not, so exact comparison in hours is not possible.
[5] A Portuguese source argued that beyond Kasang the journey 'should not be undertaken without a pilot' (see Part II below: Lemos Coelho, 31).
[6] *GT* 47.

route. Just beyond this loop is the prominent Mamayungebi Hill, noted in Portuguese and later sources because of the local tradition requiring travellers to present bared buttocks to the spirit of the hill (hence the later English name, 'Arse Hill').[1] But Jobson does not mention this (possibly because he had no Africans or Portuguese aboard the ship to draw attention to the required ritual), nor does he mention another famous locality, occurring before the northern curve, the narrowing of the river allegedly created by the African army, 'Fula's Pass'.[2] He also sails by, without noting or at least recording, the mouth of the substantial Sangedugu Creek, up which there were localities where the Portuguese traded, although the actual entrance is swamp-ridden and has generally lacked habitation. On this stretch of the middle river, from passing Kasang to 'Cuttejar', near presentday Karantabaa, Stibbs took seven sailing days, with complaints of having to be towed, little progress because of strong winds, and the river so narrow at one point that a wind turned the ship's stern 'into the bushes'.[3] From Kasang to 'Pompeton', also near Karantabaa, Jobson seems to have made the same sort of progress, taking probably 6–7 days. These times represent average speeds of 7–10 miles per day.

The river now makes further almost-circling curves, one to the south, then two to the north. This is particularly the stretch in which, despite high banks, the rainy season flood overruns the land, creating for much of the year large swamps. In this stretch is the administrative capital of the Upper River province of the modern territory, Basse, a settlement possibly non-existent in Jobson's day (although a port of this name was noted by Stibbs a century later);[4] and this stretch was to contain, further on from the site of Basse, the site of Jobson's middle river base – for the proposed exact location, see Appendix B. Creeks run in from the south and north, the main one on the north being the Tuba Kuta Creek, at whose mouth the English party most probably halted, before moving on to the site of the base. In the final northern curve lay the two 'ports' of the inland town of Sutuko, the first, Fatatenda, regularly evidenced in later sources but not mentioned by Jobson (perhaps because Stibbs found it, a century later, a mere landing-place with 'not a House near it'), the second, Sami/Same Tenda, not mentioned by name but apparently visited by Jobson in 1621 (although Stibbs found it also without habitations).[5] Important trails from north and south cross the river at both points. It would seem that it was to these ports that the Portuguese came when they spoke of trading at 'Cantor'. They took ocean-going vessels to this point, while Jobson risked only a rowed boat – Stibbs, however, took a sloop further up, to Baarakunda, and Hodges claimed that his sloop went even further.[6]

---

[1] Lemos Coelho, 36; Moore, *Inland Parts*, p. 263.
[2] Moore, *Inland Parts*, p. 258 – 'Pholey's Pass ... a Ledge of Rocks, which extends from the North Side the River quite over to within 20 or 30 Yards of the other Side, leaving just room for a Ship to pass, yet not so but the Ship brushes the Trees'. It is the 'Divil's Bridge' on the Vermuyden map.
[3] Ibid., p. 258.
[4] Ibid., p. 266.
[5] Ibid., pp. 266, 269.
[6] Stibbs's account makes it clear that the stretch of the river from Karantabaa upwards had no serious obstacles to prevent a sloop approaching Baarakunda, and Hodges passed the obstructions near Baarakunda in a sloop and went much further before continuing in a 'sciffe'. On a slack current and a high spring tide, later vessels of some size managed to pass through the Falls (for instance, a river steamer in 1906 and a yacht in 1908: [anon.], 'La Navigabilité de la Gambie an amont de la frontière

INTRODUCTION

At the middle river base Jobson was warned that the water level in the river was already falling, it being then late December.[1] An elderly marabout also warned Jobson about the hazards up-river, it being 'so full of trees, we should not be able to get our boate along'.[2] Leaving the pinnace behind, Jobson's party of eight Englishmen and two (later four) Africans made their way from the base further up-river in a shallop, a large boat. It could only be rowed in the cooler parts of the day, and not at all at night – 'for avoiding of trees sunke, rockes, and sholes'.[3] The course of the river upwards lacks the series of extreme meanders of the middle course, although beyond Baarakunda it has other severe problems for navigation; however, with the development of land transport in recent decades, even the stretch below Baarakunda is nowadays little used for traffic, so that many traditional calling places have fallen into decay. Proceeding towards Baarakunda, although Jobson failed to note (or record) the ports of Setico, he did note an unidentified town, and also Benanko (on the modern The Gambia/Senegal frontier),[4] a village abandoned in the present century. Apart from referring to the presence of hippopotamus in numbers, Jobson has little to say about this part of his journey.[5]

Towards Baarakunda, some 330 miles up-stream from the river mouth, a later Portuguese source spoke of navigation being only for vessels of very limited depth, and the presentday *Admiralty Pilot* allows for any navigation only up to Koina, twelve miles before Baarakunda.[6] Jobson's shallop encountered 'rockie passages' as it approached Baarakunda, but there is no mention of the more severe obstacle occurring a little beyond that 'port', the ledge of rocks and rapids which nineteenth-century sources christened the 'Baraconda Falls'.[7] The headman of the town of Baarakunda was contacted.[8] Jobson reported that the tides which had assisted the passage up-stream stopped at or just before Baarakunda, whereas Stibbs found a tide somewhat higher.[9] Stibbs travelled from Cuttejar to Baarakunda, some 110

---

anglaise', *L'Afrique française*, 19, 1909, pp. 330–42; Henry Fenwick Reeve, *The Gambia: its History, Ancient, Mediaeval and Modern*, London, 1912, photograph opp. 130). Jobson therefore chose to leave his pinnace half way, at the middle river base, rather than sail in it up to Baarakunda, for reasons not explained. It is possible, however, that his pinnace was larger than the sloops of the later ventures.

[1] *PJ* 922.
[2] *GT* 79. In 1724 Stibbs was similarly warned off from penetrating beyond Baarakunda by people lower down-river, but on the grounds that 'there is no sort of Eatables to be purchas'd above Barracunda', that the locality was 'look'd upon as the end of the World' with beyond 'very barbarous', and that 'no Body was ever above that Place' – but Stibbs distrusted the motives for such warnings and argued that 'hardly any Credit is to be given to what these People relate' (Moore, *Inland Parts*, pp. 268, 272–3).
[3] *GT* 15; *PJ* 923.
[4] Hence up-stream placenames are in French orthography.
[5] *PJ* 922; *GT* 11.
[6] Part II below: Lemos Coelho, 53; *Africa Pilot*, Vol. I, p. 295.
[7] *PJ* 922. For details of the Falls, see the Itinerary, p. 21, n. 7. If Purchas summarized badly, the 'rockie passages' may in fact refer to the Falls.
[8] A century later Stibbs was to report the recent destruction of the riverain town and the removal of its inhabitants to another town: Moore, *Inland Parts*, pp. 270–72. Baarakunda seems to have been later re-established, on its original or another site, before being abandoned yet again. Thus it is possible that the same cycle of change had happened earlier, and hence possible that the site in Jobson's day was not precisely where it was later.
[9] Moore, *Inland Parts*, p. 280. These reports were no doubt correct, but during the rainy season tidal influence stops at varying points down-stream.

miles, in 12 sailing days, averaging just under 10 miles per day. For comparison with Jobson, Stibbs took 19–20 sailing days from Kasang to Baarakunda, while Jobson appears to have covered the same distance in 16–19 sailing/rowing days, each covering some 160 miles at 8–10 miles per day.[1]

For the Portuguese, Baarakunda was 'the last port to which whites go',[2] and Portuguese accounts give only vague and uncertain statements about vessels which attempted higher. However, Jobson's naming of two rivers up-stream indicates that his guides knew about the upper course, presumably because canoes ascended it. Jobson, however, saw no boats on this stretch, and his account deals mainly with the animals seen on the banks, apart from references to hazards such as shoals, 'a violent current', and a whirlpool. For instance, 'on the foure and twentieth, we towed her, sometimes adding ha[u]ling by the Boats side, as sholds and trees permitted: and met with one vehement current, overthwart broken rockes, so that we were forced to hold her by force, till one taking the Anchor on his neck, waded above that quick fall, and letting it fall, we ha[u]led by our hason, and escaped that gut'.[3] The river ran alternatively narrow and wide, the latter condition producing shallow passages and extensive shoals, many dried out. In a boat with eight men and four Africans, Jobson was finally halted when a depth of only nine inches lasted for twenty yards.[4] A century later, possibly because the river was even lower in a later month, Stibbs, with fifty men, went forward with only canoes, and eventually the two least deep of these, one drawing only twelve inches unloaded, now with four whites and thirteen Africans. Yet when the tidal influence became negligible, Stibbs was halted when he found 'but 10 Inches Water in the deepest part'.[5] Thompson, Jobson's predecessor, in 1619 had gone up-stream in a boat, with two whites and an unstated number of Africans, apparently in February-March; if his boat really was rowed with only 'a payre of Oares', as Jobson stated, there must be a measure of doubt whether it got very far beyond Baarakunda, and it is possible that, to reach Tinda, Thompson thereafter went overland.[6]

In the 130 miles up-stream beyond Baarakunda that Jobson travelled, he noted much animal life in the river and on its banks, but little evidence of human presence or activity. This may have been a temporary phenomenon, caused by recent wars and devastation, or it may have represented an environmental limitation, since

---

[1] Hodges, travelling beyond Baarakunda, in difficult navigational conditions, claimed to have covered 148 miles of this stretch in thirteen days, thus at 11 miles per day, a shade faster than Jobson and Stibbs in the previous stretch, in much easier sailing conditions. A Portuguese source claimed that, 'since we sailed or rowed day and night with the [flood] tides', a 1580s journey in 'the largest launch at hand' from Kasang to 'Cantor' took only three days for a distance estimated at 60 leagues but actually some 60 miles, thus at a rate of some 20 miles per day; but the speed as well as the distance seems in error – see Part II below: Donelha, f. 27ᵛ.

[2] Part II below: Lemos Coelho, 1669, 27.

[3] *PJ* 923.

[4] *GT* 12.

[5] *GT* 12; Moore, *Inland Parts*, pp. 276–86. The Africans of Stibbs' final party included two boys and – notably – a woman, a slave, apparently as a cook.

[6] *GT* 6. However, the extra men would be available to haul the boat over shoals. Moreover, Hodges claimed that his 'sciffe', carrying only two other men, travelled even further, and rapidly, albeit later in the dry season.

INTRODUCTION

few people were found in the district in the nineteenth century.[1] Two important tributaries join River Gambia, the Grey River on the south side (first sailed by a European in 1908), and River Niériko, Jobson's 'Tinda River', on the north side. Jobson noted them both.[2] The English were halted by a shoal of sand or rocks they were unable to drag the boat over, and after dealing with an African trader and a local king, they retreated down-river. Jobson reckoned he had penetrated 960 miles up-river: the true distance was about 460 miles.

Until the English reach their final halt, near 'Tinda River', Jobson's account is sparse, nevertheless it is the earliest account of the stretch above Baarakunda, the Portuguese accounts not reaching to here. The later documentation of the Royal African Company, other than that relating to Vermuyden, Hodges and Stibbs, tends to stop at Baarakunda. Hodges in 1689 may well have gone even further up-river than Jobson, but his brief account concentrates on his subsequent crossing overland to River Senegal and does little more than list place names on the upper River Gambia. Stibbs' account is rather fuller than Jobson's for the first half of the stretch beyond Baarakunda, but the 1724 party never reached Tinda. Vermuyden's 1661–1662 map goes further up-stream but notes only river mouths, the rivers with invented names, and such vague features as 'White cliffs' and 'Rocky mountaynes'. Overall, Jobson's account is, in fact, the fullest description of the river between Baarakunda and Tinda before the mid-nineteenth century.

The African Setting: Polities and Peoples

Various local polities were encountered. Jobson boasted of meeting six 'petty kings'. But he noted that the overlords of the region were three 'great kings', those of 'Bursall', 'Wolley' and 'Cantor', who were themselves subject to a greater king in the interior. As far as it is possible to tell from the limited early sources and from time-distant oral traditions, in Jobson's period Cantor (Kantora) claimed the south bank of River Gambia from an uncertain point on the lower or middle river all the way up to 'Tinda'; Bursall (a European name derived from the title of the ruler, the *buur Saalum*) claimed the north bank from the mouth to half way up the middle river, followed by Wolley (Wuli) which claimed up to (and perhaps part of) Tinda. But Jobson met none of these rulers, their residences being in the interior. Along the river subordinate polities were encountered: Combo on the south side at the mouth, and much further up, on the north side, Niani. Jobson met the king of Combo and has a passing reference to the 'king of Nany', that is, Niani. But he

---

[1] See Itinerary (Appendix A), p. 217, n. 2. In 1724 Stibbs, who also saw much animal life, was, however, regularly contacted by individual Africans who forded the river to the canoes. He stated that 'the Country on the Cantore side is populous, with small Towns here and there, but none within a League of the River; but on the other Side are no Towns or Inhabitants till you come to Tinda' (Moore, *Inland Parts*, p. 286).

[2] Jobson calls the Grey River 'Cantore River' and notes an earlier small river on the north side, not far beyond Baarakunda, the 'Wolley' – both of these names were subsequently recorded by Hodges, who also listed 'Tanda' and the names of two later rivers (see p. 47, n. 1 above).

55

failed to record two polities of the north side, Niumi, which lay opposite Combo, and Badibu (assuming that by then it already existed under that name), which lay between Niumi and Niani.[1] With one exception, the other rulers noted appear to have been rulers of small localities: the petty kings of Mangegar (under Bursall), Kasang (under Niani and ultimately under Bursall), Oranto (at least partly under Cantor, itself perhaps under Kaabu), Pereck (almost certainly under Cantor), and Jelicot (under Wolley); also the interior king (of Kiam/Kiyang?, under uncertain suzerainty) who ruled Tendabaa, and the *farang* (a subordinate ruler or governor) of Jeraconde (perhaps under Niani).[2] The exceptional ruler was that of Cabo/Kaabu, a large if somewhat obscure polity which Jobson mentions once, correctly, as lying to the south and east of the upper River Gambia.

All but one of these polities were in the hands of a single ethnicity. The exception was Bursall, whose ruler was Wolof, as was part of its population, but this ethnicity extended only thinly towards the river, and Jobson does not seem to have met any Wolof individuals. All the other polities were immediately controlled by rulers from the majority population of Mandinka. One other ethnicity had, however, established itself in the economic and political interstices of Mandinka dominion. This was the Fulbe or Fula, still in the Gambia region largely a pastoral people. Finally, among the peoples of the river Jobson listed the Portuguese, more correctly to be defined as Afro-Portuguese, being regularly of mixed descent.

The Mandinka

The people chiefly encountered by the English in River Gambia were those whom European sources have called 'Mandingo' (or later, 'Manding'), but whose self-name, used here, is Mandinka.[3] The language of the Mandinka and a group of closely related languages evolved in the interior. The early history of the languages and peoples is therefore obscure. Arabic sources, confirming certain aspects of latterly-collected 'oral traditions', evidence, in the centuries immediately before the

---

[1] DPG suggests that Jobson may have ignored Badibu because its villages lay back from the river, up creeks and hidden behind thick masses of mangroves. The English did not need to contact this polity and may have been unaware of its existence.

[2] It is unlikely that this is a full list of polities along the river. The 1732 map names other ones, e.g. Badibu and Tomany, and DPG notes that traditions tend to claim that present-day divisions existed even further back than *c.*1620.

[3] From among a considerable literature on the Mandinka in general, including a substantial literature on the Gambian Mandinka, the following works have been especially useful for this summary account: H. Lloyd Pryce, 'The Laws and Customs of the Mandingos of the North Bank Territory of the Gambia Protectorate', unpublished report, 1907, Gambian National Archives; P. Holderer, 'Notes sur la coutume Mandingue du Ouli (Cercle de Tombacounda)', in *Coutumes juridiques de l'Afrique occidentale française: Sénégal*, Paris, 1939, pp. 323–48; Mody Cissoko and Sambou Kaoussou, *Receuil des traditions orales des Mandingues de Gambie*, Centre régional de documentation pour la tradition orale, Niamey, 1974; Winifred Galloway, 'A History of Wuli from the Thirteenth to the Nineteenth Century', Ph.D. dissertation, Indiana University, 1975; Sory Camara, *Gens de la Parole: essai sur la condition et le rôle des griots dans la société malinke*, The Hague, 1976; Peter M. Weil, 'Land Use, Labour and Intensification among the Mandinka of Eastern Gambia' [especially Wuli District], unpublished paper presented at the Annual Meeting of the African Studies Association, October 1980; Peter M. Weil, 'Slavery, Groundnuts, and European Capitalism in the Wuli Kingdom of Senegambia, 1820–1930', *Research in Economic Anthropology*, 6, 1984, pp. 77–119; David C. Conrad and Barbara E. Frank, *Status and Identity in West Africa:* nyamakalaw *of Mande*, Bloomington, 1995.

INTRODUCTION

European arrival, a polity in the interior, around the upper River Niger, known as Mali. This polity evolved as a conquest empire, with the Mandinka the ruling group, possibly to some extent as a result of Islamization at its nucleus, and at one stage it became the dominant influence over much of western Guinea. Its exact structure is, however, unclear, as is its degree of control over distant provinces, the latter no doubt varying by circumstances and over time. It is generally considered that the nucleus was in decline by the period of European arrival in Guinea, but the extent to which outer provinces any longer owed allegiance to the nucleus by the sixteenth and early seventeenth Christian centuries remains controversial. On the face of it, what the Europeans found and encountered was a series of independent Mandinka-ized 'kingdoms', apparently with elements of a hierarchical structure of political administration, certainly within the kingdom and notionally upwards towards a distant 'emperor'.

The expansion of Mandinka influence down River Gambia is generally considered to have occurred in the later pre-European centuries. The Mandinka language spread widely, partly by the migration of genetic stock, partly by the conquest and removal of the ethnicities earlier inhabiting the banks of the river, and partly by the social and cultural over-running and conversion of these same ethnicities. It would seem that the interior had in certain respects technological and possibly ideological advantages over the peoples nearer the coast. The motives for this expansion were no doubt the normal ones, for instance, population pressure, desire for hegemony, and not least economic gain. Although there was no oceanic trade and the intercoastal marine trade was almost certainly slight, there were economic attractions in a drive towards the sea.[1] Probably important was an attempt to control, as middlemen, the apparently long-standing trade moving salt from the coastal salines upriver to the interior. Another commodity most likely sought by the interior and provided downstream on the river was slaves. Certain districts of the middle stretches of the river, being less thickly wooded than the districts of the lower river, grew cotton, so that a considerable trade in both raw cotton and cotton cloths, exported both up-river and down-river, was in place by the time the first European sources reported.[2] The Mandinka influence formed new polities or reshaped old ones, but, despite the spread of the language, the rate and extent of acculturation are not entirely clear. While this influence eventually pressed Islam on polytheistic aborigines, it is clear that, even as late as Jobson's day, many of those along the river who spoke Mandinka were only lightly if at all subject to Islam, as shown by the evidence of alcoholic consumption and the distinct role of the 'marabouts' (Islamic activists). Non-Islamic Mandinka-speakers were later termed Soninke, when they appear to represent whole sections of the total Mandinka-speaking population, but

---

[1] The view expressed in George E. Brooks, *Landlords and Strangers: Ecology, Society, and Trade in Western Africa, 1000–1630*, Boulder, 1992, esp. ch. 5, that extensive coastwise navigation and commerce had existed for many centuries before the European arrival is non-evidenced and somewhat implausible; but Brooks does not lend support to the myth, widely circulated among Afro-American study programs, that a Mandinka fleet once sailed down River Gambia and made a pre-Columbian discovery of America.

[2] It is possible that before the European arrival gold from the interior was also conveyed down-river, perhaps to buy salt.

it is unclear whether the term was in use, at least with this meaning, in Jobson's time.

In so far as the developed structure of Gambian Mandinka communities around 1620 can be represented by analyses of later descendant communities, based on descriptions and research in recent times, but with a working hypothesis of close continuity over the centuries confirmed to some extent by earlier accounts such as Jobson's, the context of English contacts with the Mandinka was as follows.[1]

Speakers of the Mandinka language (possibly at first in an earlier form) moved into the Gambia valley over a number of centuries, probably both from the east and from Kaabu in the SE (apparently an earlier area of Mandinka expansion). Some were hunters or farmers, others members of warrior groups which established a series of fortified villages. In time the incomers formed small polities or 'kingdoms' along both banks of the river, traditions claiming that many received legitimization either from the Mandinka *mansa*, the ruler of Mali, or from the ruler of Kaabu, who presented them with the symbols of power, in the form of a 'crown' or a set of sacred staffs (*chonolu*). In many regions the local kingship came to rotate between different clans in specific villages ('royal towns'). However, in the polity known as Wuli, on the north side of the middle river, power seems to have remained in the hands of one clan, the Wali. As the centuries passed, the Mandinka absorbed whatever small groups of indigenous people existed in the districts along River Gambia, as well as a variety of people taken in wars and brought there as slaves, a large number being Bambara from the north and east. In addition, refugees from other regions also arrived.

In recent times – as most probably earlier – the Mandinka have lived in compact villages. The oldest man of the founding lineage is the village head, *sateyo-tiyo*, and land is allocated by him to other settlers who wish to join the community. If a village is large, it is normally divided into wards (*kaabiiloolu*), with a dominant lineage in each. The ward is composed of a number of compounds – fenced off enclosures (*kordaalu*), which house part of a lineage, and are divided into smaller family sections (*dabadalu*). The wider structure of society has latterly been considerably modified but formerly it comprised certain 'royal lineages' which provided the king (*mansa* or *farang*), a majority body of ordinary farmers who became warriors in time of war, the endogamous craft groups – the blacksmiths, leather-workers, and *jaloolu* (the 'griots', that is, musicians and story tellers) – and a slave population owned by the various freeborn families, including those in the craft groups.

Villages are generally found on higher land, but not far from the swamps bordering rivers and creeks in which rice is now grown. (Jobson identified rice but only briefly refers to riziculture.)[2] Originally most villages were non-Muslim (Soninke), but Islamic religious teachers (Jobson's 'Mary-buckes' or marabouts), often of Jaxanke origin, arrived and settled. While the number of Muslims gradually increased,

---

[1] Although statements about the distant past are here expressed as factual, many are based solely on oral traditions latterly collected, there being no written or archaeological records, and as such must be regarded as more or less conjectural.

[2] *GT* 124–5. Jobson was not on the river during the period of rice growing and harvesting, June to October. Moreover, when ashore he spent much time in the Wuli and Kantora districts where little rice is grown, its place being taken by millet and sorghum.

it seems that large scale conversion in the region was not achieved until after the Soninke/Marabout wars of the nineteenth century, inspired by the Islamic concept of *jihad*. When the European colonial powers established peace, conversion proceeded rapidly since religious teachers could move more freely. At the present time, though the Gambian Mandinka are overwhelmingly Muslim, memories of old events remain, and distinctions between the former Soninke (polytheist) and Marabout (Muslim) villages are still felt, so that even now marriage between them is uncommon.

Cutting across the contemporary kinship structure are the age sets (social groupings, each made up of all individuals sharing a limited range of birth years), existing separately for each sex, the younger male age groups performing communal works, such as clearing paths and building bridges. Close ties also link those who are circumcised on the same occasion, thus modifying rank and caste distinctions. The routine tasks of men and women are complementary, the men working the upland farms producing millet, corn, and groundnuts, while the women cultivate rice in the riverside swamps and look after small gardens. The power of certain masked figures controlled by the men, for example, the *kankurang*, which appears at circumcision ceremonies, is balanced by certain ritual offices held by the women, such as that of *ngansimbaa*, an official in charge of the girls' initiation ceremonies, or those of leaders of the *kanyelango*, a women's organization concerned with invoking fertility. When a community became Muslim, the power of the village head was tempered by the authority of the *imam*, the leader of Muslim activities. In dealing with strangers, there is nowadays generally an *alkaloo* (perhaps the same term as Jobson's 'Alcaide', or a variant), whose duty it is to collect taxes and handle suchlike external relations, the *satewo-tiyo* or village head being primarily concerned with internal affairs, family disputes, marriages, land tenure, and related lesser issues. In the past, the major clans generally looked after their own affairs, and villages were to a significant extent independent units, controlling their own land, regulating marriages, divorces, inheritance, circumcision ceremonies and disputes within the community. Matters of concern would be decided at a general village meeting (*bengo*) of adult men.

The region on the middle river within which the 1618–21 English chiefly operated was in later times unambiguously defined as that of the polity of Wuli. Although the early historical sources appear to indicate that Wuli's boundaries have shifted somewhat over time, nevertheless it seems that Wuli was the polity with which the English had most contact. Winifred Galloway, a recent student of Wuli traditions, regards this polity as having been established in the form of a confederation of a number of great families (clans), known as *kaabiiloolu*, a term also used in the sense of a 'ward' in a village (as above). The leading group, she believes, were non-Muslims (Soninke), belonging to a noble warrior class, but by virtue of an alleged conquest in the sixteenth century the Wali took the lead. Be all that as it may, Jobson's evidence suggests that the polity of Wuli was still expanding in his day and even that a district on the north bank close to the river, certainly at a later date forming part of Wuli, was still reckoned by some as part of Kantora, the polity to the south.¹ At an uncertain date, the Wali were joined by other powerful families, among them the Saane, who claimed to be related to the ruling family dominant in Kaabu and who remained powerful and proud of their supposed descent. Jobson drew attention to what historians term 'the myth of noble origins' when he observed, apparently in respect of Wuli, that 'there is no people in the world, stand more upon their antiquitie, and dignities of blood, then they doe betweene themselves'². The Wali originally had their capital near Tambakunda, but later (according to Mungo Park) the king had his residence at Madina, further to the SE.³

The *mansa* of Wuli was concerned, according to Galloway, with maintaining an army and making war, levying taxes to support a large number of dependents, dealing with serious crime, maintaining the safety of major trade routes, and entertaining visitors. An important function was that of persuading various travellors passing through Wuli to settle in what had been a sparsely populated territory. Agreements (*dankutolu*) were often made to the effect that those who settled would be able to live for ever in peace with the rulers and would not be discriminated against or mistreated. The revenue of the *mansa* was based on an annual tax on the harvests and herds of his subjects. Unless there was any urgent need, this was levied after harvest time, and paid in millet, locally-made cloth, livestock, and salt – or in any other products which the subjects possessed and which the *mansa* required. Duties were also levied on the *juloolu* (traders)⁴ and on foreign merchants. Judicial fines, and gifts from subordinate chiefs or individuals wanting favours, also added to his revenues. To keep his soldiers happy, raids were sporadically carried out against neighbouring territories to allow them to acquire plunder. The authority of the *mansa* depended on a combination of factors: seniority within the leading *kaabiiloo*, a large personal following and ostentatious generosity to clients, a sense of destiny to rule as well as actual possession of the royal insignia, and, finally, strong spiritual and magical forces – the innate mystical strength called *nyamo* – deriving from both powerful charms and the support of Muslims through their prayers. As the *mansa* of Wuli then lived far inland, Jobson did not come into contact with him, but encountered local chiefs (*farang*, plural *farangolu*) and the 'Ferambra', who seems to have been a regional representative of the king. These subordinates would have kept the *mansa* informed about the activities of the whites.

While the inland areas and the zones along the major tributary streams were firmly in the hands of the noble warrior class, the riverside area was dominated by traders and clerics. The trading community, the *juloolu* ('Julas', Jobson's 'Juliettos'),⁵ constituted an important element in Wuli society. Many of the major clans had founded, near the stretch of the river below Baarakunda Falls, settlements to which caravans came from the interior, to link up with the riverain trade. Many individual

---

¹ Although we suppose that Oranto was in Wuli, it is curious that the ruler lived on the south side of the river and, according to Jobson was a subject, not of Wuli, but of Kantora (*PJ* 922). At an earlier date Sutuko was inferred to be in 'Cantor' – see Part II below: Gomes, f. 276ᵛ.

² *GT* 56.

³ In 1724 Stibbs reported that the king lived at Cussana, 30 miles further inland than Sutuko (Moore, *Inland Parts*, p. 266).

⁴ Mandinka *jula/juloolu* 'trader/s. In the annotation, the English forms, Jula and Julas, are employed.

⁵ *GT* 92.

jula were therefore long-distance traders. The noble class and the *juloolu* treated each other with mutual respect, the latter providing the *mansa* with such items as horses, and after Jobson's time, guns and gunpowder. A large proportion of the *juloolu* were of Jaxanke origin (see below), reflecting a close association between individual *jula* and the Muslim communities, since Muslim teachers provided the traders with prayers, blessings, and charms for protection and successful journeys, as well as safe places to stay on their travels.

A third major element in Wuli society was the Muslim community, consisting of the Jaxanke (described below) and the Manding Sula Moros ('Old-Muslim' Mandinka). The latter were Mandinka of high rank who had converted to Islam at an early stage. They did not participate in secular wars and raids, which were the main occupation of the Soninke rulers. Nevertheless, they continued to be respected for their ability to provide amulets, to foresee the future, and to pray for the success of individuals. Muslim communities tended to establish themselves in the southern zone of Wuli, the leading town being Sutuko (Sutukobaa) (Jobson's Setico), which attracted numerous settlers, especially scholars, traders, and artisans.

Within the lower ranks of Mandinka society were the smiths, leatherworkers, and 'griots' (praise-singers and musicians), together termed *nyamaloolu* and generally attached to the high-ranking groups for whom they performed their services. Slaves contributed much of the agricultural labour, and were of two categories: those born in the family who were treated like junior members of the household and could not be sold unless they had committed a crime, and those taken in wars and raids who could be sold to traders. Individuals of slave origin provided a large proportion of the *mansa*'s soldiers, and also of his office holders – 'crown servants', responsible to the reigning ruler, who provided advice and intelligence about the countryside, collected taxes, delivered messages, arrested criminals, and performed other subordinate duties. Slaves who were officials could acquire their own personal slaves.

Wuli was essentially Mandinka but contained minor ethnic elements. Few Serahuli (see below) lived there until the second half of the nineteenth century, and such as came tended to be quickly absorbed into Mandinka society. The Wolof (see below), whose homeland lay to the NW, were limited in numbers, being for the most part refugees from troubled regions. In a former time, traditions claim, the Wolof rulers of Saalum and Jolof had demanded tribute from Wuli, but once the Wali clan had established control, further tribute was refused, and the Wolof could not enforce their demands. The Fulbe (see below), being cattle keepers and mobile in their constant search for water and pasture, even though at times involved in parochial agreements and disputes with their neighbours, took little part in wider political developments.

Jobson provides a great deal of information about the Mandinka of his day, most at that time apparently non-Muslim. Those near the river mouth were still suspicious of foreign vessels until they knew their intentions, since some local individuals had been captured and exported as slaves. As a result, most of Jobson's information on the peoples was gathered in the middle river, from Kasang onwards, and especially in the region of Wuli. Although as a non-African he had only a limited understanding of the structure and functions of Mandinka society, and although his information was often constricted and slanted by his having to collect it through interpreters or from other Europeans, nevertheless he was an enthusiastic and moderately sharp observer – as demonstrated by his descriptions of items of natural history – and he provided himself with a range of opportunities for social observation and inquiry (or possibly these were provided for him, more by luck than by judgement). Thus he met rulers and traders, and was around at the right moment to attend a circumcision festival and the deathbed and burial of an Islamic dignity. In consequence, not only are his keyhole glimpses of Mandinka society, in the form in which he reported them, apparently in general accurate, but they usefully mark out the broad dimensions of the society as it then existed and provide material for back-extrapolation from later and fuller descriptions. In several aspects his account is less fragmentary than any of the earlier accounts. Jobson's writings have understandably became a classic account of an African people, and they have been heavily relied on in later studies of the ethnohistory of the River Gambia region.[1]

The Jaxanke (Jakhanke)

When Jobson speaks of the people he called 'Marybuckes', that is, marabouts, he is referring to a distinct segment of the Mandinka-speaking people known as Jaxanke (Jakhanke, Jahanke), meaning people from Jaxa, distinguished principally by their social functions.[2] Jaxanke traditions state that they migrated from Dia (Ja) on River Niger to Diakha (Jaxa) in Bambuk, then to Bundu, and finally westwards to River Gambia. In earlier centuries they were renowned for spreading Islam through peaceful means, by teaching and example, and they were opposed to war, although they might be asked to pray for the success of a particular warrior chief and active defence on their own part was permissible. Their activities were supported by a large body of slaves, who worked farms and performed certain craft work, for instance, weaving cotton cloth. Jaxanke towns also attracted blacksmiths who made farming tools, and leatherworkers who made covers for amulets. Students who came to acquire religious knowledge were required to perform farming and domestic tasks, their studying being by firelight at night.

Certain clans of the Jaxanke were long-distance traders and therefore part of the

---

[1] There is thus some danger of circular argumentation, students of latterday Mandinka society using Jobson's account as a part guide to their understanding of it, and historians then using the consequent understanding to interpret Jobson's account.

[2] In some regions the Jaxanke are regarded as Serahuli (close linguistic cousins of the Mandinka), but in the Gambia region they became identified with the Mandinka. For the Jaxanke, see Park, *Travels*, pp. 516–19; Philip D. Curtin, 'Pre-Colonial Trading Networks and Traders: The Diakhanke', in Claude Meillassoux, ed., *The Development of Indigenous Trade and Markets in West Africa*, London, 1971, pp. 228–39; Lamin O. Sanneh, 'Field-work among the Jakhanke of Senegambia', *Présence africaine*, 93, 1975, pp. 92–112; Lamin O. Sanneh, 'Slavery, Islam, and the Jakhanke People of West Africa', *Africa*, 46, 1976, pp. 80–97; Thomas Charles Hunter, 'The Development of an Islamic Tradition of Learning among the Jahanke of West Africa', Ph.D. thesis, University of Chicago, 1977; Donald R. Wright, 'Darbo Jula: the Role of a Mandinka Jula Clan in the Long-distance Trade of the Gambia River and its Hinterland', *African Economic History*, 3, 1977, pp. 33–45; Lamin O. Sanneh, *The Jakhanke*, London, 1979; Lamin O. Sanneh, *The Jakhanke Muslim Clerics: a Religious and Historical Study of Islam in Senegambia*, Lanham, 1989.

class of *juloolu*.¹ In his studies of the Jaxanke, L.O. Sanneh emphasizes their religious role and plays down their trading, but nevertheless concedes that the existence of Jaxanke towns and villages along trade routes meant that long-distance caravans had safe places to halt and to rest without being plundered. To some extent the trading and religious roles appear to have been differentiated.² A few families of traders 'converted' and turned to religious scholarship; more rarely individuals from a religious family turned to trading; and intermarriage between the two groups seems to have been uncommon. Much knowledge was handed down within families. Marriages were in general arranged to ensure that children had access to as much religious and mystical knowledge as possible, while trading families also passed down detailed knowledge of particular areas, and of kinsfolk dispersed in various localities on whom they could depend for help. Jobson commented on intermarriage among the Marybuckes.³ The religious teachers sometimes looked down on the traders, and if they felt that their community was becoming too secular, they would move on and establish a new village of their own. Nevertheless, it appears that in earlier centuries there were individuals who combined both avocations, or at least practised them on different occasions, one of them being Jobson's chief guide, Fodee Careere. In a later century, Mungo Park, on his return to River Gambia in 1797, was in the care of a religious teacher, who together with other Jaxanke, was bringing slaves to the river for sale.⁴ Many of the villages at which Park stayed bore (as they continue to bear) common Jaxanke names, such as Medina, Baniserile (Bani Israel), and Kirwani (Kerewaan). The combination of religious scholars, long-distance traders, and artisans, was a powerful alliance, making for extremely prosperous communities.

Jaxanke towns were usually some distance away from the towns of the secular rulers. In general, the Jaxanke picked sites not too far from a river, with fertile lands around them. Some sites were chosen because of their nearness to woods rich in medicinal plants. But a particular marabout might be invited to settle in an already existing town. The Jaxanke exerted influence over the non-Muslim population through their ability to provide charms and amulets (these being described by

---

¹ The exact relationship between *juloolu* and Jaxanke is historically complex, since there appears to have been both convergence and separating out of these classification categories over time. Probably in Jobson's day most of the *juloolu* were Jaxanke. Non-Jaxanke *juloolu* could never become Jaxanke, but Jaxanke could move in and out of being *juloolu*. Today, however, certain clans claim to be *jula* as of origin, although not now primarily engaged in trading.

² While there is agreement that Islam was regularly spread in West Africa by traders and 'clerics', scholars disagree over the degree of distinction between the avocations. See Nehemia Levtzion, 'Merchants vs. Scholars and Clerics in West Africa: Differential and Complementary Roles', in N. Levtzion and Humphrey J. Fisher, eds, *Rural and Urban Islam in West Africa*, Boulder, 1987. A broad view is that while the roles were often combined, fully-fledged scholars and large merchants were unable to combine them. For comment in relation to a district in which Islamization is more recent than in the Gambia region and the process therefore more fully recorded, see Allen M. Howard, 'Trade and Islam in Sierra Leone, 18th–20th Centuries', in Alusine Jalloh and David E. Skinner, *Islam and Trade in Sierra Leone*, Trenton/Asmara, 1997, pp. 21–63.

³ *GT* 62.

⁴ Park, *Travels*, chs 25–6.

Jobson);[1] also to bring blessings through prayers or harm through curses, to divine the future, and to treat illnesses with remedies derived from the bark, leaves and roots of trees and shrubs, the remedies being handed down as part of the knowledge of particular kin groups. Certain renowned marabouts also treated mental illness. Jobson seems to have been genuinely interested in discussing religious ideas with the marabouts, in particular learning about their knowledge of Old Testament characters. He was impressed by their sobriety, as contrasted with the alcohol-drinking of the Soninke. He apparently believed that the marabouts might be ready for conversion to Christianity.

Caravans from the interior came to Jaxanke towns near the river to trade, the town of Sutuko (Sutukobaa) being one of these. The inhabitants were accustomed to providing hospitality for strangers, particularly those of the trading fraternity. Jobson describes the large size of Sutuko, mentioning the extensive enclosed area designed to safeguard the inhabitants' cattle, and he also notes the asses used in large numbers by the traders. Jobson had hired Jaxanke as guides for the upper river, and he found the marabouts, in general, extremely cooperative. At Sutuko, the hospitality due to strangers, and particularly traders, was freely extended to Jobson and his companions. But at the same time the marabouts were able to withhold vital information about trade routes and the sources of gold, information which might be turned to their commercial disadvantage. As Jobson noted:

> These people of Setico were the most unwilling we should proceed in the search of the River ... not onely telling us themselves of many dangers, but at all townes where we came, and amongst our familiars, had left their perswasions, if it could have prevailed to discourage us; or whether they did verely thinke, our boate could never have found passage, in regard it was never attempted by any such vessell before, or as I incline unto, fearing we might be hinderers to the Trade, they had so long followed, and whereunto they were setled, being well provided with such numbers of asses, as beasts of burthen, to proceed and follow the same, so as from them, wee could get no comfortable intelligence, wherein reason leades the way, that every mans profit is nearest to himselfe ... [2]

However, the warnings had substance and were not entirely scaremongering. By all accounts, river traffic above Baarakunda Falls was negligible, the river having many hazards. Traders from the interior normally followed land routes until they came to towns near the navigable portion of the river below the falls. To the marabouts of Sutuko, travelling far up the river in small boats must have seemed a useless and dangerous project.

The Fulbe

Jobson uses the correct self-name for this people, but the Mandinka use the variant 'Fula', a form also often used adjectivally in English with reference to the Fulbe of

[1] *GT* 51, 78, 12.
[2] *GT* 81–2. On Sutuko in general, see *GT*, p. 123, n. 1.

INTRODUCTION

the River Gambia region.[1] The physical features of sections of the Fulbe suggest to many (but not all) historians a mixing in the distant past of West African and North African stocks, and the process of emergence of Fulbe as a distinct language seems to have begun far beyond River Gambia, most probably to the NW. Be these 'origins' as they may, it is certain that over many centuries the Fulbe have spread in waves and lateral ripples, principally in an eastwards direction across the grasslands zone of West Africa, and are now found in a series of pockets of varying size from Senegal to Cameroons. The traditions of the Gambian Fulbe suggest the in-migration of separate parties from the north and east.[2] Although latterly segments of the Fulbe have become sedentary and some even urbanized (in Nigeria), their earlier predominant status was that of a pastoral, transhumant cattle-owning people, and it was in this form that they entered the lands along River Gambia and were found there, in the middle stretches of the river, by Jobson in 1620–21. Traditions of a major crossing of the river by one belligerent group, probably in the fifteenth century, were recorded in Portuguese sources, and gave rise to a toponym, 'Fula's Pass', still in use in the 1730s.[3] As evidenced in recent times, each Fula sub-group marries within itself, thus maintaining an alliance between scattered settlements.

The in-coming Fulbe tended to form Fula settlements near Mandinka villages, allowing the Fulbe to develop a semi-symbiotic relationship with their Mandinka neighbours. In many instances the association was long maintained, extending even up to the present day. Mandinka drummers may still be called on to participate in Fula ceremonies, and Mandinka smiths make farming tools for the Fulbe. In earlier centuries the Fulbe contributed to the economy of the Gambia region, and were to some extent partners of the Mandinka in this, but they did not normally share in the political process. While regulating the affairs of their own communities, and while eventually accepting Islam, the Fulbe left wider issues to the Mandinka, who greatly outnumbered them.

Jobson has important observations on the Gambian Fulbe of his day, and in the

---

[1] In this language, *pullo* = 'a single member of the ethnicity' (hence, via the Wolof version *pEl*, the French form 'Peul'); *fulbe* = 'more than one individual, the people, the whole ethnicity'. In Nigeria the Fulbe are known as Fulani. For the Fulbe of the Senegambia region see Cheikh Ba, *Les Peuls du Sénégal: étude géographique*, Dakar, 1986. The present summary account of the Gambian Fulbe is also based on the following unpublished material: W. B. Stanley, 'Notes on the Physical Distribution of the Country, and Political Organization of the Fullahs of the Gambia, their Customs, Laws, etc. [1907]', Gambian National Archives; 'Notes, compiled by Father Maloney and John Balde' [1946], Roman Catholic Mission, Basse; David P. Gamble, field notes.

[2] The Fulbe of Kantora and Fuladu East have a tradition of migration from Masina in the east; the Fulbe of Wuli, sometimes called Jeeri Fulbe, the most traditional of the Fula groups, are predominantly Roroobe (known to the Mandinka as Lorobo) who moved south from Senegal; other migrants have come west from Futa Tooro, and are considered to have been Muslim on arrival. In recent centuries, increasing numbers of migrants have come from the Fuuta Jalon region (Futankoobe) to the SE, where the Fulbe once controlled a powerful state.

[3] Part II below: Lemos Coelho, 36. Up-stream from MacCarthy Island the river makes a sharp turn south, and here a ridge of rocks leaves only a narrow navigable channel where vessels brush against trees along the bank. While this passage presents no problem to power-driven vessels, sailing vessels had a strong current to overcome and little room to manoeuvre. 'Pholey's Pass' was marked on the 1732 Leach map; it was later known as Bruko Rocks. According to the tradition related by Lemos Coelho, the crossing was made by the Fulbe throwing rocks into the river to make it shallow enough to cross, but it is more probably a natural feature.

noted respects the ethnicity seems to have changed relatively little since then. The presentday pattern of the Fula pastoral economy is that cattle are moved inland from swamp areas and cultivated lands during the rainy season, and brought back again after harvest, when the dry season begins. Pasturage is then available from new grass and crop residues, and water is to be found in numerous pools. Often the Fula villages are fixed, the herdsmen alone moving with the cattle, since only short distances are involved. In the dry season cattle are tethered in the farms round the villages to improve the fertility of the maize and millet fields, with the result that the Fulbe generally have an abundance of grain. The selling of milk and milk products is in the hands of women, who are free to travel to any nearby market. Jobson commented enthusiastically on the general cleanliness of the milk-sellers.[1]

In the past, by attaching themselves to Mandinka villages, Fulbe had pasturage and water rights, and also protection against raiders from other districts. But throughout the history of the association, local chiefs have felt they had the right to seize cattle from the Fulbe when they needed them. Even in 1950 DPG heard a Mandinka village head remark: 'God gave us the Fulbe so that we could live off them'. The cattle demanded by chiefs were regarded as a payment for grass and water, but when the demands of chiefs became excessive the best option of the Fulbe was to move elsewhere, though a few instances are recorded of resistance and revolt. Jobson stressed what he regarded as the oppression of the Fulbe by the Mandinka.

The competence of the Fulbe in cattle rearing is legendary. In the tethering area each animal knows its place, and the peg to which it is to be attached. In the middle is a platform where the herdsman sleeps overnight, and where a fire is kept burning. Songs are sung to soothe restless animals, and also to provide protection, by noise or magic incantation, against predators, the songs often being accompanied by the playing of a stringed instrument. In the morning the animals are released and guided to pasture. Cattle are trained to obey the herdsman's call – whether, in the old days, to scatter if raiders came, or, in all periods, not to cross a certain line marking someone's farm. Herdsmen still have to defend their animals against predators, though these are rare nowadays.[2]

Jobson's contact with the Fulbe was limited and the extracts from his log include no mention of this people. However, in his book Jobson noted a 'towne of the Fulbies' near his middle river base, and he made further contacts with Fulbe while travelling up-river from the base, becoming so impressed by the encounters that he devoted a section of his text to 'the wandering Fulbie'.[3] Jobson noted their physical characteristics and distinguished them from 'blacke-men'. The English visited Fula tethering places 'diverse times', in order to buy 'a beef, or beeves', and the Fulbe

---

[1] *GT* 36.
[2] In 1953, however, DPG saw a small lion which had been shot by a herdsman with a ramshackle gun. He made a little money out of it, for he stuffed the animal, and brought it to the nearest large town (Basse), where he charged spectators a penny each to see it. A section of the Fulbe still specializes in cattle raising, and whereas in Senegambia the cattle are moved to daily and seasonal pastures, in Nigeria they are also 'driven' long distances to urban markets, the Fulbe (Fula) man often walking in front and 'talking' the cattle along.
[3] *GT* 33–7, 115, 116, 132.

INTRODUCTION

men came to the English to sell other 'commodities'. Moreover, 'in most places, within the ebbing and flowing [i.e. below Baarakunda], where we did lie for Trade', Fulbe women daily came down to the river to sell milk and milk products. The women were especially eager to buy salt from the English, and probably the salt was in part intended for their husbands' cattle. Against their oppressive Mandinka neighbours, the Fulbe spoke 'many disdainfull words' – according to Jobson. But he also noted that 'higher up in the Countrey' the Fulbe had 'domination among themselves', presumably a reference to the Fulbe polity several hundred miles to the SE, in the Fuuta Jalon.

Other African peoples

Jobson did not name any of the other African peoples on the river, and refers only indirectly, if at all, to their activities. The lower south side of River Gambia, near the mouth, was occupied by the Bainunka people, an ethnicity already receding under Mandinka pressure and today almost extinct in that district. A creek leading into Bainunka territory, River Bintang, had important trading centres up-stream, much discussed in Portuguese sources and later visited by agents of the Royal African Company. Jobson ignored this district, and while noting the riverain traffic that passed between Mandinka ports, was apparently unaware of the older, probably Bainunka-led, commerce that linked River Gambia, via creeks, portages, and small streams, to rivers to the south.[1]

A large stretch of the north bank of lower and middle River Gambia was subject, at least notionally, to Wolof polities, the Wolof being the major ethnicity of Senegal.[2] Few Wolof, however, lived along the river, and it is possible that in Jobson's period Wolof influence was generally in decline. Jobson referred to the 'King of Bursall', that is, the *buur Saalum* or ruler of a polity around River Saalum, a polity of mainly Serer people under a Wolof ruling group, to whom the ruler of neighbouring Niumi/Nyoomi on the lower Gambia was seemingly subject. Jobson did not specify Nyoomi but claimed that, 'on the north side … from the sea-side, about halfe way we went up, they did acknowledge the King of Bursall'.[3] At Kasang Jobson noted the influence of the *buur Saalum*, in ordering the local king to be replaced. Jobson supplied no specific statement that the polities of Niani and Wuli were subject to Wolof control, indeed he listed the 'King of Bursall' and 'the King of Wolley' as each a 'great King'.[4] However, if Niani was the immediate overlord of the king of

---

[1] Baynunk is the self-name, Bainunka the Mandinka version. On the history of the Bainunka in relation to River Gambia, see A. W. Mitchinson, *The Expiring Continent: a Narrative of Travel in Senegambia*, London, 1881, pp. 426–8; Dr Lasnet, *Les Races du Sénégal: Sénégambie et Casamance*, Paris, 1900, pp. 173–8; Robert M. Baum, 'Incomplete Assimilation: Koonjaen and Diola in Pre-colonial Senegambia', paper, American Historical Association annual meeting, 1983; Stephan Bühnen, 'Place Names as an Historical Source: an Introduction with Examples from Southern Senegambia and Germany', *History in Africa*, 19, 1992, pp. 45–101; Peter Mark, *The Wild Bull and the Sacred Forest: Form, Meaning, and Change in Senegambian Initiation Masks*, Cambridge, 1992; Brooks, *Landlords and Strangers*, esp. pp. 87–95 (but much conjectural); and especially an unpublished thesis, Bühnen, 'Geschichte der Bainunk …', whose discussion is comprehensive.

[2] On the Wolof, see David P. Gamble, *The Wolof of Senegambia*, London, 1967.

[3] *GT* 48.

[4] *GT* 48. This was in contrast to a statement made *c.* 1500 that 'Setico and two of the other towns [in Cantor], being on the frontier of Mandinka, speak the language of Mandingua, but belong to the kingdom of Jalofo [of the Wolof]'; and that 'River Gambia divides the kingdom of Jolof from the great kingdom of Mandingua' – see Part II below: Duarte Pacheco Pereira.

Kasang, then presumably the former as well as the latter was subject to Wolof control, at least in some measure. Jobson further remarked that 'there is warres between the one side of the river and the other, and especially from the king of Bursall, in so much as the people would tell us, if hee could have any means to transport his horse on the farther side, he would in a short time overrunne great part of that Country'.[1] This again implies Wolof influence on the north bank, although it is unclear whether the remark was intended to apply to Wuli as well as to Niani. The *buur Saalum*, like other Wolof rulers, purchased horses and maintained cavalry, in his case partly from the income he derived from the salt trade up River Gambia and the export trade in slaves brought down-river. The extent to which this ruler in Jobson's day directly controlled the salt depots at Wolley Wolley and elsewhere is uncertain, but a century later the incumbent *buur Saalum* had a residence on the river (at Joar) which he at times visited.[2] It is possible that when Jobson twice called at Mansegar he was in a district directly under Saalum and therefore Wolof rule, but nothing he says confirms this, and he supplies no evidence about Wolof society.

Jobson's young assistant, Samgulley, had a name (Sanguli) common among the Serahuli, an interior people linguistically-related to the Mandinka, who moved into the Gambia valley in large numbers in the nineteenth century and are now influential, but who were probably very thinly represented there in the early seventeenth century. However, a Serahuli clan, the Jatta, who were renowned as elephant hunters, had become attached to the Wali and assimilated to the general Mandinka. Jobson records the method of hunting elephants and gifts of their meat, and possibly the hunters involved were Jatta.[3]

When Jobson traded at the furthest point up-river he reached, in Tinda, apart from the Mandinka merchant, Buckor Sano, and his followers, the English made contact with groups of non-Mandinka people from the interior who also arrived to trade. Jobson failed to name the ethnicities involved, but referred to aspects of their appearance. One group speaking a language other than Mandinka and seemingly non-Islamic was probably from the Basari/Bassari to the south, another was perhaps from the Bambara to the north. Jobson's Mandinka attendants spoke scornfully of the former group.

A final ethnic group appears implicitly but unrecognized in Jobson's account. His references to the salt trade in River Gambia predicate the activity of the Nyoominka, a mixed group mainly of Serer origin living in the delta and islands of River Saalum.[4] Seasonally a fishing people, they used their canoes to convey local

---

[1] *GT* 48.
[2] Moore, *Inland Parts*, p. 83.
[3] *GT* 149=141–142.
[4] On the Nyoominka, see E. L. Borel, 'Voyage á la Gambie: description des rives de ce fleuve et des populations qui les habitent', *Le Globe: Mémoires de la Société de géographie de Genève*, 5, 1866, pp. 5–31; Paul Pelissier, *Les Paysans du Sénégal*, Saint-Yrieix, Haute Vienne, 1966, pp. 409–26, map 444, fig. 28; Régine Van-Chi-Bonnardel, 'Exemple de migrations multiformes intégrées: les migrations des Niominka (Îles du bas Saloum, Sénégal)', *Bulletin de l'Institut fondamental d'Afrique noire*, sér. B, 19, 1977, pp. 836–9; David P. Gamble, field notes. For a conjectural history of how Nyoomi on River Gambia, as a Mandinka-led polity, came to be differentiated from an earlier Nyoominka polity centred on River Saluum, see Donald R. Wright, *The World and a Very Small Place in Africa*, New York, 1997, p. 78.

INTRODUCTION

products, particularly salt and fish (fresh or dried), into and up River Gambia (as well as northwards to the Little Coast of Senegal). Most likely they had residential settlements within the Mandinka ports. Possibly they dominated the riverain trade, and brought further down-river some of the slaves that caravans from the interior delivered at the Mandinka ports. Yet Jobson at no point records having met or seen trading canoes on the river, a puzzling omission.

The Afro-Portuguese

Under the heading 'The severall Inhabitants, &c', Jobson's first candidate as resident along River Gambia was not an African ethnicity but 'The vagrant Portingall'.[1] This is probably testimony to a belief that the Portuguese presence was the most serious obstacle the English had to overcome in penetrating River Gambia. The Portuguese had traded in the river for 170 years, at times with considerable success, and in the sixteenth century had had a regular official trade in gold, the crown-licenced ships visiting 'Cantor' at least annually. Portuguese sources also make it clear that traders from the Cape Verde Islands frequently visited the river, some settling there. Indeed, two of the Islands traders, describing the river in the 1570s and the 1640s, give an impression of lively African-Portuguese commerce.[2] This is difficult to reconcile with the comments, or at least asides, of Jobson which tend to belittle the Portuguese influence. No doubt Jobson had his motives for presenting the Portuguese presence as turning out to be of limited significance. Nevertheless, it may have been the case that around 1620 Portuguese trade with the river was in the doldrums. The Cape Verde Islands had suffered severely from enemy assaults, and the decline of Portugal in Europe had led to cut-backs in official trade with Guinea, including the gold trade of River Gambia.

Moreover, the passage of time had led to the creation, both in the islands and on the coast, of a mixed population, the Afro-Portuguese.[3] While continuing the use of the Portuguese language (albeit in an increasingly 'creole' dialect),[4] and maintaining a claim to be Portuguese in aspects of culture, the Afro-Portuguese found submission to the often muddled orders and demands of Lisbon more and more difficult to accept. Those on the mainland were regularly denounced in official circles as 'run-aways' (*lançados*) – they might otherwise have been described, in the meanings both positive and negative, as 'adventurers'. Some substance was given to the denunciation by the fact that a small number of the *lançados* were escaping or taking refuge from the official judicial processes of Church and State. These included a number of Jews and alarmed 'New Christians'. Economically the *lançados* both aided and adjusted to a commercial revolution, by switching their own activities from contact with the declining Portuguese trade network to contact with the rapidly evolving trade network of Portugal's enemies, the French, Dutch and English. Largely neglected by the Church, those of the *lançados* who stoutly called

---

[1] *GT* 27–33.

[2] Part II below: Donelha, chs 10–11; Lemos Coelho, ch. 2.

[3] On the Afro-Portuguese, otherwise Luso-Africans, see Jean Boulègue, *Les Luso-Africains de Sénégambie, XVIe-XIXe siècle*, Dakar, 1972; Brooks, *Landlords and Strangers*, passim.

[4] For the modern form of Crioulo, see W. A. Wilson, *The Crioulo of Guiné*, Johannesburg, 1962; Luigi Scantamburlo, *Gramática e Dicionário da língua Criol da Guiné-Bissau (GCr)*, Bologna, 1981; Jean-Louis Rougé, *Petit Dictionnaire étymologique du Kriol de Guiné-Bissau et Casamance*, Bissau, 1988.

themselves Christians nevertheless in practice adjusted elements of their behaviour to the social morals of their African neighbours and relatives.[1]

Jobson's view of the Afro-Portuguese in River Gambia was influenced in large part by his knowledge of earlier events. From the 1580s the English had traded in Senegal with the Afro-Portuguese resident or with bases there, but both sides had done so opportunistically, with innate suspicion at times ending in violence. The same had occurred at Kasang in River Gambia in 1619. Although the English were also guilty of violence, Jobson thought of the Afro-Portuguese in terms of the two massacres of Englishmen, in 1588 and 1619. He therefore denounced them in much the same terms as Lisbon officialdom did, as criminals, renegades, and heathen. Probably he intended to be equally derogatory when he claimed that only 'few ...[of the] 'Portingales' were other than 'Molatos, betweene blacke and white, but the most part as blacke, as the naturall inhabitants'.[2]

His actual contacts with the Afro-Portuguese were in fact limited, even surprisingly so. Before entering River Gambia, on River Saalum the English took revenge on the individual supposedly responsible for the Kasang massacre.[3] In the river, Jobson recorded a first encounter with a 'Portugall' at Tendabaa, the man informing him about the murder of Thompson up-river.[4] Both occasions testified to the network of Afro-Portuguese commerce and communication within Senegambia. Jobson stated that in the river the Portuguese were 'scattered, some two or three dwellers in a place', but all with families from unions with black women. This reference probably gave the contemporary reader an impression that the Afro-Portuguese were fewer than they actually were, but it is unlikely that in 1620–21 the number of families was more than a figure in the lower scores. Travelling up-river, Jobson seems to have avoided calling at – or at least, recording that the English called at – localities where, most probably, Afro-Portuguese traders lived. It is likely, for instance, that some would have been found at Tankular and at Wolley Wolley.[5] At Mangegar, a Portuguese was met 'by the way'.[6] He presented Jobson with a leopard skin, and it is notable that in both River Saalum and River Gambia all the Afro-Portugèse encountered were friendly and helpful, if ostentatiously so and perhaps somewhat insincerely. It is also notable that most of the friendly contacts, while recorded in Jobson's log, are not mentioned in his book. At Kasang, however, where almost certainly there was normally a fair number of resident traders, 'no Portugall

---

[1] In the first decade of the 17th century a Jesuit said of the Afro-Portuguese on the Little Coast of Senegal that, 'though Portuguese by nation and Christian by religion and baptism, they nevertheless live as if they were neither Portuguese nor Christian ... so forgetful of God and their salvation that they might be blacks themselves and the heathen of the land' (Guerreiro, *Relaçam*, pt 1605, liv. 4, cap. 8, f. 130).

[2] *GT* 28.

[3] *PJ* 921.

[4] *PJ* 922.

[5] A century after Jobson's journey, it was reported that 'Tancrowal [Tankular] is divided into two parts, one for the Portuguese and one for the Mundingoes' (Moore, *Inland Parts*, p. 50). The Portuguese were often granted a tract of land near the river or on a major creek, and this was marked by planting silk cotton trees or safeguarding trees already there, these trees serving to identify the place from the river, and also preserving memory of the site through later time. Thus, Tankular people still have the tradition that it was a ward called *Tafakunda kaabiiloo* which gave the Portuguese land.

[6] *PJ* 922.

INTRODUCTION

would now be seen', Jobson seeming to imply that they had fled after the massacre and the disapproval of the local Africans.[1] Since the traditional trade with the Afro-Portuguese must have been of more importance to the town than a possible second-only visit of the strange English the disapproval must be doubted. However, Jobson was informed that, should the English proceed further up-river, 'the Portugals had hired men of that Towne to kill us', and whether true or not the information indicated the influence of the Afro-Portuguese and presumably their presence in the vicinity. On his down-river journey, Jobson stayed at Kasang for a month, but his account of the town makes no further mention of the Afro-Portuguese.

Up-river from Kasang, the English received a very friendly welcome from a Portuguese, one of two living at Pompeton.[2] (Jobson detailed it in his account but probably only because the welcome formed part of an anecdote about African magic.) Jobson now claimed that up-river from Pompeton 'dwels no Portugall in this River', information which, whether true or not, may have been supplied to him by his host. Certainly Jobson records no further meeting with Afro-Portuguese. However, his account notes that a group of five Englishmen at an unstated locality but probably the middle river base, saw a vessel pass by travelling up-stream; and Jobson attempts a generalization about Afro-Portuguese up-river trade. 'Some few of these sorting themselves together, in one time of the yeare, have used to go up this River, in a boate or small barke, as farre as Setico [Sutuko], and there to remaine in trade, from whence it is certainely knowne they have returned much gold, above which place they never attempted, which is not halfe the way we have already gone up ...'.[3] Although Jobson manages to make it unclear whether these up-river voyages were in his own day or merely in the past, his reference to the vessel that was seen allows that it was proceeding to Sutuko. When Jobson himself reaches Sutuko, his account makes no mention of Portuguese activity, nevertheless it is unlikely that, even around 1620, Afro-Portuguese trade there was as slight as he hints. That the Portuguese had never ascended the river further than the ports of Sutuko was incorrect, but it is unlikely that there was ever much trade higher, or at least enough trade to outweigh the extreme difficulties of the passage or to undercut the goods brought from the interior by overland caravans. Beyond Baarakunda it is uncertain how far the Portuguese had ever penetrated, and Jobson's boast that he had gone higher than ever they did, while perhaps not wholly correct, was an understandable claim in terms of the information available to him.[4]

Jobson does not signal it, but the long history of Portuguese trade and residence on River Gambia meant that the African peoples had been significantly affected. Jobson does refer to the export down-river of slaves and charges the Portuguese with 'especially' dealing in this commodity. But the wide range of commodities actually bought by these European traders for decades before the English arrived must have modified several aspects of the local economy, as also must the delivery of import goods. Even if we concentrate, probably unduly, on slaves, the trade-route

---

[1] *PJ* 922.
[2] *PJ* 922; *GT* 118 – for some confusion about the visit, see the Itinerary, p. 213, n. 1.
[3] *GT* 29.
[4] For the limited evidence on Portuguese penetration of the river, see p. 7, n. 1 above.

involving the transport of captive 'heathen' from the Islamic frontier to the interior must have been to some extent reversed, when a proportion of the captives travelled in the opposite direction, to the Atlantic for export by Europeans. It is plausible that polities and populations had gradually shifted nearer to the river as Afro-European trade built on the earlier inter-African trade, and that some 'ports' had extended themselves, if not being actually new foundations. In general, the English were shadowing, rather feebly, the commercial relations with Africans initiated by the Portuguese, and in their social and cultural relations with Africans were doubtless treated as merely a peculiar and inexperienced variety of Portuguese. It seems that the Portuguese language was fairly widely known among the African traders on the river, and although Jobson, perhaps deliberately, reveals little about the problem of communication, it is clear that Portuguese was frequently one of the mediating languages between the English and the Africans – perhaps even the most used one.[1] The Portuguese had employed African assistants, mainly as pilots and interpreters, and it was no doubt the history of this relationship which enabled Jobson to easily hire, and subsequently enjoy excellent relations with, his marabout assistant.

---

[1] The borrowing of a number of Portuguese terms into Mandinka is one indication of the Portuguese cultural influence. The terms include *sumbo* 'lead' (Portuguese *chumbo*,) *kojaaroo* 'spoon' (*colher*), *furnoo* 'oven' (*forno*), *kaitoo* 'paper' (*carta*), *kadianoo* 'lock' (*cadela*), *larincoo* 'orange' (*laranja*), *patato* 'potato' (*batata*) (David P. Gamble, *Terms Found in Old Writings about Senegambia*, Gambian Studies 28, Brisbane, California, 1993). For similar borrowings from Portuguese in other African languages of western Guinea, see A. T. von S. Bradshaw, 'Vestiges of Portuguese in the Languages of Sierra Leone', *Sierra Leone Language Review*, 4, 1965, pp. 5–37.

PART I

JOBSON'S TEXTS

The Golden Trade; OR, a discovery of the River *Gambra* ... (1623)
'A true Relation of Master Richard Jobsons Voyage ...'
'The discovery of the Cuntry of King Solomon ...'

THE GOLDEN TRADE: OR, A DISCOVERY OF THE RIVER *GAMBRA*

The Epistle Dedicatory.

TO THE RIGHT WORSHIPFULL, SIR

WILLIAM ST. JOHN Knight, Governour of the Countries of *Ginney*, and Binney: Sr. *Allen Apsley* Knight, *Liuetenant of the Tower of London, and* Deputy Governour as a foresayd: Sr. *Thomas Button* Knight, *and other the Noble Knights, and Gentlemen*,[1] adventurers for the sayd Countries of Ginney, *and* Binney.[2]

Honorable Gentlemen:
IF it may please you to take into consideration, the cause of publishing this ensuing treatise, it may some-way satisfie for my presumption, in offering to bring to the publike presse, that which to you hath bin so chargeable in discovering, and therefore by all reason to you belongs the benefit of what is discovered, or at the least free will to dispose of your owne /[ii]/ priviledges as to your wisedome should be most approveable; But such are the turbulent spirits of some men, that no curtesies can win faire correspondency, but as profest enemies to the ingenious search of worthy minded Gentlemen, proclaims warres against their indevors, if they tend to merchandizing, thereby intermedling (as they terme it) to discover their secret mistery, although in their perticulars to begin such an interprize, they can no wayes paralell the meanest of your true experience and well grounded desig[n]ments, but it might have bin hop'd, that there would have beene some better respect towards you, in regard of your persons, and not to deale in that nature as to every ordinary Gentleman, or other by them imployed, by whose industry after the way is opened to any profitable businesse, and that hee hath made plaine the discovery, then they doe find occasion to cavill and turne them off, and presently imploy servants of their owne, many times very unfitting, in regard they will not requite deserts, nor allow of any society in an apparant way of gaine:[3] *All*

---

[1] Both here and in his 'Conclusion' Jobson addresses the 'Gentlemen' as if they were the only Adventurers, thus ignoring the 'Merchants'. While dedications were normally addressed only to socially and politically influential individuals, this extended selectivity reflects his attitude to the 'Merchants', whom he here goes on to criticize, at least by inference.

[2] While the Adventurers might correctly be said to be 'for the Countries of Ginney and Binney', the Governor and Deputy Governor were officers of the 'Company of Ginney and Binney', not the Countries. This is either rhetorical inflation, or, more likely, a misreading of Jobson's manuscript.

[3] Jobson is writing when there was rising complaint against the Company in respect of its monopoly of Guinea trade, but he also seems to be hinting at some discord within the Company itself, whereby certain employees were discarded. In his extant petition he complained that unnamed officers were employing their co-religionists, Roman Catholics. Although he participated in the abortive voyage of 1624 or 1625, perhaps when writing his book in 1622–3 his own employment was threatened. There may also be a hint that certain members were trading privately and if so this accusation was probably directed against John Davies.

which is indevored towards you, for whilst you have bin suffered to disburse your monies in the first discovering, and as it were beating and laying open the way, where and how this Golden Trade should rise, you have quietly past one, but now there can be no more evasions, but that the profit plainly appeares; what complaints have bin framd? what combination and plotting together? Wherein to avoyd suspition, the face of simplicity, the honest Country-man hath bin made the instrument to bring about the incroaching gaine they aymed at: That it may therefore appeare, how the first grounds of this hopefull businesse by you were layd, and how you have seconded one losse by another, /[iii]/ and how needfull and necessary it is, that you should now proceede to follow what is begun, and made use of what you so deare have payd for; I have written this Discourse out of my owne carefull observation, in the time of my imployment for you, that you might (if you please) see what you have done, and what (if it please God to blesse the courses) you are like to do, which may not onely incourage you, but invite other Gentlemen of your Ranke, to associate with you, to follow and proceede upon this hopefull enterprize, wherein intending faithfully, in demonstrating the truth, to manifest the zeale and service I must ever owe you, humbly craving pardon, I remaine

        Your devoted Servant,

        RICHARD JOBSON.

/1/   *The Invitement to this golden Trade,*
   shewing the cause of the first undertaking it, and orderly proceedings therein.

It hath beene the usuall course (for the most part,) of such as travaile Forraine parts, in the observing and setting downe such things as they see, to neglect the noting of what is held publike, in regard that after the whole Company, be they 60. 40. or but ten together, have taken perfect view, it stands conceited, the same is as well manifest to our whole Country, whereby diverse times, many things worthy of note here at home, to such as take pleasure in reading of other mens adventures, and delight in variety of other nations, are either quite left out, or sli[gh]ted in so poore a manner, as the Reader goes away unsatisfied: I having received this caveat from that worthy gentleman, Mr. *Samuel Purchus*, who is so dilligent a searcher, and setter forth of all our English travailes, of whose true industry those great volumes he hath publisht to the /2/ world, shalbe perpetuall witnesse, spending therein (as he rightly termeth it) his talent for his Countries service, and being likewise incouraged by him, after he had seene and read my journall, breefly relating each dayes particular,[1] in my travailes, into this great and spacious Country; whereof by Gods grace I entend to write, laying as it were a commande upon me, not to conceale that, which by publishing may first tend, unto the advancement of Gods glory, and next undoubtedly the honor, wealth, and preferment of our owne nation: Likewise having beene still earnestly invited by all sorts of people, and especially by some of worthy note, (as occasions have fallen out at any time, whereby I have beene drawne into discourse of thes[e] travailes) that I ought not, nor might not without offence leave unpublished, that which doth proffer so apparent hopes of so great a golden Trade, which at this time seemes so needfull, that by the generall compla[i]nt of our great want, the earth hath shut up her rich bowels toward us in other places, the rather to envite us to seeke after that, which lies as it were under our noses, in respect of other travailes, and hath been left as a concealed businesse, untell our time of neede, that then it might be more effectually followed, and more seriously regarded: For apparant proofes whereof, first there is no Historian but will accord, that in all ancient Histories discoursing of the inward parts of *Affrica*, assure[d]ly alwayes called by the name of *Ethiopia*, it hath beene noted for the golden region, in the whole conquests of *Alexander*, as *Quintus Curtius* sets it downe,[2] he onely had

---

[1] The material 'Extracted' from the Journall and published by Purchas omits many daily entries, not only for the periods of sea voyages but even during the period when Jobson was travelling on River Gambia. For this and the relationship between Jobson and Purchas, see the section on 'The Sources' in the Introduction.

[2] Quintus Curtius Rufus, *Historiae Alexandri Magni*, bk 4, ch. 3 (a very passing and dubious reference, on which see J. E. Atkinson, *A Commentary on Q. Curtius Rufus' Historiae* …, Amsterdam, 1980, p. 363).

a great desire to /3/ visit these parts of *Ethiopia*, but never came there. The *Romans* likewise, carefull Relaters of their great victories, doe speake little of the interior parts of *Affrica*, their greatest entrance being in the wars of *Jugurth*, and in pursuite of him, onely mention is made, of a great desire they had to search the South parts, in regard they were thereunto invited, by those rich and golden armes, they found those blacke people to come against them withall, where of so many golden shields, were carried to their famous City, in their so glorious triumphs, but in their discoveries they had no successe: Returning with the losse of most of their people, in regard as is alleaged, they met with diverse drie and sandy deserts to passe, wherein as many were lost and over-whelmed, so againe the parching heate, and continuall droughth was cause of the perishing of many others, and inforced their returne, without any satisfaction.

The selfe same causes continue still, for which we neede not search written bookes, but talke or discourse with any Marchant of this City of London, who have yearely trade and commerse in *Barbary*, being the nearest parts of *Affrica*, adjoyning unto us, and many times from our Country, into their principall Harbours, runne in twelve dayes, and in the like time againe, from them to us, and inquire of them whe[n]ce the *Moore* of *Barbary*[1] hath that rich gold, he makes his Chequens[2] of, and they will tell you, there is no gold growing, within the confines of *Morocco*, or *Fesse*, at least that is knowne, or made use of, but that the great aboundance of that rich gold they have, is fetcht and brought into /4/ the Country, by the naturall inhabitants, for which they undergo great travailes, onely by land wherein they do passe great desertes of sand, with much danger, as appeares by the losse they receive many yeares, of diverse both of their people, and Camels, yet so commodious is their trade, and followed with such great dilligence and government, that amongst themselves, none are admitted but principall persons, and by especiall order, without entertaining any other nation, what respect or familiarity so ever, they have gained amongst them.

The trade of the Moores in in [sic] Barbary for their gold.

This in effect hath beene the sole ground, to attaine unto that knowledg, which I presume here to write for my Countries service, wherein duety especially requires me, to manifest the care and diligence, of those noble and worthy Gentlemen, who are the grounds, and originals of this hopefull worke, unto whom these my labours, as their owne proper rights are dedicated, whose vertues ayming at good actions, in this our blessed and peacefull time, and cessation from those sea affaires, they were wont to be busied

---

[1] With a single exception, in his book Jobson reserves the term 'Moor' for North African Muslims, usually speaking of the 'Moor of Barbary', and he does not apply the term to West Africans. The exception is when he mentions children playing in a river blessed by a Muslim cleric (*GT* 24) and perhaps he wished to stress that the children were also Muslim. The Journall, however, twice terms Jobson's four hired assistants on the up-river expedition, two of whom were marabouts, 'Moors', perhaps because all were Muslims (*PJ* 923). Although they are later termed 'blackes', it is perhaps unlikely that in each instance it was Purchas summarizing who introduced the term 'Moors'.

[2] 'chequens' = chequins, sequins, etc ('chikinoes' in Richard Hakluyt, *Principall Navigations* ..., London, 1589, p. 193), a gold coin used in Turkey, the Mediterranean and Barbary (*OED*).

in[,]¹ summond them up, to inquire and make search after the goldnest hopes, and upon good grounded conferances with such principall Merchants of *Barbary* as their wisdomes could make choyse of, attaine some better satisfaction, to their former knowledge of the Moore of *Barbaries* Marchandizing, as I lightly have toucht before, wherein their practise and true understanding in the Mathematiques assured them,² the Moors unknowne travaile must be to the South-west, if other wayes our Quotidian trade,³ into all and every part of the Mediterrane sea, must needes have had /5/ some or other intelligence. And therefore uniting themselves together, concluded upon a lawfull and warrant-able course to undergo the search of this golden trade, by the South-parts, and to adventure uppon those promising rivers, that fall into the maine Ocean, on the South-west side, wherein it now requires, I should briefly relate, the manner of their proceeding.⁴

The Kings Majesties Letters Patents.

In the yeare 1618. in the month of September, they set forth a ship called the *Catherine*, burthen 120. tun, and in her imployd on[e] *George Thompson* a man about fifty yeares of age, who had lived many years a Marchant in *Barbary*, the carcazon⁵ of goods hee carried with him amounted unto 1856£. 19s. 2d. having his instructions from the Governour and Company to enter in the River of *Gambra*, and with such shallops,⁶ as hee had, and were thought convenient for him, to follow his trade, and to discover up the River, leaving the shippe in a secured Harborough: All which in his part being carefully performed, in his absence, through the overmuch trust of our English hearts, and faire familiarity wee use to all nations, with whom we are in amity, the shippe was betrayde, and every man left in her, his throat cut, by a few

The first voyage.

---

¹ Of the three leading 'Gentlemen' named by Jobson in his dedication, two, St John and Sir Thomas Button, had been commanders of fleets, as had also been at least two of the other Adventurers (Thomas Love, Sir Richard Hawkins). When Jobson wrote, the long war with Spain and Portugal had ceased, at least in European waters, although conflict with the Dutch was on the horizon.

² In contemporary usage 'Mathematiques' often referred to the science of navigation, and hence, in a general way, to knowledge of geography.

³ 'Quotidian' = daily, regular, commonplace.

⁴ The reader is warned that Jobson now tells the story of the first two voyages in long, breathless sentences, which to add to the difficulty of comprehension contain a number of obsolete terms and misprints. Briefly, the sequence of events was as follows. In September 1618 Thompson in the *Catherine* reached River Gambia; he left his ship (at Kasang, as we learn later) and with some men travelled further up-river. In (probably) early 1619 those left aboard the ship were murdered and the ship sunk or stolen. Having heard the news, Thompson remained up-river, while some of his companions passed overland to the coast and eventually reached England. In (probably) mid 1619 the *Saint John* was sent to River Gambia, to support or rescue Thompson, who declined to return and sent back an encouraging report. Later, with two of the seven other Englishmen remaining in the river, he rowed up-stream to Tinda, to gain information. In March 1620, while or after returning down-river, Thompson was killed by one of his own men; the remaining English (we learn later from Jobson) are now settled at Oranto. In October 1620, without news of Thompson's death, the Company sent out the *Sion* and *Saint John*, with Jobson, to support the enterprise.

⁵ 'Carcazon' = Spanish *cargazón* 'cargo'.

⁶ Shallop = a small boat propelled by oars or a sail, used in shallow waters or between ships, to carry persons and a light cargo.

The ship taken by the vagrant Portingall, and the men slaine.	poore dejected *Portingals* and *Molatos*, whom they gave free recourse aboord,[1] being onely banisht people, and for the most runnagados from their Country, as when I come more particularly to write of them,[2] will more fitly be delivered: *Thompson* upon intelligence, being gotten farre uppe into the River, and finding the inhabitants to use him curteously, with the Kings allowance of the Country, seated himselfe uppon /6/ the land,[3] and thorough the kindnesse of the inhabitants, neere those parts where the shippe was lost, some of the English who came downe from *Thompson*, where [sic: were] safely conveighed many dayes travaile over land, untill they found meanes, to meete with shipping to transport them home, with their woefull tidings:[4]
The second voyage.	Whereupon the noble Adventurers, with all expedition set forth a Pinnace of fifty tunnes,[5] called the S. *John*, and in her a new supply of goods, and direction to *Thompson*, either for his repaire, withall his Company home, or as he did affect his trade, or had hope of his discovery, to make use of those goods, and abide there: He utterly refused to come away, and therefore sent away the S. *John*, who for that they came in an unseasonable time,[6] which then experience made them understand, and thorough some other abuses, which more conveniently else where I shall set downe,[7] which [sic: with] losse of many of her men returned, and [brought] as little comfort of gaine to the Adventurers, onely hopefull letters from *Thompson*, inviting them to a new supply, and by the next season to send unto him a shippe and pinnace, with some especiall commodity hee made mention of, confidently affirming, they should no wayes doubt of a hopefull discovery, where the Moores of *Barbary* traded, and a valewable returne for their losses sustained, promising in the meane time, which [sic: with] such company as he had left with him, being in all onely eight persons, in his small boate to search up the River, which hee attempted in a payre of Oares,[8] takeing onely two of his owne Company with him, the rest /7/ people of the Country,[9] which [sic: with] whom hee past up

---

[1] A marginal note in Purchas's summary gives the following information. 'Hector Nunes, &c. which under colour of trade waited their time to kill the English and take their ship, Tomson and others being in their trade in the Countrey, others on shore, and divers sick: after much love and pretended kindness' (Purchas, *Pilgrimes*, 1/9/13, 1569). The Journall has the following statement. 'In the River of Borsall we entred, where we tooke a small Boat belonging in part to Hector Numez, the principall in that Treachery and Murder aforesaid and detayned some of his goods therein for satisfaction, taking thereof a publike Inventorie, that if any other could lay just clayme they might be restored. This was done for punishing Numez, and to terrifie others from like treacherous attempts, not without effect' (*PJ* 921).

[2] See *GT* 28 ff.

[3] Thompson's base is not named but it was probably at or near Oranto where Jobson was later to find the surviving English.

[4] They made their way overland to 'Cape Verde' (*GT* 31), probably meaning the 'Little Coast' of Senegal (south of Cape Verde up to almost River Saalum).

[5] pinnace = a small, light vessel, generally two-masted and schooner rigged.

[6] Presumably during the rainy season, that is, in June–October 1619.

[7] Jobson blamed the crew of the *Saint John* for taking drinking water from the river, *GT* 126–8, 164/156. If there were other 'abuses' he forgot to mention them later.

[8] 'a payre of Oares' = a boat rowed by two men (*OED*).

[9] 'people of the Country' = Africans.

the River, and got to *Tinda*,¹ a place hee aymed at; in hope to have had conference with a blacke Merchant, called *Buckor Sano*, (of whom I shall have cause to speake in the Relation of my owne travailes)[;] fayling of [meeting] him, for that hee was then in his travailes within the land, hee stayd not many houres above, how-be-it in that time, hee received such intelligence of the trade hee lookt after, that such an extasie of joy possest him, as it is and hath beene aleadged against him, that growing more peremptory then he was wont, and seeming to governe with more contempt, by a quarrell falling out amongst them one of his Company slew him,² to the utter losse of what he had attaind unto, who in regard of emulation in striving to keepe others hee affected not³ in ignorance, committed nothing to paper, so as all his endeavours and labours were lost with him.⁴ These things I have presumed to write, that it may appeare, what rubs [*sc*. mishaps] have beene in the infancy of this discovery, and may partly make answere to the question may bee propounded, by any that shalbee pleased to read over my insuing discourse, why so hopeful and promising a businesse should bee neglected.

<small>Captaine Thompson slain.</small>

And now I returne to the worthy Adventurers who little distrusting this mishap, notwithstanding *Thompson* was slaine in March, whereof they could have no intelligence: In October after[,] beeing a convenient season[, they] set forth againe a shippe and Pinnace, the shippe called the *Syon*, burthen 200. tunne, and the S. *John* a Pinnace of 50 tunne: /8/ In this shippe it

<small>The third voyage.</small>

---

¹ Tinda/Tenda/Tanda was a vaguely defined region of the interior encountered up-river around the confluence of River Gambia and River Niériko, where the former, after allowing penetration broadly eastwards, has its early course from more or less due south. Jobson later refers to 'the Towne' of Tinda (*GT* 83), but no town of that name is known today or recorded in other earlier sources, the chief town of Tinda being Bady/Badeyo. Jobson supplies the earliest record of the name 'Tinda'.

² The reference in the Journall reads - '... killed by one of his Company, and that the rest were in health' (*PJ* 922), this making it clear that the murderer was one of the English. Jobson passes rapidly over an episode embarrassing for the Company (which may have disapproved of it being publicized), damaging to the good standing of its agents on the river, and awkward for a narrator who has just deplored the murder of Englishmen on the river by Portuguese – not least because he was informed of this English scandal by a Portuguese (*PJ* 922). At this point it is left unclear whether the murder of Thompson was committed while he was up-river with two men or after they had rejoined the other five, but later Jobson seems to indicate that it was after he 'returned' (*GT* 85), which fits the fact, soon to be given, that he was killed in March. Jobson eventually met all seven survivors, but he can only have learned about the 'not many houres above' from Thompson's two companions, one of whom, Brewer, Jobson mentions (*PJ* 922). Jobson does not name the killer. If Thompson behaved 'more peremptory than he was wont', possibly he had acquired a fever on his travels. All the survivors rescued by Jobson returned to England (*GT* 6, 128), but there is no suggestion that the murderer should be punished, and no evidence that he was. Presumably it was because the Portuguese and the local Africans learned of the murder that the Englishmen did not conspire to produce a different explanation of the loss of Thompson, a killing by Africans, by Portuguese or by wild animals.

³ 'others hee affected not' = others he had no affection for.

⁴ Somewhat of an exaggeration. From the two companions Jobson patently learned something about the up-river conditions, and possibly the name of Buckor Sano, although Thompson may have announced his intention of meeting this merchant when he sent a written report back in 1619. However Jobson is thinking principally of the information Thompson may have gained about the interior gold trade.

pleased them to imploy mee the present wrighter,[1] and now what doeth insewe of this discourse; is written from mee either as an eye witnesse, or what I have received from the Country people,[2] and none but such, as were of esteeme, and as my confidence assures, would deliver no false thing, as where I come to speake of the blacke people in particular, may be more aptly conceived. The 25. of October 1620. wee set sayle from *Dart-mouth*, the 4. of November, when the day appeared we were up with the Iland of *Launcerot*, and the next day by noone, past the Canary Iland, and had layd all that land [?;] the 17. of November, we came to an anckor in the River of *Gambra*, having had some occasion of stay by the way, to the losse of neere three dayes, so as our whole travaile from *Dart-mouth* thither was in 20. dayes,[3] we anckored some foure leagues within the mouth of the River.

The whole way from England to the River, runne in 20. dayes.

And to avoide inconveniences, by intermingling one thing with another, to set downe each particular as they presented themselves: I have thought it most acceptable to the Reader, to divide my discourse into particular heads,[4] the more aptly to bee understood, wherein I thinke it fit to beginne with the description of the River, with the limit and bounds thereof, so farre as we have seene, likewise what opinion experience makes mee hold for the continuance thereof, and how necessary it is, to bee searcht into for advancing the *Golden Trade*, with a relation what we find living therein, which may serve for sustenance, and maintaine the Traveller, next the several sorts of people, inhabiting /9/ upon the land, *Blackmen alias Mandingos*, or *Ethiopians, Fulbies*, and the vagrant *Portingall*, with the manner of their lives, buildings, and fortifications, the state of their Kings, and the title of other Commaunders, and their manner of life. The government of the Mary-bucke[s] [*sc.* marabouts] or Bissareas,[5] the discourse of their Religion, and seperations from the rest,

The particulers handled in this booke.

---

[1] It is possible that the wording is intended to convey that Jobson held a post aboard the ship as a mariner, presumably as an officer, but not certain since he nowhere confirms this or states a post.

[2] The claim is misleading, since Jobson also received information from the English survivors and, almost certainly, from some of the Afro-Portuguese on the river. For his 'eye witnesse' observations, see the section on 'The English in River Gambia 1620–21' in the Introduction.

[3] Jobson is anxious to stress that River Gambia is conveniently close to England in sailing time. For the dates, see the Itinerary (Appendix A), p. 208, n. 2.

[4] See below, as regards the subjects about to be mentioned, for the river 10 ff; the people 27 ff; the Mandingoes 37 ff; the Fulbies 33 ff; the Portingall 28 ff; the buildings, etc 42 ff; the kings 56 ff; the Mary-bucke 61 ff; the Merchants 80 ff; Buckor Sano 82 ff; the Juddies 105 ff; circumcision 109 ff; trades 123 ff; tilling and plants 119 ff; the seasons 125 ff; the wild beasts 136 ff; the fowl 146 ff; conclusion 152 ff.

[5] 'Mary-bucke' = 'marabout', Islamic cleric. It is not clear why Jobson uses his peculiar and unique corruption of the term (later sources employing it were copying Jobson). Marabout (an orthographic form originating and continuing in sources in French, e.g. Alexis de S. Lo, *Relation du voyage du Cap-Verd*, Paris, 1637, p. 27), from the Arabic for a Muslim 'holy man' (*al-murābiṭūn* 'religious frontiersman, activist'), was used by Europeans to distinguish in Senegambia (a) Muslims from 'pagan' animists; (b) Muslim communities at a time when the majority of the population was still pagan (e.g., the 'Soninke–Marabout' wars); and (c) in Jobson's day, especially the Jaxanke, itinerant Muslims, acting either as traders or as clerics, or as both, who served as exemplars of Islam and to at least that extent as active 'missionaries' (for whom, see the section of the Introduction on 'Polities and Peoples'). An economic and social distinction was that marabouts were literate, at least in Arabic, hence the term also connoted a learned ('educated') man. The term 'Bissarreas', from

and course of trading, and therein speaking of their Juliettos[1] or Merchants, with the Relation of my meeting with Buckor Sano, a great blacke Merchant, and commerce with him: Their Juddies or Fidlers,[2] and manner of meeting, with the discou[r]se of circumcision, and report of their divell Ho-re,[3] what manner of trad[e]s are amongst the common people, their order for tilling the ground, and severall sorts of graine, and other plants in use amongst them, and therewith an ample Relation of the times and seasons of the yeare, when those great stormes of thunder and lightning, with aboundance of raine do fal, the unwholsomen[e]s of the ayre in those times and what naturall reasons may be alledged aswell for those contagious times, as also to avoyde the inconveniences that have formerly by most of our nations [? our natione] beene fallen unto: Againe what variety of wilde beasts as well offensive, and ravenous, as also such as are for the sustenance and comfort of those as travaile, we find the Country replenished with, thereunto adjoyning what land foule is likewise there, the aboundance of both, which kinds are always at hand to mend the dyet of any ingenious looker out; and with a briefe conclusion from my selfe, I shut up my discourse unto which severally I now proceed: & first to the River.

Arabic *mubecherin* 'preacher of Islam', appears regularly in early Portuguese accounts of Senegambia, usually in the form '*bexerin(s)*', and appears to approximate to 'marabout'. The transition from 'bexerin' to 'marabout' occurred with the earliest French accounts of Senegal, notably S. Lo, *Relation*, p. 27 'marabou'; Claude Jannequin, *Le Voyage de Lybie au Royaume de Senega ...*, Paris, 1643, p. 111. Jobson's version, Mary-bucke, appears to be the earliest appearance of the term 'marabout' in printed Guinea sources or extant manuscript records of the region. However, the term 'Morabite', defined as a Muslim 'of the Hermitage', appeared in the section added to the 1600 English translation of Leo Africanus (Pory, *Historie*, in Brown, *The History*, 3, p. 1009); and no doubt in this or cognate forms the term had appeared earlier in European literature dealing with the advance of Islam.

[1] 'Juliettos' = Mandinka *juloo/jula* 'trader' + ?. The term in this extended form is peculiar to Jobson. The ending may involve an English misunderstanding, or conceivably a humourous extension (perhaps by relating the term to the English forename 'Juliet'); or it may be intended to represent a diminutive, i.e. 'petty trader'. Finally, it is just possible that Jobson wished to distinguish the term from a word in contemporary usage, 'julio' (an Italian silver coin, *OED*).

[2] 'Juddies', an English (and only-Jobsonian) corruption of Portuguese (and Portuguese Crioulo) *Judeus* 'Jews' – therefore 'Juddies' was probably to be pronounced as a plural of English 'Judy'. The term was applied in early Portuguese sources on Senegambia to a caste of musicians and praise-singers, often itinerant and to some extent beggars (cf. Part II below: Lemos Coelho, 1669, 11). In later and modern European sources these individuals were known as '*griots*' (a term first appearing as *guiriots* 'Basteleurs' [jugglers, buffoons] in S. Lo, *Relation*, p. 70.) The general population regarded the caste as useful, yet contemptibly parasitical, and members were discriminated against in certain respects (for instance, denied access to certain wells, and also burial in the ground, their corpses instead being deposited in the hollows of baobab trees). The Portuguese therefore believed, or affected to believe, that they were Jews (there may also have been some confusion with 'gypsies'). In fact, there is no convincing evidence that they were of Jewish origin. Jobson's use of the term indicates his reliance on Portuguese-speaking informants.

[3] In recent times, the name *Chore, Chore Mama* 'Chore, grandfather Chore!' was called out by men when the sound of a hidden bullroarer was heard during circumcision ceremonies, the sound being regarded as the voice of a spirit (oral information supplied to DPG, Kerewaan (Badibu), 1947; and Kundam (Fuladu East), 1954).

/10/ *The description of the River.*

This River, whereof I now take in hand, (by Gods Grace) to write, is scituate in the latitude of 13 degrees and ½, by all or the most part of Mappes and Cardes [*sc.* charts], and by some called, by the name of *Gambia*, by others *Gamba*, and by another sort set downe *Gambra*,[1] to which latter name being most frequent, I doe apply my selfe, for by the naturall inhabitants, either belowe in the mouth of it, neither it above to the farthest I have travelled, being upon the truest accompt I could keepe, some 320 leagues, or 960 miles,[2] could I ever heare any proper name, but only the word *Gee*, which in their language, they use to all rivers, and waters:[3] It hath one sole entrance, which in the very mouth, is about some 4 leagues broad, and in the channell 3 faddome water, at the least, without any barre, contrary to the setting of it formerly forth, where it is generally noted to have a barre, and much sho[a]ller water then we have found:[4] After we are run some 4 leagues in, it doth spread it selfe, into so many rivers, bayes, and creekes, that for the

---

[1] The latitude is almost exactly correct, the mouth being at 13°28–29′N. 'Gambia' was the form of the name in contemporary Portuguese sources, but 'Gambra', a form ultimately derived from Cadamosto, as well as appearing in Hakluyt, was the form on some non-Portuguese maps (e.g., Mercator's 1595 'Africa', and the Mercator/Hondius 'Africae Descriptio' of 1607 included in Purchas, *Pilgrimes*).

[2] The meandering of the river no doubt made the exact distance difficult for a voyager to reckon but Jobson exaggerates grossly, the true distance he travelled up-river being only some 460 miles.

[3] '*gee*' = *jiyoo*, the Mandinka term for water (simply the liquid), hence *san-jiyoo* 'sky-water, *sc.* rain'; *teng-jiyoo* 'palm water, *sc.* palm wine'. Possibly Jobson heard all these terms and memorized *jiyoo*, but it is unlikely that *jiyoo* would be given as the name of the river rather than as a description of its substance. On a journey down-river from Basse in 1979, DPG enquired at intervals what name the inhabitants had for the river, and in all cases individuals replied with the general Mandinka term for 'river', *baa*. In Western Kiyang, a town name could be added to further distinguish the main river from creeks, so that when *Bintang Bolong* was given for Bintam creek, *Tankular Baa* was given for the river. The term *Kambi Bolong* used for the river by Alex Haley in *Roots* was heard only in the estuary. The origin of the name 'Gambia' is uncertain. A recently collected 'tradition' avers that the Portuguese on arrival met a man named Kambi who gave the name of his locality as *kambi-yaa*, Mandinka for 'Kambi's place' (D. P. Gamble, with Louise Sperling, *A General Bibliography of The Gambia*, Boston, Mass., 1979, p. xiii). Although this is most likely a latterday folk etymology, it may contain some historical substance, inasmuch as the Bainunk, who in earlier times inhabited the lower River Gambia, have a clan-name 'Kambi': Bühnen, 'Geschichte der Bainunk', p. 177.

[4] The widest part of the river entrance, from Bunyadu Point on the north to Cape Point on the south, measures just over 10 miles, not far from Jobson's estimate. The estuary narrows to 3 miles between Banjul and Barra, and then opens up again to a width of 8 miles. As regards the depths, Cadamosto's account described ships entering River Gambia cautiously because of shallows and sandbanks, the caution no doubt because of the difficulties encountered by early European explorers in the previous rivers, River Senegal and River Saalum; and Jobson may have known Cadamosto's account, at least at second hand. But his reference to a 'generally noted' bar is curious. His assertion that River Gambia is 'without any barre' seems to mean that he was using the term 'bar' in the sense in which the mouth of River Senegal is partly blocked by a sand 'bar' – which indeed River Gambia lacks. But as regards a bar being nevertheless 'generally noted', since the only navigational guide to the coast in print by this period was in Portuguese, Manuel de Figueiredo, *Hydrografia ... Com os Roteiros do Brasil, ...Guiné, ...* (Lisbon, 1614), and therefore unlikely to have been seen by Jobson, he must have been referring either to maps or to oral information. Possibly he was confused by the appearance on maps of the toponym Barra (Portuguese for 'bar'), which he must have mistakenly assumed to refer to the

space of some 30 leagues, unto a Towne called *Tauckro valley*,[1] it is so intricate, that many months might be spent to search each particuler within that limit; but for that my occasion of writing, is grounded upon the great hopes, and expectations, that are from above [*sc.* up-river] I entend not to make any stay there,[2] but refer what is to be sayd, until I speake of the inhabitants only, as I proceed to let you know, that the maine channell, is not to be mistaken, except within the limit /11/ aforesayde, and then also thorough great neglect, or rather some wilfull ignorance.

Thus with a faire streame, this brave river shooteth in flowing from his mouth, into the land, neare upon 200 leagues, unto a Towne called *Baraconda*, or some little above, & that is the uttermost bounds of his flowing, even in the lowest season of the yeare:[3] For as in all rivers, running into the sea, the increase of the inland waters, occasioned by raines, or snowes, doe abate of the seas in draught; so much more, in this great River, who swels upright 30 foote, observing one due time, and season of the yeare for ever, must the seas force in those swelling times, be mightely driven backe, whereby a certaine knowledge is attaind, which are the setled times, to be followed earnestly, to meete with no impediment, in passing up, which impediment, is onely want of water to passe over flats, which in the lowest season of the yeare, in certaine places, are met withall, as is commonly seene, in all rivers, of such mighty inlets, which bankes as it were being past, presently a faire passable River continues, for many leagues;[4] and as we expect our seasons of Winter, and Sommer, so do these inhabitants these times of floods, occasioned by aboundance of raine, which raines always proceed forth of the South-east, and have their beginnings, much sooner in the inland, then at the Rivers mouth, so as in those parts, where we have had aboade, they begin to fall in the latter end of May, and at the Rivers mouth, not untill the

---

river rather than to a locality on land. Almost certainly the name had been applied by the Portuguese because of the proximity of the locality to the bar of the river, using 'bar' in its sense of merely the zone or line where the river water meets the sea ('crossing the bar').

[1] '*Tauckro valley*' = Tankular, on the south bank in present-day Western Kiyang District: see the Itinerary (Appendix A), p. 210, n. 1. Jobson exaggerates the distance travelled, Tankular being 52 miles up-river, not 90; and the wording gives the impression that the town was called at, whereas it was 'over-shot' (*PJ* 921). A century after Jobson's voyage, Tankular was said to possess a church, visited annually by a priest to baptize and perform other services for the Afro-Portuguese residents (Francis Moore, *Travels into the Inland Parts of Africa* ..., London, 1738, p. 51).

[2] After this remark, Jobson ignores the first part of his journey up-river, resuming only when he comes to describe Kasang. He fails to tell the reader that the *Sion* remained at Tendabaa and that he ascended the river in the *Saint John* (*PJ* 922; and see Itinerary (Appendix A), p. 211, n. 1).

[3] Baarakunda is some 330 miles up-river, not 'neare upon' 600. For the course to Baarakunda, see the section in the Introduction on 'The River'.

[4] The argument for ascending the river in a particular season is not as clear-cut as Jobson supposed. While in the dry season the inflowing tides carry ships higher, in the wet season the winds blow more favorably up-river, as Jobson himself noted (*PJ* 921), albeit in dangerous gusts. The deeper water covers the rocks and shoals which on the higher reaches caused Jobson and his party so much trouble, although with a faster down-flow which at times and places vessels could not overcome, even when towed.

These are all more largly written of, where the tillage of the ground is handled.

end of June:¹ These raines continue very violent, for three moneths, comming downe with great winds, and very much /12/ thunder and lightning, not perpetually, but as we say, in suddaine gustes, and stormes, the violence whereof being overpassed, the people continue their labour, as where I write of their manner of Tillage, is more largely set downe, as also a more free discription of these contagious times.² The increase of the River likewise, in the beginning of the yeare, before any raine is seene to fall where we aboade, did make it propable [*sic:* probable], that raine was fallen, within the land, before we tasted any, all which affirmes the great inlet [*sc.* penetration inland] of this hopefull River, and gives an assurance, that it is passable, if times and seasons be observed, and with dilligence followed, with boates, and vessels fitted accordingly, as experience in travelling it already so farre, may some wayes warrant a sufficient director.³

We were 10 of our owne company, that went up in a shallop, and 4 Blacke[s] that I hired to carry up a Canoe.

Next to shew a continuance of this great streame, when we had rowed beyond the ebbing and flowing,⁴ and 12. dayes against the currant, which wee travailed in the moneth of January, when the water was at the lowest of his nourishment, and then the shole we met withall, and stopped our further proceeding,⁵ had 9. inches water, which shallownesse continued not above 20 yards, wherein if we had beene an able company together, being onely 10. of us,⁶ and likewise had had provision of tooles wherewithall, and beene

---

¹ Purchas's summary introduced an error: 'These Raines alwaies proceed from the Southeast, beginning within Land, where wee abode in the later end of May, and at the Rivers mouth in the end of June' (Purchas, *Pilgrimes*, 1/9/13, 1567). But in the second half of May Jobson was careening the ship in Combo district, near the mouth of River Gambia, and was not 'within Land, where wee abode', i.e. up-river. And by the end of June his ship had left the river. Only the months November to May were spent on River Gambia. Jobson's information about weather conditions throughout the year, and about matters related to the seasons, must have come only in part from personal observation. The rest was probably obtained from the English survivors of the previous voyage, they having spent more than two whole years living up-river, that is, 'where wee have had aboade', as *GT* more correctly states.

² See *GT* 123 ff. below.

³ As detailed in the section on 'The River' in the Introduction, River Gambia is tidal in the dry season as far as Baarakunda (beyond the easternmost boundary of The Gambia, in Senegal). In the rainy season, June to October, the flow of the river increases, and the tidal effect is diminished in the middle reaches of the river. In this region, the banks are high, 20–30 feet at Basse and Fatoto, but in years of heavy rainfall, the water overflows the banks, creating swamps and pools. However, even when the winds are favourable (p. 85, n. 4), frequent and sharp changes in the direction of the river mean that travelling beyond Kasang is difficult for large vessels, the river channel narrowing and producing on each side banks of sand and mud, and thus giving little room for tacking. Hence sailing vessels commonly make use of the tide to progress up-river, anchoring when the tide turns.

⁴ That is, beyond Baarakunda.

⁵ The shoal is mentioned again (*GT* 84) and Buckor Sano states that the party had been stopped by want of water (*GT* 98). But in Purchas's extracts from the Journall nothing is said about being stopped by a shoal, it being instead implied that the English simply halted when they had ascended the river far enough to meet Buckor Sano (*PJ* 923).

⁶ In the extracts from the Journall no mention is made of a canoe, and none is made at this point in the present text. But Jobson later notes that a flying fish hit the canoe (*GT* 25). Elsewhere the Africans are said to have been hired to help to row the boat, that is, the shallop (*GT* 18). The canoe, used to enable landings on the banks and to explore shallows, was no doubt towed behind the shallop.

assured of a commodius trade, and so friendly a people to converse withall, as after we found, any encouragement would have made us worke a gut [*sc.* narrow channel] thorough that little distance, and being past that place, the river shewed himselfe againe, with faire promising, so farre as wee had occasion to looke, neare a league, and how far /13/ he might so continue, we are ignorant, and in those places above did we see sea-horses [*sc.* hippopotamuses], whose nature requires deep waters, as where I write at large of him,[1] you may better perceive, likewise the higher still, more store of Crocodiles, which addes incouragement of the largnesse of the River; and likewise a faire breadth between the shores: I follow these probabilities, to encourage the farther search of the River, which dilligently followed, may even in one season, give a full satisfaction to the forward Adventurer, and if it so fall out, we can meete with any towne above, standing by the River side, it will assuredly prove a commodious place, to make our aboade in, to take the advantage of the seasonable times, and to make returnes, to, and againe, as experience must leade, to the greatest advantage. And for trade there is no question, but a marvailous recourse would be unto us, which is already testified, in that so many hundreds of them came downe unto us, to the remote place where we were enforced to stay, building them houses of reads on both sides the shore, and the recourse still more, and more increasing, in so much as we had intelligence, the people were comming, from a great Towne called *Jaye* in their language, and wee doe conceave it to be *Gago*,[2] if wee had beene furnisht with commodities enough for them, and likewise knowne the seasonable times for our passage in the River, and convenient Harborough for our safer aboade; and why may not the towne they call *Mumbar*, which they say is but 6. dayes journey from the place we stayed at, according to their travaile, which in the discourse of the people I after /14/ lay downe, be likewise upon the River,[4] if so, how great an advantage, might it bring unto us, if wee

<small>These places are more largly written of when I set downe the manner of our trade, at the highest we went in the River.[3]</small>

---

[1] See *GT* 19 ff. below.

[2] Gago is Gao, a distant town on River Niger. Jobson follows the form of the name in Hakluyt (*Principal Navigations*, II/2, p. 193), in an English letter of 1600 (Purchas, *Pilgrimes*, 1/6/2, 852), and in the 1600 English translation of Leo Africanus (Pory, *Historie*, in Brown, *The History*, II, p. 826) – hence in Purchas, *Pilgrimes*, 1/6/1, 829, this form derived from Portuguese. (Although Pory's additions to Leo Africanus include passing references to 'Gambra or Gambea', Elephants' Isle, and the 'factorie of Cantor' [Pory, *Historie*, I, pp. 18, 81], it does not seem that Jobson borrowed from this work, not even for references to Barbary.) For a further reference to Jaye and Gago, see *GT* 102 below; see also the Introduction, p. 22, n. 8.

[3] See *GT* 102 below.

[4] The name '*Mumbar*' (later '*Mumbarre*', 93) appears in seemingly only one other source. Pina, writing *c.*1500, spoke of River Senegal penetrating a hinterland extending '*pera Cidade de Tombucutu, e per Mombare, em que sam os mais ricos tratos, e feiras d'ouro que ha no mundo*' ('to the City of Timbuktu and to Mombare, in which are the richest trades and markets for gold in the whole world': Ruy de Pina, *Cronica de el-Rei D. João II*, ed. Alberto Martins de Carvalho, Coimbra, 1950, p. 95). Pina's account was unpublished in Jobson's day and the published writings which borrowed from it, notably Barros, did not mention Mombare. Pina gives no source for the name, but presumably it had been acquired in Guinea by a contemporary Portuguese. Since the name did not reach Jobson from anything he read, it appears that, as he states, it was supplied to him by an informant on the river. Given that the name emerges on two occasions over a century apart, it may well be a genuine African toponym. Untraced as the name of any recorded town, it is possibly a variant of 'Bambara', the name of an ethnicity immediately NE

were minded to stay there, when the Moore of *Barbary* come, for at this towne the Caravan from *Barbary* doth stay and abide,[1] we know their whole trade is for gold, but what quantity they have here, or what people it is, they trade withall, we are as yet ignorant, and this adventure up the river, would undoubtedly discover, that the gold is there, wee are assured of, having bartered, and had trade for some,[2] and upon triall the same in goodnesse, that *Barbary* affords our Countries, having the river to friend, we should be able, though but few of us, to defend our selves, from the rage of the *Barbary* Moore, if he should attempt any thing against us, for undoubtedly, when he shall see us entered into his trade, he will appose what may be, to affront us; And although I have beene promised safe defence, by the country people, yet a boat is a certaine retreate, and the River a constant friend, to trust unto.

Againe, what know we, whither the River may bring us within the confines of those people, who will not be seene, and are those to whom our salte doeth passe, of whom in the relation of the Country, in his place, as it followeth I write,[3] and if it be as in all descriptions that are set out, it is layd downe, that the River of *Senega*, and this River do meete, yet cannot be in any probability, but a few dayes journey, above the place, and heigh[t], we have already beene at, must needes reach to it, and no doubt afterwards, that which affordes two such branches, must containe within himselfe, a faire /15/ and promising streame, which may take head from some great and large lake, above, such as is described, to be about *Gago*,[4] and if any such place

---

of the Mandinka. The Bambara lie across the upper River Niger, perhaps the river indicated, although their territory begins at more than six days' distance; and the trading town might therefore be Jenne. Jobson, however, clearly believes that the locality is higher up River Gambia (which he assumes to be part of an inland network of rivers).

[1] Contemporary accounts of the trans-Saharan trade – admittedly slight and distant – reported the North African merchants making their exchanges at Timbuktu or Gao (e.g. Purchas, *Pilgrimes*, 1/6/2, 852–3), and there must be some doubt whether they (and their camels: *GT* 3) penetrated further south to 'Mumbar'. It is perhaps more likely that trade between the auriferous districts and the Saharan-exchange towns was conducted, probably via Jenne, by merchants from these three towns (see J. D. Fage in Roland Oliver, ed., *The Cambridge History of Africa*, III, from c.1050 to c.1600, Cambridge, 1977, pp. 487–8). But as most would be Muslims, and many would have, and boast, some Arab ancestry, they might be mistaken for 'Arabeckes', that is, North Africans.

[2] Cf. *PJ* 923. Since gold was not specified in the statement of financial returns on the voyage, probably the amount was tiny.

[3] See *GT* 101 below. This is a reference to the famous but perhaps mythical 'silent trade'. Jobson may have learned about it from any one of three different sets of sources: local-Islamic, Portuguese, English. English sources normally derived the story from Cadamosto, although one English source of 1609 (the relevant section reprinted in Purchas, *Pilgrimes*, 1/6/2, 872), may have derived it from information supplied by Moroccan merchants. Cadamosto claimed that he had been told the story by Saharan Arab merchants, and it probably therefore circulated among the Islamic trading community on River Gambia. One early Portuguese source, Pacheco Pereira, repeated the story, but it was ignored by the Cape Verde Islands traders who wrote about River Gambia in more or less Jobson's day. See P. F. de Moraes Farias, 'Silent Trade: Myth and Historical Evidence', *History in Africa*, 1, 1974, pp. 9–24.

[4] The 'lake about Gago', if factual information, would refer to River Niger, but it more likely represents the contemporary belief, perhaps deduced from knowledge of River Nile, that the inland waterways of Africa all began in lakes.

should be found, what use or profit might arise, cannot but promise a hopefull expectation.

And lastly, if the inhabitants above, be enemies amongst themselves, as we see in the mouth of the River, and heare likewise of them, what advantage our force in the River, may worke, will easily be considered, in regard they have not the use of any boates, above where it ebbs and flowes, so farre as we have hetherunto beene, which is about 120. leagues, or 360. miles,[1] which we were travelling, as I have sayd before, onely 12. dayes, wherein is to be understood, we laboured not the whole day, but setting forward so soone as it was daylight, we continued working untill 9. or 10. of the clocke, resting the heate of the day, and againe from 3 until the evening shut in, and not at all in the night when it was coole and convenient, for avoiding of trees sunke, rockes, and sholes, which in the day time we could see, and have now taken notice of, and perfectly writ downe,[2] that upon any second attempt, we may be much bolder, and thereby aske lesser time for performance; howbeit our returne downeward, for those 12. dayes travaile, was in 6. and God be praised, both going, comming, and staying there, without sicknesse, or losse of any one man; Nay more (to our great comfort wee found) the higher we went, the more healthfull our bodies. And it is likely, if townes were found againe, neare the River, they do so continue, for from *Baraconda*, whither the River flowed, we never hard, nor saw /16/ of any Towne, or plantation, nor recourse of any people unto us, but what we sent for, neither shew of any boate, onely some two, or three bundles of Palmeta leaves, we found bound uptogether, which our Blackes would tell unto us, some of the people had made shift to passe the river upon, so as our passage then must needes afford more discouragement to the Actors, then any that can, (by Gods grace) happen hereafter, for we were discouraged, that the people above were of a bad condition, if we could passe unto them, which the inhabitants held, as impossible, in regard they did affirme, the river was full of trees suncke, and drifts, wee should meete withall, and our time in passing, being uncertaine, our provision which was small, might faile us, and poorely (God knowes) we were provided of those materials, that would have helpt to maintaine that principall, in respect the place, and way affordeth it; and what experience hereafter, can direct in that kind to doe, which being good comforts, and encouragers to the Adventurer, I will not by any meanes leave unwritten.

There is abounding in this River, who are bred and live therein, two sorts especiall, as I may terme them monstrous, the one devouring as the people report, and the other daungerous, as I have found: The devouring is the

The Country people have no boates or Canoos above the ebbing and flowing.

We laboured to get up the River onely 7. houres in 24.

No townes neare the River side after wee past the ebbing and flowing.

---

[1] Whereas Jobson previously stated (*GT* 10 above), that he had travelled up-river 320 leagues or 960 miles, the present figures, as he clumsily expresses, relate only to the distance above the tidal stretch, that is, above Baarakunda. But the distance is much exaggerated, being actually only some 130 miles.

[2] Jobson appears to be saying that he composed a river chart, showing obstacles, but if so, it must have been a very rough one, since there is no mention of his taking bearings (although he did carry a compass), and the maps available to him merely sketched a notional course for the river above Kasang.

Crocodile or *Alegatha*, because they carry one, and the same resemblance, but doubtlesse, I am perswaded, there is no other Crocodile, but such as wee have seene in this River, whom the people call by the name of *Bumbo*,[1] sundry times when we have driven them from the shore, where they have beene lying in the morning, or /17/ otherwise forth of the water, when wee have observed the print they leave behind them, upon the soft sand, we have found by measure of rule, his whole length, from the point of his nose, to the end of his tayle, containe thirty three foote; The people of the Country, stand in such dread of these, that they dare not wash their hands in the great River, much lesse, offer to swimme, or wade therein, reporting unto us many lamentable stories, how many of their friends, and acquaintance have beene devoured by them: neither do they at any time bring any of their Cattle, to passe the River, as within ebbing, and flowing, they have diverse occasions to doe, but with great dread, and ceremony: for at all Townes within that compasse, they have small boats, which we call Canoos, to ferry over withall, which cannot receive a live beefe, onely some five or six of the people: but when they passe a beefe over, he is led into the water, with a rope to his hornes, whereby one holds him close to the boate, and another taking up his tayle, holds in the like manner; the Priest, or Mary-bucke, stands over the middle of the beast, praying and spitting upon him, according to their ceremonies, charming the Crocodile, and another againe by him, with his bow and arrowes ready drawne, to expect when the Crocodile will ceaze, and in this manner, if there be twenty at a time, the[y] passe them one after another, never thinking them safe, untill they be on the toppe of the River bancke: One thing more, to shew the feare they have of him, when I was going in my discovery up the River; having as I sayde onely nine of our owne people with me, I did hire /18/ Blacke-men, as I had occasion to use them, to serve as Interpreters, likewise to send abroade, and to helpe to row, and get up the boate, so that when I came to passe the flowing, and to goe all against the currant, I did furnish my selfe, of foure able Blackmen: the first place we found a stiffe gut to resist us, the water being not above foure foote deepe, for speedier and more easier passing, our men went into the water, and laying hands, some one [on] the one side of the boate, and some likewise on the other, waded along, and led her through, which we found a good refreshing; the River being sweete and cleare, was comfortable in the heate, by no meanes I could not make any of my blacke people, go out of the boate, denying flatly to go into the water, saying that *Bumbo* would have them; after some two of these passages, there was another streight where was a necessity of more hands, so that striping my selfe, I leapt into the water, the Blackes seeing me prepare, seeme much to diswade me, but when they saw me in the

---

[1] *Bumbo* = Mandinka *bamboo* 'crocodile'. The commonest species in River Gambia is the Nile crocodile (*Crocodilus niloticus chamses*) (Chris M. Loiser and Anthony D. Barber, 'The Crocodile Pools of the Western Division, The Gambia', *British Herpetological Bulletin*, 47, 1994, pp. 16–22). Early sources regularly commented on the river's crocodiles, see Part II below: Cadamosto, [32]; Pacheco Pereira; Fernandes, f. 112; Barros; Almada, 5/7; Donelha, f. 23; Lemos Coelho, 67.

water, they presently consulting together, stript themselves, and came likewise in, the businesse ended, and we all aboord againe, I askt of them the cause made them come in, having so earnestly denied it before, they made answere, they had considered amongst themselves, the white man, shine more in the water,[1] then they did, and therefore if *Bumbo* come, hee would surely take us first, so that after they never refused to go in, yet in all our whole passage, did we never receive any ass[a]ult, but to the contrary, where we have seene great companies of them, lying upon the sands, they have perpetually avoyded us, with the same /19/ shines [shyness], that Snakes doe use, to avoyde the noyse, and sight of men here, onely boldest to shew himselfe where the water was deepest, and the Blacke people, do not sticke to say, that since the white men have had to doe [? goe] in the River, the Crocodile is not so daungerous, as in former times: Againe, whereby it doth appeare, they are more aboundantly above, whereas he doth naturally smel exceeding sweete, after the manner of muske, so as in all places, where they use to come one [on] shore they leave a sent behind them, that many times we are not able to receive, but inforced to stoppe our nostrils: some three dayes, before wee came to the highest place we stayd at, we beganne to find the River water, which was our daily drinke, to change his relish, but after we came there, it had such a sweete musky tast, that we not onely refused, to drinke of it, but also could not endure our meate to be drest therewith, but sought out springs and freshes, upon the land, nay more, those great fish which with our hookes, we tooke in that place, lost the savor they had below, and did tast and rellish as the Crocodile smelt, that we utterly refused, to eate them ourselves, but bestowed them upon the people of the Country, which received them thankfully; and likewise the cry and noyse of them in this place, was more then we had heard al the way, for the noyse he makes is resembled right to the sound of a deepe great well, with which the great ones call one to another, and may be distinctly heard a league,[2] which surely argues, the continuance of this hopefull river, and that some great lake above may bee the nourisher of them. /20/

The other is the Sea-horse, who in this River do wounderfuly abound, and for that the name of Sea-horse is a common word, in regard of the Greeneland voyages, where they use the same to the Sea-mosses[3] they kill there, who are of contrary shapes, I thinke it fit to describe this fish or beast, or what I may call him, because questionlesse, there was never beast, nor any thing in that kinde, set forth to shewe in these our Countries, that would produce more admiration. He is in fashion of body, a complete horse, as round buttock'd as a horse of service, and in his whole body answerable: his head like

---

[1] The curious construction at this point may be Jobson's imitation of Africans speaking broken English – although to the modern reader many of his other constructions are either obsolete or idiosyncratic.

[2] The distance over which the noise of the crocodile can be heard is greatly exaggerated. DPG has heard the sound at an estimated distance of one quarter of a mile.

[3] 'Sea-mosses' = sea-morses or walruses ('morse' ultimately from the Lappish). The term 'morse' is frequently found in Hakluyt, where Jobson may have encountered it.

unto a horse with short eares, but palpably appearing which he wags, and stirres, as he shewes himselfe, onely toward his mouth, he growes broade downe like a Bull, and hath two teeth standing right before upon his lower chopp [*sc.* jaw], which are great and dangerous in regard he strikes with them: his crye, or neighing, directly like a great horse, and hath in the same manner foure legges, answerable to his body, whereupon hee goes, and wherewith hee likewise swimmeth, as a horse doeth, yet in these is his greatest difference, for they are somewhat shorter in proportion, then horses are, and where they should be round hoofte, it devides it selfe into five pawes, upon every which hee hath a hoofe, the whole foote, containing a compasse of great breadth, as the beast is in growth, insomuch as I have taken the measure of some prints they leave, where they walke, of twenty ynches over: His manner of feeding likewise, resembles the horse, for although he live /21/ all day in the River, yet every night he goeth duely on shore, in divers places feeding upon their Rice, and Corne, doing the Country people much spoyle, but his general feeding, is upon low marish [marsh] grounds, where the grasse or sedge is greene, to which they resort in great companies, & in those reaches of the River, which have deepest water, and lie nearest, and convenienst, to such manner of grounds, do wee always finde greatest store: in some places, they go a mile from the shore side to their feede, having trackes that are beaten as hard and palpable, as London high way; he returnes by the breake of day to the River, where he is very bold, when our boates come by, hee will hold his head above the water, many times store of them together, and so neare as within Pistolls shot, snorting, neighing, and tossing the water, making shewes of great displeasure, and sometimes attempting it, for in my passage too and againe in the River, my boate was stricken by them three times, and one of the blowes was very daungerous, for he stroke his tooth quite through, which I was enforced, with a great deale of dilligence to stoppe, or it had daungered our sinking;[1] but the hazard of them may be well avoyded, if men be provided to shoote at them, when they presse over-bold, which wee could not do, in regard our allowance of powder was small, and we were driven to put it to other uses, neither had wee peeces accordingly, thorough the neglect of some ill wishing persons, who deceive /22/ the trust the worthy Adventurers apposed upon them: In the night, while wee had candle burning, some of them, disturbed by us, would remaine in the River, and would come staring up the streame, snoring, and pressing neere upon us,

---

[1] In recent times, the hippopotamus population in the MacCarthy Island stretch of the river having apparently increased, 108 incidents involving the animal were reported in one year (1980); but when these were attacks on individuals in canoes, it was suspected that the hippopotamus had been wounded, or was protecting a young one hidden in the vicinity (Cherno A. Jallow, 'Hippo Lands and Islands', *The Gambia News Bulletin*, 18.6.1982, 2). Hunting hippopotamuses is now forbidden by law in The Gambia. Traditionally, no one under forty, and unless initiated by certain rites known in hunting communities, would hunt the animal. Early sources comment regularly and extensively on the river's hippopotamuses, see Part II below: Cadamosto, [32]; Pacheco Pereira; Valentim Fernandes, f. 112; Barros; Almada, 6/7; Lemos Coelho, 66.

but wee found meanes to send them packing, for breaking a small peece of wood, we would sticke a short candle lighted upon it, and let it drive with the streame upon them, from which they would flie, and make way, with a great deale of horrour, and one note we observed amongst them, they were always most dangerous, when they had their young with them, which they sometimes leave on shore, but being in the water, every female carries her young upon her backe, so as when she puts up her head, the young head likewise will looke his share, and where they appeare many heads together, there is asmuch variety, as from the great horse, to the hunting nagge: the Sea-horse, we found greatest store: when we were likewise past the flowing of the tide, and continued above the highest place we were, which still argues a large and constant River: The people do account of these for an excellent meate, not refusing to eate them, if they be taken up dead in the River, as they are many times found swimming, howsoever they come killed, howbeit I conceive, the Crocodile and they agree, for that I have stood upon the bancke, and see[n] them swimme, one by another without offence. *The Country people esteeme the sea-horse, for excellent meate.*

Having spoken of these, I now returne to matter of sustenance, which the River affordeth, there is variety of good fish, among which great store of Mullet, if men have nets, and provision to take /23/ them, which in some places, within the ebbing and flowing, the shore lies convenient to make use of, and above, that in most places, howbeit we never made use above any place where our shippe ridde, who always kept the net with her, wherewith we made diverse draughts, most especially at a Towne called *Cassan*, and against which the shippe did ride, and was the highest place in the River she went,[1] where our convenientst drawing was close to the Towne, and when the people at any time saw us bring our net on the shore, and provide to fish, as the net came neare the shore, they would come rudely in and many times with their uncivilnes, indanger the breaking, and spoyling of our net, with their greedinesse to lay hold on the fish, that wee were inforced to speake unto the King, dwelling in the Towne, to command them to forbeare troubling us, promising when we had taken for our own present use, and reserved some for him, the residue should be taken out and remaine amongst them, and his Commaund being given, they wete [were] carefull to observe it.

Amongst the rest, one time having made a draught, we had not such plenty as usually, onely some fish, in the cod of the net, which being taken up, were shakt into a basket standing in the boate, with which we rowed aboord, & the basket being handed in as the custome is, the fish were powred upon the Decke, whereof many rude Saylers will be their owne carvers, amongst which fish, there was one, much like unto our English breame, but of a great thicknes, which one of the Saylers thinking for his turne, thought to take away, putting therefore /24/ his hands unto him, so soone as he

---

[1] This was the *Sion*, which was eventually met there on Jobson's journey down-river (*PJ* 925). The ship stayed first at Tendabaa, and when it went up-river to Kasang, and why, are not recorded.

<small>A strange operation of a fish.</small>

toucht, the fellow presently cried out, he had lost the use both of his hands, and armes: another standing by sayd, what with touching this fish? and in speaking, put thereto his foote, he being bare-legged, who presently cried out in the like manner, the sence of his leg was gone: this gave others, of better rancke, occasion to come forth, and looke upon them, who perceiving the sence to come againe, called up for the Cooke, who was in his roome below, knowing nothing what had hapned, & being come wild [willed] him to take that fish, and dresse, which he being a plaine stayd fellow, orderly stooping to take up, as his hands were on him, suncke presently upon his hinder parts, and in the like manner, made grievous mone: he felt not his hands, which bred a wonderfull admiration amongst us: from the shore at the same time was comming a Canoe aboord us, in which was a Blacke man called *Sandie*, who in regard he had some small knowledge of the Portingall tongue, had great recourse amongst us, we brought him to the fish, and shewed it unto him, upon sight whereof, he fell into a laughter, and told us, it was a fish they much feared in the water, for what he toucht hee num'd, his nature being to stroke himselfe upon another fish, whom presently he likewise num'b [*sic*: num'd], and then pray'd upon him, but bid us cut of his head, and being dead, his vertue was gone, and he very good to eate:[1] At this place onely we should see many Moores sporting, playing, and comming boldly into the water, a good distance from the shore, where lay a sandy banke, but they never went beyond their heights, and they /25/ would tell us, there was a blessing granted to that place, by some great Mary-bucke, that Bumbo should never hurt them; and on that side the Towne stood, as our ship did ride in the middle of the River, and we have observed, we never saw any Crocodile, but on the contrary side, many times very great ones; And this being assuredly true, for varieties sake I have placed here.

<small>The running fish.</small>

In the upper part of the River, there are store of fish, and more conveniently to be come by, if men go provided. Amongst which we note one little fish, which may well be called the running fish, and is much like our English Roach, with a red tayle, who is inforced to runne above the water, and will continue a great way, but only touching of it, to save his life from his pursuing enemy, who comes chopping after him, just like the Trout after the flie, and is of that bignesse the Trout is, that somtimes the little fish hath runne into our Canoe to avoyde the pursuer:[2] Likewise of foule, the higher we go,

---

[1] The fish involved was an 'electric cat-fish' (Genus *malopterurus*), called *tingoo* by the Mandinka. A marginal note in Purchas's summary reads 'Torpedo, Tremedor, or Thinta' (Purchas, *Pilgrimes*, 1/9/13, 1568), 'torpedo' being a name for the electric eel and 'tremedor' being Portuguese for 'trembling, the trembler'. In spite of what local people told Jobson about the use of the power to attack other fish, modern authorities hold that the ability to give an electric shock is solely a defence mechanism (M. Holden and W. Reed, *West African Freshwater Fish*, London, 1972, p. 49). The fish plays an important part in the mythology of the Bambara people of the interior (and therefore probably of the related River Gambia peoples), and is also linked to successful childbirth, the dried skins providing a medicine for a difficult labour (Sarah C. Brett-Smith, *The Making of Bamana Sculpture: Creativity and Gender*, Cambridge, 1994, pp. 127–9). No earlier source had mentioned the fish.

[2] A freshwater flying fish (*Pantodon bucholzi*).

we find plenty, and much variety, but this we have ever observed, that in the maine River, we never see them swimming, but as they are in sholes together be they Ducke, and Mallard, or any other in their kind, they sit upon the shore, close to the River side, and dare not surely venter in, for feare of the Crocodile, but have their principall feeding upon the marish grounds, and ponds, which lie from the River, whereof the Country is very full, and you can finde no such place, but is aboundantly furnished, among which are many geese, of colour white, and blacke, rather bigger then our English tame goose, who hath upon /26/ each pinion of his wing, a sharpe spur, in every point resembling a Cockes spurre of the largest size, with which they are apt, not being shot dead, to give offence:[1] but for foule that live naturally by the shore side, as Hernes, Corlews, Storkes, Pluffer and the like, it doth yeeld plenty, so that whomsoever shall go well furnished of peeces, and powder, shalbe sure to mend their fare, and light upon many a dainty dish.

<small>The nature of the river foule.</small>

The people of the Country have likewise divers wares, which they make use of in the times of raines, and when the River is over-floude, at which times, they kill much fish, and they have also a strange maner of fishing, in their lakes and ponds, of which there are many that are very broade, and containe much circute, but are not deepe, to which they resort as they desire to fish, a whole towne or plantation together, only the men, every one having a kind of basket, with the mouth open, which hee holds downeward, and so going into the water, close on[e] by another, they over spread the pond, whereby the fish is moved; and so clapping downe the mouth of their baskets before them, they hit upon the fish, and in this manner they take so many that most of them go loaden home, and if at any time we be neare those places, they will lovingly impart them to us, upon returnes from us of poore valew:[2] These things which now we know, and can tell how to provide for, may serve as incouragements, to proceede upon a farther discovery, but in that which followes, concerning the love of the people, what trade we already have found, and what reliefe they bring us, and at what rates, as /27/ also what Deare, and wild Cattle the land affords throughout, with such variety of land foule, and other necessaries, wherof in their due place I meane to speake, I hope (as I desire) may be some furtherances, to invite Adventurers, to advise of some few dayes search further into this hopefull River: Wherein the very River, if we had nothing else to friend, proving but as we have hetherunto found it, will afford that comfortable reliefe, men neede not stand in dread of starving, which considered, and the probable good that may and wlll [sic: will] rise, in obtaining the Golden Trade, I conclude it most necessary, to follow dilligently a farther search, for which if I should be thereunto required, in place convenient, I could yeeld some other speciall account for

<small>The manner of the peoples fishing.</small>

---

[1] Cf. Part II below: Almada, 5/3.

[2] Communal fishing in ponds with baskets and barbed spears is still common in the middle river districts. This takes place towards the end of the dry season when the pools are beginning to dry up. For such fishing near Panchang (Upper Saalum) in 1947, see Plate II.

Man fishing with a barbed spear

which for some respects I forbeare to publish,[1] and following my order, proceede to the inhabitants.

The severall Inhabitants, &c.

To speake of the Country, and the inhabitants,[2] I take my beginning from the mouth of the River, whereat our first entrance, we find the Black men called *Mandingos*, and that they do continue amongst themselves, still one and selfe same language: Those of them who are inhabiting, or dwelling in the mouth of the River, or within certain leagues of the first enterance, are very fearefull to speake with any shipping, except they have perfect knowledge of them, in regard they have beene many times, by severall nations, surprized, taken and carried away,[3] but upon some knowledge /28/ they will resort to the shore neare unto us, and bring with them Beeves, Goates, Hennes, and aboundance of Bonanos, in the West Indies called Plantanos, a most excellent good, and wholesome fruit; likewise of their Country pease, and other graine, and in way of Trade some hides:[4] they there alone have the domination, their Kings and Governors being their seated, as in the upper parts, of whom

---

[1] It is not clear what Jobson is hinting at, but unless he was suggesting that he had additional information about the gold trade, it may be that he was threatening an account of the lapses of his fellow travellers and the inadequacy of certain of the arrangements.

[2] For the peoples mentioned in this section of the book, see the section of the Introduction on 'Polities and Peoples'.

[3] Any slowness of the villagers of the estuary to trade may have been for the reason given by Jobson (although the district was rarely visited by vessels other than Portuguese ones, which are unlikely to have spoiled their regular trading by seizing individuals). But perhaps it was instead due to their uncertainty whether the visitors had paid customs dues to their ruler.

[4] No contacts with the Africans of the estuary on the up-river voyage are detailed in either text, but it is possible that contacts were in fact made at Barra: see the discussion in the Itinerary (Appendix A), p. 209, n. 3.

especially my discourse is intended, howbeit for the more playner proceeding, I must breake of[f] a while from them, and acquainte you first, of another sort of people we finde dwelling, or rather lurking, amongst these *Mandingos*, onely some certaine way up the River.

And these are, as they call themselves, *Portingales*, and some few of them seeme the same; others of them are *Molatoes*, betweene blacke and white, but the most part as blacke, as the naturall inhabitants : they are scattered, some two or three dwellers in a place, and all are married, or rather keepe with them the countrey blackewomen, of whom they beget children, howbeit they have amongst them, neither Church, nor Frier, nor any other religious order.[1] It doth manifestly appeare, that they are such, as have beene banished, or fled away, from forth either of Portingall, or the Iles belonging unto that governement,[2] they doe generally imploy themselves in buying such commodities the countrey affords, wherein especially they covet the country people, who are sold unto them, when they commit offences, as you shall reade where I write of the generall governement: all which things they are ready to vent, unto such as come into the river, /29/ but the blacke people are bought away by their owne nation, and by them either carried or solde unto the Spaniard, for him to carry into the West *Indies*, to remaine as slaves, either in their Mines, or in any other servile uses, they in those countries put them to: Some few of these sorting themselves together, in one time of the yeare, have used to go up this River, in a boate or small barke, as farre as Setico, and there to remaine in trade, from whence it is certainely knowne they have returned much gold, above which place they never attempted,[3] which is not halfe the way, we have already gone up, since our trading there. With these, in their places of dwelling, wee are very conversant, notwithstanding, we received such a horrible treachery from them, as is set downe in

The vagrant Portingall.

---

[1] Two generations after Jobson, a French visitor to River Gambia described the Afro-Portuguese as follows. 'They always have a large chaplet hanging from their neck and they bear the names of saints, although neither baptized nor having any trace of the Christian religion. They normally wear a hat, shirt and breeches like Europeans, and although blacks they nevertheless assert that they are whites, meaning that they are Christians like the whites. Most of them say neither Christian nor Muslim prayers and others say both, when with blacks saying the latter, and when with whites wearing their chaplet and doing as the whites do.' He added that they spoke not metropolitan but Creole Portuguese (Prosper Cultru, ed., *Premier voyage de Sieur de la Courbe fait á la coste d'Afrique en 1685*, Paris, 1913, pp. 192–3). This source may have exaggerated the degree of assimilation which, in any case, may have been much less in 1620.

[2] 'the Iles' = the Cape Verde Islands. Jobson was of course biassed against the local Portuguese. It is true that some of them were 'run-aways' from the civil or ecclesiastical government in the Cape Verde Islands or further afield – some Jews in Senegal had come from Europe – and were refugees from persecution or individuals escaping from legal justice. But many of those in River Gambia were merely economic adventurers, or their Portuguese ancestors had been such, who had settled there to better themselves, given that conditions in the islands were harsh. Jobson was, however, repeating an English view expressed in the 1590s – most of the Portuguese in Senegambia were 'banished men or fugitives, for committing most heinous crimes and incestuous acts' (Hakluyt, *Principal Navigations*, II/2, p. 192). Similarly a Dutchman in 1606 opined that the Portuguese on the Little Coast of Senegal were 'mostly men banished from Portugal' (Moraes, *Petite Côte*, p. 138).

[3] At best, a half truth: the Portuguese had probably never attempted regular trade above 'Cantor' but had certainly on occasions explored higher up-river.

my beginning, in regarde they tell us, those that were the Actors thereof, are banished from amongst them, as being hated and detested for the fact. Howsoever, wee hope, and desire it may stand, for all our Nations warning, never to let them have the like occasion, but beleeve, ever they will doe as they say, in telling us they do love and wish us wel, provided they may never have us under their power, to be able to doe us ill, which it behooveth us to take especiall care of.

<small>An especiall Caveat.</small>

The conditions they live subject unto, under the blacke Kings, makes it appeare, they have litle comfort in any Christian countrey, or else themselves are very carelesse what becommeth of their posteritie; for whensoever the husband, father, or maister of the familie dies, if hee be of any worth, the King seizeth upon what hee hath, without respect, /30/ either to wife, children, or servant, except they have warning to provide before, or are capable of themselves, to looke out for the future time;[1] whereby we finde in some those few places we trade with them, poore distressed children left, who as it were exposed to the charitie of the country, become in a manner naturalized, and as they grow up, apply themselves to buy and sell one thing for another as the whole country doth, still reserving carefully, the use of the *Portingall* tongue, and with a kinde of an affectionate zeale, the name of Christians, taking it in a great disdaine, be they never so blacke, to be called a *Negro*: and these, for the most part, are the *Portingalls*, which live within this River, who since they see we have followed a trade, and begunne to settle upon it, in regard they much doubt, wee waite but an oportunitie (as they say amongst themselves) to have a valuable satisfaction, for the wrong their Nation began with, knowing the Englishmen doe not ordinarily digest such horrible abuses, it hath made such as were of worth, and dwelling upon the coast, who were woont to looke into the River, forbeare that recourse, and also those, that were of the best and most ablest estates, to quit their dwellings, and to seeke out else-where, leaving none but a few poore snakes, who for feare, rather then love, offer themselves, to do us any maner of service: which feare of theirs, is the more increased, because the naturall blacke people, out of their morall understanding, and were some of them spectators of their bloody murther, the shippe then riding before the Towne, when the fact was done, and by them rightly understood, to /31/ be treacherously done in betraying our faithfull trust, contrary to the great protestations and obligements before these inhabitants made and confirmed, did not onely utterly disallow of the fact, but exclaiming against them, caused them to forsake

---

[1] In western Guinea it was common for African rulers to welcome resident Europeans and protect them, but to operate the principle that on the death of a stranger within the polity his goods became the property of the ruler. The early Portuguese accounts of the coast complained about this principle, in particular its operation on the 'Little Coast' of Senegal (e.g. Donelha, *Descrição*, pp. 129, 131), but it was also referred to at Kasang – see Part II below: Lemos Coelho, 29. The Portuguese had themselves an elaborate legal and administrative procedure for winding up the estates of those deceased in Guinea and conveying the proceeds, or part of them, to dependants. The two sets of customs clashed.

their dwellings in that Towne, neither have they at this time any habitations there, notwithstanding they had had continuance for many yeares before.[1]

And further, when some of our people, who were above in the River, not knowing of this evill accident, and were upon occasions returning to the shippe, whom they found so miserably lost, and carried away, the people of the Towne, especially some principall, and most powerfull men, tooke such compassion upon them, that they fed them, and lodged them, with a great deale of loving care, and that for no small time, until they had devised and concluded amongst themselves, what course to take; and having resolved, to take a tedious journey by Land, in seeking to crosse the country to the North-ward, untill they came to *Cape de Verde*, where they were sure to meete with shipping, they not onely fitted them, with such necessaries as they could, but also sent of their owne people, as guides with them: and being in that manner commended from one King to another, were lovingly entertained, lodged, and fed, and with new guides still conveyed, never leaving them, untill their desire was satisfied, and they safely arrived, where they found convenient shipping, and still the commendations that went alongst with them, from one blacke King to another, was, in regard their shippe was betraied, /32/ and taken away by the *Portingall*, whereby they found such compassion, that in some places they had horses to ride on, and in other places were entreated to rest, and recreate themselves, longer then they were willing.

The curteous usage of the natural inhabitants.

And thus much is said, for those people of the country, amongst whom the *Portingalls* dwelt, had their aboade, and all familiar commerce; but for those blacke people who are dwelling above in the River, where these *Portingalls* never had any habitation, onely as I sayd, a trade, in their boates up some part of the River, and amongest whom wee have setled our selves, with great league and testification of much amitie (as I must deliver when I come unto them,) these I say, when there was only five of our men dwelling amongst them, their houses seated by the River side; and that cer[t]aine *Portingals*, in a smal bark or Boat, were to passe by them in following their Trade to *Setico*, being a matter of some 16 leagues, above the place our men lived at:[2] these people when they saw our men make ready their armes, & prepare their peeces, to stand on their guard, being so few of then, not daring to trust the *Portingals* flattering promises, did not only put themselves in companies for their defence, but likewise animated our men, to set upon them, promising if

---

[1] According to a later writer, at Kasang the spilling of blood was regarded as spiritual pollution and those responsible were penalized – see Part II below: Lemos Coelho, 30. Possibly this was why the Portuguese left. Although one might have expected the offence to relate to all blood-spilling, it perhaps only related to blood-spilling by non-natives, since there were several instances of large-scale local attacks on whites, apparently unpenalized.

[2] When this incident occurred is unclear. It may have been when Jobson and Lowe were both in boats trading on the river and only five men were left at the middle river base. Yet the *Saint John* was lying there and one might have expected the defence of this from the Portuguese to have been mentioned. Alternatively, and perhaps more likely, it may have been in 1619 when Thompson was up-river, when probably only five other survivors were left in the middle river. For the English middle river base, see Appendix B.

they would give the on-set, they would prosecute it, to the confusion of all and every man of them, in the same manner, as they before had dealt with us, with great vehemency pressing them, as a thing they were especially bound to do, which our men refusing they in themselves did carry towards them a kind of sullen, and insolent behaviour: so as their /33/ bloody act, wherewith they thought to daunt, and discourage us, in seeking or following of any trade here, and more securely to settle themselves, hath no doubt (by Gods providence) if it be carefully considered and dilligently observed, by a timely following of what doth offer it selfe, turned to the cleane contrary, and through their owne guilt, enforc'd them to avoyde the place, leaving it of their owne accords, whereby if wee imbrace the occasion, many good and profitable ends may bee made, and this have I truely related: the *Portingal*, who as he sees we prepare with earnestnesse, to follow this Trade, with the like earnestnesse, will prepare to leave the River, which preparations as I hope and desire, may speedily and earnestly in our Countries behalfe be under gone, and followed, so in his preparation I would be no hinderer, but thinke it a faire riddance, of a false friend, and so I leave him.

The reward of treachery.

The wandering Fulbie.

There is one people more, dwelling and abiding among these Maudingos [*sic:* Mandingos], and under their subjection, of whom it is necessary for me to speak, before I come to the principall. These are called Fulbies, being a Tawny people, and have a resemblance right unto those we call Egiptians:[1] the women amongst them, are streight, upright, and excellently well bodied, having very good features, with a long blacke haire, much more loose then the blacke women have,[2] wherewith they attire themselves very neatly, but in their apparell they goe /34/ clothed and weare the same habite, the blacke woemen do, the men are not in their kinds, so generally handsome, as the women are, which may be imputed to their course of lives, whereof I proceede to tell you; Their profession is keeping of Cattle, some Goats they have, but the Heards they tend are Beefes, wherof they are aboundantly stored: In some places they have setled Townes, but for the most part they are still wandering, uniting themselves in kindred and families, and so drive their heards together; where they find the ground and soyle most fitte for their Cattle, there, with the Kings allowance of the Country, they sit downe, building themselves houses, as the season of the yeare serves, and in such

[1] 'Egiptians' = gypsies. Purchas's summary adds this curious marginal note: 'These being tawny, and the others blacke, sheweth that this colour comes from the seed which takes hold on the Portugals borne of Negro women, and not of the Sunne-burned Fulby, in the same place' (Purchas, *Pilgrimes*, 1/9/13, 1569).
[2] Apart from Jobson's masculine appreciation of these females, the Fulbe have an element of North African ancestry and their features therefore tend to appear more in conformity with conventional European notions of beauty and handsomeness than do those of other West African peoples, especially in respect of skin colour, shape of nose, and quality of hair.

places as lies most convenient, for preservation of their Heards they looke unto: during the times of the raines, they retire to the mountaines, and higher grounds, and againe as they grow drie, and barraine to the low plaines and bottomes, even to the River side; that in the times of our chiefest Trade, their cattle are feeding by us, and the women with their commodities daily customers to us. These mens labour and toyle is continuall, for in the day time, they watch and keepe them together, from straying, and especially from comming to neare the River, where the Crocodile doth haunt, and in the night time, they bring them home about their howses, and parting them in severall Heards, they make fires round about them, and likewise in the middle of them, about which they lie themselves, ready uppon any occasion to defend them from their roring enemies, which are Lyons, Ounces,[1] and such devouring /35/ beasts, whereof the Country is full, as when I speak of them wilbe perceived.

This is the poore *Fulbies* life, whereunto he is so enured, that in a manner he is become bestiall, for I have noted diverse times, when we have come up in the morning, before his Cattle had beene disperst, or gone to feede, when we have called for the Mr [*sc.* Master], or chiefe of them, to make a bargaine for a beefe, or beeves, as we had occasion, hee would come unto us, from forth the middle of the heard, and those parts of him which were bare, as his face and hands, but especially his face, would stand so thicke of flyes, as they use to sit in the hot Sumner time upon our horses, and teemes here in England, and they were the same manner of flye wee have, which the *Fulbie* would let alone, not offring to put up his hand, to drive them away, therein seeming more senslesse, then our Country beasts, who will wiske with their tayles, and seeke any other defence, to avoyde or be rid of them,[2] but for our own parts we were faine, during our parley with them, to hold a greene bow [bough] to beat of the flie, finding his stay never so little, very offensive. These people live in great subjection to the *Maudingo*, under which they seeme to groane, for he cannot at any time kill a beefe but if they know it, the blackmen will have the greatest share, neither can hee sell or barter with us for any commodity hee hath, but if it be knowne the other will be his partner, in so much as when the men come unto us, they will watch the blackemans absence, or hiding their commodities, draw us covertly to see it, that

The misery of the Fulbie.

---

[1] The term 'ounce' normally referred to leopards. Jobson later writes of 'Ounces and Leopards' (*GT* 145/137), but was either confused or was referring to another local feline, the cerval, or wild cat (Mandinka *bambango*) – a third feline, the civet cat, he separately mentions and identifies. Until recently, when Gambian petty traders offered visiting non-Africans handbags made from animal skins, they would describe any skin with spots as 'leopard'. If not merely acknowledging the limited zoological knowledge of the visitors, they were perhaps following locally an older common usage. Early sources frequently mention leopards (e.g., Part II below: Gomes, f. 278v; Fernandes, f. 111; Barros).

[2] On the daily movement of Fulbe cattle, see the section of the Introduction on 'The Fulbe'. Jobson notes the Fulbe rapport with cattle but fails to specify their expertise in giving the cattle verbal commands (cf. Part II below: Almada, 6/4), perhaps because a degree of this skill is not unknown among stock-herders in Britain. As for the flies, presumably in the working conditions any attempt to disperse them was wasted labour and energy, so the Fulbe man learned to disregard them.

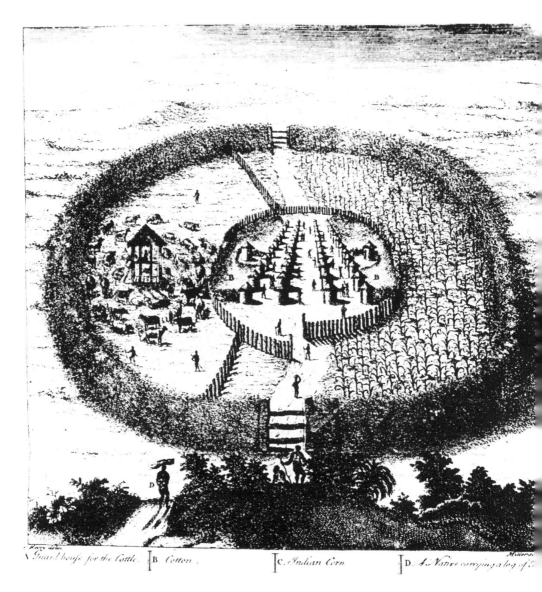

'Draught of a Pholey Town and Plantation about' (Moore, 1738)

they may have their returne private, and not sticke /36/ many times, when he knowes the other out of hearing, to speak many disdainfull words against him: And of these people the Country is very full, being disperst and spread in such a manner of families, as I sayd before, over the whole Country; and higher up in the Country, as we here [hear], and I shall shew hereafter, they are in on[e] part principall, and have excluded the Blackes, holding domination amongst themselves, and for the most part continually in warre. The language the *Fulbie* speakes, is different from the black-men, the women are our chiefest customers, for in most places, within the ebbing and flowing, where we did lie for Trade, we should be sure to have their custome every day, which was to bring us new milke, sowre milke, and curdes, and two sorts of butter, the one new and white, the other hard and of an excellent colour, which we called refined butter, and is without question, but for a little freshnes, as good as any we have at home: all which they brought unto us, in great and small gourds like dishes, made up very handsomely, and one thing let me not forget to give them due praise in, that in what somever you received from them, you should have it so neate and cleane that in your milke you shold not perceive a mote, nor in her butter any uncleanlines : nay the gourds, or dishes, they brought it in, on the very outsides would shine with cleanlinesse, and one [on] the inward parts, without any nastinesse, and if at any time, by any mischance there had beene a mote, or haire, which you had shewed unto her, she would have seemed to blush, in defence of her cleanely meaning. The cleanelines of the Fulby women.

In noting of which, I have diverse times sayd, /37/ there was great difference betweene them, and the Irish Calios although their manner of lives had great resemblance in following of their Cattle,[1] and as they were out of heart in one ground, to remoove whole Townes together, which but few yeares since was the Irish Kernes true course of life;[2] but with cleanlinesse your Irish women hath no acquaintance, and therefore I returne backe to my Tawny *Fulbie*: the commodities shee askt for were small beades, and poore knifes of 16 d. a dozen, with other trifling things, but after they once saw, and tasted of salt, which in their language they called *Ram-Dam*,[3] there was no other thing could so well please them, although it were never so little; we found the variety of these things many times agreeable to our natures, and therefore gave

---

[1] 'Irish Calios' = Irish *cailleacha* 'old women' – cf. 'the men of war could not be maintained without their churls and "Calliackes", or women who milked their "Creates" [Irish *creata* 'herds']' (D. B. Quinn, *The Elizabethans and the Irish*, Ithaca, 1966, p. 127, citing a 1579 pamphlet – the Jobson reference also noted, p. 25). For a spelling 'calliots' in writings of c. 1620, see ibid., p. 166, and also James Carty, *Ireland from the Flight of the Earls to Grattan's Parliament (1607–1782)*, Dublin, 1949, p. 7.

[2] 'Kernes' = corruption of Irish *ceithearn/aigh* 'foot soldier/s' – in practice lightly-armed 'feudal' infantry, in English eyes devoted to itinerant cattle-raiding (see many references in Quinn, *Elizabethans and the Irish*). Jobson seems to envisage 'Kernes' as herders of cattle, moving them between summer and winter pastures, and armed to protect their herds.

[3] 'Ram-Dam' = Gambian Fula *lamdam* 'salt'. The Fulbe have special ceremonies in which salt is fed to their cattle (Almada, *Tratado breve*, ch. 1/23; Ousmane Diallo, 'Le "Tuppal"', *Notes africaines*, 21, January 1944, 19).

faire recourse unto the people, for if we denide but one day to buy of them, we should want their company a weeke after, what earnest occasion somever we had to use them; and these things were to be had from none but them, because the *Maudingo* [Mandingo], or Blacke-man applies himselfe, at no time, in keeping or preserving of Cattle, but leaves it to this painefull *Fulby*, whom I likewise leave looking to his Heards, and come to the commanding Blacke.

The Maudingo *or* Ethiopian, *being the naturall Inhabitants, distinguished by the name of the* Maudingos.

The people, who are Lords, and Commaunders of this country, and professe themselves the naturall Inhabitants,[1] are perfectly blacke, both men /38/ and women. The men for their parts, do live a most idle kind of life, imploying themselves (I meane the greater part) to no kinde of trade nor exercise, except it be onely some two monehts of the yeate [*sic:* yeare], which is in tilling, and bringing home their countrey corne, and graine, wherein the preservation of their lives consists, and in that time their labour is sore, as when I come to shew the manner, you may easily conceive; All other times of the yeare, they live wandring up and downe, from one to another, having little understanding, either to hunt in the woods, or fish in the waters; notwithstanding, both the one and the other, in their kindes, are infinitely replenished, that to their very doores wild beasts doe resort, and about their houses in every corner, abundance of *Ginny* hennes, and excellent partridges.[2] In the heat of the day, the men will come forth, and sit themselves in companies, under the shady trees, to receive the fresh aire, and there passe the time in communication, having only one kind of game to recreate themselves withall, and that is in a peece of wood, certaine great holes cut, which they set upon the ground betwixt two of them, and with a number of some thirtie pibble stones, after a manner of counting, they take one from the other, untill one is possessed of all, whereat some of them are wondrous nimble:[3] we do perceive amongst them, that the ordinary people eate but one meale a day, and especially the younger sort, of what kinde soever; their houre of feeding being onely after the day light is in, and then with fires of

<small>The time and manner of the peoples feeding.</small>

---

[1] 'naturall' = native, aboriginal.

[2] DPG, when in 1946 he first lived among the Mandinka, conceived much the same sentiment as Jobson. During the 'hungry season' there was little effort to catch fish in the river, or utilize game. In return for some cartridges a local hunter would bring him bush fowl, guinea fowl, pigeons, etc. with which to vary his diet, but the villagers made no use of such foodstuffs. In the case of pigeons, their eating by children was believed to slow the healing of circumcision wounds. However, on afterthought, DPG appreciates that the supply of game birds would have done little to alleviate the hunger in a large village. As for domestic fowl, the villagers rarely ate hens, except at the time of major Muslim festivals, or at naming ceremonies. The hens were reserved for feeding distinguished strangers, or for sale to Europeans.

[3] The game is widespread in West Africa, under the name of *warri* (or *wori* or *oware*). Boards are now to be found in the tourist markets, and in The Gambia are common in villages of the Kombo region and in the capital, Banjul. Up-river, the game is less commonly played by adults, but adolescents and children play it by making holes in the sandy soil. The game has been described as follows:

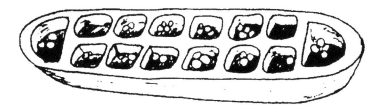

The game board

Reedes, without the doore, they sit them round, and fall to their viands, which for the most /39/ part, is either Rice, or some other graine, boyled, which being brought unto them by the women in goardes, hot, putting in their hands, they rowle up into balles, and cast into their mouthes, and this is their manner of feeding: they doe seldome eate either flesh or fish, the rather because they cannot get it, then out of any will to refuse it: and although they are great breeders of such very poultry as are our Cockes and Hennes, and have understanding to cut Capons, yet they are great sparers thereof, and preserve them to sell unto us, for small peeces of Iron, beades, and such like commodities, wherof if we be furnished, we can want none of that provision.

They will say, their feeding so seldome, is a great preservation of their healths, & at that time, when the Sunne is downe, a fittest time for nourishment, avoyding especially, to eate in the heate of the day, as a thing wonderfull unwholesome, wherein it may please you to give me leave to utter my opinion, which happly comming to be scanned by men of knowledge, may

---

The board consists of two rows of four, or six, holes or cups on each side. Players may use pebbles, seeds, or other small objects. To start, in each cup are placed four pebbles. Players take turns in distributing the pebbles from one cup to the adjoining cups, one by one. When the pebbles from one cup are all gone the player takes up the pebbles from the cup into which he last played a pebble. He continues until he lands the first pebble in an empty space. If in the distribution of the pebbles he gets four in one cup on his own side, he picks up the pebbles. If he finishes his distribution in a cup of four on the opponent's side, he takes these too. The object is to try to make cups containing four pebbles and take the pebbles, since the person who gets the most pebbles is the winner' (from Amadou Traoré, 'African Games', paper presented at the Manding Studies Conference, London, 1972)

A fuller description, 'How to Play Oware', appeared on the Mind Sports Olympiad internet site, August 1998, with the claim that the game was 4,000 years old. The game resembles chess in that success lies in forecasting the future moves of both parties, and it certainly calls for much concentration. For illustrations of *warri*, see Frederic Shoberl, ed., *The World in Miniature* (translation of Geoffroy de Villeneuve, *L'Afrique, ou histoire, moeurs, usages et coutumes des Africains*, Paris, 1814), 4 vols, London, 1821–5, in vol. 3, 'A Negro Girl studying the game of Ourri'; Anne-Marie Bouttiaux, *Senegal Behind Glass*, Munich/New York, 1994, p. 114 (painting by Arona Diarra, four women playing and arguing); Robert Zimmerman, *Enchantment of the World: The Gambia*, Chicago, 1994, p. 112 (two Banjul men playing); *The Times*, 20.8.1997 (photograph of the game being played at a Mind Sports Olympiad in London). Jobson appears to supply the earliest reference to this game.

A digression, by the Writer, for the better preserving of mens lives and healths.	produce some rules of better order, then hath hither unto beene kept amongest our Nation, which hath caused the losse of so many lives, and the dangerous sickenesse of others, therewithall laying a generall scandall upon the countrey it selfe, to be both infectious, and unwholesome for our bodies, whereas indeede it is our owne disorders: For the custome that hath beene hither-unto held, especially into this River, hath beene without any diligence, to make choise of such Sea-men for governours, as were men of temperance, and commaund, who being able to /40/ temper well themselves, might the better governe the rest, whose ill carriages may be thought to bee great provokers for shortning other mens dayes: I will example it in this last voyage of ours. The Maister was a man knowne, for an excellent Artsman,[1] but in the governement of himselfe, so farre from knowledge, that after our passage from *Dartmouth*, which was in October, untill the middle of March after, about which time he dyed, it will be justified, he was never twenty dayes sober, in which time he went not alone, but our Chirurgion, with sundry officers that were of his societie, with their lives payed for their riotous order. And further, whereas wee were divided into two shallops, to goe up the River, the bigger whereof, the principall Factor was to follow his trade in, and carried therefore in the same a Butte of Sacke, and a Hogges-head of *Aqua vitæ* [*sc.* brandy], making choice of such men as were the most able, and likeliest bodies to hold out, and he in himselfe carefull enough, as his experience might well advise him, having spent many yeares, and made many voyages upon that continent, to observe both in his diet of eating and drinking; yet towards his people, hee carried at sometimes, such an over-sparing hand, that they fell to practising how they might deceive him, making use of all advantages, to steale those hot drinkes from him, which being purchased, as it were from his niggardly nature, they would divide as a spoyle with great greedinesse amongest themselves, and thereby wrought their owne confusions, that of those people he carried with him, they eyther died before hee came backe to the shippe, or /41/ shortly after, some two at the most excepted, who escaped with dangerous sickenesse, whereas to the contrary (with thankefulnesse to God be it spoken) my selfe going up in the other shallop, and wherein I must take such men as were given me, not such as I desired, observing amongst our selves, a loving and orderly course of diet, wherein everie man had his equall share, notwithstanding I went one hundred and fortie leagues above the other shallop, returned without the losse of any one man; nay, in all my going to the highest, and in my returne to the Pinnace, I never had any man sicke, but upon a second returne up some part of the. River, some of my olde men being changed, two or three fell sicke, howbeit (with comfort be it spoken) there was not one man died that went with mee, and for my owne part, through the whole voyage, I was never one quarter of an houre sicke (blessed be the name of God.)

Now for my opinion concerning our diet, I hold well with the Blacks, that

---

[1] 'Artsman' = man skilled in arts or crafts, in this case the art of navigation.

to feed at noone, is an unholesome thing, for that the Sunne, being then in his extremitie of heate, and by his neerenes having such power over us, the moisture that lies within the body, is exhaled to the exterior parts, to comfort, and refresh that, which the heate doth drie, and then are the interior parts most cold, and unapt for nutriment, wherein experience makes us see, that in the height and heate of the day we can with great facilitie, and without offence, drinke off such a draught, or quantitie of *Aqua vitæ*, or hot waters, as if we should drinke heere in our native countrey at one time, would certainly burne out our harts; /42/ nay more, wee finde our bodies naturally desiring, and longing for the same, (wherein I might heere shew some reason, partly to blame a neglect in our owne provisions, but that I assure my selfe, it hath beene rather ignorance, to know what was good, then want of wil to provide it) whereas in the coole of the morning, and againe in the evening, wee receive it with much more temperance, and a little giveth satisfaction, so that my conclusion is, that to us that have able, and working bodies, and in our occasions ore stirring, and labouring in the morning earely, and after the heate of the day, are the fittest & convenientst times to receive our sustenance, wherein I shal ever submit my selfe to those of more able judgement; and returning to the Blackes, let you know, that their usuall and ordinary drinke, is, either the River water, or from some Spring, howbeit they have growne from trees, several sorts of wine, or drinkes, as also the making of a kinde of liquor they call *Bullo*, made and compounded of their countrey corne;[1] whereof more conveniently else-where I shall effectually satisfie you;[2] and now goe on to shew you, the manner of their building, and fortifications.

The Writers opinion concerning dyet.

The Caveat must be lookt carefully to, in the setting forth.

They place themselves in their habitations round together and for the most part have a wall, though it be but of Reede, platted and made up together, some sixe foot in height, circling and going round their Towne, with doores of the same, in the night time to be orderly shut, some of the houses within their walls, likewise are made of the same Reedes; but the better sort do build the walles of their houses, of loame, which after it is tempered, and layde /43/ up together, carrieth a kinde of red colour with it, and doth remayne with an extraordinary hardnesse, that doubtlesse (as I have carefully divers times observed) it would make the most excellent and durablest Bricke in the world, the whole countrey, except upon the mountains, yeelding the same earth, whereof I will not forget to report one thing,

The manner of their building.

---

[1] '*Bullo*', a misprint for '*dullo*' (given correctly, *GT* 132) = Mandinka *doloo*, 'beer', a brew made from maize or millet. An account of Senegal and the Wolof of *c*.1510 described beer-making as follows. 'Millet beer is made in this way. They take the millet and pound it finely, to make flour, and in this flour they put boiling water. Then they filter it through a special palm leaf matting. This liquid they put into pots, and let it ferment several days. The older the beer is, the better it is. And they drink much more of this beer than any other' (Valentim Fernandes, *Description de la côte occidentale d'Afrique (1506–10)*, ed. T. Monod, A. Teixeira da Mota, and R. Mauny, Bissau, 1951, p. 17, translated). See also Almada, *Tratado breve*, ch. 4/15. DPG has not heard of beer being made within The Gambia in modern times.

[2] If this refers to the beer, see *GT* 132 below; if to the corn, for its cultivation see *GT* 124 below.

Strange Ante-hils.

which (in my opinion) deserves admiration: we doe finde in most places, hills cast up by Ants or Emmets, which we heere call Ant-hills,[1] some of them twenty foote of height, of such compasses, as will hold or containe a dozen men, which with the heate of the Sunne is growne to that hardnesse, as wee doe use to hide and conceale our selves in the ragged tops of them, when wee take up stands, to shoote at the country deere, or any other manner of wilde beast; the forme of their houses, whether it be loame, or Reed, is alwayes round, and the round roofes made lowe, ever covered with reedes, and tyed fast to rafters, that they may be able to abide, and lie fast, in the outragious windes and gusts, that come in the times of raine; for which purpose also they build their houses round, that the winde may have the lesser force against them; and the walles enclosing and keeping them in, is to avoyde those ravening and devouring beasts, which in the night time range and bustle about, wherewith divers times notwithstanding, they are much affrighted, and by making fires, and raising cries at midnight, to chase and drive them from their mansion dwellings. This for the meaner Townes or countrey Villages, but they have likewise Townes of force, according unto the manner of warre, they use amongst them, fortified, /44/ and trencht in, after a strong and defencible nature; whereof (they say) the countrey within is full, especially where the Kings are seated, the maner whereof we have seene in some two or three places, wherof I will instance onely one: which is the Towne of *Cassan*, against which (as I sayd before) the shippe which was betrayed did ride, and we in our last voyage, did make it our highest port for

The towne of Cassan with the manner of fortification.

our bigger shippe. This Towne is the Kings seate, and by the name of the Towne hee holdes his title, King of *Cassan*; It is seated upon the Rivers side and inclosed round neare to the houses, with hurdles, such as our shepheards use, but they are above ten foot high, and fastned to strong and able poles, the toppes whereof remaine above the hurdle; on the inside in divers places, they have rooms, and buildings, made up like Turrets, from whence they within may shoot their arrowes, and throw their darts over the wall, against their approaching enemies; on the out-side likewise, round the wall, they have cast a ditch or trench, of a great breadth, & beyond that againe a pretty distance, the whole Towne is circled with posts and peeces of trees, set close and fast into the ground, some five foot high, so thicke, that except in stiles, or places made of purpose, a single man cannot get through, and in the like manner, a small distance off againe, the like defence, and this is as they do signifie unto us, to keepe off the force of horse, to which purpose, it seemes to be very strong and availeable; considering what armes and Weapons they have in use, which in this place is necessary to be knowne.[2]

---

[1] Although still commonly so described, these are actually termite nest-mounds.
[2] Donelha, who was in the river in 1585 and perhaps later, wrote at great length about the town, including the following:

> The town of Casan stands a pistol-shot away from the port. The port is handy, we disembark on dry land, and there is a certain amount of sand. [Here] ... boats are caulked, masts are shaped which are cut on Cabopa Island, and rigging and ropes for the ships are

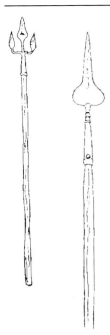

Traditional spears now used as ceremonial staffs by village heads

They doe usually walke, with a Staffe or Javelin /45/ in their hands, which they call an *Assegie*,[1] being a Reede of some sixe foote long, the head whereof is an Iron pike, much like our Javelines, but most of them very artificially made, and full of danger; others they have also made, to throw like Irishmens darts, with heads all barbed, full of crueltie to the receiver: each man likewise, about his necke doth weare, in a Bandeleere of red or yellow cloth, a short Sword of some two foot long, with an open handle, which Swords they make of the Iron is brought unto them, as you shall heare when I speak of their Trades, and also the better sort of them, doe carry their bowe in their hands, and at their backe a case, very artificially made, which may hold within it some twenty foure of their arrowes, it is the smallest arrow used by any Nation, made of a Reed, about the bignesse of a Swans quill, and some two foote in length, there is fastned in the end, a small Iron with a barbed head, all which Iron is dangerously poisoned; the arrow hath neither nock[2] nor feather, but is shot from the bowe, which is also made of a Reede, by a flat string, or rather sticke smoothed and

The armes or weapons the people have in use.

> repaired. The town is small, and built in a circle, with round houses of baked brick, whitewashed with a white clay that resembles lime. ... The town is entirely surrounded with high timber stakes, called *tabanca*: outside this, a deep and wide ditch which runs around the whole town is full of water in winter-time. There are four bridges and four gates: the bridges being of palm trees. ... There is great trade in cloths, cotton, wax, ivory, gold and hides of various animals ... (Part II below: Donelha, ff. 25–27v).

Lemos Coelho, who lived at Kasang for three years, probably in the 1650s, three decades after Jobson's visit, also described the town at some length, including the following:

> The port of Cação is very attractive, the houses of the village being visible from the river. Whites have always lived there, but in a village separate from the heathen, which is not the case in any other port. ... The land has much trade in hides, as well as in blacks, cloths, and some ivory. Cola sells there extremely well. The blacks have many foodstuffs ... Thus everything for human existence is found in this land in great plenty and sumptuousness. There are very good facilities for repairing ships, because of the timber available in bulk at every point along the bank of the river ... Above all, this is a very healthy land, well cleansed by the wind, and since it is sixty leagues from the mouth of the river seldom sultry, ... (Part II below: Lemos Coelho, 29).

For *cabopa* 'a tree, probably *Mitragyna stipulosa*' and *tabanca* 'earthworks, fortifications', see P. E. H. Hair, *Africa Encountered: European Contacts and Evidence 1450–1700*, Aldershot, 1997, VIII, pp. 42, 46. Both terms appear to have been incorporated into Portuguese Crioulo from the Temne language of Sierra Leone.

[1] '*Assegie*' = a lance or javelin. Apparently derived from Berber, this term appears in the early Portuguese texts on Guinea as *azagaia*, and from Portuguese was borrowed into other European languages. It is not used in presentday Gambian speech. But chiefs and village heads often possess spears (*tambolu*) which nowadays have solely a ceremonial use.

[2] 'nock' = normally the notch at each end of a bow for holding the string but here the notch at the end of the arrow for inserting the string.

109

made even, and fastned to the bow. so that the bowe and the string, are one and the selfe-same wood, whose force is small, and therefore the offence lies in the poyson, which neare hand upon their cotten garments, may make an entry, but to a Buffe Jerkin [*sc.* leather jacket] or any other sleight garment of defence, except it be very neare, can be little offensive: we have seene of them likewise on horse backe, the horses being of a small stature, bridled and sadled after the Spanish fashion, each man having his *Assegie* in his hand, and upon the right side of his /46/ horse a broad Buckler hanging; and this is all the weapon in use amongst them.

The Kings house is in the middle of the Towne, inclosed by it selfe, onely his wives severall houses about him, to which you cannot come, but as it were through a Court of Guard, passing through an open house,[1] where stands his chaire empty, unlawfull for any but himselfe to sit in, by which hangs his drummes, the onely instruments of warre which we see amongst them, neither are these drummes without dayly imployment, for this is their continuall custome every night after it seemes they have filled their bellies, they repaire to this Court of Guard, making fires both in the middle of the house, and in the open yard, about which they doe continue drumming, hooping [whooping], singing, and makeing a hethenish noyse, most commonly untill the day beginnes to breake, when as we conceive dead-sleepes take them, by which meanes sleeping one part of the day, it makes the other part seeme shorter, untill the time of feeding come againe, otherwise it is done to that purpose in the night, to feare and keepe away the Lyons, and ravening beasts from about their dwellings, who are at that season ranging and looking out; for this manner of course is held amongst them, not only in their fortified Townes, but also in every particular village, and habitation, whereof few of them is without such poore drums they use, and if they be, yet they continue the custome, through hooping, singing, and using their voyces, but when it happens musicke is amongst them, then is the horrible din, as I shall signifie when I overtake their fidlers. /47/

The inhabitan[t]s custome in the night.

But first I am to meddle with matter of state, and acquaint you concerning their Kings, and Governors, for so I have promist, the better to bring my worke together: In following of which, I shall entreate your patience to observe with me, that the better to distinguish of their governments, I must devide the Country by the River, that howsoever the River trends, which in his windings is surely all points of the compasse, I shall always call that part, which lies to the southward in our intrance, the southside, and the other the northside, of both which sides, although we have seene divers petty Kings, and other Commanders, to whom we payd a kind of poore custome, which in the mouth of the River, where the *Portingall* hath used, is not onely greater, but perremptorily demaunded, whereas above it is lesse, and rather taken as

---

[1] The compounds of chiefs and village heads were often entered through a structure whose open entrances could, if necessary, be closed off. Many instances can still be seen at the present day.

a curtesie presented, which morall kindnesse requires all strangers, comming in the way of amity, to begin withall to a principall person, then any manner of custome, that is (as we say) exacted, but howsomever both below and especially above, it is of that poore quantity, it hardly deserves the paines of so long a rehersall, whereby you may be drawne to conceit of a greater valuation: Those petty Kings I say, whereof I both saw, had conference, and did eate and drinke within [with in] sixe severall places, who had the title of *Mansa*, which in their language, is the proper name for the King, have all reference to their greater Kings, who live farther from those places; on the southside, the whole Country we past, even to the highest we went, which you must needes conceive to be very spacious, had all /48/ reference to the great King of *Cantore*:[1] on the northside likewise, from the sea-side, about halfe the way we went up, they did acknowledge the King of *Bursall*,[2] and after him, to the highest wee went on that side, the great King of *Wolley*:[3]

The great King of Cantore.
The great King of Bursall.
The great King of Wolley.

---

[1] The polity of Kantora, on the south bank of the middle River Gambia, a Mandinka 'kingdom' at this date. Much of the currently-presented history of the regions surrounding River Gambia is conjectural, being based on latterday oral traditions, but early Portuguese sources seem to indicate that Kantora was at times a part of, or subject to, Kaabu (Cabo), a larger Mandinka polity to the SE – see Part II below: Almada, 6/8; Donelha, f. 15.

[2] 'Bursall' = *buur Saalum* 'ruler of Saalum'. Jobson gives a variant of a term employed in the early Portuguese texts but probably communicated to him by Portuguese-speaking contacts on River Gambia. The term combines Wolof *buur* 'ruler', and part of the district name Saalum, which itself may be derived from a personal name. The exact distribution and inter-relationship of Wolof polities at this period is not known with any precision, but for an attempt to synthesize the limited evidence, see Jean Boulègue, *Les Anciens Royaumes Wolof (Sénégal): Le Grand Jolof (XIIIe–XVIe siècle)*, Blois/Paris, 1987. (Note that 'Wolof' is the modern self-name for the whole ethnicity, but 'Jolof', apparently a name for only one section, has been consistently used, or misused, both by the early sources and by modern historians, to describe the whole people and also their language.) Exactly what polity of the 1620s was represented by 'Bursall' is obscure, and it is possible that the term involved an outdated concept. Nevertheless, there is no doubt that not only did a Wolof-dominated polity centred on the district of the lower River Saalum, to the NW of River Gambia, exercise some measure of control over the Mandinka polities of the lower part of the north bank of River Gambia, but also that this Wolof polity, or an antecedent one, had at one time exercised control of the north bank as far up-river as Wuli. At the very least, the Wolof attempted to exact tribute from the peoples along the river, partly through subordinate rulers. Thus Jobson describes how the 'King of Bursall' successfully ordered the replacement of the 'King of Cassan' (*GT* 59). For more on 'Bursall', see the section of the Introduction on 'Polities and Peoples'.

[3] 'Wolley' = Wuli, a Mandinka kingdom on the north side of middle River Gambia. The early sources, including Jobson, evidence it being visited by long-distance caravans from the interior, some of which apparently passed through en route to the coast to collect salt – see Part II below: Lemos Coelho, 2, 55. It appears that the area was renowned for cotton growing and the production of cloths, and that from these activities the ruler derived a large proportion of his wealth. Oral traditions assert that the Wali clan established the polity and provided the rulers. At first the Wolof demanded tribute, but when the Wali had consolidated their power, they ceased to pay: Winifred Galloway, 'A History of Wuli from the Thirteenth to the Nineteenth Century', Ph.D. dissertation, Indiana University, 1975, p. 57. It is unclear how far this process had gone in Jobson's day. Eventually *jula* (trading) families from Tinda, further east, settled in Wuli, representing a powerful, but perhaps additional, Islamic influence, as evident by the 1620s. Jobson did not encounter the ruler, the *mansa*, who lived inland (the capital changed its location over time), but only his representatives, the *farang-baa* ('Ferambra'), his executive representative, and several *farang*, local rulers or governors. For more on Wuli, see the section of the Introduction on 'Polities and Peoples'.

These great Kings are likewise tributaries to one great King far ahove [*sic:* above] in the land, as is reported to us.	These three Kings we hard of, but saw none of them; the report going that they were such as shewed not themselves abroad, but in a manner of pompe, and that they were not seene to hunt, but with great number of horse, and especially on the northside, whereof diverse English have beene eye witnesse;[1] concerning *Bursall*, whose continuall aboode is neare the seaside, whereby some recourse hath beene unto him, there is warres betweene the one side of the River, and the other, and especially from this King of *Bursall*, in so much as the people would tell us, if hee could have any meanes to transport his horse on the farther side, hee would in short time over-runne great part of that Country:[2] the state of the great Kings, we may conjecture at, by the observances those small ones we see, doe assume unto themselves, and those people they governe per-
The reverence of the people to the petty Kings.	forme unto them, for there is no man, but at his first approach before them, where they sit commonly in their houses, onely upon a mat which is spread upon the ground, but presents himselfe with a great deale of reverence in kneeling on his knee, and comming nearer, layes first his hand upon the bare ground, and then on the toppe of his owne uncovered head, many of them taking up the dust, and laying it upon his bare-head, which action he useth twise or thrise, before hee come at him, where with a great deale of submission, he /49/ layes his hand upon the Kings thigh, and so retireth himselfe a good distance backe, and if it chance in any company or resort unto him, that there be a Mary-bucke, which be their priests, as soone as they have made their maner of
The Religious ceremony of these people.	salutation, they al kneele downe, and hee fals to praying, the substance of his prayer, being for the preservation of the King, and in the same blessing him, to which himselfe crossing his armes, and laying his right hand over his left shoulder, and his left hand, over his right shoulder, useth the word *Amena, Amena*, many times over, which signifies the same as we say Amen, or so be it:[3] Nay more even among the common people, when they meete in the high way, and are of acquantance, having beene absent from one another, any distance of time, if there be a Mary-bucke amongst them, they put themselves into a round ring, and before any salutation fall on their knees to prayer: The Kings respect unto them againe, is onely nodding of his head, which is acceptably received,

[1] This seems to say that some of the English had seen a king in a hunting party. Which king? Perhaps the party of 1618 survivors making their way overland to Cape Verde saw the ruler of Saalum. Alternatively, the king may have been seen by the survivors remaining at Oranto. If the survivors, no doubt themselves out hunting, had wandered some distance to the west, this might have been the king of Niani; but if close to Oranto or to the east, then presumably the king of Wuli.

[2] The *buur Saalum*, a Wolof ruler over a population of mixed Wolof and Serer, appears to have derived a great deal of his income from the export to the interior, partly overland by caravan and partly by canoe up River Gambia, of the salt produced in the coastal regions of his polity. (For salt manufacturing on the Little Coast of Senegal, see G. Balandier, 'Notes sur l'exploitation de sel par les vieilles femmes de Bargny (environs de Rufisque)', *Notes africaines*, 32, October 1946, p. 22.) A second source of income was the sale of slaves. With their income, Wolof rulers were able to purchase horses from more northern parts and these provided them with a formidable cavalry, effective in areas where the countryside was relatively open.

[3] In Islamic prayer the reiteration of *amīne* is standard but the crossing of the arms as described is not (p.c. Louis Brenner); the Wolof and Mandinka say *amiin*.

how be it in manner of habite [*sc.* dress], there is betweene the King, and his people, little or no manner of difference; which may be imputed to the necessity of the Country, because it yeeldeth but one onely materiall, to make apparell of, which is a Cotten wooll, whereof they plant great fields, and it growes up as it were our rose bushes, yeellding a cod, that in his full maturity, breakes in some part, and shewes a perfect white cotten, of which you shall find more written, when I come to rehearse what trees and plants wee finde amongst them.

Now for the manner of their apparell, it is soone /50/ related, they being for the most part bare-head, only bedecked or hang'd over with gregories, as they are likewise over their bodies, legges, and armes, which word I will presently expound unto you, but first tell you their onely garments are a shirt and a paire of breeches, their shirts are made downe to their knees, wide in the manner of a Sirplace, and with great sleaves, the which when he commeth to use his bowe or armes, he rowleth up and it continueth fast at the shoulder, his breeches are made with so much stuffe gathered just on his buttockes, that he seemeth to carry a cushion, and after a manner makes him stradle as he goes, bare-legged, and without shooes, except it be some few of them, who have a peece of leather under their foot, cut like a shooe-sole, butned about the great toe, and againe about the instoppe.[1] The Gregories bee things of great esteeme amongst them, for the most part they are made of leather of severall fashions, wounderous neatly, they are hollow, and within them is placed, and sowed up close, certaine writings, or spels which they receive from their Mary-buckes,[2] whereof they conceive such a religious respect, that they do confidently beleeve no hurt can betide them, whilst these Gregories are about them, and it seemes to encrease their superstition; the Mary-buckes do devide [? devise] these blessings for every severall and particular part, for uppon their heads they weare them, in manner of a crosse, aswell from the fore-head to the necke, as from one eare to another, likewise about their neckes, and [a]crosse both shoulders about their bodies, round their middles, great /51/ store, as also uppon their armes, both above and below the elbow, so that in a manner, they seeme as it were laden, and carriyng [carrying] an outward burthen of religious blessings, whereof there is none so throughly laden as the Kings, although of all sorts they are furnished with some, both men and weomen, and this more I have taken notice of, that if any of them be possest of any malady, or have any swelling or sore

The manner of their apparrel.

The description of their Gregories, which are charmes they receive from their Mary-buckes.

---

[1] Jobson is describing the local sandals.

[2] 'Gregories'= amulets. Jobson gives an unique version of a term usually written 'gris-gris'. (The term appeared in European sources from the mid-16th century: its derivation is uncertain and may not be from an African language: Hair, 'An Ethnolinguistic Inventory of the Upper Guinea Coast before 1700', in *Africa Encountered*, p. 70). As Jobson describes, these were amulets, their cover made by a leatherworker, the contents a scrap of paper on which a religious cleric had written a charm, sometimes a verse from the Koran (cf. Part II below: Donelha, f. 30ᵛ; also Almada, 2/5). Present day Mandinka call them *safolu*, from *safe* 'to write'. Their extreme popularity was a feature of a transition period, communally and individually, from a belief in animist 'magic' to a belief in 'pure' Islam; and the sale of amulets, with their accompanying message of the functionality of literacy and universal religion, was an initial and deliberate device in the process of Islamization.

upon them, the remedy they have, is onely by placing one of these blessed Gregories, where the griefe lies, which they conceite will helpe them: and for ought I can perceive, this is all the Physicke they have amongst them, and they do not onely observe this for themselves, but their horses doe usually weare of these about their neckes, and most of their bowes are hanged and furnished with them.

To countenance his state, he hath many times two of his wives sitting by him, supporting his body, and laying their hands upon his naked skin, above the wast, stroking, and gently pulling the same, wherein he seemes to receive content, and because I have named two of his wives, before I proceed to any thing else, I will acquaint you with the manner of their women, the multiplicity of their wives, and the wonderfull great subjection they live under.

The number of their wives. The King hath an orderly allowance of seven women, which are called wives; that is which are esteemed, and acknowledged, and with a setled ceremony amongst them, distinguisht from other women which he hath use of, being absolutely tied, to attend his only pleasure, and therefore /52/ in relating of them, according to our proper phrase, can give them no other title but wife, of which it seemes he cannot exceede the number of seven: for he hath the

Allowance of other women for necessitie sake. use of other women, who are not of that esteeme, but rather as we may terme them Concubines, who are of a lower birth then his wives, and these likewise are tyed unto him, but not with that manner of strictnes the other are, so as it may appeare, they are rather taken for necessity then that it is a setled course amongst them, which word necessity I must better explaine, and therefore tell you, that it may and doth diverse times fall out, that of his seven wives he hath none to accompany him in the nature of a wife; For undoubtedly these people originally sprung from the race of *Canaan*, the sonne of *Ham*, who discovered his father *Noahs* secrets, for which *Noah* awakeing cursed *Canaan* as our holy Scripture testifieth, the curse as by Scholemen hath been disputed, extended to his ensuing race, in laying hold upon the same place, where the originall cause began, whereof these people are witnesse, who are

The reason of that necessity. furnisht with such members as are after a sort burthensome unto them,[1] whereby their women being once conceived with child, so soone as it is perfectly discerned, accompanies the man no longer, because he shall not destroy what is conceived to the losse of that, and danger of the bearer, neither until she hath brought up the child, to a full and fitting time to be weaned, which every woman doth to her owne childe[,] is she allowed, in that

---

[1] Jobson argues that the alleged large penis size of African males (which Europeans claimed to observe because of African 'nakedness') is a divine curse. The text he cites is Genesis 9: 21–5, the curse on Ham, because he saw his father's genitalia, falling on Ham's son, Canaan. In Jobson's period, the curse on Ham was regularly interpreted by Europeans as being on Ham's descendants, understood as the Africans, but normally in relation to their hard way of life. Jobson, however relates it, seemingly claiming an untraced 'Scholeman' source, instead to an alleged physical characteristic; and makes this characteristic a curse because of his belief, curious but no doubt genuine, that intercourse with this organ would therefore damage the child in the womb (perhaps by causing, as he elsewhere suggested, women to have 'bloody issues': *GT* 81).

nature, the mans society,[1] so that many times it falles out, he hath not a wife to lie withall, and therefore as I said, hath allowance of /53/ other women, for necessities sake, which may seeme not over-strange unto us, in that our holy Writ doth make mention therof, as you may reade in the 23.chap. of the Prophet *Ezechiel*, where Jerusalem and Samaria, being called by the names of the two Sisters, *Aholah* and *Aholibah*, being charged with fornication, are in the twentie verse of the same chapter, said to doate upon those people, whose members were as the members of Asses, and whose issue was like the issue of horses, therein right and amply explaining these people.[2]

And for the chastitie of the wife, their lawes and customes are in that kinde very severe, for, being taken or found an offendor, both she and the man shee shall offend withall, are without redemption sold away, in this sort they punish all great offences, putting none at all to death, and such as these are the people the *Portingalls* buy and ttansport [transport] for the west *Indies*, as before I shewed you; and this is the course held amongst them all, howbeit every man cannot have so many wives, but according as he hath means to keep them, and wherwithall to buy them: for first, every man must compound or have the Kings or chiefe Governors consent, for any wife or wives he shall take, to whom he must give some gratification; and next he doth buy,

Strict punishment for unchastity.

The men buy their wives.

---

[1] The substance of Jobson's explanation is the custom, still widespread throughout West Africa, that after giving birth a woman does not resume sexual intercourse until the the child is weaned, generally when it is aged between eighteen months and two years, on the grounds that a further, immediate pregnancy would stop lactation and the unweaned child therefore be harmed. In the Gambia region it is also the custom that intercourse ceases once a woman acknowledges herself to be pregnant (perhaps the explanation of Jobson's belief that in the local circumstances intercourse would damage the unborn child – see the previous note). That these customs require men to practice polygamy is a view expressed nowadays by polygamists when challenged, but it is unclear from Jobson's account whether open expression of this rationalization was current at the time and he collected it, or whether he thought it up himself.

[2] Jobson accepts the reasonableness of polygamy, given the custom referred to in the previous note, on the premise of the 'necessitie' of regular sexual intercourse for males, but in turn justifies this on a very curious and confused argument. He cites Ezekiel 23: 20, which in the Authorized Version translation reads - 'For she doted upon their paramours whose flesh is as the flesh of asses and whose issue is like the issue of horses'; the version in the earlier Bishops' Bible, which Jobson may have known better, is not dissimilar - 'She burnt in lust upon their lechers, whose flesh was lyke the fleshe of asses and their issue like the issue of horses'. Ezekiel, in English translation, was certainly saying that the women in question 'doted upon' passionate sexual activity on the part of male partners, who are compared with equines, apparently in terms of their penis size. (The *OED* indicates that the terms 'flesh' and 'issue' could at the time be given an anatomical connotation, and Purchas summarized Jobson by reference to 'their members Monstrosity' and added a marginal note, 'Priapeian Stallions': Purchas, *Pilgrimes*, 1/9/13, 1571.) Yet the text relates penis size primarily to excessive female satisfaction rather than to excessive male libido. Thus the text condemns female lust but does not justify male 'necessitie', the notion for which Jobson was attempting to find Scriptural support. (In fact, unlike many other early European observers, Jobson does not impute an excessive libido to African women.) However, irrespective of the text, he probably thought it commonplace wisdom that equine behaviour – and animal behaviour at rut in general – could be considered proof of the 'necessitie' of regular sexual activity for males, including men.

with some commoditie, the woman of her friends, and what hee giveth, doth remayne as we say in banke, if he should die, which shee hath for her better maintenance, or if shee please to buy a husband, for as every man when he takes a maid, must buy her, so every widow, if shee will have a husband, must buy him,[1] through which occasion /54/ of buying the women, may be conjectured, they yeeld themselves to that subjection, but whence soever it growes, I am sure there is no woman can be under more servitude, for first they doe in morters, with such great staves wee call Coole-staves,[2] beate and cleanse both the Rice, and all manner of other graine they eate, which is onely womens worke, and very painefull: next they dresse both that and all other manner of victuall the men doe eate; and when it is so ordered, they do bring and set it downe upon the Matte before them, presently withdrawing themselves, and are never admitted to sit and eate with them; nay I can confidently affirme, that notwithstanding I have eaten sundry times, both at the Kings, and other mens houses, where the men and wee have put our hands in a gourd and fed together, yet never could I see any woman allowed to eate, albeit I have earnestly entreated the same, for amongst their many wives, there is none of them, but have one especially accounted of, which hath a greater priviledge by being about him, and more conversant then the rest, which by us (though strangers) is easily discerned, and therefore amongst us we call her his hand wife, to whom wee use always to be more free, in those gifts we bestow, yet this wife is not allowed to eate in sight, but in another house, although she be priviledged of many other labours:[3] neither are the men ever seene to use any manner of familiar dalliance with them, insomuch as I thinke, there is hardly any Englishman can say, he ever saw the Blackeman kisse a women. Againe, which is to be noted, notwithstanding this partialitie of affection,[4] and so /55/ many of them of an equalitie living together, it is never heard, that they doe brawle, or scold, or fall out among theemselves, howbeit it may be thought in [a] matter that neare concernes them, they cannot chuse but have many aggreevances, contrary to our English proverb that sayes, *Two women in one house,* &c. but there indeede I will make answer, That although they have recourse all day together, in his great or more spacious haunt, yet for the night every woman hath a severall house, whether [whither] she retires herselfe, to give attendance as his desires shall

---

[1] As normally the case throughout West Africa today, in the Gambia region a marriage 'payment', in money or kind, is made by a man (or his relatives) to the parents of his bride. But DPG has never heard of a widow having to 'buy a husband'.

[2] 'Coole-staves': 'coole' = cowl, a large tub carried by two men by means of a projecting stave or pole. Presumably Jobson thought that the stave was the nearest English equivalent to the long heavy pestle used in mortars in West Africa, today as earlier, normally by women (as Jobson says), in beating out and preparing various foodstuffs.

[3] It is normal practice today in the Gambia region, as throughout West Africa, for men and women to eat separately.

[4] Lip kissing between the sexes is, of course, a social practice only in certain cultures, and remains not normal in Black Africa. Jobson interprets this as a lack of marital affection. It is of interest that, as a mark of such affection, he – and presumably his English contemporaries – expected kissing, as one form of 'familiar dalliance', to be publicly performed.

Woman pounding grain with mortar and pestle

leade him:[1] and as they appeare before him, in the morning, their salutations are uppon their knees, laying their hands upon his thigh.[2] And lastly, for her apparrell, it is loose clothes party-coloured, blew and white, of the same cotton stuffe the men weare, whereof the women commonly weare but one tuckt about their middles, and from the waste upward, bare, in regard they are, for the most part, wrought, or rather printed upon the backe, especially in the higher parts we were at, insomuch as we have seene some of them, with workes all over their backes, resembling right the printed lids and covers which wee see layd and set uppon our baked meates:[3] whereof it appeares they take extraordinarie pride, because they will turne themselves, wee should take notice of it, and be very well pleased, wee should touch or handle it, as a matter to bee esteemed or set by; otherwise they cast another like cloth, as they weare below, uppon their shoulders, which hangeth loose, and those be their garments, and without question, many, or the most part of them, very chary or nice [*sc.* fastidious, particular] in shewing of their /56/ secrecie [*sc.* secret parts]; and to conclude, concerning women which [in] the country use they serve in marriages: the man doth commonly bespeake the women while she is young, and not of abilitie, which is done and confirmed by the friends consent, yet when the time of her full age is come, or afterwards, as they agree amongst them, the men getting his friends, which are all the youth, and younger sort of men hee can procure come to the Towne where the Maide is dwelling, in the beginning of the night, when the Moone shines, and as it were by violence lift her up amongst them, and carry her away, who makes a noyse, skrieking [shrieking], and crying out, which is seconded by the other young maides of the place, and thereupon the younger sort of men that are dwellers in the place gather themselves together, and (as it were) in reskew of her, while the other with great shouts and cries of rejoycing, carry her quite away, to his place of abode,[4] where she remaines unseene

The womens clothing.

The manner of taking away their wives, which in some sort is used in Ireland at this day.

---

[1] It is still normal in rural West Africa for a man and each of his wives to have a separate sleeping hut within a family compound.

[2] Today, the morning greeting of a Gambian wife to her husband is a deep curtsey. Among the Wolof in rural villages, a curtsey may be made when passing a group of elders seated at the *dat* (a sitting-place under a shady tree). A curtsey is also appropriate when presenting a drink to a stranger.

[3] Many West African peoples raise designs on the skin by means of scarification, produced by cutting the skin and often by the insertion of keloids. This is normally for body decoration, but in the Gambia region in recent times the insertion of small keloids has been considered a form of preventive medicine. For the earliest reference to the local practice, see Part II below: Cadamosto, [26].

[4] In comparatively recent times, the ritual of wife-capture was occasionally found among the Wolof, when the parents of a bride had been reluctant to let their daughter go to her future husband, usually because they needed her to work at home. A mock capture was then carried out by the companions and relatives of the groom, and, after an interval, any remaining regular marriage formalities were performed. For symbolic bride capture among the Serer people neighbouring River Gambia, see Bérenger-Féraud, 'Le Mariage chez les nègres Sénégambiens', *Revue d'anthropologie*, 12, 1883, pp. 284–97. There appears to be only a single report of wife-capture among the Mandinka, and the instance, in 1818, seems to be represented as an exceptional episode involving an unconsenting female, rather than as a recognized custom – but the European observers may have misunderstood: William Gray and - Dochard, *Travels in Western Africa ...*, London 1825, p. 56. If Jobson was correct and wife-capture was a normal or

for a certaine time, and when she doth come first forth, for certaine Moones, she doth not shew her open face, but with a cloth cast over her head, covers all but one eye, after the maner of the Spanish vaile,[1] observing herein a shamefast modestie, not to be looked for, among such a kinde of blacke or barbarous people.

The modesty of a new married woman

I proceede againe in the state of their Kings, there is no people in the world, stand more upon their antiquitie, and dignitie of bloud, then they doe betweene themselves, insomuch, as once I had a quarrell grew in my house, betweene one of the Kings sonnes, under whose governement our habitation was, and another Blacke, who was a very lusty and /57/ able man, called *Boo John*,[2] unto whom for some respects, we carried a more kinde of extraordinary affection, wherein they braved one another, concerning their dignities of birth, the one was better, and the other was better, insomuch as *Boo John* with his hand stroke the other in the face, and thereuppon their weapons were drawne, and parties making on both sides, danger was like to ensue, but calling more of my company, with a small gunne I stept betwixt them, and made them sever; howbeit the Kings sonne promised to returne the next morning, and take revenge if he durst abide him, which the other promised to doe, and accordingly, notwithstanding all the entreaties I could use, would not depart, but withall sent out for more people, who in the morning betimes came into him, all of them fitted with their countrey armes. And the Kings sonne also performed his word, for in the morning he came to the Rivers side, and called for a boate to passe over, bringing a live biefe with him, as a token unto me from his father, as he was many times accustomed to send, and there came likewise with him diverse people, and all with armes. I was very doubtfull some hurt would arise, and laboured *Boo John* to goe away: but all I could gaine of him, was, he would only sit downe in our yard, under the shadow of our houses, with his people about him, and if no wrong were offered him, hee would not stirre nor give no offence; and by meanes I wrought so with the Kings sonne, carrying him and his companie into my house, and using them courteously, that for that time I pacified them, and they parted quietly, howbeit not without threatening upon /58/ another occasion; They doe distinctly know every governement, who shall be King, and how the succession shall hold, for in their temporall governements, one brother doth ever succeed another, untill that race be extinct, & then the

These people stand much upon their dignity.

A dangerous quarrell betwixt them.

The certaine knowledge of their Kings & Governours and their successors.

---

common practice among the Mandinka in his day, it has completely disappeared in modern times, possibly due to Islamization.

[1] Today, brides being taken to their husband's house normally wear a shawl of white cotton cloth over their heads. But Serahuli brides wear a dark muslin veil over their faces; and Mandinka women wear a shawl over their heads during the first rainy season after marriage (see Plate VI).

[2] Although *Bo John* is here used as a personal name, on the next page Jobson wrongly describes it as a title. Bojang is a common clan name now found in the Kombo region on the south bank of the lower River Gambia. According to oral tradition, the Bojang lived in Wuli for a time, before proceeding to Kaabu, and from there one branch then moved to become an important ruling dynasty in the polity of Kombo (Galloway, 'History of Wuli', p. 43).

eldest brothers sonne beginnes: and likewise they doe distinguish of governments, as they are in age.[1] For there were foure brethren, the eldest whereof was the great King of *Cantore*, whom we never saw; the second was *Summaway*,[2] King of the next place, and he came downe and was aboard our boate; the third brother was King of the place where our Land-dwelling was, being a blinde man, at whose house I have sundry times beene; and the fourth brother was called by the name of *Ferran*, and had the government of a countrey, where we had much and often trade, and for the most part, kept a Factor lying, and this, notwithstanding hee was of great age, was the youngest brother, and as any of his brethren died, they were all to remoove still, giving him the latter place:[3] And this may suffice to shew their manner of government; and for their severall Titles, they have onely foure, which is *Mansa* for the King, *Ferran* a second name, *Ferambra* the third, and *Boo John* the last;[4] every one of these foure being in their places commanders aod [and] governors: their greatest Riches consists, in having of most slaves, and from the King to the slave, they are all perpetuall beggars from us, howbeit small matters will satisfie them, except it be in *Aqua vitæ*, for which they sell all things they have, and the Kings and all will drinke, until they be starke drunke and fall fast asleepe, so that to describe the life of the /59/ Kings truly, is, that they doe eate, drinke and sleep, and keepe company with their women, and in this manner consume their time, untill Time consumes them; with their great bloud and dignitie, whereof they so much eesteme [esteeme]: and with one example I will rehearse unto you, I shal conclude concerning their kingly priviledges.

The King of *Cassan*, who was dwelling in the Towne, as wee came up the River, with whom I spake, drunke two or three bottles of *Aquae vitæ* and had

Their titles of honor.

Wherein their Riches consists.

Great Beggers.

The temporall people great drinkers of Aquavitæ.

The life of their Kings truly described.

---

[1] Jobson describes the normal traditional succession to office (among many peoples of West Africa), from oldest brother to next surviving oldest, and so on down the fraternity, before passing to the oldest of the surviving sons of the (deceased) oldest brother.

[2] A *suma* is the heir apparent, often in charge of an outlying province of a Mandinka kingdom until his time comes to succeed. His claim can, however, be contested by rivals. Numerous places are called Sumakunda, meaning 'the *suma*'s town'.

[3] This is a confused passage. It is possible that Jobson misunderstood the use of the term 'brethren', since, among the Mandinka, rulers referred to as 'brothers' may be honorific 'brothers' rather than biological ones. The 'King of *Cantore*' ruled the important territory of that name, on the SE of the middle river (see p. 111, n. 1 above); *Summaway* was a dependent ruler (described in the Journall as 'King of Bereck under the great King of Cantore': *PJ* 922); the blind ruler was the King of Oranto, also a dependant ('his name Summa Tumba, a blind man and subject to the great King of Cantore', ibid.); the identity and location of the fourth, a mere *Ferran*, are uncertain, and how he could be 'of great age' and yet the youngest brother (in any sense) is puzzling. The final clauses seem garbled but presumably were intended to confirm the statement about fraternal succession, the 'youngest' moving up the hierarchy on the death of each senior.

[4] The term *mansa* is the normal Mandinka word for 'king': *ferran, farang, fari*, and other variants, are titles for a ruler owing allegiance to a *mansa* (but are also used to designate a warrior ruler). Jobson's 'Ferambra' is presumably *farang* + *baa* 'great, senior', and if so, Jobson may have been wrong in making Ferambra third in the hierarchy rather than second. It has been stated that, in Wuli, 'Faramba ... meant something like "lesser ruler", or "arbiter" who acts on behalf of the *Mansa* – in other words a person to whom the *mansa* could delegate certain duties' (Galloway, 'History of Wuli', p. 244). In Wolof polities, the term *farba* was often used as a title for an official of slave status. For *Boo John*, see p. 119, n. 2 above.

much familiar conference, who had lived in that place many yeares, howbeit hee was a lame man; after we returned downe the River, and made stay at that Towne we found him nothing so merrily inclined, as he was at our going up: whereupon demanding among some of the rest the reason, they told us, hee was to be put out of his kingdome; enquiring wherefore, they replied, there was another to come who had more right then hee, for sayd they, this was the Kings sonne, but begotten of a base woman, such as I described their concubines to be: howbeit the King left behinde him a sonne, who was right borne, but very young, which childe, being now come to yeares, craves his rightfull inheritance, and the King of *Bursall*, under whom they hold, could not deny, but had taken order to send him forward, so as he was lookd for every day, and indeed, the second day after I was gone from the Towne,[1] he came thither, bringing abundance of people with him, to whom the people of the Town resorted, leaving the old King, to whom notwithstanding the yong King sent, willing him to depart, and goe away with his wives and family before he /60/ came, or if hee found him there, hee should suffer death, which at the first he refused to doe, saying he would enclose himselfe in his dwelling, and die ther: but upon better advice, when he saw himselfe quite forsaken, he passed himselfe and wives crosse the River, and left the Towne to this yong Commander, who after his entrance, our pinnace riding there, and being ready to come away, he sent for the Factor, to whom after some familiar conference, hee sayd, this towne and kingdome hath been lost, but hereafter you shall see it found againe, promising at all occasions his friendly assistance.[2]

    But before I leave them, it were necessary I should acquaint you, what manner of ceremony doth passe betwixt them and us as we meet together; the King is commonly sitting on his mat, laid on the ground, which in our entrance he observeth, not offring to rise; and in regard he hath nothing but his Gregories on his head, which are fast and cannot stirre, We do not use to moove our hatts, or uncover our heads when we come to him, but drawing neare, somewhat bending our bodies, wee lay our hands upon our breasts, which he also performs to us, and when we come neare, he holding foorth his hand, we first take hold on the upper part of one anothers hand, next on the lower part, and the third time joyne palmes, and with a full hand shaking one another, downe wee sit by him, and after some small parlee concerning the cause of our coming, wherein the kings part is performed in state, whatsoever he speakes being related over by another:[3] out goes our bottle of *Aqua*

Marginalia: The deposing of Kings. — The ceremony used betwixt the King and us, when first we meet.

---

[1] Jobson sailed down-river in the *Sion* but the *Saint John* followed later, hence Jobson heard about events after he left Kasang.

[2] The king's comment is puzzling – what had been 'lost'? The previous king had been friendly, up to a point, but perhaps the new king was alluding to the 1619 massacre of the English, by the Afro-Portuguese, with or without local assistance, for which the old king, since he had ruled 'many yeares', may have been deemed to have had some responsibility. But perhaps the new king was simply applauding his own new regime and the English misunderstood.

[3] In formal events West African rulers normally speak in low voices, their words being relayed to the audience by a speaker with a powerful voice and oratorical skills, in the Gambian region usually a griot (cf. Part II below: Fernandes, f. 108ᵛ).

vitæ which must not be wanting, and a bottle of Sacke too, it is so much the better, & /61/ calling for a small gourd to drinke in, which is their richest Plate; I first beginne, and drinking off a cup, present both the cup and bottle unto the king, the botle he delivers presently to some one of principall regard with him, who by his appoyntment, after the king hath first drunke, and he himselfe not failing of his next turne, distributes it to the whole attendance, who for the most part stand round by the wall of the house, one after another taking his cup, and then to the king againe. In their drinking I observe one thing, that in regard of the goodnesse or the strangenesse of the liquor, when he receiveth the first cup, before hee drinke himselfe, with the same liquor, he wets one of his principal Gregories: The king many times calling for a cup, breakes the square, otherwise they never leave untill the bottle is out, and so all the bottles we bring, wherein we have one priviledge, for after we have tasted the first cup, which must be of every bottle to secure them of danger; we need not unles we please, take any more, but shaking of our heads when they offer a cup, it stands for a faire refusall, and so passeth to another; neither doe they suffer their women publiquely to drinke amongst them, except some especially respected, who may sometimes have a cup, but even divide it among themselves, so long as the bottle will run, and their brains hold out, which being the true and proper element, they delight in, I will heere in [herein] in their hearts desires leave them, and proceed to a more civill; I am sure soberer sort of people.

He gives his chiefe Gregory drinke first.

Their women not allowed to drinke in publike, although they love it well.

The discourse of their Maribuckes or religious men.

And so by order, I am now come to speake of /62/ their *Marybuckes* or *Bissareas*, which we in our language, may call religious Persons, or Priests of the country. The Mary-buckes are separated from the common people, both in their habitations & course of lives, concerning whom I have with dilligence observed, that in their whole proceeding they have a wonderous reference to the leviticall law, as it is in our holy Bible related; the principalls whereof they are not ignorant in, for they do report concerning *Adam* and *Eve*, whom they call *Adama* and *Evahaha*, talking of *Noahs* flood, and of *Moses*, with many other things our sacred History makes mention of;[1] their houses or dwellings are seperated from the common people, having their Townes and lands set out in severall within themselves, wherein no common people have dwelling, except such as are their slaves, that worke and labour for them, which slaves they suffer to marry and cherish the race that comes of them, which race remaines to them, and their heires or posterity as perpetuall bond-men; they marry likewise in their owne tribe or kindred, taking no wives, but the daughters of Mary buckes, and all the children they have, are nourished and bred up, unto the ceremonies of their fathers.

They observe the leviticall Law.

And have great knowledge of the old Testament.

They marry in their own tribe and breede up their children in their owne sects.

---

[1] Jobson was wrong in thinking that the names and episodes showed knowledge of the Old Testament since they also occur in the Koran. 'Eve' is Arabic *Hawā*.

But for the number of their wives and women, they have the selfe course, that I described before among the Kings, and temporall people, in the like manner amongst them, every man in his dignity, and precedence having more or lesse: wherein there is no severed [sc. separate] towne but hath a principall, for better relation whereof, I will declare unto you the towne and place, where there [their] especiall, or, as I /63/ may say, high Priest doth dwell.

The Marybucks have the same allowance of women the Kings or temporall people have.

The Towne is called *Setico*, lying from the River side some three miles:[1] to this Towne I went, having occasion in following of our Trade, to lye with my boate so neere as I could come, my Guide or Conductor, was one of my blacke people I hired, called *Fodee Careere*,[2] who is his profession was a Marybucke, and they are for our commodities to be hired, and will put their hand to any needfull occasion, like as any of the temporall sort would do.

This Fodee Kareere was my Alchade and bought & sold for me.

This was the first of the Country, who ever I entertained and continued with mee,[3] both up into the highest part I went, as likewise all the time I followed any trade in the River, with whom we doe agree by the moone, how

---

[1] Setico = Sutuko, north of the river in Wuli territory (see the Itinerary (Appendix A), p. 220, n. 1). The presentday town is called Sutukobaa, that is, 'Big Sutuko'. (A more recent town called Sutukung, in the form Sutukonding, that is, 'Little Sutukung', is located on the north bank opposite Basse. As the Jaxanke spread, they carried the names of original villages with them, so one finds many communities with names like Bani, Bani Israel, Gunjuur (Kunjuur), Jaxa (Diakha), Kerewaan, Sutukung, Tuba, etc.) A modern account of Sutukobaa based on current oral traditions claims that the town was originally founded by animist Mandinka (the Jabi family), and that they were joined by a Mandinka trading family (the Nyaballis), followed by a trader from Ja, on the Niger, who had also spent some time in Tinda, and by other trading families. Families of blacksmiths and leatherworkers were also persuaded to join the settlement: Galloway, 'History of Wuli', pp. 64–6. Be all that as it may, by Jobson's time Jaxanke religious scholars had settled in the area and become the dominant element (for a larger West African town dominated by Islamic scholars see Elias N. Saad, *Social history of Timbuktu: the Role of Muslim Scholars and Notables 1400–1900*, Cambridge, 1983). This created an extremely influential centre inhabited by long-distance traders, religious scholars, and artisans, as well as by a large slave population employed both in agriculture, to grow food and cotton, and in the weaving of cotton cloth. The town had connections to further inland and to the interior, that is, to Tinda and the auriferous district of Bondu on the east and NE, as well as connections to the south and SW, to River Gambia and beyond, and even to the coast.

[2] Mandinka *fodee* is a title for a religious instructor (cf. *fodige*, Part II below: Lemos Coelho, 17), and 'Kareere/Careere' possibly represents the Arabic name Khalīl, or a local variant. The man is not named in the Journall. 'Alchade' ('Alcaide' *PJ* 924)) = Portuguese *alcaide* (derived from Arabic *al-qâid* 'chief') 'official of various kinds', but in Senegambia at this period a representative of either a ruler or a notable trader (cf. Bühnen, 'Geschichte der Bainunk', p. 418, 'alkaid'). Before reaching River Gambia Jobson may have learned the term from the account of the English in Senegal in the 1590s (Hakluyt, *Principal Navigations*, II/2, p. 189, *alcaide* 'governor'); but otherwise from local informants. A term in modern Mandinka, *alkaali*, usually denoting a village head, the representative of the district head and/or the central government, may derive from 'alcaide'. Alternatively it may derive from Arabic *al-qādī* 'judge', but since Islamic law has even as yet barely superseded customary law in the general countryside this is perhaps less likely, although it is possible that Jaxanke towns may have appointed judges earlier.

[3] Jobson says that he 'did hire four Blacke-men … to serve as Interpreters, likewise to send abroade, and to help to rowe' (*GT* 17–18, cf. 118). Two were hired at Baato, the youth Sanguli and a 'a young Mary-bucke' (*PJ* 922). Two more, 'Selyman' another marabout, and 'Bacan Tombo his kinsman, who he said had lived at Tinda' – where Jobson was heading – were engaged at Baarakunda (*GT* 83). Earlier, at Kasang, 'we could not get any Blackman to goe with us to be our Pilot and Linguist [interpreter]' (*PJ* 922). It is not clear whether these remarks rule

[63/4] THE DISCOVERY OF RIVER GAMBRA

much hee is to have, which agreement he receives when the moone is ended, in some commodity of ours which he desireth, the valuation whereof, amounts unto a poore summe: our continuance together, had bred such an affectionate league, betweene us, that we were united as people of one place, and in those courses of trade we followed, I did not onely aske and require his advise, but in most things allowed and followed the same: with that Marybucke, I had diverse and sundry communications, concerning their Religion,[1] wherein many times he would wish, that I might once come to converse with their chiefe man, whom he called *Fodee Bram*, who would as he sayd give me full satisfaction, concerning their religious orders; and being now come to the Port of *Setico*, for so we called it, he was very importunate I should goe up to see the towne and visit this religious person, I received his /64/ direction, what Present it was fittest, I should carry him up, wherwith furnished, taking two of our men with me, one of them carrying a fowling peece on his necke, we came to the towne, but meeting of some of the dwellers by the way, they had told us that this *Fodee Bram* was very il, and dangerously sicke, whereat they seemed much to mourne.

Fodee Bram was the chiefe Marybucke of all the Country.

Being entered the Towne, and come unto his house, I found without many people, demaunding of him what they were, he told me they were all as he was, Mary-buckes: for by their habite they are not to be discerned, being all clothed in one and the same manner, as the common people are: I was intreated to sit downe under their open shades, which are made by the better sort of people, on the out-side of their houses, to take the ayre in; and in the meane time my *Alchade*, for by that name my hired Mary-bucke was called, went into his house, and in my name presented my present, acquainting of him that I was the Captaine, and Commander of our people, whereupon he caused himselfe to be lifted up from his bed, or mat whereon he lay, sitting on the side whereof, supported and helde up by three of his wives, he sent out to have me brought unto him, and after our salutations past, he held me fast by the hand, giving me many thankes, for that great present he had received, bemoneing much his sicknesse hindered him, he could not accompany me, thereby to shew his respect unto me, during our conference he caused a dinner to be made ready: I did conceive him to be daungerously sicke, for his hand wherewith hee held me, did burne with that extremity, as

Both Priest & people weare one manner of apparrell.

The chiefe Mary-bucke, daungerously sicke.

---

out any engagement of Africans in the lower river (or even at River Saalum), to act as pilots as well as interpreters, which might be thought to have been a necessary move. Fodee Careere/Kareere was, however, undoubtedly hired at Baato and was therefore the 'young Marybucke', since on the return down-river Jobson attended a circumcision festival at Baato accompanied by 'our blacke Alchade, who wee hired from this town, wherein his mother likewise dwels' (*GT* 109). As Jobson goes on to explain, at Sutuko, Fodee Careere – 'my *Alchade*, for by that name my hired Mary-bucke was called' – introduces him to the senior marabout (*GT* 64, cf. 83). On the journey up-river one of the marabouts had to be rescued from a whirlpool but refused an alcohol-restorative, and almost certainly this was Fodee Careere (*GT* 76, cf. *PJ* 923).

[1] Although Jobson refers to his having obtained much information about Islam from Fodee Careere, and to his acting regularly on the marabout's trading advice, he nowhere explains how they communicated. This matter is discussed in the section of the Introduction on 'The English in River Gambia 1620–21'.

the heat /65/ gave such offence, that I wisht very willingly I might be loosed; I tooke notice of those women who held him, and to every of them gave a pewter ring, which both from them, and him was thankefully received: dinner provided, he intreated me with my Company: to passe into another house adjoyning, which after their Country manner was prepared, and we had Hennes, and other provision brought in, amongst which one sort of sustenance I never saw before, nor after in the Country, which was compounded of their Country graine, made up in round cakes, resembling very much our English Jelly, and as our Alchard told me, was one of the principall dainties, esteemed amongst them; while I was at dinner, a messenger came from him, bringing word he was very sory to heare I fed not, and likewise by him sent unto me a large hide, and an Elephants tooth of a good bignesse, as a gratuity for the present I had brought him, (wherein may be judged what losse I received, when the valuation of what I delivered, and by him so much esteemed, did not cost here at home, according to our Merchants accompt, above the great summe of xviijd): After I had eate my desire was to go see the Towne, and view their dwelling, he sent certaine people with mee, who brought me through their streetes or housing into the plaine fields, where I might throughly discerne the whole Scituation, I did heedfully regarde it, for it did appeare the greatest Towne, or place, that I had seene, and the manner thereof in my opinion, was worthy the observation: The towne was built round, after the manner of a Circle, whereof the front of the /66/ houses, did not contain any great thicknesse, but as we may say, the breedth of a reasonable faire street, joyning their houses or walles of their yards and barnes close together, the diameter whereof, that is from the North, to the South, or likewise from any one point to his opposite, we did conceive to be neere an English mile, within which Circute was much Cattle, especially store of Asses, whereby it may be conjectured, that they contrived their towne in that sort, to keepe out the ravening beasts, and securing those Cattle they had about them, whereof at this place they had the greatest use, I meane of their Asses, as I will relate unto you, but first I must take leave of the high Priest:[1] after I had satisfied my selfe in looking abroad, I returned unto him into his house, where I found him layd along, in extremity of paine, he seemed to labour much in his desire to have had conference with me, saying

His manner of entertaining me.

The valuation of the present I gave him, which was so highly esteemed.

The description of the town called Setico.

---

[1] A description of Setico/Sutuko (modern Sutukobaa) a century or so earlier included the following passage:

> At Sutucoo is held a great fair to which the Mandinguas bring many asses; these same Mandinguas, when the country is at peace and there are no wars, come to our ships.....and buy common red, blue and green cloths, kerchiefs, thin coloured silk, brass bracelets, caps, hats, the stones called '*alaquequas*' [bloodstones], and much more merchandise, so that in time of peace .... five and six thousand doubloons of good gold are brought thence to Portugal. Sutucoo and these other towns belong to the kingdom of Jalofo, but being on the frontier of Mandingua they speak the language of Mandingua (Part II below: Duarte Pacheco Pereira).

Sutuko remained a trading centre, and retained a reputation for hospitality, down to the 19th century.

> We reached the village of Soutouko, inhabited by Mandinka marabouts, who understood

The Chiefe Mary-bucke wonderfull desirous to confer with me about our Religion.	he had earnestly wished to see me, that we might have spoken together, concerning our lawes, and something he spake, but what came from him, was very unperfect, which I imputed to the paine he endured, only I noted when he wold have spoken of *Adam* and *Eve*, and *Moses*, & so I tooke notice of the names, in that manner he might well perceive, we had knowledge of them, he seemed to take pleasure therein, but the concluslon with him, was to no purpose, nor worth rehersall: So that after he had made me to eate by him, a dish of excellent creame drest with corne after their best fashion, & a solemne farewel past betwixt us, I left him. Notwithstanding before I proceed further, I will relate what I have gathered of their /67/ profession, and what they conceit of us. They do worship the same as we do, the true and
They worship the true God above, whom they call Alle.	only God, to whom they pray, and on his name they call, in their language, expressed by the word *Alle*, insomuch as if he see any thing which begets in him admiration, casting his eyes to the Heavens, hee cries *Alle Alle*, neither have they amongst them, any manner of image, or picture, or resemblance
They have no manner of Image.	of any divine things but as far as we can perceive, such things are distastastfull unto them: They doe ackowledge *Mahomet*, and all are circumcised, the manner of their circumcision, I refer to speak of in another place: their Sabboth or the seventh day is uppon the friday, and [they] have distinctions, and proper names for seven dayes in their weeke, they reckon their age or times, by the raines, in saying hee hath lived so many raines, as we say so many yeares, howbeit, we doe never heare them call upon the name of *Mahomet*,
They have no Churches.	neither have they amongst them any Churches, nor places they dedicate to holy uses, so farre as we can perceive, neither do we finde that they celebrate
They observe not their Sabboth.	or solemnize their Sabboth day, for even on that day will they follow any Trade, they will have with us, and their owne occasions: without any intermission, they have certaine round houses built open, and are spacious, according to their fashions, wherein they teach their youth,[1] the manner
The manner of teaching their male children to write, and reade.	whereof is this: All the male children that proceedes from these Mary-buckes, are taught to write and reade, and in regard they have no paper amongst them, but what wee or others bring them in the way of Trade; and therefore is of esteeme, they have for their bookes a small smooth boord, fit to /68/ hold in their hands, on with [which] the childrens lessons are written with a kinde of blacke incke they make, and the pen is in a manner of a pensill.[2]

---

hospitality better than the fierce inhabitants of the interior. Soutouko, situated in a charming place next to a rich green forest, presents the appearance of civilization. Its houses are better constructed, the roofs higher, and the interiors more spacious. It also has an elegant fort, and gardens enclosed with bamboo, in which are rich cultivations. We obtained milk and a sheep ... Soutouko is also a depôt for English goods (Anne Raffenel, *Voyages dans l'Afrique occidentale... et de la Gambie, depuis Baracounda jusqu'à l'Océan en 1841 et 1844*, Paris, 1846, p. 481, translated).

[1] Round open houses used by students are found today among the Serahuli of the middle River Gambia region: see Plate V. But village Mandinka now have square houses, with corrugated metal roofing.

[2] For a present-day student with a writing board, see Plate IV. The ink is made from soot from the bottom of cooking pots, but for permanent records a more elaborate form of ink

The Character they use, being much like the Hebrewe, which in regard I understood not,[1] I caused my Mary-bucke to write in paper, some part of their law, which I brought home with me, that some of our learned Schollers might peruse, if wee might by that meanes come to any better knowledge, then the small practise we have hetherto had, and by men of our capacities cannot so easily be attained, onely this much wee discerne, that the religion and law they teach, is not writ in the same tongue, they publickly speake,[2] and moreover, that none of the temporall people, of what dignity soever, are traded [*sic:* traned, i.e. trained] up to write, or reade, or have any use of bookes or letters amongst them.[3] And whether these open houses, they teach their children in, be places for their religious ceremonies, and for their publike meetings, in their holy exercises, because they are ever placed, neere the principall mans dwelling, and as it were joyning unto him, wee cannot well resolve, but rather conceite the contrary, in regard they lie open, and are not swept, and kept with any manner of decency, and therefore do rather thinke they make use of the open fields, where under some spacious shady tree, they observe their meetings,[4] some part of the manner whereof, as an eye witnesse, I will readily recite.

The place where we had houses built, and walled with straw for our owne uses, was seated by the River side, upon the top of the banke; and by the

---

is prepared: Ousmane Diallo, 'Encres et teintures au Fouta Djallon', *Notes africaines*, 28, October 1945, p. 17; Roland Portères, 'Encres et tablettes à écrire de fabrication et d'utilization locales à Dalaba (Fouta-Djalon, République de Guinée)', ibid., 101, 1964, pp. 28–9). Children would normally write Koranic texts.

[1] The script is, of course, Arabic, and it is strange that Jobson did not realize this. Exhibiting some but not full knowledge, Purchas's summary added a marginal note: 'It seemes they speake some Arabike words: in which also their Law is written' (Purchas, *Pilgrimes*, 1/9/13, 1572).

[2] Religious texts are written in Arabic, not the vernacular language (but see below). For the training of a student in Islamic studies in this region, see L. O. Sanneh, 'The Muslim Education of an African Child: Stresses and Tensions', in *Papers of the Conference on Manding Studies*, School of Oriental and African Studies, University of London, 1972; 'A Childhood Muslim Education: *Barakah*, Identity, and the Roots of Change', in *The Crown and the Turban: Muslims and West African Pluralism*, New York, 1997, pp. 121–46. At the end of the 17th century a visitor to River Gambia claimed that the Arabic script was used to write 'the natural language', that is, Mandinka: Cultru, *La Courbe*, p. 191. But there must be some doubt whether this was at all common – even today it is rare, and the Mandinka say that if a letter is sent in which the Mandinka is written in Arabic script, then the writer had better accompany the letter to read it out (DPG). However, it is not inconceivable that Mandinka in Arabic script was already in use, not for general communication, but by marabouts for glossing religious texts and translating parts of the Koran (p.c. Louis Brenner).

[3] It was no doubt correct that education, or at least schooling, was fundamentally in terms of Islamic learning, and that, apart from the instruction given to a probably limited number of children in reading and writing (but only Arabic, not their own language), other children only learned to recite the Koran (again in Arabic). In similar situations, 'Muslim' and 'literate' are equated, therefore non-Muslim and illiterate. Whether there existed non-Muslim autodidacts who managed to acquire a limited capacity in letters and numbers, in order to facilitate their commercial affairs, in analogy to the position in other cultures with restricted literacy, is perhaps doubtful, inasmuch as Muslims tended to regard the Arabic script as sacred and therefore forbidden to non-Muslims.

[4] The practice today is that major religious gatherings, for instance at the end of the Fast Month (*Ramadan*) or at the festival of Tobaski (*id-el-Kabir*), are usually held in an open space on the east side of a village.

[margin: They called our dwelling the white mens towne.]

people /69/ of the Country, called *Tobabo Condo*, the whitemans towne:[1] some hundred paces within us, from the River, was a small towne of these religious people, wherein lived an ancient Mary-bucke, called *Mahome*, who could not be lesse in apparance then an hundred yeares of age, from whom we received much relation concerning the countrey above, and of the abundance of gold there, which himselfe had travelled and seene, as it is their profession to doe, and I shall have occasion foorthwith to shew you: This ancient *Mahome*, was

[margin: In any occasion of falling out betweene the people and us, this old man would come with his Assegy presently to ayde us.]

ever a faithfull and loving neighbour unto us, howbeit in regard of his age, he did not teach the children, nor was not the eminentest man of the Towne, but one *Hammet*, who was not so trusty a neighbour. The olde *Mahome* did diverse times lodge and entertaine strangers, that came, especially of his owne profession, amongest which, there had laine one night at his house, a *Mary-bucke*, who in the morning, coming downe to the River side, close without our wall; having his slaves [*sic*: slave] to follow him, who brought in his hand a great gourd, in the River he filled it full of faire water, and brought unto him, whereinto presently pulling forth all his privie members, hee put them, without any nicenesse of being seene what he did, and after hee had well washed them, hee made him to throw away the water; which done, and the gourd well washed or rinced, he brought him another, filled also with water, wherein he washed and rubbed his hands, and in the like manner it was throwne away, and a third brought, wherewith he washed and cleansed his face, all which performed, he making a kind of lowe reverence with his bodie, /70/ and laying

[margin: Their manner of devotion.]

his hand on his breast, his face directed towards the East, kneeled downe, and there mumbled or uttered foorth, after a decent manner, it should seeme, certaine prayers, wherein, after hee had continued for a space, kissing the ground, hee rose up, and turning himself about, with his face directed to the West, hee performed the like ceremoney, which ended, after hee had stayed, and looked a while upon us, hee returned to his place of lodging.[2]

[margin: The death of the chiefe Mary-bucke.]

One ceremony more of their Religion, I will relate, if you please to remember, where and how I left the chiefe *Mary-bucke* sicke and full of danger, it did manifest no lesse, for in the evening, the day after I came from him, he died, the report whereof was immediatly spread over the whole countrey, who

---

[1] Mandinka *tubaboo* 'white man', *kunda* 'village/town'.

[2] Jobson was witnessing the purification before prayers, and the prescribed pattern for daily prayers. (For general ritual ablutions, see Part II below: Donelha, f. 25.) Prayers are made facing east, towards Mecca. The ritual does not include 'kissing the ground', as Jobson writes, but touching the forehead to the ground. There is also a problem when Jobson writes of the marabout turning to the west to continue the prayer. A commentator on this passage has noted that 'in Muslim canon law the worshipper may adopt as *qiblah* any direction if he or she is in any doubt about the exact location [of Mecca]' (Lamin O. Sanneh, *The Jakhanke Muslim Clerics: A Religious and Historical Study of Islam in Senegambia*, Lanham, 1989, p. 185). But it is difficult to see that in these particular circumstances there would have been any doubt about east and west; moreover, any doubt does not, presumably, call for repetition in each direction. The only explanation that occurs to us is that the man felt a sudden need to urinate, turned away from facing Mecca, moved a short distance, squatted down, and then stood up. If Jobson was watching from a distance, he might have interpreted the actions as a continuation of the prayer. However, it is perhaps more likely that in recollection Jobson invented the westward ceremony.

from all parts came in, after that abundant manner, to solemnize his funerall, so many thousands of men and women gathered together, as in such a desart and scattered countrey might breed admiration, which I thinke was rather increased in regard at that time he died, the moone was high, and gave her light, and they in whole troupes travelled, eyther the whole night, or most part of the same together; the place or port whereat my boat did ride, was a Passage or Ferry to the towne, from the whole countrey, on the further side, whereunto belonged a great Canoe, which I had hired, having likewise another of my owne, both which never stood still, but were used, night and day in passing the people, none of them came emptie, some brought beeves, others goates, and cockes and hennes, with rice, and all sort of graine the country /71/ yeelded, so as there came in a wonderfull deale of provision, my Mary-bucke entreated mee, to send something of sweet savour, to be cast upon his body, which the people much esteeme of: I sent some *Spica Romana*, and some Orras,[1] which by his sonne was thankefully received: the manner of his buriall, was after this sort, hee was layed in a house, where a grave was digged,[2] and a great pot of water set in the roome, and just after the same manner, as the Irish doe use,[3] with a wonderfull noyse of cries and lamentations, hee was layed into the ground; the people, especially the women, running about the house, and from place to place, with their armes spread, after a lunaticke fashion, seemd with great sorrow to bewaile his departure. They also assembled themselves, in the most convenient place, to receive the multitude, and nearest unto the grave, and sitting down in a round ring, in the middle came foorth a Mary-bucke, who betwixt saying and singing, did rehearse as it were certaine verses, in the praise and remembrance of him departed, which it should seeme was done *extempore*, or provided for that assembly, because upon divers words or sentences hee spake, the people would make such sodaine exultations, by clapping of their hands, and every one running in, to give and present unto him, some one or other manner of thing, might be thought acceptable, that one after another, every severall Mary-bucke would have his speech, wherein they onely went away with the gratifications, who had the pleasingest stile, or as we terme it, the most eloquente phrase, in setting forth the praises of him departed, in which the people /72/ were so much delighted; another ceremonie was, that every principall Mary-bucke and men of note amongst them, would take of the earth, which came forth of the place his grave was digged, and with the same water, which was in the pot, standing in, the same roome, would moysten the said earth, and so forme therof a round ball, which they would carry away with them, and esteeme of as a great Relique: whereof my *Alchade* or Mary-bucke, because of those perfumes I sent, was admitted to have one, which he

---

[1] 'orras' = a fragrant root of iris, a perfume.
[2] Burial within a house was usual in the case of prominent people among the Mandinka until modern times, when Islamization and colonial views regarding hygiene have led to its total disappearance.
[3] That is, at wakes.

[72/3] THE DISCOVERY OF RIVER GAMBRA

so highly esteemed, I could not at any rate purchase it from him, although I made him offers, of more then I meant to give.[1]

The investing of the eldest sonne in the fathers place.

This Assembly held, for the space of ten dayes, with a continuall recourse, of comming and going, but not altogether for the buriall of the dead; for after certayne dayes were spent in the celebrating of his Obsequies,[2] then beganne a great solemnitie, for the establishing and investing of his eldest sonne in his place and dignitie: whereunto came agayne many gifts, and presents: amongst those that passed by me, I tooke notice of a great Ramme,

A Ramme for sacrifice.

which was carried betweene two, bound fast and layd uppon a hurdle. In the whole time I was in the Country, I never saw any Ramme, or Sheepe, but that which was brought very farre, his wooll might more properly be called haire, it was of that hardnesse:[3] I did understand by my Mary-bucke, he was to be used, after some manner of sacrifice, and I understood likewise, that in their high Priesthood, the sonne succeeded the father, & this course is held amongst their Religious orders, wherein they differ from the temporall governerments. /73/

It followes, I should now deliver their poore opinion they should [showed] concerning us and our profession: wherein, with humble reverence, I crave pardon, that my hand hold [*sic:* should] in the least sort, be made an instrument, to shew or set downe, any thing opposite unto my Lord and Saviour, but by shewing the weaknes of naturall man, and the wisedome that remaines in rotten flesh, the glory of God more perfectly appeares, to the confirming and comforting of every true and perfect establisht Christian; when wee shew unto them we honour and serve God above, and likewise his Sonne, who was sent upon the earth, and suffered death for us, who was

They cal Christ by the name of Nale.

called Jesus, by that name they doe not know him, but by the name of *Nale*, they speake of a great Prophet, who did many and great miracles, whereof they have amongst them diverse repetitions, and that his mothers name was *Maria*, and him they doe acknowledge, to be a wondrous good man, but to be Gods sonne, they say it is impossible, for say they God was never seene, and who can see God and live, and much more, for God to have the knowl-

The opinion they hold concerning him.

edge of woman, in that kind that we should beleeve it, they do wonder at us;[4] the rather they say, because they see God loves us, better then them, in giving us such good things, they see we have and are able to bring unto them; and likewise they do admire our knowledge, being able to make such vessells, as can carry us through such great waters, and how we should finde our way, more especially higher up in the River; when we talk of the Sea,

---

[1] The burial rites and ceremonies described appear to combine Muslim practice and local custom (p.c. Louis Brenner).

[2] In Islamic and Mandinka tradition, ceremonies (involving 'charities', Arabic *sadaqa*, the ritual distribution of alms, generally in the form of rice cakes, kola nuts, etc) are held on seven and then on forty days after a death.

[3] The type of sheep referred to is the Sahelian breed.

[4] This is an accurate rendering of the Muslim view of Jesus (*Īsā*) as a prophet but not as divine. 'Nale' is therefore likely to be a mistaken version of *nabī* 'prophet'. 'Maria' would be *Maryam* (p.c. Louis Brenner).

whereof they are altogether ignorant, onely by the name, or word Fancassa, which signifieth /74/ great waters;[1] thus like humaine creatures in darknesse they argue, being barred from that glorious light, which shines in the east, wh[e]reof though they have heard, they have not yet made use but no doubt when the fulnesse of time is come they shall; for amongst themselves a prophesy remaines, that they shall be subdued, and remaine subject to a white people: And what know we, but that determinate time of God is at hand, and that it shall be his Almighty pleasure, to make our nation his instruments, whereof in my part I am strongly comforted in regard of the familiar conversation wee find amongst them, and the faire acceptance I received: in the upper parts I attained, where I had a people came down unto me, who had never seene white men before,[2] with whom we traded with a faire commerce, and some favor of a golden sequell: the relation whereof, will follow very speedily: onely it is necessary, I part not obruptly from my religious company, and to acquaint you that they have great bookes, all manuscripts of their Religion,[3] and that we have seene, when companies of Mary-buckes have travelled by us, some of their people laden therewith, many of them being very great, and of a large volume, which travell of theirs, it is most necessary I acquaint you withall, in regard from thence proceedes, a great deale of intelligence we have, and I may not let passe one vertue of theirs, the narration whereof, may make their intelligence somewhat more respected, and in my poore opinion carry alongst a better esteeme.

They have bookes of great volumes all manuscripts.

It may please you to call to minde, when I left the /75/ Kings in the middest of their cups, I promised to shew you a soberer people, which are these Marybucks, betwixt whom and the temporal people, is a wonderfull difference, the rather in regard they live upon one and the same ground, the temperature of the day being the same, wherein the desires of those common people, is for Aqua-vitæ, and hot drinkes that they will many times pawne their armes, both their bowes and arrowes, and swords from their neckes for that hot liquor, yea many times their clothes from their backes, to satiat and glut their earnest desires, which seeme to us never to be satisfied: Now to the contrary the Marybucke, will by no meanes take or touch on[e] droppe thereof, of what kind somever it be, tying himselfe strictly to no manner of drinke but water, and not onely himselfe, that is the men, or malekind, but likewise their wives and women, neither will or can at any time be drawne to

The wonderfull sobernesse of these Marybuckes.

---

[1] Mandinka *fankasoo* 'the ocean'. The term was recorded on a Benincasa map of the 1460s (assumed to be based on information from Cadamosto although the term is not in his account as we have it), and also recorded by Diogo Gomes. However, it seems to have related, not to River Gambia, but to a further river on the coast (A. Teixeira da Mota, *Mar, além Mar*, Lisbon, 1972, pp. 241–3).

[2] See p. 143, n. 2 below.

[3] Although one source speaks of a marabout preaching 'legends on parchments which he unrolls' (Álvares, *Ethiopia Minor*, f. 11ᵛ), the fact that the marabouts regularly sought paper from European traders indicates that Jobson's 'books' were not normally scrolls but manuscript sheets tied together and sometimes gathered within locally-made leather-covered portfolios (as witness the older Islamic books displayed in the region today). For the books of marabouts, cf. Almada, *Tratado breve*, Prologue, f. 2.

tast or receive any jot of this our comfortable liquor, nay more, they will not suffer none of their children, not so much as the little infant, who in the place we lived at, through daily recourse one with another, were growne to such familiarity with us, that they would many times steale from their homes, and come and hang about us, these smal ones we might not give any wine, no nor any maner of fruit as reasons [raisins], or sugar, or any sweete things, without great offence unto the parents, and if hee [*sic:* it] hapned they found it with them, they would take it away with great displeasure, and although themselves were never so sicke, and in those times we would perswade them, how comfortable it would be unto /76/ them, we could by no meanes prevaile, to gaine any manner of inclination towards it, for example, as I was travelling up the River in my boate, upon some occasions our people being in the water, and in the shallow, leading up our boate, a suddaine deepnesse, occasioned by a steepe banke, brought them beyond their reaches, and enforced them to shift for themselves by swimming: my Alchade or Mary-bucke, being one of them, who could reasonably use his armes, was notwithstanding taken in a whirle poole, and in great danger of drowning, having beene twise at the bottome, but at the second rise, one of our men tooke hold upon him, and with helpe, we presently got him aboord, being almost spent, and his senses gone, we earnest to recover him, fearing the agony we saw him in, got rosa-solis[1] to put in his mouth, the sent whereof, as it appeared, made him hold close his lippes, that we gave him none, but within a while he came perfectly to himselfe, and as it seemed retained the savor, so as he askt whether he had taken any or no: He was answered no: I had rather (sayth hee) have died, then any should have come within me, although I am verily perswaded, the very savor refresht, and did him good, wherein they have a great resemblance to the *Rechabites,* spoken of in the thirty five Chapter of the Prophet *Jeremy,* who kept zealously the Commaund of *Jonadab* their father, from whence these may be lineally discended, in regard it is sayd, they proceeded from *Hobab,* the father in Law of *Moses,* and *Moses* wife is noted to be an Ethiopian: And this is the principall marke, we know /77/ these Mary-buckes by, that howsomever they cannot by their habite be discerned from the common people yet in offering them to tast, or drinke our foresayde liquors, they are presently to bee distinguished, which sobernesse of their, being an evident signe, that they are always themselves:[2] To which I adde, that as they do not love, wee should promise them any thing, but be sure of performance, so in any thing we can discerne, we receive no false reports, or untruthes from them, with which confidence, I goe forward with the relation of their trade, and travaile.

These Mary-buckes are a people, who dispose themselves in generall, when they are in their able age to travaile, going in whole families together,

---

[1] 'rosa solis' = 'rose of the sun', a cordial flavoured with the juice of the sundew plant (a bog plant of genus *Drosera*), and fortified with brandy.

[2] However, a Portuguese source described the dress of *bixirins* and implied that it was distinctive: see Part II, Almada, 5/10.

and carrying along with them their bookes, and manuscripts, and their boyes or younger race with them, whom they teach, and instruct in any place they rest, or repose themselves, for which the whole Country is open before them, to harbour and sit downe as night or necessity over-taketh them, always disposing themselves to some Towne whereunto they are not overchargeable, but only to rest their bodies, in regard we see them alwayes carry provision for the belly with them, which we conceite is renewed, as they meete with some principall persons, or make their Rendevow[1] in some eminent place, this wee are sure that there is not any of them passe us, but they will use the custome of the whole Country, which is to begge without any deniall, and although to us it is but a poore matter, in respect of the Trade we have, much more what we hope and looke for, to give unto /78/ them, or amongst a whole company, a quier of paper, which cost three pence,[2] yet to them it is a rich reward, out of which they questionlesse doe rayse the greater part of sustenance to travell withall, and what else may be availeable unto them, making thereof, by writing in the paper their blessed Gregories, which they give and bestow as they finde occasion, and to confirme us herein, this wee note, that if wee have occasion to send any of the Countrey people, of any message or employment for us, after he hath agreed for his reward, he will looke to have a sheete or two of paper given him, which is to buy him sustenance, as hee passeth from towne to towne: so as you shall never meete with any of this profession, but in discourse they can speake of more Countries then their owne native places: one chiefe reason to encourage their travell, we have learned, which is, that they have free recourse through all places, so that howsoever the Kings and Countries are at warres, and up in armes, the one against the other, yet still the Mary-bucke is a priviledged person, and many [*sic*: may] follow his trade, or course of travelling, without any let or interruption of either side. Notwithstanding there is none of these Mary-buckes but goe armed, and are as compleatly furnished, as any of the other people, and have the manner of use and exercise of their weapons, in as ample manner as they have, whereunto I thinke they are rather invited, in regard of those wilde and ravening beasts, the countrey is stored withall, that upon any occasion, they may be able to defend themselves, and offend their offensive enemies. To particularize heerein I may /79/ tell you of those two ancients [*sic*: ancient] Mary-buckes, who were our neighbours, in the towne where our housing stood, who both of them would relate unto us, of infinite store of gold, which they had seene the Countrey above to abound withall, wherein the most auntient [ancient] man, whom wee found so loving a friend, would speake marvellous confidently, howbeit he would tell us, there were a dangerous people to passe, before wee came unto them, and that the River was so full of trees, we should not be able to get our boate along; and in token of feare, when I was to beginne my journey upward, and came in the evening to

---

[1] An earlier appearance in English writing of 'rendezvous' than noted in *OED*.
[2] In the 1940s DPG also found that white paper was a form of gift acceptable to Koranic teachers.

take my leave of him; taking my right hand betwixt both his, hee uttered over it, diverse unknowne words, and ever and anone, would sparingly spatter, with his spettle uppon it, after which laying his mouth close to my necke, over my right shoulder, hee would after the like manner performe there:[1] which his superstitious zeale being assuredly done in love, I did not contemptuously refuse, because I was ignorant of any offence therein, but with a friendly curtesie parted with him, and my returne backe was to him as joyfull; the other who was a more, or as I may say, most subtile fellow, promised to be my guide along, and to passe in the boate with me, and thereby wrought upon my willingnes, to embrace his company, to the serving his owne turne, and getting from me many such gifts and curtesies, as otherwise hee could never have attained, holding me in hand, hee would meet me at a Port above, but there deceived mee, to my further trouble, and at our returne prevented [? pretended], his cause was feare of the people /80/ above, which (God be praised) fell out to the contrary, whereof forthwith it will fall right to tell you; onely I must first say something, concerning the great Towne of *Setico*, and of the trade they follow, with those same number of Asses, whereof before I told you.

The inhabitants heere, who are all Mary-buckes, are the onely people, who follow a continuall trade from their owne houses downe to the King of *Bursall*, whose dwellings (as you may remember) is sayd to be by the Sea side; at which place, the Sea shoare doth naturally yeeld great store of Salt, but it is a course and durty kinde, insomuch as the greatest part, which we have seene, and taken notice of, doth rather looke like durt, or Sea-coale ashes,[2] then resemble the Salt we have in use, or make our trade withall, to buy which they carry downe, as their chiefest commoditie, the slaves or people of the Countrey, whereof the King of *Bursall* doth make such profit, as it is supposed to be a principall of the revenew, wherewith he maintaines his greatnesse. This commoditie the people doe carry likewise farre up unto the Country,[3] for amongst themselves, we can perceive they make little use thereof, so as their travell is long and tedious: the returne they make, is not discernd to be any thing but gold, and a kinde of Nuts they call *Cola*,[4] which

---

[1] The spittle of religious teachers was believed to be special through their recitations of the Koran, their prayers, and their blessings, and the virtue of sanctity could be conveyed to another person through the spittle. For an instance in Senegal in 1697 when a marabout took the hand of an important woman, spat on it, and applied the spittle to her forehead, eyes, nose, mouth and ears, while murmuring in Arabic, and after the ceremony assured his visitors that their journey would be successful, see Jean-Baptiste Labat, *Nouvelle relation de l'Afrique occidentale* ..., 5 vols, Paris, 1728, 4, p. 173. The Senegambian view is that 'like honey in water, speech, good or bad, dissolves in the saliva which keeps part of its power' (DPG).

[2] 'Sea-coale' = ordinary coal, as distinct from charcoal, so called because brought to London and southern England from the rivers Tyne and Wear by sea.

[3] On this salt, its source and the salt trade, see the Introduction, p. 29, n. 11.

[4] Kola nuts, the fruit of *Cola nitida*, are grown in limited districts within the forest belt of West Africa, and at this period those distributed in western Guinea came mainly from the Scarcies district of Sierra Leone. Kola was – and is – highly valued in West Africa as a stimulant, alleviating hunger and thirst when working or travelling, having an effect equivalent to a cup of strong coffee. (It is purportedly one of the 'secret ingredients' of presentday worldwide 'cola' drinks.) Giving and sharing kola nuts symbolizes friendship and trust (see *GT* 97 below), and they continue to be

is in great esteeme amongst them, the vertue whereof I shall hereafter tell you: and for that it may be here demanded, what becoms of the gold by them brought downe, I will shew you what by report is told us; These Mary-buckes doe hold an opinion, that after their death they shall appeare in another world, /81/ wherein this gold wil be of great esteem, and therefore strive to furnish themselves all they can therewith, which either in their life time, they secretly in the ground doe hide, or by their dearest friends cause to bee buried with them, esteeming themselves happiest, that can with greatest quantitie be furnished: another use they make, is, to buy from the *Portingals*, a sort of faire, long & square blew stones, which stones their women weare about their middle, to keepe them from bloudy issues, unto which they are generally subject, the occasion rising from the men, as may be well supposed, if you but remember or call to your minde, after what sort they are discribed,[1] and this is seene, by that esteeme the *Portingalls* make of that commoditie which brings (as I observed) so great a store and quantitie of gold amongst them; other use within themselves they have none, but that the women weare it hanging in their eares, in rings, and pendants, made up with little Arte, and as unhansome workmanship. These people of *Setico* were the most unwilling we should proceed in the search of the River, of [*sic: or*] any other, not onely telling us themselves of many dangers, but at all townes where we came, and amongst our familiars, had left their perswasions, if it could have prevailed to discourage us; or whether they did verily thinke, our boate could never have found passage, in regard it was never attempted by any such vessell before, or as I incline unto, fearing we might be hinderers to the Trade, they had so long followed, and whereunto they were setled, being wel provided with such numbers of Asses, as beasts of burthen, to proceed and /82/ follow, the same, so as from them, wee could get no comfortable intelligence,[2] wherein reason leades the way, that every mans profit is nearest

An ill opinion of the Marybuckes, to bury their golde.

A good commoditie.

The reason of looking after Buckor Sano.

---

used on many formal occasions, for instance, when greeting strangers, in naming ceremonies, and during marriage negociations. Jobson refers again to kola nuts in *GT* 134–5. For other 17th-century descriptions of kola, its uses and the trade in the River Gambia region, see Cultru, *La Courbe*, pp. 202–3; and Part II below: Almada, 6/9; Lemos Coelho, 23, 29.

[1] Jobson's insistence on 'bloody issues' which are other than menstrual may have been influenced by the Biblical reference to a woman with a 12-year history of an 'issue of blood' (Mark 5: 25). For alleged male responsibility, see note p. 115, n. 2 above.

[2] This sentence reads as if the English were advised against proceeding higher up-river by direct contact with the 'people of *Setico*'. But according to Jobson's Journall, Sutuko, in the interior, was only visited after the the English party had returned from the up-river journey. Perhaps Jobson meant to say that Sutuko, when at length visited, discouraged further journeys up-river. Or perhaps Jobson met Sutuko merchants lower down the river before ascending, and obtained their opinion at this stage. It is unlikely that, despite it never being mentioned, the party proceeding up-river halted at one of the ports of Sutuko and spoke to merchants encountered there. However, Jobson adds that Sutuko, by 'perswasions', influenced other towns on the river, that is, down-river, to discourage the English from proceeding. This implies that the Sutuko merchant community had learned about the English intention before the party set out, which may or may not have been the case. In claiming that never earlier had 'any such vessel' attempted the upper river, the discouragers were showing ignorance of, or at least ignoring or forgetting, Thompson's venture. For Buckor Sano, see p. 137, n. 1 below.

Note.	to himselfe; but as it shall please God, to encourage you the noble Governour and Company, to prepare and settle your selves, with a serious resolution, to follow the farther search of this rich expectation: These people of *Setico*, of all others are the likeliest, and dwell the most convenientest, to be brought to a more setled, and commodious trade, which will fall upon them with a greate deale lesse trouble, and infinite lesse travell, and withall be made especiall instruments of our good, whereunto as yet their grosse understandings cannot ascend, and ancient customes are harsh to be altred, howbeit these were the considerations that made me endevour to settle a league with the high Priest, and establish a perfect course of amitie betwixt us; which course of mine, I shall more boldly commend to your faire acceptance, when you shall see it grounded upon the experience of my whole travel and trade in the RIver, and after my discourse and conference with that great blacke Marchant *Buckor Sano*, concerning whom and all my proceedings above, I now am come to make a full relation.

Our travell up the River.

The Marybuckes name was Selyman, the other Tombo. Samgulley a blacke boy.	When I was come to *Baraconda*, which is the highest Towne the River flowes unto [*sc*. has tidal flow to], and notwithstanding all the discouragement I received, was absolutely bent to proceed up the River, by means of *Bacay Tombo*, a principall man /83/ of that Countrey, who brought mee two beeves.¹ I was furnished with two Blacke-men more to go up with mee, the one a Mary-bucke, the other *Bacan Tombo* his kinsman, who he said had lived at *Tinda*, which was the place wee aymed at, not to the Towne it selfe, but to the mouth of a little River, which was said to runne neare unto the place, and from thence fell into the maine River where wee were:² Two Blacke-men I had before, that was my *Alchade*, of whom I talked, and a pretty youth called *Samgulley*, who from the first coming of *George Tompson* into the River, had always lived with the English, and followed their affaires, so as hee was come to speake our tongue, very handsomely, and him I used many times as an Interpreter: so as in all we had foure Blacke-men, whose help we could not misse, in regard wee carried with us a small *Cano*, that was ready at all times to put ashoare: and when wee came to an anker, to fetch wood, or any other provisions, as likewise to carry us ashoare, and bring others to us, reserving our boate in the middest of the River, as a castle and refuge for us.

Thus we being ten Englishmen, and these foure Blackes, went the fifteenth of January in the evening from *Baraconda*, and were going against the streame, untill friday the twenty sixt in the morning, which was eleven dayes travel, wherein I desire to be rightly understood, that our labour was about

---

¹ The punctuation should probably be corrected, to read - '... up the River. By means of *Bacay Tombo* ... I was furnished ...'. '*Bacay*' may represent the commmon forename 'Bakari'.

² In the Journall this river is called Tinda River (*PJ* 923): for its identification, and for 'Tinda', see the Itinerary (Appendix A), p. 218, nn. 1, 3.

foure houres in the morning, and foure in the evening, so as our whole time spent therein, amounted to but foure score and eight houres, in which time, our Sabboth day was observed, onely two houres in the evening, whenas my men /84/ earnestly entreated to be going: in which time we recovered within half a league of the place or Rivers mouth, we intended to goe unto, and further wee would not have passed, nor endangered our selves, and what wee carried into the little River, untill we had made triall of the peoples dispositions, and how they should stand affected to us; we met here with a shole, as I have said in the description of the River, which stayed us we could passe no higher; we concluded therefore in the after-noone, to send away three of our Blacke-men, who were willing to goe, directing them to *Buckor Sano* of *Tinda*, to whom we sent a Present, and likewise I sent to the King of the place, as the manner is;[1] and gave our Blacke-men wherewithall to buy them victualls, demanding of them when they thought we should expect their returne, they said Sonday night, wherin I speake after our owne phrase: and while they are travelling, it will be necessary I acquaint you, what were the grounds or reasons we sought after this *Buckor Sano*, and laboured to get neare to this *Tinda*, making more especiall choice of him, then of any other man.

George Tompson, in his diligence, while hee lived, hearing of diverse Caravans, that past in the country, and went downe to the King of *Bursals* dominions for salt, had learned, that the onely and principallest man that maintained the greatest Trade, was that *Buckor Sano*, whose dwelling was at *Tinda*, who maintained and kept 300. Asses following that tedious travell. *Tompsons* desire led him forthwith, to goe finde this Marchant, and in a paire of Oares, as I spake in the beginning,[2] went up the /85/ River, and travelling some way by land recovered *Tinda*, but found not his blacke Merchant, in regard he was travelled higher into the Country, in the sale and uttering of his salt Commodity: *Thompson* returned, but found his expectation so satisfied, in that he had hard of the Moores of *Barbary*, and was come so neere where they frequented, that hee talkt of nothing, but how to settle habitations, and fortefie the River to defende themselves, and keepe out other nations; but these his desires died in his unhappy end, and this was all our acquaintance; which now I came to second, by sending unto this *Buckor Sano* to come downe unto the River to us, as the onely man we were willing to sell, and commend our commodities unto.

And by this time Sunday night is come, and none of my blackmen returned, monday likewise all spent in expectation, on twesday our men began to grumble, and my especiall consort to speake out, there was no reason wee should hazard our selves by staying any longer, in regard it was

Eleven dayes travell against the streame, wherein wee wrought eighty eight houres.

Some of our men grew fearfull.

---

[1] 'Buckor Sano', that is (most likely), Bakari Sanuo. Sanuo was a distinctive Jaxanke surname (Galloway, 'History of Wuli', p. 155). But Bakari Sanuo was not a rigid Muslim, as shown by his drinking of alcohol. Jobson repeats Thompson's enterprise which, although Thompson is dead, he has learned about from the English survivors. The 'King of the place' was probably the king of Jalakoto.

[2] *GT* 6–7.

fallen out, as we were told below, that they were a bloody, and dangerous people, and therefore those people we sent up were murdered; and if we stayed our turnes would be next, and likewise that we had no flesh left, and our other provisions were very scanty. I gave them content with faire words, that the place might be farther of then they conceited, in regard we had beene on the toppe of the mountaines, and could discerne no likelihood, of Townes or habitations, of which we had had no acquaintance, since we came from *Baraconde*, and in regard the Country about us was aboundantly replenished /86/ with all manner of wild beasts, we would try our indeavors, and on the wednesday morning I went out with two more, and killed a great and goodly beast,[1] which was no sooner brought downe & cut out, & hanged to coole under the shady trees on shore, but there appeared in sight three blackmen, the one was one of those we sent, who had brought with him *Buckor Sanus* brother, and the King of *Tindas* servant, and they came before to see us, and what commodities wee had, bringing word that the next day *Buckor Sano* would be there himselfe; I had them aboord my boate, and made them curteous entertainment, giving them some small commodities, and when the evening came, we feasted with our Venison.

    The next day about noone, came *Buckor Sano* with his musicke playing before him, with great solemnity, and his best clothes on, and about some 40. more, armed with their bows and arrowes with him, hee shewed no more at first, bowbeit [*sic:* howbeit] within two houres after, there were two hundred men and women come thither: he sat downe upon the banke under a shady tree: after a little stay, I went ashore to him, and our salutations being past, I desired him to go aboord, whereof he kindly accepted; and withall shewed me a beefe he had brought to give me for the present I had sent him, diverse goates the people had likewise brought, and corne, and cockes, and hens, so as there was no neede to doubt any more want of victuall: He carried no more aboord with him, but two; after he was in the boate, I shot off three such guns as I had to welcome him, at the noyse /87/ whereof he seemed much to rejoyce, calling the report of the powder, by the name of the white mens thunder, and taking notice of the head, and the hide of the Deare which we had killed, which we shewed him was slaine by one of our guns, they sent, with admiration, from one place to another, and certified, that there was a people come, who with thunder killed the wild beasts in the wood, and the fowles in the ayre: Which for it was our dayly use to kill one sort of fowle called a Stalker, which is as high as a man, and hath as much meate of his body, as is in a Lambe, which diverse times we used to kill, and eate, more especiall we desired to have his feathers, which grew on his tayle, which are of use, and such as are worne, and esteemed of here at home

---

[1] For similar descriptions of the terrain in 1848 and 1881, see the Itinerary (Appendix A), p. 217, n. 2. The antelope may have been a hartebeest (*Bubalis major*, Mandinka *tonkongo*); but the region has many other large antelopes, the cob (*Cobus kob*, Mandinka *wontoo*), roan antelope (*Hippotragus equinus gambianus*, Mandinka *daa-koyo*), and Derby Eland (*Taurotragus derbianus*, Mandinka *jinki-janko*).

amongst us: I had of my owne provision good Rosa-solis, taking forth a glasse, I dranke unto him, after he had dranke he tooke off his sword and gave it me to lay up, saying defend me here in your boate, and I will secure you on shore, he liked our drinke so well, he suckt it in, and as it seems not knowing the strength of it, took more then he would have done, insomuch as he fell asleepe, the people that came with him, in the meane time cutting of reedes, made them houses, others [after] fetching in wood, made fires every where about them, so as it seemed a little towne; *Buckor Sano* slept soundly upon my bed by me in the boate, and in the morning complained of his head, and this much I must justifie in his behalfe, that during the time we were together, he was never overtaken by drinking after, but observed the course he saw we used, to take a small cup before meate, and /88/ another after, and this ever gave him satisfaction: He desired to see all the Commodities we had, which he liked very well of, and whereas we thought our Iron would have beene greedily desired, we found it not so, for they told us, there was a people neighbours unto them who had knowledge to make it,[1] howbeit they were diverse times in wars together, but some of our Iron we put away at better rates then below, by one third, and might have done away all we had, if we would have accepted of hides, which for the reason I shall presently shew was refused; howsomever this was the maine businesse, that after they saw our salt, no other thing was esteemed amongst them, which at first seemed strange unto them, forasmuch as they had never seene any of that fashion before: the salt we had, was onely bay salt,[2] which after they put in their mouthes, and tasted, they would looke up and cry, *Alle*, in token of the good esteeme they had of it; After two houres of the morning spent, my Merchant went on shore, keeping my gowne about him, which when the evening shut in, the night before I had put upon him, and in a manner of state, he went one shore withall, wearing of it in that manner, it might well appeare, they were not used to such kind of ornaments.

    The first thing he did, after he came on shore, he caused on [one] to make a lowed outcry, in manner of a proclamation prohib[it]ing any of the people, to buy or barter with us, but as he bargaind.

Marginalia: The saying of Buckor Sano aboord the boate. / Hee was but once overtaken with our strong drinkes. / The great esteeme of salt. / He makes a proclamation.

---

[1] In 1795, during his first journey to the Niger, which skirted the Tinda region and came back through it, Mungo Park noted iron smelting in the village of Kamalia, west of Bamako, north of River Gambia (Park, *Travels to the Interior*, ch. 21). But it is on record that, at a much earlier date, the Susu in the Futa Jalon to the east also smelted iron (Fernandes, *Description*, ff. 110^v, 125, 135; Claude Francis-Boeuf, 'L'industrie autochtone du fer en Afrique Occidentale Française', *Bulletin du Comité d'études historiques et scientifiques de l'A.O.F.*, 20, 1937, pp. 403–64). However, since local methods of smelting produced small quantities of iron and both districts were some hundreds of miles distant from River Gambia, it is likely that the iron imported from Europe was cheaper on the river, as well as perhaps of better quality. For iron working in the river, cf. Part II below: Fernandes, f. 110^v; Almada, 5/6, 6/3. The imported iron bars had to be cut into smaller pieces by a local blacksmith (120–21 below; cf. Part II: Almada, 6/3).

[2] 'bay salt' = salt from the evaporation of sea water, not mined salt. The Portuguese and later the Dutch, obtained naturally evaporated sea salt from extensive deposits in the Cape Verde Islands, but since the English ships did not stop there when en route to River Gambia Jobson's salt must have come from elsewhere. Since the salt normally sold in the river came either from River Saalum or the lower River Gambia, it would seem that Jobson's different salt must have come from England, where much salt was still 'bay salt'.

	All that day hee found himselfe so sicke, after his drinking, that hee told me hee could tend no businesse, onely hee shewed unto mee, certaine young /89/ blacke women, who were standing by themselves, and had white strings
He offers women to sell unto us.	crosse their bodies, which hee told me were slaves, brought for me to buy, I made answer, We were a people, who did not deale in any such commodities, neither did wee buy or sell one another, or any that had our owne shapes; he seemed to marvell much at it, and told us, it was the only marchandize, they carried downe into the countrey, where they fetch all their salt, and that they were solde there to white men, who earnestly desired them, especially such young women, as hee had brought for us:[1] we answered, They were another kinde of people different from us, but for our part, if they had no other com-
Their commodi- ties.	modities, we would returne againe: he made reply, that they had hides and Elephants teeth, cotton yarne, and the clothes of the country, which in our trade we call Negroes clothes:[2] he was answerd, for their hides, we would not buy, in regard our boate was little, and wee could not conveniently carry them, but if they would bring them lower downe the River, where our bigger vessels could come, we would buy them all, but for their teeth, cotton, and clothes, wee would deale for them: so against the next morning, being Sat-
A markethouse made a shore.	terday, we had a house built by the water side, open round about, and cov- ered with reeds on the toppe, to shadow us from the Sunne: and this was our market house; when we came to trade, we asked which should be the Staple commoditie, to pitch the price upon, to value other things by, they shewed us one of their clothes,[3] and for that they onely desired our salt, we fell to love- ing [*sc.* bargaining] and bidding upon the proportion, wherein we had such /90/ difference, and held so long, that many of them seemed to dislike, and made shew, that they would goe away, but after we concluded, there was no more difference, every man bringing his commodities, our salt went away, and as they dispatcht, they likewise returned in companies together, and still others came, that we had the place continually furnished: We never talked unto them of golde, the principall we came for, but wayted opportunitie, and
Warning not to take notice of their gold.	notwithstanding we saw it worne in their womens eares, warning was given, none of our people, should take any great notice of it, as a thing wee should greatly desire, untill occasion was given, by *Buckor Sano* himselfe, who taking

---

[1] While this may have been essentially a commercial offer on the part of the African, as Jobson interprets it (or affects to do so), the inclusion of Buckor Sano's response to Jobson's refusal makes it clear that, as might be assumed, since only young women were presented it was intended, at least in part, as an act of courtesy, on the assumption that the Englishmen would welcome the sexual services of young women. The episode is discussed in the section of the Introduction on 'The English in River Gambia 1620–21', including p. 32, n. 1.

[2] Cotton was widely grown in the middle river area, Wuli being renowned for its cloth pro- duction.

[3] While West Africa lacked coinage, currencies of set amounts of local products serving as intermediate exchange tokens were not uncommon. (Similarly, in the later Afro-European marine trade bars of iron served as a currency.) Strips, pieces, or bundles of locally woven coarse cloth were used as a measure of value in certain parts, including Senegambia (D. W. Ames, 'The Use of a Transitional Cloth-money Token among the Wolof', *American Anthropolo- gist*, 57, 1955, pp. 1016–24).

note of our guilt swords, and some other things wee had, although but poorely set out, with some shew of gold trimming, did aske if that were gold: hee was answered, Yes: it should seeme sayth he, you have much of this in your Countrey. Wee affirmed the same, and that it was a thing our men did all use to weare, and therefore if they had any, wee would buy it of them, because we had more use then they for it, you shall have sayd he, what is amongst our women here; but if I did know you would esteeme of that, I would be provided, to bring you such quantitie, as should buy all things you brought: and if you would be sure to come still unto us, I would not faile to meete you. And proceeding further hee sayd: This Countrey above doth abound therewith, insomuch as these eyes of mine (poynting two of his fingers to his eyes, as the Countrey manner in speaking is,) hath beene foure severall times, at a great Towne above, the /91/ houses whereof are covered onely with gold:[1] wee demaunded of him, how long he was going, and comming thither: he answered foure Moones; we asked him, if hee would carry some of us thither, hee answered: Yes, but they had enemies by the way, sometimes to fight with them,[2] wee shewed him presently our gunnes, and tolde him wee would carry them with us, and kill them all, at which he seemed to take a great deale of content.[3]

Buckor Sanos report of gold and of the houses above covered therewith.

He seem'd wonderous willing of our companies.

Before I goe further, I will take occasion heere to set downe their manner of travell. They goe in companies together, and drive before them their Asses, whose ordinary pace they follow, beginning their dayes journey, when the day appeares, which is even at the Sunne rising, (for so neare the Equinoctiall, there is a short dawning, eyther before the Sunne riseth, or after shee sets)[4] and continue travelling some three houres, then are they enforced to rest all the heate of the day, some two houres before the Sunne setteth, going forward againe, and so continue untill night comes, whenas they are sure to harbour themselves, for feare of wilde beasts, except in some Moone light nights, and then they will travell the better, likewise when they come to some speciall Townes, they will rest themselves and their Asses, 2. or 3. daies together, laying all their burdens, under some shadie trees, close to the town, [and] set forth such things as they have to sale, maintaining in the time they are ther a kind of market, & their asses being spanseld, which is their 2. forelegs tied together, feed by them, the people themselves, lodging among their burdens, upon such matts as they ever carry with them; of which /92/ kind of Innes or lodging places, they can seldome misse, the Country

---

[1] While the African trader had an interest in exaggerating and no doubt used hyperbole, it is possible that his remark was mistranslated, and that he was not being deliberately deceitful.

[2] Jobson's petition, summarizing this episode, adds that the merchant had had 'divers encounters, & received many wounds shewing for a witnesse severall hurts upon his owne bodie', Petition, f. 3.

[3] The English boast that with so few men and guns they could give realistic assistance adds a further note of unreality to the whole reported conversation. In his petition, Jobson added that the people of the gold country wore in their ears and lips rings of iron, this metal being more valuable than gold there (Petition, f. 3).

[4] As first noted in print by Cadamosto, see Part II below: Cadamosto, [15].

being well replenished: So as if you please to observe, although the time seeme much which is spent in this journ[ey]ing, yet the way cannot be much, if you consider the maner of journ[ey]ing, wherein leaving to speake farther, untill I come to a more ample application, I returne againe to *Buckor Sano* my blacke Merchant.

Buckor Sano his subtill speech.

In our time of trading together, if it were his owne goods he bartred for, he would tell us, this is for my selfe, and you must deale better with me, then either with the Kings of the Country or any others, because I am as you are, a Julietto, which signifies a Merchant, that goes from place to place,[1] neither do I, as the Kings of our Country do which is to eate, and drinke, and lye still at home amongst their women, but I seeke abroad as you doe; and therefore am nearer unto you,[2] neither was I unwilling to answere somewayes his expectations, in hope I should better forward our owne endes. In our course of familiarity, after time I tooke some speciall note of the blade of his sword, and a paire of brasse bracelts [bracelets] one of his wives had upon her armes, both which things did appeare to me, to be such as might very well be brought in their beginnings, either from London, or some other part of this our native Country, I demanded of him where he had them, he made answere there was a people used to come amongst them, whom they called *Arabecks*,[3] who brought them these, and diverse other commodities;[4] we askt what manner of people, he described the Tawny Moore[5] unto us, and sayde they came in great companies together, and with many Cammels: How acceptable this /93/ report was unto me, may be conjectured by any such, who are seriously enclined, to give a faire and just accompt of any such imployments they are interest[ed] in, and whose desires, with affection, labours the full satisfaction of the trust imposed upon them;

His declaration of the Moores of Barbary.

Pleasing intelligence, being the maine businesse wee ayme at.

This his relation made it certaine, that these were the Moores of *Barbary*, the discovery of whose trade and trafficke, was the ground of this our being so high in the river: we grew to question him, how neare those people came to the place we were now at, he answered, within 6 dayes journey there is a towne called *Mumbarre*,[6] unto which towne, the next Moone, these *Arabeckes* will come: we askt againe, what commodities they brought with them, he answered much salt and divers other things, wee desired then to know what

---

[1] For 'Julietto', see p. 83, n. 1 above.
[2] It is not inconceivable that a trader would speak disparagingly of his political rulers, but the problem of interpretation into English, and the fact that the sentiment of the speech smacks of Jobson's own prejudices, throws a measure of doubt on its authenticity.
[3] '*Arabecks*' = 'Arabs', presumably. It is unclear whether the ending is Jobson's attempt to represent a terminal sound in another language (cf. Mary-bucke), or confusion on his part between 'Arab' and 'Arabic'. The *OED* notes 'gum-arabecke' in a work of 1616.
[4] By way of the trans-Saharan trade, a sword blade, although probably not mere brass bracelets, may well have come from North Africa, especially Morocco ('Barbary'). But if any of the commodities reaching the upper stretches of River Gambia from the North were indeed of European manufacture, it is more likely that their source was Iberia than that it was England.
[5] By 'tawny', Jobson means that the North Africans were neither 'white' nor 'black', but 'brown', as indeed are the Berbers and Arabized Berbers of North Africa.
[6] See p. 87, n. 4 above.

they exchange for, and carryed backe: he answered nothing but gold, and that they onely desire to have, and returned nothing else: wee questioned him farther, whether hee would undertake to carry any of us safe to see those *Arabeckes*, and that wee might returne without danger; hee stopt his nose betweene his finger and his thumbe: and cryed *Hore, Hore*, which is the greatest oath they use amongst them that he would performe it: some other conference past betwixt us at this time, howbeit by reason of a disaster that fell in the way betwixt mee and my chiefe interpreter,[1] I was hindred from understanding divers particulars, wherein *Buckor Sano*, seemed very desirous to give me full satisfaction, so as from him these were the principals I gathered, howbeit, another occasion fell, whereby I had some farther relation, as in his due place shall follow; for the conveniency whereof, I must once againe with your favour returne to a great company /94/ on shore, who expect their trade, for this our conference with *Buckor Sano*, was aboard our boate at dinner.

Their course of trading.

An oath they observe carefully.

An unhappy accident.

The people who came unto us for the first foure dayes [we] were staide here, came onely upon that side our Marchant came; but one of our hyred black men, spake unto mee, to give him some paper, and beades to buy him, and two other [blacks, their] provision as they went, and they would goe seeke other inhabitants likewise who dwell on the other part of the River, wherewith furnished they went away, and two dayes after returned, and brought with them divers people, who in the like manner, made them houses of Reedes, to harbour themselves under: These people had never seene white men before;[2] and the woemen that came with them were very shye, and fearefull of us, insomuch as they would runne behind the men, and into the houses to hide from us; when we offered to come neare them: I sent therefore into the boate for some beades and such things, and went unto some of

A people that never saw white men before.

---

[1] Presumably the 'chiefe interpreter' was the English-speaking youth Sanguli. Jobson does not disclose the nature of the 'disaster' and 'unhappy accident'. For a further discussion of the episode, see the section of the Introduction on 'The English in River Gambia 1620–21'. But if within a few days Jobson obtained further information from the trader (as he goes on to say) the disaster cannot have been serious or other than temporary. The 'conference' with the trader seems to have been at the regular dinner aboard – could it have been that on this occasion it was the youthful interpreter who collapsed drunk (perhaps after being plied with liquor by the other Englishmen, and perhaps even to annoy the abstemious marabout, Fodee Careere, whose relationship with Jobson they may have resented)?

[2] It was perhaps not completely correct that none of them had ever seen white men before. Since the messengers were away only two days, the new arrivals lived within one day of the river. Indeed, as stated later, some were within earshot of the firing of guns by the English. Moreover, 'the better sort' could speak Mandinka. Their instant attraction to the river, when news came of a trading opportunity, suggests that among them were a number who had visited the river before, or travelled to trade elsewhere. These individuals may well have encountered, at least once, representatives of either the Afro-Portuguese who traded up-river or the 'Arabeckes' who traded in the north, both being of mixed 'white' and darker ancestry. However, it is possible, perhaps likely, that these, although alien, were not considered 'white'. Jobson, for his part, tended not to regard the Afro-Portuguese as 'white'. In his petition Jobson boasted that 'myselfe was the first white man made tryall of their conditon & brought them such commodities [as] were very acceptable. I speake of those higher parts of the river …' (Petition, f. 2). Even allowing that only the English were white, this claim was only partially true, since Thompson had preceded Jobson to the upper river, but (as far as we know) he had not actually traded, as Jobson did.

the boldest, giving them thereof into their hands, which they were willing to receive, and with these curtesies imboldned them, that they soone became familiar, and in requitall gave me againe, Tobacco, and fine neate Canes they had to take Tobacco with:[1] these woemen were the deepest, and largest printed upon the backe that ever wee saw. The men likewise shewed [themselves] a more savage kinde of people then we had seene; many having breeches made of rawe hydes, either of Deare, or other cattle, the grace whereof was the taile of the beast which remaineth on the skinne, did sticke right forth upon the hinder part of the mans buttocks, resembling the manner as the beast wore it, whereat those blacke men, who were /95/ our hyrelings, would in scorne jest and deride.[2]

Strange breeches the common people did weare.

The principall man of these parts, called *Baiage Dinggo*[3] I had into my boate, and curteously used, demaunding if hee did not knowe of our being there, in reguard we thought he might heare our gunnes, he answered they had hard the noyse, and were much amazed, supposing it to bee thunder, being the more fearefull unto them, because the sound came that waies, it was not accustomed to doe; they were very desirous also to buy of our salt, and some of that little store we brought they had; which the people on the other side did unwillingly spare; we bought of them such commodities as the rest had, and wonderfull desirous they were wee should come againe unto them; the best of these women likewise did weare in their eares gold, and many commodities the women askt for, which we had not, but the men especially for salt: there was not so little, as 500. men and woemen, who came downe on this side the river, and it did plainely appeare, both the one and the other side were of familiar acquaintance, in regard they did passe in our Canoe, the one to the other, and had neighbourly salutations together: and further wee noted, that the naturall tongue there spoken, was another different language: yet all the better sort did talke together in the same speech we broght

An encouragement to search further up the River.

Those people had another language.

---

[1] The habit of tobacco-smoking spread with extreme rapidity in Africa, as well as in Europe and Asia, after the late 16th-century conveyance of tobacco from the Americas to England and France. However, in West Africa certain fragrant plants other than genuine tobacco were also smoked, tobacco not being imported in quantities or grown locally until probably the 18th century, and it is therefore just possible that smoking preceded the export of tobacco from the New World. Finally, tobacco was sometimes chewed rather than smoked, a development from chewing or at least sucking other plant products, notably kola. Tobacco-smoking on the higher River Gambia could have been a cultural import from North Africa and the Middle East rather than from Europe. For tobacco pipes, see *GT* 122 below and p. 162, n. 1.

[2] Apart from its Mandinka and Jaxanke communities, the area of Tinda and the nearby region of Damentan were inhabited by peoples such as the Basari, Koniaji, Bedik, and Badyaranke, who (even in later evidenced times) used animal skins for clothing. (For 'Bachares' = Basari as captives, see Part II below: Lemos Coelho, 61.) It may be noted that the Islamized Africans, like the Europeans, entertained a poor opinion of 'breeds without the Law'.

[3] Or 'Bajay Dinko' in the Journall and 'Bajay Dinko usko' in a side note (*PJ* 924). The Mandinka term *dingo* 'child' added to a name means either 'lesser, junior' or 'child (of the place)'. The side note asserts that the man was 'called by the name of his Countrey' and it seems likely that 'Biagge/Bajay' is Hodge's 'Badjady', which in turn seems to represent modern Pajady, an ethnic group of the region south of the confluence, known in writings on Portuguese Guinea as 'Pajadincas' (Teixeira da Mota, *Mar*, pp. 309 ff.). See also *PJ*, p. 192, n. 5; Itinerary (Appendix A), p. 219, n. 3. Jobson's is the earliest reference to this ethnonym.

with us, & is from the mouth of the River;[1] their familiaritie did assure us that they had commerce together,[2] which was an encourager to the confidence I have of the rivers continuance, and that they have the use of canoes above, to ferry over, in regard in this place where we were, although it was so shallow they might wade over, they were afeard to attempt it, dreading their neighbout [*sic:* neighbour] Bombo [*sc.* crocodile] would bee in the /96/ way and hinder their passage: leaving them therefore safe, on their one [own] side, with a faithful expectation of our second returne, which we promised, and they earnestly desired, I crost the river to our first acquaintance; to whom was come downe the King of the countrey, called by the name of the King of *Jelicot*;[3] who was the immediate King of that land whereon wee kept our trade; I went on shore, intreating him to come into our boate, and there we dranke: I gave him a present which he accepted, and brought him likewise ashore, where he had houses built for himselfe and his wives, on whom also some thing must bee bestowed; they brought with them commodities to barter, in doing of which, and looking upon us, they continued foure dayes.

These people expect our returne.

The King comes unto us.

Each night after wee had supt aboard our boate, where *Buckor Sano* did ever eate with me, nor during the w[h]ole time of my stay there, did hee faile mee one meale: notwithstanding I did divers times send both fish and fowle unto his wives, so did I likewise unto the King while hee remained by us, which wee tooke with our hookes, and kild with our gunne. Our manner was to go on shoare, and either at the Kings house, or *Buckor Sanos*, before the doore [,] fires being made, and mats spread to sit downe upon; the Fidlers plaide on their Musicke, and in dawncing and singing, and sometimes the men marching with their bowes and arrowes to shew their warlike exercises we spent great part of the evening, wherein with the small shew we could make, wee were not behind hand to let them heare our powder, and see our manner of marching ; wherein they tooke much pleasure, and commonly these evening sports, cost us a quart bottle of Aquavita, /97/ which made their tongues freer, and their bodies fuller of agility; wherein alwayes they did strive to shew a desire they had to give content unto us: I may not omit one principall manifestation: some three nights before we had finished all our occasions in this place, *Buckor Sano* being aboard at supper, by way of desire, spake that hee might bee called by the name of the white mens *Alchade*,[4] in regard hee did performe the same manner of office for us; that is, to make bargaines, to deliver and receive, according to the trust reposed upon him: I tooke it very kindly hee would accept the name, and to invest him therein I tooke a string of Currall [coral], and another of Christall, and put about his necke: he had likewise a small chaine of silver put upon him; then causing three gunnes to bee shot off, with a cup of Rosasolis drunke

Buckor Sano did alwayes eate with us in the boate.

These exercises did commonly hold three houres in the night.

---

[1] Unfortunately Jobson does not name the new language, but the language spoken 'by the better sort' and 'from the mouth of the River' was undoubtedly Mandinka.

[2] Since some of the visitors from the southern side had previously been in commercial contact with those on the northern side they must have been acquainted with the river and its trade, perhaps weakening Jobson's previous assertion that none of them had ever seen whites before.

[3] Present-day Jalakoto/Dialacoto.

[4] See p. 123, n. 2 above.

THE DISCOVERY OF RIVER GAMBRA

<small>Buckor Sano made the white mens Alchade.</small>

unto him, with a great shot [shout] we cryed out, Alchade Buckor Sano, Alchade Buckor Sano, that all the shore did ring of us: hee presently calling to the shore, commaunded the Fidlers to meete him at the water side, and that his wives should bring him thither, all the nuts he had, (which be of great esteeme amongst them.) I brought him on shore with as much grace as wee could, as soone as he landed, the people being all come forth to meete him, he gave his nuts frankely, Distributing them about,[1] in *teste* [*sc.* Latin *in teste* 'in witness'] of

<small>The acknowledgment of his new title.</small>

his new honour, his musicke playing before him, he went up to the kings house, whom wee found sitting upon a matt without doores, with a fire of reeds before him, which is still maintained, by the light whereof they performe their ceremonies, and have use of no other candles: he placed me upon the matt by the King, and going himselfe a certaine distance off, made a speech unto him, the effect whereof was, that /98/ in regard we were staide, through want of water,

<small>His mediation to the King in our behalfes.</small>

from passing higher, and had setled our trade upon his land, the King would deale lovingly with us, which he would not onely acknowledge with much curtesie himselfe, but speake unto his King, whether we would have gone to give thankes unto him, all which speech he performed standing:[2] The King as hee

<small>The Kings answere.</small>

sat made answre [answer] unto him, that hee did very well like of us, and did desire much we should come againe, and that we should have free liberty to kill any beastes, as also to trade or travaile in any part of his countrey; upon his curteous answer, Buckor Sano kneeled downe upon both his knees, giving him thankes, and taking from about his necke, the string of Currall, and the string of

<small>Buckor Sanos gratification.</small>

Chrystall, which I had given unto him, sent them unto the King, as a courteous gratification in our behalfe, which hee received and put about his necke: Buckor

<small>The Kings acceptance and faire reply.</small>

Sano still remayning on his knees, to whom the King spake, that for his sake, and to shew his love to us, he would freely give us all that countrey we were in, and would deliver it unto him for us, as the great King had given it unto him:

<small>He gives us the Country.</small>

upon these words, Buckor Sano puld his shirt over his head, which he gave to one of his woemen that stood by, kneeling naked from the wast upward untill the Mary-bucks I had with me, and another that was with the King, had scrapt together a great quantity of dust, sand, and small gravell such as the ground in that very place afforded; whereupon Buckor Sano lying with his face downe flat along, they lightly covered him, from the head to the heele, when presently hee taking his hands full of the earth, throwes it this way, and that way, after a carelesse manner, every where round about him. /99/

Which done, and kneeling up againe, with his face towards us as we sat in the mid way between us, the Marybucks gathered together a round heape of the same dust, and with their hands made likewise of dust, a Circle, a small

---

[1] For the distribution of kola nuts, see notes p. 130, n. 2; p. 134, n. 4 above.

[2] In Buckor Sano's speech to the king, he seems to be apologizing that trading began before the English found and met the king, but he explains that they were prevented from travelling further by shallow water and sandbanks. The king proved amenable, and formally granted land to Jobson as a trading site, giving him freedom to trade and to travel, and to kill wild animals on his land, an interesting testimony to the legal rights of African rulers, not dissimilar to the corresponding rights of contemporary European rulers.

distance of, round about the heape, in which Circle, one of them did write with his finger, in the Character they use, so much as the Circle would containe: this performed, *Buckor Sano* came upon his knees, and hands to the heape, and stooping with his mouth, tooke a full mouthfull, which presently he spet, and spattered foorth againe, then taking both his hands full of the loose earth, hee came forward on his knees, and both our Marybuckes, with their handsfull in the like manner, and threw the earth into my lappe: which done, they presently rose, and two women came with loose cloathes, therewith fanning, cooling, and wiping the body of *Buckor Sano*,[1] who retiring himselfe into his house, put on his best cloathes, arming himselfe with his bowe and arrowes, and some forty more in the same manner following him, came againe before us, every man an arrowe or two, ready in his bowe, marcht round about us, twice or thrice: which done, *Buckor Sano* came to mee, and turning his face from me, kneeled downe upon his right knee, putting his legge betweene mine as I sat, which his body seemed to shadow mine, presenting his bowe, and drawing his arrowe up, signifying, that so he would fight, and oppose his body, in defence of mine: then rising up, he gave away his bowe, and arrowes to another, and sat downe by me, so all the rest as their turnes came, kneeling either before mee, or him, performed the like action, and this was the manner of giving us the /100/ Countrey, and delivering a full possession; which bounteous gift, and great solemnitie, could not require lesse, then two or three bottles of our best liquor, which was accordingly performed, & set them into their ceremonious discourses, wherein because the night was farre spent, wee were willing to leave them, and betoke our selves to our owne Castle, beeing little the richer, for all that great gift, and spatious scope of land, we were in that ample manner indowed withall.

The next day amongst many that came unto the place, we did observe some, who were markt under both their eyes, with three blewe stroakes, resembling stripes, all after one manner, demaunding therefore what they were, it was answered a people who belonged to a King further up in the Countrey, and by those markes they were distinguished and knowne from other people:[2] This day likewise came unto us, an ancient Marybucke, taking acquaintance of our people, to whom he was very well knowne, I enquired what hee was, and they told me, it was a Marybucke, who did use to Ferambra,[3] and was acquainted with *George Thompson*, and that he had a wife in Ferambras towne:[4] which was within three miles of *Tobabo Conda*, the place

A strange recemoney [sic: ceremony] in takeing of possession.

The possession given unto me.

A great protestation of defence.

A people markt in the face.

Observe this Mary-bucke.

---

[1] For another ceremonial gesture involving the throwing of earth, see Part II below: Fernandes, f. 108. A large cloth (Mandinka *faanoo*) is used by young women and girls to fan elders, both to create a breeze and to drive off mosquitoes.

[2] Three vertical stripes on the face are characteristic of adult males among the Bambara, a people nowadays occupying a homeland north and NE of the Mandinka, mainly within the modern state of Mali, and beginning about 200 miles from Jobson's location on River Gambia.

[3] 'Nany' = Niani, a kingdom (in recent administrations a chiefdom), on the north side of River Gambia, below Wuli. The important information in the sidenote (overleaf) does not appear elsewhere in *GT* and is not in *PJ*.

[4] It was probably not uncommon for such itinerant Jaxanke to have a wife and household in several of the towns they regularly visited.

Ferambra was Lord of his Country, and when the Portingals had got the King of Nany to send horsmen to kill Thompson and his small company, hee did preserve them, and put himselfe & Country in armes for their defence.

The marybucks first tale.

where our habitation upon the land was: and *George Thompson* did use to lye at this *Ferambras* house, who had shewed himselfe a faithfull friend, in time of neede unto him, and his company, and likewise confirmed the same unto me, by divers Curtesies, which past betwixt us, during my time of trade in the river: I askt the Marybucke from whence he came, and he answered from Jaye, a towne nine dayes travell higher in the /101/ countrey,[1] whether [whither] he was sent by *George Thompson*,[2] and seemed to lament much for his death, hee tolde us there was great store of gold, at that place, and that the Arabecks, which are the Moores of Barbary, came thither: we askt him, if he would undertake to carry fome [*sic:* some] of us thither, he answered, yes, if wee were but past these bad people, pointing to our consorts ashore, we told him they were a good people, and very loving to us, shaking his head, he said *immane, immane,* which signifies naught, or the thing they like not:[3] hee had two or three other consorts with him, and [they] made hast to be gone, keeping themselves severed, from the other people: during their time, of staye, he told us, he was going downe to *Ferambras*, and by him I sent letters to our Marchants below, which were very orderly delivered, by reason he was in some feare, and came in the night aboard of us, making such hast away, we had little conference; but meeting of him afterwards in my trade at *Setico*, I

His second discourse.

had a full and large discourse, and received the intelligence which I will now recite: Hee told me that not farre from Jaye, there were a people who would not be seene, and that the salt was carryed unto them, and how the Arabecks, had all their gold from them, although they did never see them: demanding the cause, hee made a signe unto his lippe, and could receive no farther answer,[4] he saide likewise, if we could have gone further up the River with our boate, many people would have come unto us, and brought great store of gold; and to the place where we did staye, hee saide, many people were comming downe, both from Jaye, and many other townes, but that they were sent backe againe, by /102/ such as had gone from us, and told our salt was endded, we askt him, as we had done *Buckor Sano* and others, if they had a

---

[1] Probably not the town of Jenne on the upper River Niger, some 600 miles east of Jobson's location, this being too far for only 'nine dayes travell'; and therefore perhaps Diakha, in Bambouk, an auriferous district about 100 miles NE of the location. However, it was suggested earlier (p. 87, n. 4 above) that 'Mumbar' might represent Jenne and this was stated to be only six days journey from the river. The Journall extracts appear to contradict *GT* in some respects: the marabout was 'borne in Jaye' and not stated to come from there, while the text can be read as indicating that he arrived in fact from Tambakunda ('Combacunda').

[2] If Thompson was indeed only in Tinda for 'not many hours', *GT* 7, it is unlikely that he had time to organize the sending of this man to 'Jaye', and so must have met him and made the arrangement when the man was at Ferambra's town. It is of note that the marabout was travelling between 'Jaye' and this locality, a considerable distance, and was shortly afterwards again met (as Jobson goes on to say) at Sutuko, an intermediate point.

[3] The expression is not identified, but Mandinka *mang/man* is a negative indicator.

[4] This is a reference to the famous, albeit possibly mythical 'silent trade' (see p. 83, n. 3 above), which is expounded in more detail in the next two pages. Unlike the further details (see p. 149, n. 4 below), the present reference represents the story as it appeared in Arabic literature and tradition and therefore may be accepted as having been what Jobson was told by the marabout.

THE GOLDEN TRADE [102/3]

towne called *Tombutto*, that name they knew not, but a towne called *Tomboconda*, they saide was neerer where we were then Fay:[1] the pronuntiation of which two places, as *Tomboconda* for *Tombutto*, and *Fay* for *Gago*, may if there be such two places, carry some resemblance,[2] wherein I stand to put forth this question, who should hee be that directs these names, being it is apparent never white men either by land or water were up to this countrey so farre but us;[3] this man likewise tolde us of the houses covered with gold, and many strong incouragements to invite us on, especially to goe further up the River.

Our opinion cercerning [*sic*: concerning] Trombutto [*sic*: Tombutto] and Gago.

More incouragment to go further up the River.

To this let me now set downe, although not able to name particular authours, what is the generall report for the Moore of *Barbary* his trade. That it is certaine when they come up into the country where they have their chiefest trade; they do observe one set time and day, to be at a certaine place, whereas houses are appointed for them, wherein they finde no body, nor have sight of any persons. At this place they doe unlade their commodities; and laying their salt in severall heapes, and likewise setting their beades, bracelets, and any other commodities in parcells together, they depart, and remaine away a whole day, in which day comes the people they trade withall, and to each severall layes downe a proportion of gold, as he valewes it, and leaving both the commodity and the gold goes his wayes: the Merchant returning againe, as he accepts of the bargaine, takes away the gold and lets the commodity /103/ remaine, or if he finde there is to[o] little left, divides his commodity into another part, for which he will have more, at the unknowne peoples returne, they take to themselves, where they see the gold is gone, and either lay more gold, or take away what was laid before, and remaines in suspence: so that at the Marchants third time, his bargaine is finished, for either he findes more gold, or the first taken away, and his commodity left, and this [? thus] it is saide, they have a just manner of trading and never see one another: to which is added, that the reason why these people will not be seene, is for that they are naturally borne, with their lower lippe of that greatnesse, it turnes againe, and covers the greater part of their bosome, and remaines with that rawnesse on the side that hangs downe, that through occasion of the Sunnes extreame heate, it is subject to putrifaction, so as they have no means to preserve themselves, but by continuall casting salt upon it, and this is the reason, salt is so pretious amongst them: their countrey beeing so farre up in the land, naturally yeeldes none.[4] And this

The manner of merchandising without speech or sight one of the other.

The report of the people with the great lippe.

---

[1] Jobson was most probably referring to Timbuktu, far away to the north, but the local people were naming Tambakunda, only some 30 miles from Jobson's present location.

[2] 'Fay' is a misprint for Jay(e) (as given a few lines above), and this makes more sense of Jobson's claim of a pronunciation resemblance between Gago and 'Jay'.

[3] Except, apparently, Thompson.

[4] Certain West African peoples, notably the Lobi and Sara, in recent times enlarged their lips and wore labrets in them. But none of the known peoples live in western Guinea and there is no record of the practice near River Gambia. In fact, the reference to enlarged lips was added to earlier accounts of the silent trade in Arabic sources by Cadamosto, and the deformity then appeared in all European accounts of the trade based on Cadamosto (Farias, 'Silent Trade', p. 13). Since it is unlikely that these European sources had affected the Arabic version of the myth, Jobson's account in this paragraph was not based on anything he was told in River Gambia, but, as he himself states,

carryes some appearance by what wee are able to say, for first out of our owne experience wee find that these people, who trade with us for our salt, have for their owne occasions little or no use thereof, and being demanded what they doe with it, they doe not deny to tell us they carry it up further into the countrey, unto another people, to whom they do sell and vent the same: which still are good inducements to follow a further search, and therein to neglect no time, but diligently to follow the occasions already obtained, which is the love of these people, who were held so dangerous, who were so earnest for our /104/ comming againe, and to hold a certaine course of trade with them, which by us was faithfully promised, wherein the Adventurers have this strong encouragement, that if we should attaine no further then amongst these people the gaine is knowne to bee great upon our exchanges, and fitted accordingly to carry a good proportion will yeeld a valuable returne to the full satisfaction a reasonable desire may aime at: but if it please God to prosper the discovery, and that we meete with any place of habitation againe by the river side, which may bee a convenient seate of residence to maintaine a setled trade in, there can be no opposition to gainesay it must be the greatest and gainfullest trade, considering the short returne that ever fell into our little Iland: which commending to your worthy consideration I will conclude my discourse of trading, with the curteous farewell that past betwixt our blacke Marchant *Buckor Sano* and us. They were earnest we would give a name unto the place wee traded in, that might remaine as a memoriall of of [sic] our being there, I called it by the name of Saint *Johns* Mart,[1] which they repeated diverse times over to be perfect in: and when our salt was gone, seeing us hast away, which wee were carefull to doe, fearing lesse water in the river, hee in his affection would needes desire to goe some little way along in our great boate, passing about a mile with us, with curteous embracings we parted, shooting off three gunnes for his farewell, not forgetting the drinking of three or foure cuppes, and so put him on shore: from whence by holding up his armes, he againe saluted us, and with his hopefull expectation to see us there againe, I will let him rest, and /105/ according to my course propounded fall uppon a merrier company, which is their Juddies, or as wee may terme them, Fidlers of the Countrey, neither the musike they make or instruments they play uppon, deserving to have a better title: and may sort also reasonable well to the company, because at all especiall meetings their divell *Ho-re*[2] makes [? ,] on the relation whereof I proceed unto.

There is, without doubt, no people on the earth more naturally affected to the sound of musicke then these people; which the principall persons do hold as an ornament of their state, so as when wee come to see them, their

---

on 'the general report', although he is not able 'to name particular authors', presumably because he could not remember which printed sources he had read.

[1] Most likely so named after the ship but perhaps with indirect reference to the name of the Governor of the Company.

[2] For *Hore*, see *GT* 105–6 below.

musicke will seldome be wanting, wherein they have a perfect resemblance to the Irish Rimer sitting in the same maner as they doe upon the ground, somewhat remote from the company; and as they use singing of Songs unto their musicke, the ground and effect whereof is the rehearsall of the auncient stocke of the King, exalting his antientry, and recounting over all the worthy and famous acts by him or them hath been atchieved: singing likewise *extempore* upon any occasion is offered, whereby the principall may bee pleased; wherein diverse times they will not forget in our presence to sing in the praise of us white men, for which he will expect from us some manner of gratification. Also, if at any time the Kings or principall persons come unto us trading in the River, they will have their musicke playing before them, and will follow in order after their manner, presenting a shew of State. They have little varietie of instruments, that which is most common in use, is made of a great gourd, and a necke thereunto /106/ fastned, resembling, in some sort, our Bandora; but they have no manner of fret, and the strings they are either such as the place yeeldes, or their invention can attaine to make, being very unapt to yeeld a sweete and musicall sound, notwithstanding with pinnes they winde and bring to agree in tunable notes, having not above six strings upon their greatest instrument:[1] In consortship with this they have many times another who playes upon a little drumme which he holds under his left arme, and with a crooked sticke in his right hand, and his naked fingers on the left he strikes the drumme,[2] & with his mouth gaping open, makes a rude noyse, resembling much the manner and countenance of those kinde of distressed people which amongst us are called Changelings;[3] I do the rather recite this that it may please you to marke, what opinion the people have of the men of this profession, and how they dispose of them after they are dead; but first I would acquaint you of their most principall instrument, which is called Ballards[,] made to stand a foot above the ground, hollow under, and hath uppon the top some seventeene woodden keyes standing like the Organ, upon which hee that playes sitting upon the ground, just against the middle of the instrument, strikes with a sticke in either hand, about a foote long, at the end whereof is made fast a round ball, covered with some soft stuffe, to avoyd the clattering noyse the bare stickes would make: and upon either arme hee hath great rings of Iron: out of which are wrought pretty hansomly smaller Irons to stand out, who hold upon them smaller rings and juggling toyes, which as hee stirreth his /107/ armes, makes a kinde of musicall sound agreeing to their barbarous content: the sound that proceeds

The fashion of the Irish Rimer.

Upon this instrument only they play with their fingers.

A strange consortship.

Their chiefest instrument.

---

[1] Jobson provides an accurate account of the griot's skills and functions. The 'bandora' was a European four- or six-stringed instrument resembling a guitar or lute. The six-stringed instrument seen by Jobson was probably a *bolongo* (*bolombato*) – see Plate XI. The better-known *kora* with twenty-one strings is a more recent development. The *bolongo* was used to inspire warriors in battle.

[2] The underarm drum is a *tama* (Wolof and Mandinka), a widespread type of drum in West Africa – see Plate XIII.

[3] 'Changelings' = creatures believed to have been substituted by the fairies for certain babies, to explain why these babies did not develop normally, the term coming to mean 'half-witted'.

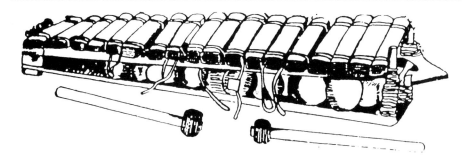

Mandinka xylophone

from this instrument is worth the observing, for we can heare it a good English mile, the making of this instrument being one of the most ingenious things amongst them: for to every one of these keyes there belongs a small Iron the bignesse of a quill, and is a foote long, the breadth of the instrument, upon which hangs two gourdes under the hollow, like bottles, who receives the sound, and returnes it againe with that extraordinary loudnesse;[1] there are not many of these, as we can perceive, because they are not common, but when they doe come to any place, the resort unto them is to be admired; for both day and night, more especially all the night the people continue daunsing, untill he that playes be quite tyred out; the most desirous of dancing are the women, who dance without men, and but one alone, with crooked knees and bended bodies they foot it nimbly, while the standers by seeme to grace the dancer, by clapping their hands together after the manner of keeping time;[2] and when the men dance they doe it with their swords naked in their hands, with which they use some action,[3] and both men and women when they have ended their first dance, do give somewhat unto the player: whereby they are held and esteemed amongst them to be rich; and their wives have more Cristall blew stones and beades about them, then the Kings wives: but if there be any licentious libertie, it is unto these women, whose outward carriage is such wee may well conceit it: and this one

The manner of this instrument

Their manner of daunsing.

Their Fidlers rich.

---

[1] The instrument Jobson describes at length, and accurately, is the local xylophone, known as *baloo* or *balafongo* in Mandinka, and as *balafon* in European writings. For a good description in the 1680s (said to be of a *bala* played by a *balafon*, the wrong term), see Cultru, *La Courbe*, pp. 195–6; and for an early illustration of a *balafon*, see F. Froger, *Relation du voyage de Mr. de Gennes ...*, Paris, 1698, opp. 45 (repeated in John Barbot, *A Description of the Coast of North and South Guinea*, London, 1732, Plate 3). The area in the modern Gambia in which the instrument is today most frequently found is Wuli, approximately where Jobson encountered it, the Suso family in Kanape being particularly renowned. For a detailed modern study, with a description of the instrument, traditions about its origin, and the repertoire of the musicians who play it, see Lynne Jessup, *The Mandinka Balafon: an Introduction with Notation for Teaching*, La Mesa [California], 1983.

[2] For a modern account of drumming and women dancing, see Roderic Knight, 'Mandinka Drumming', *African Arts*, 8/4, 1974, pp. 24–35.

[3] Men have a special dance performed at circumcision ceremonies when they pass a cutlass back and forwards across their chests without cutting themselves (as observed by DPG at Kerewaan in 1947).

especiall note, howsoever the people affect /108/ musicke, yet so basely doe they esteeme of the player, that when any of them die, they doe not vouchsafe them buriall, as other people have; but set his dead corps upright in a hollow tree, where hee is left to consume:[1] when they have beene demanded a reason for so doing, they will answer, they are a people, who have always a familiar conversation with their divell *Ho-re*: and therefore they doe so dispose of them: which opinion of theirs caused us to neglect and especially in their hearing to play upon any Lute or Instrument which some of us for our private exercise did carry with us, in regard if they had hapned to see us, they would in a manner of scorne say, hee that played was a Juddy: The greatest resort of people, with the most aboundance of these Juddies, is at their times of Cricumcision [*sic:* Circumcision], wherein they observe one due season, and for that I desire heerein to give a full relation, I will follow my discourse with what I saw, and as an eye-witnes am able to deliver.

They are basely esteemed of, and being dead are not buried.

I have set downe before, where I report the manner of our going up the river, that I carried with me foure blackes; whereof the one was a boy, or young youth, whom I call by the name of *Samgully*; who in reguard of his continuance with *George Tompson*, and after him with the rest of our company, had learnd to speake pretty English, and withall had taken such an affection towards us; that he did seeme even hartely to neglect father and mother, and his owne home, in his desires to follow us; he was about the age of 17. yeeres, a straight youth and of a handsome growth; yet was he not circumcised: howbeit hee should have beene the yeere /109/ before, but his absence with the white people, which was some of our company, when the time of circumcision came, was the only cause he mist cutting them [*sic:* then]: and this yeere hee was to bee circumcised, or else there was some great penalty to light upon his friends, or danger to himselfe; which appeareth in their earnestnesse to keepe him from going up with us; notwithstanding, hee was stollen beyond the towne his friends dwelt in with us, as farre as our boate would goe in two tides; and there overtooke us his mother: who on the shore made grievous moane to have him sent backe, the boy had spide her, and hid himselfe in the boate, bidding us say that he was gone backe, and albeit her moane was great, because shee saide, he would be absent againe in the time of circumcision, which would bee the next Moone; and if wee would not put him on shore, she would throw her selfe from the banke into the river; the boy lying along in the boate, said, she will not drowne, shee will not drowne, let us bee gone; and alongst he went with us: It was the eight day of *January* when his mother made this moane, and the ninth day of *February* after we came by that place againe, and that evening, as the Sunne set, came to an anker at the port that belonged to the towne where the boyes friends dwelt,

The affection of Samgulley our blacke boy unto us.

---

[1] Griots were buried in the hollows of baobab trees, particularly in Siin and Saalum, among the Wolof and Serer. See the references to *Judeus* ('Jews') in Almada, *Tratado breve*, ch. 4/9–10; also see Part II below: Lemos Coelho, 1669, 31, confirming what Jobson writes.

|[109/11]| THE DISCOVERY OF RIVER GAMBRA

This Bo John was brother to Ferambra.

which was called *Boo Johns* towne, a man whom we did well affect; the towne stood some mile from the water side[;] to carry the boy home, and refresh our selves, wee were willing to walke on shore.

So taking our Chirurgion, and one more of my consorts, with our blacke Alchade, who wee hired from this town, wherein his mother likewise dwelt, /110/ and our blacke boy, ashore we went, the bakne [*sic:* banke] was high from the river; which the boy first gat up, where presently ascended, he began to leape and sing, making great shewes of joy, holding up his hand, and pointing towards the towne, which as I said was a mile from us; but when wee likewise ascended, wee heard a great noyce of musicke and shooting [shouting], where at the boy so much rejoyced, and said it was the cutting of Prickes; for so hee cald it, and that hee was come time enough; we walkt towards the towne, and as the evening went in came thither, I had an intent to have gone to the Maister of the towne his house; but my Marybucke told mee, hee had a world of strangers, and was earnest with mee to goe to his mothers house; we were also to passe by the house where our blacke boies father and mother dwelt: the father in regard he was blind, and kept house, we saw not; but his mother being within, and hearing one call her sons name, came forth and met him, and presently turning her head to the side of the house, fell into a bitter weeping, calling only upon his name, *Samgulley, Samgulley*; I would have had the boy alongst where I lay, but hee was taken from us, and not suffered to goe; howbeit I charged him he shold not be cut, which as they told us, was to be done in the morning, untill I came to see him, which he promised, and so we departed to our lodging: at which place likewise was many people, and much musicke; but after a while that wee had beene there, they all quitted the place, shewing a kind of modesttie, not to disturbe us. There was no housing, nor dwellings, but was full of people; nay likewise, under every shady and convenient /111/ tree, there was great fyres, whereas [? where at] there was [? people], their pots a seething, and their victuals adressing, and also their mats laid, to take up their lodging, sorting themselves together in great companies, and in most places, having musicke, drumming, and dauncing; making such a noyse and din, as might well proceede from such kinde of Actors: and amongst them likewise they had commerce, one thing for another, so as it had a manner of resemblance to our fayres here in England; neither was there want of any manner of provision, for as much as all kind of people that came thither, brought some manner of sustenance with them, and the people of the place, did provide and reserve themselves against this time, so as I may well say it had a right resemblance to our countrey martes:

Amongst the rest of these dispersed companies; I tooke specially notice of one, who stood more remote, and was closde and severed in, under the shady trees, which [*sic:* with] reedes, and bowes set up together like a hedge, from whence proceeded, a greater noyse of voyces; as also drumming, and thumping, mor[e] clamorosly: demaunding what it meant, I was answered,

The feast of their Circumcision.

Samgulley taken from us to be circumsised.

The great resort to this solemnity.

154

in that place remained those youthes that were cut, and they were to continue until such time as they were recovered of their sorenes, and that the greatnesse of the noyse did come from those people who kept them company, which were the yonger sort of people, above their age; who had already past, and received their circumcision:[1] I went likewise that night, after we had supt, to the maister of the townes house; who had sent unto mee to mend my supper, a brace of Part[r]idges, and finding /112/ there the Ballards, or best musicke, and the younger sort of women gathered together beheld their dancing, and for that they might see we had such pleasures amongst us; I tooke one of them by the hand and daunced with her, whereof they gave testimony of great gladness, inviting the rest of my company to do the like: *Boo John,* the maister of the place, excusing himselfe that we lay not at his house, in regard of much company and noyse: but more especially, because of one his wives was lately delivered of a child, unto whom he carried me within a house by her selfe, where she lay after their fashion upon a mat handsomely; I gave unto the mother for the childe, a few poore beades, which were very thankfully taken, and he said, if it had beene a man child, it should have had one of our companies name, with whome he had beene longest acquainted; but saith he, my wife *Dowry*[2] is with child, and if shee bring a man child, it shall carry your name; for so shee earnestly desires; these familiarities past betwixt us, after which wee betooke our selves to our severall lodgings, and were nothing frighted with the roaring cry of their divell; who at these ceremonious meetings so soone as evening comes is conversant by his roaring voyce amongst them, and so continues all, or the most part of the night, whereof I shall presently give you relation: But first, I must conclude of their circumcision; for the sight whereof, as even now I told you, we did reserve our selves to receive advertisement of our Samgulleyes cutting, which was to be done in the morning. And accordingly, the Sunne some two houres high, we had a messenger came to entreate we would send him a white cloath, and that hee would pray us to come /113/ & see him. As soon as we came, he was broght forth into the open field between the houses, and the place where they remained, who were cut the day before [:] they had taken away his clothes, they broght him ashore in; which was shirt, breeches, and a cap of stript stuffe, after the bravest fashion of the country, and onely brought him with a white cloath close about him; whereas we did expect some great ceremony after a religious manner to be performed: He was first set downe upon a little mole-hill,[3] divers people comming foorth to see him, amongst the which were most women, who stood directly a little distance off, looking on: the maister of the towne was likewise there, and three of us;

They that were cut, kept all together.

The curtesie and mirth that past betwixt us.

women looke upon the circumcision.

---

[1] For an earlier and briefer reference to a male initiation ceremony, see Part II below: Valentim Fernandes, f. 109; and for male circumcision (in Senegambia), see Almada, 4/17. Today newly circumcised boys are lodged in a 'shed' or enclosure, and looked after by youths who have been circumcised on earlier occasions: see Plates VII, VIII.
[2] The name is unidentified.
[3] Presumably a small termite mound.

amongst which our Chirurgion was one to comfort him not to feare: hee was very confident, entreating me to lay my hand upon his shoulder; from amongst the blackes came foorth an ordinary man, with a short knife in his hand, which he whetted as he came, like one of our butchers unto a beast; and causing the boy to stand up, he tooke of his cloath, giving it to a stander by to hold, so as he was starke naked, and set his hands upon his sides, being neither bound nor held: howbeit there were some by, who offered to hold his armes, but because hee promised not to move, they let him alone, the executioner taking hold of his members [*sic:* member], drawing the skinne over very farre, as we conceived, cut him largly, and had three severall cuts afore hee had done, whereat the boy shrunke very little; insomuch as the maister of the towne who stood by told us, hee had very seldome seen any abide it with so great a courage: to our thinking it was exceeding feareful and full of terror, /114/ insomuch as I told the doer in a very angry manner he had utterly spoyled him; when he askt wherein, I replyed, in cutting him so deepe: His answere was, it is so much the better for him, and without any curiosity taking up his cloath shewed his owne members, that it might appeare he was cut as farre; howbeit my distaste was such upon him, that I could not yeeld to give him any thing in the way of gratuitie to wash his hands withall, and as the manner of the countrey is to doe by such as are friends, to the party circumcised [*sc.* friends to the party circumcised]: the thing performed, the boyes white cloath was cast over him, and by two men, who held his armes, he was hurried apace to the same quarter, where the other that were cut remained: We made first a request that they would let us goe along to the place with him, and were going with some of the people; but presently in hast overtooke us four ancient men, who did not onely stay our going, but made shew of much displeasure to such as were going with us, and would by no meanes suffer wee should come amongst them: then we desired we might have the boy away with us telling them wee had better meanes to cure him, and to make him sooner well, then they had, shewing our Chirurgion unto them, who they knew had healed wounds and sores amongst them; but wee could not prevaile, by the interposing of these auncient men some of the rest seeming to consent unto us: so as wee were there driven to leave our boy, who amongst the rest of his consorts, had without doubt no other chirurgery to cure their tender maladies, but onely to attend the expectation of time, who by the helpe of their youth & nature might weare it out, which appears the rather to us, in regard at these times, there is unto these /115/ youthes allowed a certaine licentious liberty, whereby they may steale and take away peoples hennes, or powltry; nay from the Fulbies, a biefe or cattle to eate and banquet withall amongst themselves without any offence to the lawes, or government of the countrey; which at other times is strict in that behalfe, thereby animating, and encouraging their spirits to more alacritie, and according to the condition of their wanton age by these stollen delights to draw them more willingly from the thing, and make the time of their recovery lesse tedious to themselves, and discourageable to others.[1] And if I might be worthy to deliver my opinion, considering this their circumcision, as I have carefully observed I should conclude, it were done of meere necessitie, as a morrall law for the preservation of their lives and healthes, and so found out by their precedent auncestors, and by strict observations laid peremptorily upon them, wherein I should submit my selfe in the account I could give to more able judgement: onely this you may please to note, it is done without any religious ceremony, and the word in their language is expressed unto us by no other signification, then cutting of prickes;[2] and this is done in certaine bigger townes of the countrey, whether [whither] the smaller townes and habitations make their resorts, bringing their youth to be all cut together.

And to make up the number at all these meetings, there is one sure card that never failes, which is their roaring devill, that before I spake of, whose attendance may seeme to keep the youth in awe, and he is called by the name of *Ho-re*,[3] whose strange report I /116/ proceede unto: There is at all these meetings, some distance of from the place, heard the noyse of a roaring

The discourse of their divell Ho-re.

---

[1] While Islamization further sanctified male circumcision, it appears to have been a practice originating in pre-Islamic times. As far as can be judged from later 20th-century practice, Jobson covers the essential points of Gambian circumcision ceremonies (though there are minor variations in different regions): thus, the taking out of the boys in the early morning for the operation; the wearing of a white cloth; the bravery expected during the actual operation, which is generally carried out by a blacksmith, not a religious official; the separation of the boys from the village during the healing process; the reluctance of the elders to allow any non-initiate to visit the circumcision shed; the silence demanded of the initiates when outside the shed (see *GT* 117 below). In 1907 it was noted that 'during the time of confinement, they go out into the bush to bring in grass, wood, etc., but always with an escort, neither are they allowed to speak to any one except the attendants', and that they are later allowed to carry out raids to seize chickens, and other animals (H. Lloyd Pryce, 'The Laws and Customs of the Mandingos of the North Bank Territory of the Gambia Protectorate', unpublished report, 1907, Gambian National Archives). As Jobson goes on to state (*GT* 116), it is believed that the initiates die and are re-born, having been swallowed by a monster, who needs to be fed vast quantities of food, lest harm come to the boys. Some groups hold a ceremony at night which represents the swallowing. There are various mechanisms for producing the alarming sounds which represent the voice of the spirits.

[2] Medical opinion remains divided about the advantages and disadvantages of male circumcision (debated in 1997 issues of the electronic journal *Circumcision*). That in the past it was practised for medical rather than social reasons – marking either distinctiveness of the community, e.g. among Semitic peoples, or of adulthood by traumatic initiation – is a view only latterly put forward by Africans. The common view expressed by village West Africans is that it is a social ritual sanctified by ancestral command, without any other precise rationalization. Jobson's adoption of a modern view is curious. He may well have considered circumcision previously, in respect of the Jews and its mention in both the Old and New Testaments, but his plumping for medical reasons in Africa seems original. His reference to 'the preservation of their lives and healthes' is not further explained. A modern medical view is that male circumcision reduces the (small) risk of cancer of the penis, but it seems unlikely that this view had emerged in Jobson's day. However, an inexplicit view apparently not uncommon among Europeans, to the effect that circumcision both increases the pleasure of intercourse and the likelihood of conception – a view in part derived from a baleful envy of Jews – may well have existed in earlier times and influenced Jobson.

[3] The name associated with the sound of the bullroarer, said to be the voice of a spirit. (Perhaps inspired by reading about Jobson's interest in village spirits, a century later another English observer gave to a Gambian spirit, visible on ceremonial occasions in the form of a masked figure uttering in a strange language, the name of 'Mumbo-Jumbo' [from Mandinka *mama Jombo* 'grandfather Jombo'] (Moore, *Inland Parts*, pp. 40, 116–18), a term which passed into the English language.)

voice, resembling the greatest base of a mans voice; when we demand of them what it is, they will answer, with a kinde of feare, it is Ho-re, and then describe him to be a fearefull spirit, that none may come neere, without danger of being destroyde, carryed away, or torne in pieces: there is at all their meetings, upon the first notice of his voice, a preparation for him of all manner of victuals, they have amongst them, every one imparting some-what, all which is carryed towards the voyce, and there under a tree set downe, and within small time, bee it of what quantitie soever; it will be found devoured, and not so much as a bone to bee seene, uneaten, or left behind, and if they be not ready forthwith to carry him such provisions, as shall content him, some of their uncircumcised sons are instantly taken away; females he meddles not with,[1] and saide to remaine in Ho-reyes belly, some of them nine or tenne daies, from whence they must be redeemed with some belly provision: and it is strange to heare, how confidently they will report unto you, that they have beene carryed away, and beene abiding there: wherein this is observed, that looke how many dayes he hath beene kept away, or remaining, as they say in Ho-reyes belly, so many dayes after they returne, it must be, before they will, or dare open their mouths, to speake a word. For confirmation of which, this I have seene: as I walkt one day into the countrey from our dwelling to *Ferambras* house, distant some foure mile, in the way we were to passe through a towne of the Fulbies, aamong the people that lookt upon us, I was shewed a /117/ youth of some eighteene yeares of age, who they said, came but the night before out of Ho-reyes belly: I went towards him, and urged him to speake unto me, but still he went backe from mee, and kept his finger before his mouth, and notwithstanding I made what meanes I could, by pulling and pinching of him, and more to terrifie him, making proffers with a false fyer to shute at him, beeing naturally exceeding fearefull of our gunnes, I could not prevaile, neither make him open his mouth: notwithstanding afterwards, the same fellow did often come, and have commerce amongst us: nay our people, who were lying, and dwelling in the countrey, had beene at severall times frighted with the voyce of this Ho-rey, for having staide in their fowling or being abroade, untill night hath overtaken them, in their comming home, as they have saide, they have heard the voyce of Ho-re, as they might conceive, some mile from them, and before they could passe tenne steppes, hee hath seemed to be in their very backes, with fright whereof, maintained by their imagination, of [? as] their report went of him, they have not, without a gastly dread, recovered home: unto which place of dwelling, he never was so bold to make any attempt: and verily my opinion is, that it is onely some illusion, either by the Marybuckes, or among the elder sort, to forme and keepe in obedience those younger sort:[2] for better approbation of what I suppose, I will crave the patience, to set downe what I observed at the circumcision of our blacke boy: The nights

---

[1] For female circumcision, see the Introduction, p. 42, n. 8.
[2] A correct assessment.

were very light, the Moone being then about the full towards midnight, comming from *Bo Johns* house to the place at Faye,¹ Ho-reyes voyce was wondrous /118/ busie, as it seemed to me, not farre of, I spake unto my consorts, we would secretly take our armes, and steale downe, to see what it was, one of our three was backeward and unwilling, whereby it came to passe, our Marybucke understood what we intended, who came earnestly unto me, intreating, I would give over that dangerous attempt, saying, I could not finde him, for one cry would be hard by me, and another instantly beyond the river, which was a mile of, and there was great danger, he would carry me into the River with him: when hee perceived, he could not alter my resolution, he held mee by the arme, and pointing to a blacke, not farre from mee, held downe his head. I went to that man, being a very lusty fellow, to speake unto him, whose voyce was growne so horse, by crying like Ho-re, he had no utterance, whereupon I returned to my Marybucke, and saide, there is one of your Devils; who with a smile went his way from me. How he was partly discovered.

But that the divell hath great recourse aamongst them, is without question, especially as I noted before, with the Rimers or Juddyes;² I will specifie one intelligence we had, and so leave him there amongst them: When wee came first up the River, we were uncertaine of our owne times, much lesse then any other, could fixe [? much less could any other fix our] houres of going, staying, or comming to a place: howbeit wee were to come to a towne called *Pompetane*, at which place dwelt a Portingall, called *Jasper Consalvos* [Gonçalves], who had a young kinsman with him, called *Marko*, but no women but blacks; this dwelling of his was the highest by many leagues of any Portugall in the River; and very faire quarter, [? as] ever past betwixt us,[:] we came /119/ to this place, the 14. of December, between eight and nine in the morning, and notwithstanding, the dwelling houses were somewhat remote from the river, we found standing upon the banke at the landing place this *Consalvos*, who in friendly sort saluted us, and carried us up to his housing, where presently we found ready a very good breakefast of hens, and other good refreshing, which he said, was provided for us: we seemed to marvell he should know of our comming thither, but after he told us, that the evening before, he was at another towne within the land, and had no meaning to come home, when as there came unto him a Juddy or Fidler, which dwelt in the towne with him, and did likewise shew us the man, who told him that Ho-re had acquainted him, that the next morning, and at such an houre, there would be so many white men at *Pompetane*, naming the number that were in our boates, and that there they would land: but to what An example of the divells converse with the Fidlers.

---

¹ The only mention of this placename on the river. On one interpretation of Jobson's obscure references, it may have been the name of the English middle river base.

² In Jobson's period, a general belief across western Europe in the active intervention of the devil in human affairs – as sometimes represented in witchcraft accusations – made it difficult for European observers to dissociate African rituals which seemed strange to them from a suspicion of diabolic origins. Jesuit missionaries in Sierra Leone *c.* 1610 shared Jobson's approach (P. E. H. Hair, 'Heretics, Slaves, and Witches – as seen by Guinea Jesuits *c.*1610', *Journal of Religion in Africa*, 28, 1998, pp. 131–44).

[119/21] THE DISCOVERY OF RIVER GAMBRA

The Divell could not tell the Portingall where [whether] we were friends or foes.

purpose, either to doe good or hurt, the Devil was ignorant; and upon this intelligence, I retyred my selfe, and came away hither to meete you, whereat wee seemed much to wonder, being altogether our selves uncertaine of any houre, in regard, we divers times went a shoare, and shooting at fowle and such like occasions divers times lighted upon us; notwithstanding the divells intelligence, we were no wayes discomforted, for he did confesse, hee was altogether ignorant of our intended actions, and the conclusion was, how by his intelligence, the worst hurt we had, was a better, and more readier breakefast.[1]

The trades or occupations they have in use, their painfull season of thunder and lightning, also what fruites & plants the Countrey yeeldes, and are growing there amongst them.

It followes, concerning what trades, or occupations is in use amongst them, whereof wee note only /120/ three the first and chiefest is the Ferraro or Smith,[2] who holds a good repute: notwithstanding, they have no Iron of their owne making, but what is brought unto them, whereof they have most needfull use, and neither may, or can live without it: for first of the Iron we bring unto them, they doe fashion and make all those short swords they were [wear], next the heads of their Assigies, or Javelings, as also the heads of their throwing darts, and the barbed heads of their shooting arrowes, which are covered over with their deadly poyson: in many of these the Smith doth shew a pretty kind of art and making: but the most needfull use amongst

The Smith.

them, is the toole or instrument, wherewith they till their ground: without use of which, they could hardly have their being, and therefore Iron, a principall commodity, that they doe call upon; at the lower part of the River, where the Portingall frequents, they have more for exchange then above, whereas we are upon a certeine trade: for wee cut our Iron of twelve ynches, and that is the proportion lookt after, and so high as it flowes, the Kings and Governors will call for that length: but passing above eight ynches will goe as friendly; which in either of them, is gaine enough, the returne even in the worst, yeelding tenne for one, carrying our yron in barres, we are inforced to make use of their Smiths, to cut it to the proportion, wee must use, and therefore sending for him, he comes to the water side, bringing his shop with him, that is his bellowes, and a small Anvill, which hee strikes into the ground under a shady tree, and onely of one kind of red wood, amongst

An excellent charcole to worke their Iron.

them, they can make artificiall Charcoale, which will give our Iron his true heate, /121/ as any seacoale, his boy blowing the bellowes, that lye on the ground, the nose of them, through a hard earth, made of purpose with

---

[1] An alternative scenario is this. The previous day the Portuguese heard from an African that the English intended to leave the next day in the morning, this most probably being no secret, and knowing the river the former could guess an estimated time of arrival. Despite Jobson's implied conviction that on the day news of the departure could not have travelled overland, given the twists in the river and the slow progress of the ship, broken by halts while members of the crew went ashore, it is most likely that this is actually what happened, the arrival of the messenger allowing the Portuguese to confirm and fine-tune his prediction. In some parts of Black Africa such news might have travelled by talking drums but Gambian drums only signal limited specific instructions, e.g. to dancers.

[2] 'Ferrero' = Portuguese *ferreiro* 'iron-worker, blacksmith'. The Mandinka term is *numoo*. It is evidence of Jobson's dependence on information obtained through Portuguese-speakers that he gives Portuguese terms for Gambian occupations, not the Mandinka terms.

a hole in it,[1] and in this manner with a hammer and a toole, they cut it for us, receiving satisfaction, to us easie enough, but what it is consists of Iron; and chary we must be to looke to our measuring, or he will use his best understanding to purloyne;[2] and this for the Smith and his esteeme amongst them [is enough]: The next is he whom we call a Sepatero;[3] one that doth make all their Gregories, wherein truely is a great deale of art shewen, they being made and fashioned of leather into all shapes, both round and square, and triangle, after that neate manner as might be allowed for workemanship, even amongst our curious handicrafts: these men are likewise they that make their saddles and bridles; of which bridles I have seene [some] so neately made up, as with leather, even here in our owne countrey, could hardly be mended: whereby appeares, they have knowledge to dresse their leather.

_{The Sepatero they of this trade are most ingenious.}

Pipe smoking

Howbeit I conceive, onely their goats and deare skinnes, which they can colour and dye: but to greater beasts hides, their apprehension cannot attaine, and some of these [sepateros] are held for curious persons, and deepe capacities: for they will bee feeling of some stuffe garments we weare, and do thinke, and will boldly say, that wee doe make them of the hides, we buy from them, and will not doe it in their sight, because they shall not learne; and for our paper, we bring, they absolutely conclude, it is made of the hyde, and likewise many other things they see us use, they will say, comes and is framed of those Elephants teeth we carry from them, allowing much of a more deeper knowledge in us, then themselves in /122/ many things[,] applying it amisse, and to impossibilities.

Another profession we finde, and those are they who temper the earth, and makes [*sic:* make] the walles of their houses, and likewise earthen pots

---

[1] Charcoal is often made from the timber of *Prosopis africana* (Mandinka *kemboo*). For the present-day Gambian blacksmith's tools, his bellows, the clay cone through which the air is pumped into the charcoal fire, and his portable anvils fixed into the ground, see Plate X.

[2] These crabbed clauses seem to mean that the blacksmith was paid in pieces of the iron bars but stole additional ones.

[3] 'Sapatero' = Portuguese *sapateiro* 'leather worker, shoemaker'. The Mandinka term is *karankewo*. For leather working at Kasang, see Part II below: Donelha, f. 27; and cf. Fernandes, f. 109^v.

The Potter & tobacco pipe maker.	they set to the fire, to boyle and dresse their food in[;] for all other occasions, they use no other mettle, but serve themselves with the gourd, which performs it very neatly; onely one principall thing, they canoot [*sic:* cannot] misse, and that is their Tabacco pipes, whereof there is few or none of them, be they men or women doth walke or go without, they do make onely the bowle of earth, with a necke of the same, about two inches long, very neatly, and artificially colouring or glasing the earth, very hansomly, all the bowles being very great, and for the most part will hold halfe an ounce of Tabacco; they put into the necke a long kane, many times a yard in length, and in that maner draw their smoake, whereof they are great takers, and cannot of all other things live without it.[1] These are the 3 professed trad[e]s, other things they need, and that are in use amongst them, are commom to every man, to doe or make, as his occasion requires, whereof the most especiall in use is
They have in the highest of the River, excellent mattes.	matts, such as they eate their meate upon, sit upon, and also make their beddes, having no other thing indeede to lie uppon, and therefore, as wee rightly terme it, is the Staple commoditie, they have amongst them: while we were in the River, at a place called *Mangegar*, against which we had occasion to ride with our ship, both up and downe, in the open fields, about a mile
A market kept every monday.	distant from any housing, is every monday a market kept; which is in the middle /123/ of the weeke,[2] unto which would come great resort of people, from round about, as heere in our countrey, who would disperse and settle themselves, with their commodities under the shady trees, and take up a good space of ground, & any thing what the Countrey did yeelde, was there brought in, and bought and sold amongst them. Now through the whole
No mony or coyne amongst them.	Countrey there is no use of any coyne, or money, neither have they any, but every man to choppe [*sc.* exchange] and barter one thing for another, and the onely nominated things is matts, as in asking the price of this, or that I desire, the word is, how many mats shall I give you ? so as they are still in use;[3] and these are the severall Trades, and manner of course the common people follow, or have among them.
	And so I passe to their laborious travell [travail], and generall trade amongst them, from which none are exempted, but the Kings and principall persons themselves, or such as by age are past their labour, otherwise all, the Mary-bucke, both Priest, and people, and of all sizes, as they are able, put to
All labour to till the earth and sow their graine.	their hands to till the earth, and sowe their corne. And for that the goodnes of God unto us may the more appeare, and the Reader stirred up the rather, to acknowledge his mercies, let us call to minde, the words which God sayd

---

[1] The types of pipes described by Jobson are no longer to be found in the Gambia region, although they appear in 19th-century drawings, e.g. in Schoberl, *World in Miniature*, I, opp.63, II, opp.199; and P. D. Boilat, *Esquisses Sénégalaises*, Paris, 1853.

[2] Presumably an Islamic week, beginning on Friday.

[3] Whereas in the higher river area cloth was used as a measure of value (p. 140, n. 3 above), Jobson records that at Mansegar mats were used. Mat-making from raffia palm remains characteristic of the central river area. However, the Kau-ur area, that is, around Mansegar, was later regarded chiefly as a place where excellent cloths could be obtained (Moore, *Inland Parts*, p. 102).

unto *Adam*, after his fall in Paradise; *In the sweate of thy browes shalt thou eate thy bread*: and with care and sorrow shalt thou eate it, and acknowledge these people, to abide the curse indeed; and our selves mittigated, through his mercifull favour: For the earth likewise receiving a curse, doth naturally bring foorth unprofitable /124/ things, whereby man is forced for his necessary sustenance, to till and plant the same: now God hath lent and given unto us, the beasts of the field, (which likewise they enjoy) but he hath endued [endowed] us with an understanding and knowledge, to make the beasts, and cattell, to serve and obey our wills in plowing and opening the earth, thereby easing, and as it were taking the sweat of our browes, which knowledge hee hath denied unto them, and notwithstanding they have so many heards of fitting cattell, they understand not to make use of them, but even with their owne hands, in the true sweat of their browes, doe they follow their painefull labour, as I heere relate it: They reserve great fields to sowe their corne in, which they raise up in furrowes, as decently as we doe here, but all their labour, is with their hands, having therein a short sticke, of some yard in length, upon the end wherof is put a broad Iron, like unto our paddle staves, which Iron set into the ground, one leading the way, carries up the earth before him, so many others following after him, with their severall Irons, doing as he leadeth, as will raise up a sufficient furrow, which followed to the end of the ground, they beginne againe in this painfulle and laborious manner,[1] fitting the earth for the graine, wherein our old proverb is to be allowed of, Many hands make light worke; otherwise it would appeare a most tedious kinde of labour. They have six severall sorts of graine, they doe feede upon, amongest which none is knowne to us by name (I meane heere in *England*) but onely Rice; the other may rather be called a kinde of seed then corne, being of as small /125/ a graine as mustard seed,[2] neither do they make any bread, but boyling their graine, rowle it up in balls (as I have said before) and so eate it warme: in like sort they boyle their Rice, and eate it warme; and even to us it is a very good and able sustenance: all other graines being sowed, the ground is with their Irons spadled over, and so left to his growth: but in Rice they do set it first in smal patches of low marish [marshy]

They understand not to make their cattle worke.

The manner of their painefull labours.

Their corne, or graine.

The manner of their Rice.

---

[1] Jobson's description does not make entirely clear the shape and use of the farming tool. The tool now normally used in planting millet and certain other crops has a handle measuring about six foot and a small blade at the end. At each stroke into the earth, the tool is twisted to turn aside the soil. The small pile of soil is then moved back by the foot of the sower after the seeds have been dropped into the hole. This operation is not particularly toilsome. But the *daramboo*, or *pafungo*, a tool with a handle at a sharp angle to the blade, was formerly used for ridging, a truly back-breaking task. (It is now rarely used, ploughing with animal traction having been widely adopted.) Planting a crop of digitaria (*findoo*) was a simpler task: the seed was broadcast and the earth loosely hoed over it with a 'weeding hoe' (or 'spadled', as Jobson goes on to say). For the earliest description of a farming tool (in Senegambia), see Crone, *Cadamosto*, ch. xxvi.

[2] The small grain is probably the seed of bulrush millet (*Pennisetum sp.*, Mandinka *sanyo/saanyoo*), still a common crop in Senegambia. The seed of digitaria (*Digitaria sp.*, Mandinka *findoo*) is like grass seed. The remaining four 'severall sorts of graine' were probably varieties of sorghum distinguished by size of grain and period of ripening, e.g. *sunoo*, *basoo*, *kintoo*. Purchas' summary has a marginal note: 'Pannike, Millet, &c.' (Purchas, *Pilgrimes*, 1/9/13, 1574), 'panic' being also a millet (i.e. of genus *Panicum*).

Using the ridging hoe (pafungo)   *The ridging hoe* (pafungo)

The planting of cotton.

grounds, and after it doth come up, disperse the plants, and set them in more spacious places, which they prepare for it, and it doth yeeld a great increase;[1] they do likewise observe their seasons, to set other plants, as Tobacco, which is ever growing about their houses; and likewise, with great carefulnesse, they prepare the ground, to set the seedes of the Cotten wooll, whereof they plant whole fields, and comming up, as Roses grow, it beareth coddes, and as they ripen, the codde breaketh, and the wooll appeareth, which shewes the time of gathering.

The misery of the people.

And before I passe to speake of other naturall plants, that proceed and come forth without labour, I must not omit to relate heere, the farther misery of this labouring people, that thereby wee may discerne, the greater mercy we doe enjoy, for whereas it hath pleased God, to affoord unto us seasonable times, to plant and sowe, and againe to reape, and enjoy our labour, sending likewise gentle showres and raines, wherby we receive them in a due season; he hath not dealt so with all Nations, whereof these are witnesses; for although their seasons are certaine, yet they are violent and fearefull: For from September, unto the latter end of /126/ Maie following, almost nine monethes, they never taste any showres of raine, so as their ground is so hard, through the extreme heat of the Sunne, that they can make no use thereof, but are compelled to stay untill raine doth fall, to moysten the earth, that their instruments may enter, to prepare the same, which raines, at the first come gently, now and then a showre, but not without thunder, and lightning;

The times of their raines & the fearefulnes thereof.

Towards the end of June, it then groweth more forcible, powring it selfe violently foorth with such horrible stormes, and gusts of winde, and with such fearefull flashes of lightning, and claps of thunder, as if (according to our

---

[1] For earlier references to rice, see Part II below: Cadamosto, [26]; Fernandes, ff. 110, 114v; Andrade; Donelha, f. 22^v; and for riziculture, Part II below: Almada, 6/7, 6/13.

phrase) heaven and earth would meet together: in all which notwithstanding, the miserable people are driven to worke and labour, in the open field, for loosing the season of the grounds softnesse; and as it doth beginne, after a more gentle manner, in the same nature and distance of time it passeth away, the most extreame force being from the middle of July, untill the middle of August, and the abundance of raine that then doth fall, may bee supposed, in that it doth raise the River from his usuall height, directly upright thirty foote, and where it hath not banke to defend it [it] over-flowes the shoares, and therefore they prepare their habitations, in their owne discretions, accordingly, and in some yeares not without danger:[1] Now in regard many people of our Country, have beene lost, and that our Seamen directly charge the unholsomnesse of the ayre, to be the sole cause, I would presume a little to argue it, delivering my opinion, hoping it may invite some abler understanding, to search into it, /127/ and produce some better assistance, to avoyde the inconvenience, then I am able to deliver. *A faire intreaty to men of judgment.*

It is certaine, in regard of the grounds hardnesse in those nine moneths when the raines are past, that the superficies, or upper part of the earth, doth receive all that venome, or poysonous humours which distill either from trees or plants, whereof there is store, as we see by the aboundance they use in poysoning their Arrowes, and some of their Launces; and likewise, what doth issue from their venemous Serpents, and Snakes, of which kindes there are very many, both great, and exceeding long; also Toades and Scorpions: the poyson doth remaine and continue in the drynesse of the ground, and rakte up in the dust and sand, which upon the first raines, being moystned, and the earth wet, by the exhalation of the hot Sunne is drawne up, and in short time in the next showres fals downe againe; in my poore judgement, some reason appeares that those first times must be very pestilent, and full of danger; which in some sort testifies it selfe, in regard those first raines, lighting upon the naked body, doe make blaines [*sc.* sores] and spots, which remaine after them, much more then after the raines have continued, and more perfectly washt, and cleard the superficies; and not onely upon the bodies, but in the garments, or clothes worne; who being laide by, after they are wet with the first raines, doe sooner, and in greater number breed and bring foorth untoward wormes; whereas other wayes, after the raines are more common, it doth not produce any such effect, or if it doe, very little. To this I say, that it is a thing to be especially observed, as much as men may, to avoide the being in those first raines, and more especially /128/ to be provided of water, either to drinke, or dresse meate withall, before these seasons fall; except it bee those who dwell and abide upon the land, and may have meanes to cover and keepe close their springs; but for men to water, in those pestilent times, and in the open Rivers, as the Saint *Johns* men in their first voyage did, I say it was a desperate attempt, and might have *The great aboundance of poyson. The nature of the first raines. An observation to be kept.*

---

[1] On the rainy season ('winter') and its effects on agriculture and the river, cf. Part II below: Cadamosto, [14]; Fernandes, f. 114ᵛ; Almada, 6/2, 6/12–13; Lemos Coelho, 1669, 11, 39.

A note of experience.	beene the confusion of them all, as indeed there were but few of them escaped; and that the countrey is not so contagious, as they would have their reports to make it, those people of ours may be witnesse, who being willing to stay behind, and remayning there almost three yeeres, there was not one of them dyed, but returned into their owne countrey, being eight of them in number, except onely Captaine *Tompson*, who as I repeate before, was slaine by unhappy accident.
	I would willingly also venter here, and speake my opinion, what naturally may bee saide, concerning these contagious times; but with this proviso, it is done to animate others, who if they knew the certaine course and season, with the true manner of each particular circumstance, would be able to demonstrate better, and so rectifie me in that where I shall doe amisse. These seasons I say, begin gently in the end of *May*, when the Sunne drawes to the end of his Northerne progresse, in the Tropicke of *Cancer*, whose power, as it may appeare, draweth up after him those great and clowdy vapours, which directly, come perpetually out of the Southeast, and from no other place or point, which following after the force of the Sunne, as they rise higher, and neerer the heate, begin to dissolve; but as the Sun turnes backe againe, and comes /129/ in his reverse to meete with these massie vapours, sending in his
An observation of the tempestuous times.	forceable raynes amongst those clowdy substances, compels them to give way, and breake in sunder, the violence whereof produceth that terrible thunder, and fearefull lightning which followeth, and great abundance of raine which falleth: which as it doth appeare, is most terrible, when the Sunne, and those vapours are as it were incorporate; for from the middle of *July*, untill the middle of *August*, the extremity is, and by that time in *September*, the Sunne is againe in his equinoctiall[,] the aire doth cleare, and all the stormes doe end; and so it appeares, that as the Sunne, after his comming from the Equinoctiall, in his whole Northern progresse is raysing, and drawing these vapours after, so in his reverse againe from the Tropicke, untill he comes to the Equinoctiall he is dissolving, and clearing the same againe, all which observing as a naturall man, I commend to the ingenious practitioner, either to amend, or make use of; And in my selfe, with humble thankefulnesse give glory to God, who shewes his almighty power to these unbeleeving
They heare & speak of Christ but will not beleeve.	people, that in regard, they will not accept of that pleasing, and peacefull intelligence of our loving and meeke Saviour his blessed Sonne; they shall feele and feare his omnipotent power, in trembling under those incomprehensible terrours, which as hee saith in *Job*, are prepared for his enemies:[1] Againe, if it hath pleased him to appoint certaine places upon the earth, where more especially those great and fearefull workes of his shall appeare, thereby to daunt and keepe down the hawty aspirings of sinnefull man; how
Gods mercy to us.	much are wee bound to praise, and acknowledge his everlasting goodnesse, in not seating us and our habitations under /130/ those contagious clymates,

---

[1] In the book of Job there are many references to 'terrors' launched against the unrighteous, for instance Job 27: 20, 'Terrors take hold on him as waters, a tempest stealeth him away in the night'. But no single verse refers to both 'terrors' and 'his enemies'.

and how much more is his great power manifested, that hath appointed bounds and lymits, as hee saith himselfe of the swelling Seas,[1] so likewise of these fearefull seasons; hethereto shall you come, and shall exceed no further.

And now to adde comfort unto us that are, or shalbe called to travell these parts; first, the times and seasons are certaine, that men may either avoide them by leaving the countrey, when they are to come, or by preparing themselves with things necessary, bee the better able to endure them, when they are come; of which now wee have had such experience, as wee can expound things outwardly, by Gods permission requisit and availeable, and inwardly frame our bodies and dispositions to the countrey and seasons agreeable: and this is encouraged with a comfortable resolution, that the continuance is not long, and that wee know the ends, and termination of the season, which before experience, was a fearefull discourager: So I returne backe, to speake of the naturall plants, which following the laborious courses, I was driven to omit. *A comfort to the traveller.*

They have naturally growing, which is but onely neare the mouth of the river, Bononos [bananas] a very excellent fruit, and they are as delicious, good, and great, as any that are in the West Indies; likewise within that lymit, store of small Lemmons, or Lymes, and for Orenges, wee have seene, and had brought unto us, farre up in the river, at some times good store, that shewes there are trees in the countrey, and that they might be stored, if the people were ingenious, and either would or could knowe how to plant them: but to speake of things that the whole countrey yeeldeth plentifully, and what is esteemed and set by amongst them, whereof /131/ especially, wee note Palmeta trees, and in some places there are whole grounds or groves of them, the use whereof is to draw from them a most sweete and pleasant drinke, which we call Palmeta Wine, and as wee approve and like of it to bee toothsome, so likewise in operation, wee find it wholesome; the manner whereof is this, they do cut into the body of the tree holes, in some more, in some lesse, as the tree is in substance, to which holes, they place a hollow cane cut sloping to goe the neatlier in, into which the juyce of the tree distilleth, and is conveid, as in pipes, unto gourdes set handsomely into the ground reddy to receive it, which is in lesse then twenty foure houres taken away, and as they please disposed of: now this is of that esteeme, that the vulgar sort may not meddle with, but the principall persons, and therefore they will send of this unto us, foure or five miles distance, as a curteous present; the tast whereof, doth truely resemble white Wine when it comes first over into England, having the same sweetnes of tast, and in colour, if they were together, not to be distinguished; only this is the misery, it will not keepe above one day, for if you reserve of it untill the morning it will grow sowre, notwithstanding any dilligence that can be used; and of this kind there are severall sortes and tastes, as there are in white Wines, which the people themselves distinguish *Plantans. Limes. Orenges.* *Good wine forth of a tree.*

---

[1] Jobson perhaps recollects Job 26: 10, 'He hath compassed the waters with bounds ...'.

Severall sorts thereof.	by severall names; calling some *Sabbagee, Bangee, Tangee*,[1] and other names, as the trees are from whence it comes.[2] Some Palmeta trees, doe likewise carry great store of Apples, which the countrey people will feede upon, especially the yonger sort.[3]
Palmeta apples.	

And being entred into their good liquor, I must not forget, to speake of the knowledge they have in /132/ making a compounded drinke, which wee can afford to tast, and accept of; and it is made of some corne, boild and ordered as wee doe our Ale, they call it *Dullo*,[4] it is not common amongst them, but when the King or principall person will make a feast, he calles all the inhabitants about him, and having a great gourd or two, sometimes three, of this liquour in his presence they drinke round, and it is devided amongst them, making an end of all before they part, and it is of that operation, it will warme their braines, and set their tongues a working: the poore *Fulbye* finding that wee affect it, will many times watch for a private conveyance, but if the Blackes meete with him, they will surely drinke it, and send him home againe, having lost his market; Now because I speake of gourdes, which are growing things, it is fit I tell you, they doe grow, and resemble just that wee call our Pumpion [*sc.* pumpkin], and in that manner are placed, and carried upon their walles and houses, being of all manner of different sorts;[5] from no bigger then an egge, to those that will hold a bushell, and the necessary use they have of them, to eate, and drinke, and wash their clothes in, with divers other very fit occasions, gives them just cause to preserve them although the meate, or substance that growes within them is to bee throwne away, in regard of the extreame bitternesse, whereof the shell it selfe so savours, as no use can be made, untill it be perfectly seasoned; and they have likewise growing Pumpions in the selfe same manner wee have, and in like case they doe convert to sustenance: But to rise higher from the ground; they have likewise great store of Locust trees, which growing in clusters of long cod[s], together in the beginning of *May*, growes to his ripenes, /133/ which the people will feede upon, especially the younger sort, if they can make shift to get them downe, the trees beeing bigge, and of a good heigth;[6] with this I must joyne hony, which doth appeare likewise to growe; and the countrey is very full, wherein the people use one of the ingenious parts I see amongst them, for upon those great trees, which are growing about their houses, in many places you shall have them make baskets of

[1] 'Bangee' = Mandinka *bango* 'raffia palm (*Raphia vinifera*)' + *ji(yoo)* 'liquid' = raffia wine; 'Tangee' = *tengo* 'oil palm (*Elaeis Guineensis*)' + *ji(yoo)* 'liquid' = palm wine; 'Sabbagee' = *sibi* 'rhun palm' + *ji(yoo)* 'liquid' = rhun palm wine (although DPG has never heard of this tree being tapped in Senegambia in recent times).
[2] On palm wine, cf. Part II below; Gomes, f. 278; Fernandes, f. 114. For the earliest description, see Crone, *Cadamosto*, ch. xxvi.
[3] Unidentified.
[4] See p. 107, n. 1 above.
[5] Gourds growing on a house roof can be seen today.
[6] The locust bean tree, *Parkia biglobosa*, Mandinka *netoo*. The pod contains a yellow powdery substance, which, sprinkled on food or drink, has a laxative effect.

reedes and sedge, which they will make fast, on the out bowes of the tree, and in those the Bees will come and breede, whereof in time they receive the profit, having so many baskets on some convenient trees, that in our ignorance, before wee knew it, being distant of, we might conceit it was some fruit the tree had yeelded;[1] also in holes of hollow trees, amongst the woods still bees are plenty, so that another *John Baptist*, if any were, might in this place and that with plenty, receive his full of Locust and wild honey.

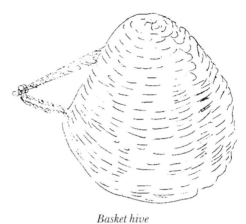

Basket hive

And for trees of great and large bodies, they be here, especially one sort, which doth carry upon a long stalke, a great and round fruite, yeelding a kinde of pleasing pith within: Whereupon the Baboones and Munkyes use to feede; whereof there had neede be store, in regard of their number, as I shall declare hereafter: and some of these trees retaine that bignesse, that sixe men by fadoming [*sc.* spanning] can hardly compasse:[2] there are other huge trees, one whereof doth carry a stony apple, which being th[o]rough ripe, to eate is tolerable, and serves if hee fall to feede the wilde swine, but that is in place, where the Baboone is a stranger:[3] And I will conclude their fruites, with that /134/ which is in most esteeme amongst them; which is a fruite in proportion, much like our bigger sort of chesnuts, flat on both sides, but hath no hard shell on the outside, they call them *Gola* [cola/kola], and we Nuts: the tast of him, when he is bitten, is extreame bitter, but the operation of him, is with them so set by, that ten is a present for a King: this operation we finde, that after we have eaten of him downe, notwithstanding his bitter taste, the water wee drinke presently after, although it he [be] out of the River, shall have a relish like white wine, carrying that sweetnesse, as if it were mixt with suger; and likewise the Tobacco wee take presently after, shall have that sweetenesse one would much admire: other operation we finde none, yet so doe they esteeme them that the auncient persons having lost their teeth, and not able to bite it, have morters wherewithall carryed to bruise it, that they may not be bard of the Juice, and comfort of it; neyther are they

---

[1] Coiled baskets are still in use as beehives. The use of hives was probably a traditional procedure and not one learned from the early Portuguese, since an early source referred to honey, wax and hives – see Part II below: Fernandes, f. 113ᵛ; cf. Almada, 5/4. Collecting wild honey was certainly traditional. Jobson does not mention wax, although beeswax formed an important element in Afro-Portuguese trade on the lower river, the wax being required in Europe for the better sort of candles, not least large altar candles.

[2] The tree is the baobab, *Adansonia digitata*, whose fruit is known in English as 'monkey's bread'.

[3] Possibly *Salacia senegalensis*, known today in Freetown as 'stone-fruit'; in Mandinka *sinjang-koyo* (a rare fruit today).

	for the common people: Fifty, of these nuts in the habitation where we dwelt, presented to the King, would buy a wife, and many times as a wonderfull great present, I have had sixe of them sent me, howbeit, we never sawe any of them

They are not growing within the limit we saw.

grow, neither are they, as they say amongst them, but brought from another people, and they are of most valew, still the lower, and neerer the mouth, yet there they bee, and the Portingals will make, as if they bring them into the River, by a trade they have in a great baye, beyond *Cacho* [Cacheo], where they meete with a people, that brings them gold, and many of these nuts; and this we can say, that when we were at the highest part of the river, the people brought them

Great store brought us, when we were above.

abundantly unto us,[1] and did wonder much, we made no more esteeme /134 [bis = 135]/ or care to buy them:[2] but where they grow, or whence they had them, wee are yet ignorant, although the Portingall affirmes, they come from the golden countrey, neither will they last, or continue by any knowledge we have to keepe them, being subject to wither, or be eate with wormes, as by tryall I prooved, keeping of them, to have shewed in England, as I much desired. They have neither Onyons, nor Garlike: howbeit, Garlike is a thing they much desire, wherewith we see them rather rub their heads and bodies, then affect to feede on:[3] neither have they any herbs, or flowers, which either for taste or

This is like our water Lilly.

smell they esteeme, but onely one called *Binning*,[4] which carrying a sharpe, or sowre taste like Samphire, we used for sauce, and that they seeing, would ordinarily bring unto us: howbeit when we came upon divers mountaines, and sundry woods, wee should retaine such sweete smells, as would be very pleasing; from whence we concluded the Bees did gather, and make up theit [*sic:* their] hony. And I will make my conclusion of the Plants amongst them, with that which unto me brought admiration; which was a tree, or bush, commonly growing upon the toppe of the *River* banke, resembling much our great Barbery bushes,[5] onely having a little ragged leafe, whereunto comming, with all the diligence might be devised, not to touch or moove it, but onely with all gentlenesse, betweene your fingers and thumbe, touching a leafe, the whole bowe

---

[1] For kola nuts, see p. 134, n. 4 above. Contemporary Portuguese sources discussed the trade in kola, the nuts being purchased mainly in the Scarcies Rivers of Sierra Leone – actually far 'beyond Cacho' – and conveyed annually in Portuguese ships up-coast to the northern rivers, including River Gambia. Portuguese marine intervention in the distribution of kola seems to have operated for about a century, between the 1550s and the 1650s. The previous distribution was entirely overland, and Jobson has earlier indicated that the upper river still received kola overland (*GT* 80); if he is correct in reporting that kola was more expensive on the lower river, the African and Portuguese merchants involved in the marine trade had higher costs or higher profits than the overland merchants. That the trade in kola at Sierra Leone was associated with a trade in gold was fictional, there having never been a gold trade on the Scarcies. Either Jobson misunderstood what he was told, presumably by the Afro-Portuguese, or was deliberately misinformed.

[2] Europeans never gained a taste for kola, and no nuts, other than as samples such as Jobson's, were ever exported to Europe, at least before the 20th century. The source of the kola purported to be an ingredient of modern 'cola' beverages is supposedly 'secret'.

[3] On the middle river, near Basse, in 1953 DPG was asked for onions, not for eating, but as a rub to counter headaches.

[4] The root of the water lily (*Nymphaea spp.*).

[5] The shrub *Berberis vulgaris*.

[bough] should presently close up every leafe together, as if they feared and found themselves offended: but if you toucht or stird a little sprigge, the whole tree sooud close his leaves after a most sensible [*sc.* sensitive] manner.[1] Whereof taking especially notice, wee did allow it to /136/ be the sensible tree, of which aunciant authours have written; which wee did observe to carry a kinde of yellow flowers like our hedge roses: with the strangenesse whereof concerning plants, I here conclude my story.

The sensible tree.

The discourse of the wild beasts.

AND now I am to speake of the wild beasts the countrey is stored withall, whereof I am to begin with those that are ravenous and offensive, keeping the people in dread, and as his pleasure requires the Lyon is first, whereof there are very many, which we can speake by perfect knowledge, although we cannot say, [as] eye-witnesses, for our gracious God hath so ordained, that those beasts which are most tyrannous to others, and boldest against man, as ashamed of their bloody actions, notwithstanding their ablenes and strength, doe shut themselves up in the bowels of the earth all the day time, as it were confinde from the glorious light of the Sunne, beeing one of Gods comfortablest creatures; So that in the night, is their times of walking, and our knowledge of them, is by their roarings and noyse they make, whereby one beast is distinguisht from another, but the Lyon is more especially to be taken notice of, besides his owne voice, in regard of a servant hee hath[,] sometimes two or three[,] that doe attend him, which we doe call the little Jacke All: it is a little blacke shagge-heard [haired] beast, about the bignesse of a small spaniell, which so soone as the Evening comes, hunts and busles about for the preye, and comming on the foot, followes the sent with /145=137/ open crye, to which the Lyon being master huntsman gives diligent eare, and applyes himselfe to follow, for his owne ease and advantage; if it so happen the Jacke All, beweary, or set up his chace before the Lyon come in, he howles mainely out, to shew the estate he stands in, and then comes the haughty Lyon, and ceazeth one [on] the weary prey: for as it is written of the Lyon in his pride, if hee faile of his prey at three jumps, he scornes to pursue, or toyle himselfe after it: and being ceazd, he remaines feeding, making a kind of grumbling noyse: whilest his small servant stands barking, and yalping by, attending untill his Master hath feasted, and then hee falls upon the remainder. And this, as we heare, and receive from the countrey people; so likewise it is affirmed unto us by our owne travailes: for as we had occasion when the tydes fell out to travaile up the River in the night; and likewise many times to ride all night at an Anker in the River against desert places, we did observe the noyse, and hunting of this Jacke All, and likewise note the reply, and answer of the Lyon, insomuch, as it was a commonn word amongst us, who will goe on shore, and accompany the Master huntsman.

All ravenous beasts in the day time keepe their dennes.

The Lyon.

His small servant.

His manner of hunting.

The causes of our knowledge.

---

[1] The plant is *Mimosa pudica*, known as 'the sensitive plant', which behaves as Jobson states.

_{Ounces, and Leopards.}

There are Ounces and Leopards great store, whereof by reason of the many dennes wee see upon the land, we may discerne the print of the foot, remaining upon the holes mouth, beeing able to assure us what is within, as also the countrey people doe bring many of their skinnes unto us to sell, how ever they light upon them; for by their owne valour, sure they dare not, and by their ingenious capacities, I beleeve they cannot devise any course to lessen their /138/ company. The Ounce doth seeme to bee more ravenous,

_{The Ounce dangerous.}

or dangerous unto them, then either Lyon or any others, and makes more spoyle upon them, as they doe complaine, I was shewed a child there, which

_{A true tale of a Child.}

the mother gave sucke unto, who early in a morning going neare to her house to a spring to fetch water, had laid her child wrapt in a Cloath without her dore upon a matte, as they use to doe, and there came a hungry Ounce, who it seemed had mist his nights prey, and tooke up the cloath and childe, and runne his wayes, the mother met him, and with wofull outcry pursued him, and as it chanced he tooke the way to come right upon the place, where the Father of the child with other people were labouring in the field, who with roaring voyces run after him, the Ounce still ran away, keeping his hold, but as it chanced, the child dropt forth of the cloath, and the father running after, recovered it, and tooke it up, the Ounce carryed cleane away the cloath, and the man brought backe the child to the mother; the which [child] wee our selves have both seene and handled: and so bold and fierce is the Ounce, that many times in the night, hee hath driven a small dogge wee had, where we dwelt on shore to our bedsides by a hole he had through our straw walles, barking and running under our beds, not daring to looke out, howsoever we encouraged him, untill we were faine with firebrands in our hands to go abroad, and so feare him away: and many more are there of night enemies, which watch and looke after their carefull huswifery, amongst which espe-

_{Civit Cat. Porcupine.}

cially is the great Civit Cat,[1] and the Porcupin, who are carefull purveighers for any outlying poultrey, whose view early in the /139/ morning is their discoverie, the Cattes by the print of their feete left in the sand, and the Porcupine by his quills, which are shed, and many times taken up in plenty;[2] and so I end with their night enemies, and as I stand conceited, cruell acquaintance, because what after I deliver, is uppon such beasts, as walke, and shew themselves by day: and howsoever, they stand in feare of them, it is rather out of a timorousnesse, in the people, then any willingnesse in the beast.

_{The Elephant.}

The first whereof is the Elephant, whose presence indeed, as he is a wilde beast, may even to a strong person give a just amazement; and such is the feare the countrey people in generall, have of them, that by all possible meanes, they seeke to shunne and flie from them, yet such is the great abundance the Countrey doth yeeld of them, that they are over all places, and

---

[1] In his petition Jobson noted that 'the woods and mountains are full of muske & civitt catts & very many codds wee bring home, which wee have of the people for poore commodities', Petition, f. 4. Civet cats and their product, musk, were regularly noted in early sources, the fullest reference being Part II below: Cadamosto, [22].

[2] Jobson was mistaken in saying that porcupines eat fowl.

wheresoever you come, you shall find the footing and apparent shew they have been there, though not presently to be seene:[1] and notwithstanding those great abundance of wilde ones, they have not any of them tame, or under commaund: as in other places of the world they have, which certainly proceeds from the feare they conceive of them: much, and great is the spoiles they doe them, both among their corne, and especially in their Cotton grounds, going in small companies together, whereof I have seene sixteene verie great ones, besides young ones that suckt, and others that were of middle statures; the proportion of the greatest I leave you thus to conjecture of: the reeds or sedge, that grows naturally in every place, is higher above our heads, then the arme of a tall /148 = 140/ man can well reach; and halfe the body of those Elephants, is above the reedes, their naturall feeding is amongst this sedge, but more especially, they doe browse upon trees, whereof in the woods you shall find store by them pulld downe, and that of bigge bodies, and round substance; the manner whereof I must relate, to correct the mistaking, which is most common in picturing the Elephant, whose two great teeth are commonly set in his lower jaw, carrying them upward, as a Bore doth his tusks, which is contrary, for he carries them downward, and with them breakes downe the trees, for after with his truncke he hath bended the toppe, he haspes over his two teeth, so as one or other must needs give way, and that is the reason, that among those multitude of teeth, that are brought over, so many broken teeth, and crackt and shaken are amongst them, for if the tree be too strong, the tooth gives way, and so the people find many junkes [chunks] and peeces, which they sell unto us, & the abundance of those teeth, that are yearely brought from thence, may satisfie what store of these beasts are in the countrey; for as I have spoken with many, who considering the great store are brought away, have conceived, they had shed their teeth, as Stagges doe mew [*sc.* shed] their hornes, which directly is nothing so, but by the death of the beast, the teeth are gotten; what casuall deaths they are subject unto wee are ignorant: and for any practize of the peoples, too much feare possesseth them, so farre as we have seene, one place alone excepted, which I will manifest unto you: within foure miles where our habitation was, there stood a good spatious plantation, /149 = 141/ the Commander whereof we called *Ferambra*, who was alwayes a friend of ours; as we were in our dwelling, upon our Christmasse day, at dinner; where (God be praised) wee had varieties of meate: to mend our fare, just in the dinner time, there came foure blacke people unto us, whereof two were laden, and had great gourds uppon their heads, as much as they could stand under, the one full of Palmeta wine, the other of raw flesh, which were Presents sent me from this *Ferambra*, who sent me word, hee had killed an Elephant, and had sent me some part thereof; our daintie stomacks looked

side-notes: The nature of the great ones. — He browses like a deare. — A false opinion. — This was that *Ferambra* I noted before. — Elephants flesh good to feede upon.

---

[1] In the modern state of The Gambia there are no longer any elephants, although some can be found in the Niokolo Game Park in neighbouring Senegal (G. Roure, *La Haute-Gambie et le Parc national du Niokolo Koba*, Dakar, 1956; André R. Dupuy, *Le Niokolo-Koba, premier grand Parc National de la République du Sénégal*, Dakar, 1971). Although there was earlier knowledge of the

asquash at such grosse flesh, yet I received it kindly, and gave it away to our blacke neighbours, who eat it very merrily. The next day I went to *Ferambras* house. & the fashion of the Country is to entertain us, with the best provision of diet, amongst which we had Elephants flesh, whereupon both my selfe, and consorts that were with me, fed very heartily, and found it good and savoury meate:[1] I desired to know how he killed them; And he shewed me one of his blacke people, and sayd, There was none but hee alone durst doe it; and taking downe a Javelin, which hung in the house, the staffe some ten foote long; the Iron or head whereof was bound up in a cloth, which he opened and shewed me, and it was laid with poyson all over; he sayd, his manner was, when hee saw the Elephants feeding in the high sedge, he would steale in amongst them, & by creeping, still keeping himselfe behinde them, he would recover so neare, as to strike his Javelin into the body of the beast, and leaving it there, take to his heeles, and through the long reeds scape away: and /142/ the warme bloud dissolving the poyson uppon the Javelin, it presently spreads it selfe, to the cruell torture of the beast, the extremitie whereof killes him; the people in the meane time, upon trees, and places of advantage, being set round about to watch him, and so soone as he is downe, come to him, presently cutting away so much of the flesh as is inflamed with the poyson, which they throw away, reserving the rest for their owne sustenance: and in this manner he hath killed mee so many, as you see I have tailes heere hanging up; And except in this place, I never heard but the people were wondrous fearefull of them: the experience whereof, was in those blacke people, I had in my boate when I went up the River: It was my manner, as I could with conveniencie, to adventure and set uppon such as wee met withall, but my Blacks would alwayes tremble, and runne away; and many severall attempts I had upon them, wherein I must say, as I found, that they were as fearefull as a forrest Stagge, and according to their greatnesse, went as swift from us as they could, which pace was faster, then a good able man could runne, whereof I had triall in one great beast, who notwithstanding wee had shot three times, the bloud running downe his sides, escapt away from us, that we lost him, whom afterwards the people found dead, and brought his teeth to sell unto us; and had wee beene provided accordingly, we might have made divers preys upon them: but what wee did, was held in admiration amongest the people, for many would come downe on purpose to looke upon us, and demaund of our Blackes, which was he that durst set upon an Elephant. /143/

There are also in the Countrey *Buffelos*, which are wilde Bulles, and cattell of that sort; also wilde Boores, very huge and great, their colour being a darke blew, and without doubt he is a very dangerous beast, for hee shewes more boldnesse then any other, being armed with great and large tusks, and

---

elephant in Europe, Cadamosto supplied a long account – see Part II below: Cadamosto, [28]–[31]; and cf. Fernandes, f. 111; Almada, 5/4.

[1] Cadamosto, who ate elephant meat 170 years earlier, disagreed; see Part II below: Cadamosto, [30].

carrying up his tufted taile, of a great length, boult upright, in a scornefull manner, will walke from us.

There are likewise large Antelops, and Deare of all manner of sorts spread over the whole countrey, with beasts of that kind, whose names wee are ignorant of; and many strange hydes they doe bring unto us, amongst which there is one beast, whose hide is fourteene foote of length, of a dunne colour, and strokt with white. Another sort I must needes remember, whose great abundance may well put me in minde, besides their society and neighbourhood, which in our travell up the River we were often acquainted withall, which are, the Babownes and Munkeys, whereof the countrey hath innumerable store, and where they are, they doe goe in heards, and companies, but are of two societies: the Munkeys alwaies keepe by themselves, and great and little as they are; onely of that kind consort together, and even in Ilands that lye within the River, they are as frequent as on the mayne, which condemnes the report is of them, that they cannot swimme, but being in the water will drowne presently, and in my owne knowledge I can affirme, that having bought a Monkey from the countrey people, who use to bring them unto us, and sel them for poore things; being got loose in my boate that /152 = 144/ rid in the middle of the River, hee leapt into the water to swimme on shore, and being pursued by one of our men, who swamme after him, hee did dive under water, divers and sundry times, before he could recover him.[1]

<small>Antelops. Deare of all sorts.</small>

<small>Munkies.</small>

But to speake of the Babowne, I must say, it is a wonderfull thing, to observe a kind of common-wealth that is amongst them; they have none but their owne kind together, and are in heardes, of three or foure thousand in a company;[2] as they travell, they goe in rancke, whereof the leaders are certaine of the greater sort, and there is as great, and large of them, as a Lyon, the smaller following, and ever now and then as a Commaunder a great one walkes;[3] the females carry their yong under their bellies, except shee have two, and then one under, the other above. In the rear comes up a great company of the biggest sort, as a guard, against any persuing enemy, and in this manner doe they march along: they are very bold, and as we passe in the river, when we come neare their troupes, they will get up into the trees, and stand in gaze upon us, and in a kind of collericke humour the great ones will shake the trees, and with his hands clatter the boughes in that fashion, as it doth exceed the strength of a man, to doe the like, barking, and making a noyse at us, as if they were much offended, and in this manner, many times they will follow us along, and in the night time, where wee ride at an ancker

<small>Babownes a strange story.</small>

---

[1] On baboons and monkeys, cf. Part II below: Cadamosto, [22]; Almada, 6/6.

[2] Jobson greatly exaggerates the size of baboon troops. A locality alongside the river known as Monkey's Court (on the north bank about ten miles up from Bansang) has over the centuries been noted as an area where large numbers of dog-faced baboons congregate: Almada, 6/6 (Part II below); A. W. Mitchinson, *The Expiring Continent*, London, 1881, pp. 45–7; Rex Hardinge, *Gambia and Beyond*, London, 1934, 120; Sire So, 'Le Tribunal des singes de Monkey Kote', *Notes africaines*, 40, October 1948, pp. 4–5).

[3] Purchas's summary adds this marginal note: 'He told me, he never saw greater Lion then of them: their height (standing up) most admirable!' (Purchas, *Pilgrimes*, 1/9/13, 1575).

A government amongst them.	take up their stands, or lodgings on the mountaine toppes, or on the trees that are above us, whereas [? whence] we heare their government: for many times in the night you shall heare such a noyse of many of their voyces together, when instantly one great voyce exalts it selfe, and presently all are /153 = 145/ hush, and the noyse is dasht,[1] so as we were wont to say, Maister Constable speakes; likewise when wee are a shore and meete with these troupes, on a sudden the great ones will come forward, and seeme to grin in our faces; but offer up a gunne, and away they packe. One of our people one day as we came neare the shore in our boate, and a troupe of these shavers [*sc.* rascals], being gazing on us, made a shot and kild one of them, which before the bote could get on shore, the others had taken up betwixt them and carried quite away, but we have kild of them, which the countrey people doe much desire, and wil eate
The people of the Country eate them.	very heartely; wherein I hope never to take their part: And lastly let mee tell you that wee have seene in the desert places they use, trees and plants, wound and made up together in that artificiall manner, and wrought together with that thicknes over head, to keepe away the sun, and shade the ground, which hath bin beaten, & smoothed under neath, and all things in the manner and shape of an excellent arbour, which place they have only used and kept for their dancing and recreation;[2] that no, man living that should have come by chance, and seene the same, without knowledge of these unlucky things, but would have confidently supposed, it had, and must have beene the handy worke of man; which some wayes confirmes the opinion the Spaniard holds of
The Spaniards opinion of them.	them, and doth not sticke to write it,[3] that they are absolutely a race and kind of people, who in regard they will not bee brought to worke, and live under subjection, refuse to speake, and so he reports of them.

And to conclude, amongst their multitude of wilde beasts, we have enquired amongst them: /154 = 146/ especially, when I was at the highest in

The pleoples [*sic:* peoples] report of a Unicorne.	the countrey, whether they could tell or report of a Unycorne, setting foorth unto them a beast, with one onely horne in his forehead, and certainely they have told me, that higher within the land there is a beast, which hath one onely horne in the same manner, but describe the beast, to be both about the colour and bignesse of a vallow [fallow] Deare, and the horne to be about the length of their arme, and no otherwise, which is nothing like to the description of a Unicorne, as he is with us set out, if there bee any such beast;[4] whereof indeed I am very doubtfull, and so I am come to the last, which is to deliver, what land fowle, and of that nature, there doe remaine wilde as we have seene amongst them.

[1] Purchas's summary reads 'husht' (ibid.).
[2] Although most of this description is undoubtedly of baboons, the assembling of 'plants' on trees suggests the nest-building of the chimpanzees which live in parts of the forest belt of West Africa, and the 'dancing' also suggests these animals rather than baboons.
[3] The source of the Spanish opinion that baboons (? monkeys) are renegade humans has not been traced.
[4] This appears to be a mythical beast. It is too small for a rhinoceros; moreover, no rhinoceros are to be found in West Africa today and probably none were to be found in Jobson's time.

The discourse of land fowle.

Amongst such fowle and birds that remaine and live upon the land, in our travels up the River, and our daily walkes and travels upon the shore, our indraught being so many hundred miles, wee never saw any Estriches, neither did any of the cuntrey people, ever bring any of their feathers to barter, or sell unto us, so as it appeares plainly there is none of them in these parts; notwithstanding in the River of Senega which is to the Northward, and likewise againe more Southerly, upon the Sea coast of Affrica great store:[1] Therefore the greatest bird or fowle we see, is called a Stalker; who by reason of his long legs and necke, when he stands up right, is in height taller then a man, his body in substance is more then an indifferent lambe which wee doe feede upon, and finde it somewhat a dry meate, but well allowed for nourishment, and by the countrey people much esteemed of: The especiall desire we have to kill them, is in regard of some feathers he hath, which being taken in due time, and so preserved, are heere at home esteemed and worne.[2] *The Stalker.*

The next in greatnesse, is called a Wake, in regard of the great noyse hee makes when hee flyeth, which resembleth what he is called by, and of these there is great abundance, who for the most part live upon their Rice grounds, and in those times do them great spoyles: they are very good to eate, and is a bird of a great stature, having the upper part of his head carrying a beautifull shew, with a pleasing tuft on his Crowne, which I have seene worne by great personages here at home.[3] *The wake.*

There is infinite store of another sort of excellent birds which wee call Ginney Hennes, in bignesse much about our Phesants, and in beauty answerable; his feathers being all laid over him like unto eyes, in a pleasing fashion, they are all the countrey over, and in flockes of many hundreds together; their food is upon their corne grounds, keeping close together, insomuch as we have killed eight of them at one shoote, they are an excellent meate, and many of these are brought into England, and given as presents to those of note, and worthy persons who preserve and keepe them for their rarenesse, as birds of pleasure: And in the very like abundance they have Patridges, whose colour is not beautifull, so much as our Patridges here; but onely of a darke feather, and these are likewise all the countrey over, where it is planted, for the most remaining neare their houses, and in the middle of their dwellings, the great plenty of both which kinds, of Gynney Hennes and Patridges, are some manifest tokens there are no Foxes at all in the countrey, who are in these our parts great enemies to both kindes of Phesant and Patridge: and the cause the[y] keepe so neare the houses, is to preserve them from as subtill enemies, which are the Babownes, and Munkeys, *Ginney hens.* *Patridges.*

---

[1] Purchas's summary adds a knowledgeable marginal note: 'Estriches & Emes use sandie Deserts' (Purchas, *Pilgrimes*, 1/9/13, 1576).
[2] Perhaps a stork (*Leptoptilos crumeniferus*).
[3] The bird is undoubtedly the Crowned Crane (*Balearica pavonina*).

Quailes.	who are no night walkers, and in the day time the recourse of people makes them keepe further off, whereas otherwayes they would not faile to be sharers. There are also great store of Quailes, who are in bignesse as great as a Woodcocke, and from whence it is derived I cannot avouch, onely it is saide, they are of those kind of Quailes as fell among the children of *Israels* tents:[1] thus much I can affirme, they are a pleasing and delightful meate, and in many places, where we have made abode, they have accustomed to fall about us: so as [*sic:* ? if] provided wherewithall to shoote them, mens dyets are mended, even in a short warning [*sc.* at short notice]: In all their townes and dwellings likewise
Pigeons.	store of Pidgeons, which feede upon the offall of their Corne, in the very doores, yet all are wilde, and of tame Pidgeons they have no knowledge: I have with my stone-bow or pelletbow in two houres killed twenty Pidegeons, even among their houses, which manner of shooting they have had in wonderfull admiration: And these birds or fowle nominated, are such that are at all times, and in most places ever neare at hand, and always ready for sustenance if men bee provided, and will take small paine to looke them. /157 = 149/
Parrats.	There are likewise in the countrey Parats, but none good for ought, except the dun Parat with the red tayle, of which sort you have some few that come to
Paraquetos.	speake well: but of Paraquetos there are very many, and beautifull birds, which are often brought home, and some few attaine to perfection.[2] Also of smaller
Variety of smal birds.	birds great varietie & sundry strange shapes, amongst which many are in colours, delightful to the eye, many in notes very pleasing to the eare: there is amongst the variety one smal bird, which for his strangnesse we observe, hee
A small bird without legges.	hath no legges, but two strings like the bird of *Arabia*, with which he hangs with his head downeward, and hath such resemblance to a dead leafe, as it hangs on the tree, being direct of that colour, whereby unlesse hee be seene too [to] light, you can hardly discover him, and he doth seeme to take pleasure to deceive mens eye-sight, hanging wonderous steddy, without motion, whilest hee is lookt after, and very neare the touching:[3] Likewise another strange bird there
A bird with foure wings, about the bignes of a turtle Dove.	is, which flyeth with foure wings [:] wee see him not all the day, but an houre before night, his two foremost wings are largest, the other are a pretty distance backward, and beares his body betweene foure palpably.[4]
	As I speake of these birds, it is very necessary I should set downe how nature teacheth these little creatures to provide for the safety of themselves,

[1] Exodus 16: 13; Numbers 11: 31.
[2] On parrots in Senegambia, see Crone, *Cadamosto*, ch. xxx.
[3] The bird which 'hangs with his head downeward' is undoubtedly the Gambian Fruit Bat (*Epomorphosus gambianus*), which is dormant during the day, but flies around at dusk in great numbers. Purchas took note (Purchas, *Pilgrimes*, 2/5/2, 965). But it is curious that Jobson did not recognize a bat. The 'bird of *Arabia*' perhaps means the mythical Phoenix.
[4] This is almost certainly the Pennant-winged Nightjar, described as follows: '*Macrodipteryx longipennis*, a large bird, in which the male dons a remarkable ornament for the breeding season, one feather shaft in each wing prolonged to end in a racket-shaped enlargement. They are nocturnal, coming out at dusk singly or in small parties, while during the day they sleep on the ground with the pennants stretched out at right angles to the body' (Reeve, *The Gambia*, pp. 214, 223).

and the young they bring forth, I have shewed before, what troupes and multitudes of Babownes and Monkeys the countrey is stored withall, which are profest enemies to feathered fowle, and therefore in these little poore creatures, who can make no resistance, Nature hath directed them by Art to prevent /158 = 150/ cruelty: Amongst the great variety of strange trees, and woods, which the countrey affordeth, whereof there is not any, that I can know, or call by an English name, by saying, this tree doth grow in England: there is especially one, who doth exceed in prickles, both upon the body, branches, and armes: even to the outermost small sprigges, many of these grow distant from the water, and many of them grow upon the banke side, hanging their toppes over the water: we observe, that of this onely tree, the little Bird makes choyce, and not content with his defence of prickles, makes use likewise of his growing over the water, and on that side which bends to the river on the very outside doe they winde their nestes with an owse, or neck, which is hollow, made of reeds and sedge, the whole neast hanging like a bottle, made fast by the necke, in some places so thicke together, that the same side of the tree, seemes as it were all covered with thatch;[1] unto which, if notwithstanding the prickles, the Babown or Munkey durst approach, the feare hee shall have, that the boughes will not beare him, and the fall hee is in daunger of, together with the fright of the water underneath him, is able to daunt him by which natural care he preserves his increase, and speedes better then many times the Parrat doth, for he likewise is provident to make his neast on the outermost smallest twigge of a tree; but on the land winding it about the twigge, so neare, as it will not beare any of his unhappy enemies, who notwithstanding are vigilant for their owne ends, and by getting upon upper boughes, will overlooke his desired prey, and when hee sees they are growne to fill up the neast, will hazard charily as the bough may beare him, and sitting fast with /159 = 151/ his two hinder feet, with his two hands [will] take up the bough, and shake it in that manner, that either some or all forth of the nest shal tumble, and being down, the gaines them for his labour. Another kind of art, nature hath taught these birds[:] in the high banke, which is steepest over the River, whose steepenesse hinders the accesse of these devourers, they will make holes so artifically round like augor holes, and of that equall distance the one by the other, so thicke as the banke will beare, carrying them at least a yard within the ground, by which places they preserve themselves and their young.[2]

But there are birds of defence such as are hawkes, whereof there is one sort, as large as our Jerfauchon,[3] and these as the people tell us, will of their owne accords, kill the wild deare by ceazing upon his head, and hanging fast,

---

[1] The description is of the nest of the weaver bird (*Plesiositagra cucullatus*, the Village Weaver), an item of natural history that fascinated many European observers (e.g. Barbot, *Barbot on Guinea*, p. 74).

[2] Among the birds which make their nests in the banks of rivers are bee-eaters and certain varieties of wagtails and kingfishers.

[3] In Purchas's summary 'Ger-falcon'.

Hawkes that will kill a Vallow deare.	[he] doth continue beating with his wings, untill the deare faintes, and then he preyes upon him. And likewise of other sorts that live upon prey, whose manner of breeding, is in the open trees, and by the continually watching and attending the nest, they are ready to defend and save their young. There are no great
Bastard Eagles.	Eagles, but of a kind of small bastard Eagles infinite store, and likewise severall sorts of ravening Kites and Buzzards, whereof the skin of one sort smells wondrous sweet and strong, after the savour of the Crocodile: These sorts are easily to bee discerned: for if at any time, wee hapned to kill a beast in the woods, whereby any blood were discovered, although there were scarce any one of these ravening birds to be seene, almost instantly, you should have such troupes of all sorts come in, as were able to devoure the whole carkas, if wee were not present to /160 = 152/ affront them:[1] And the onely meanes the people have to finde out either Elephant, or any other beast, as they dye, or come to an untimely
How the people finde the dead beasts.	end, amongst the thicke woods, or high reedes, is by observing and keeping watch to looke out where these ravening birds gather together, which is easily discerned.[,] the nature of them being to sore, and flye in the aire aloft over the place where their prey remaineth, to which place the people repaire, and many times are sharers in the booty: And to shut up this discourse, that it may appeare
The Inhabitants want knowledge to take them.	how likely it is, these birdes and fowles may well increase, wee doe not see that the people have any ingenious conceites, either by gins or otherwise to kill or take of them:[2] but upon any especiall time, when the King is determined to make a feast, they observe a course to take them, with the rehearsall whereof I
The Kings manner of Hawking.	will make an end. The greate command is sent, that all people come abroad, and being in the fields, are set, and placed severally, of an indifferent distance, the one unto the other, when the Patridge, and gynny hennes being sprung or put up, as their natures are to flye but an indifferent flight, so soone as he lights againe they are ready to runne in, and put them up, and in this manner still pursuing them, [so] that they are wearyed out, and the people with their hands take them up, and bring them to the King, even to that number as may content him, with which their Princely pastime I heere conclude my story.

The Conclusion.

And for a finall end doe earnestly desire, that what /161 = 153/ is written may be taken into consideration, thereby to stirre up a more willing affection to prosecute and goe on in a timely proceeding upon this hopefull

---

[1] The various birds of prey in West Africa include kites and buzzards, but they prey only on small creatures (or they scavenge), and the attack on a deer seems far-fetched. The 'troupes' arriving to scavenge were probably of vultures.

[2] It is probable that the Jaxanke, with whom Jobson had most contact, paid little attention to hunting, but the peoples of Senegambia in general did use snares, traps and hunting techniques (see P. P. Vinche, M. Singleton, and P. S. D. Diouf, 'Techniques et instruments de chasse chez les Sérer du Sine (Sénégal)', *Notes africaines*, 138, October 1985, pp. 113–18; Youssouf Tata Cissé, *La Confrérie des chasseurs Malinké et Bambara*, Paris, 1994). It would not have been easy for Jobson, who spent most of his time on the river, to have witnessed this activity.

trade, which will crave expedition in regard of these reasons following: First there is, as it were, a certaine combination made betwixt the people above and us, never to faile them of a yearely trade, which they in their parts, (without all doubt) will carefully expect, and as they have faithfully promised, will accordingly provide for, and if in our parts, it should be neglected, may justly cause them to take a great distrust of our fidelities, which in regard, we are now the first white people they have seene,[1] and have from them received such faire approbation, may settle a distast for the present very prejudiciall, and among such a barbarous people, wee know not whether it may be easily remooved. Againe the course we run, is allowable by our Lawes, fitting and agreeing with the peacefull time we live in, opposite to no neighbourly love or amity, either confronting any forraine Prince, by entring, or intermedling within any forbidden territories,[2] neither is it done in any warlike, or hostile manner, but by the aunciant and free Commerse, that uniteth nations, the course of marchandizing, a commodious exchange answering to either side, wherein an especiall animation is, the certaine knowledge we have gained in discovering the golden trade of the Moores in Barbary, which was the first incourager and beginning of this businesse, and for which the Adventure[r]s hitherto have beene laide, through the uncertainety whereof, those losses and mischances that have hapned, fell out, and there fore now should with a more setled resolution be /162 = 154/ followed to regaine, by knowledge, what ignorance miscarryed in. And I may joyne with this, the familiar conversation, faire acceptance, and mutual amitie, we finde the natives to embrace us withall, not onely celaring [sic: clearing] our owne doubts, which before knowledge must of necessity be, but likewise disprooving, and altogether confounding the report and speeches of all those, who, to serve their owne ends, gave out, the people above to bee a bloody and dangerous nation.[3]

Againe to advance the Adventurer, let the already knowne and certaine trade be remembred, which in my owne perfect knowledge I will make good, (against all Maligners, and secret opposers) that in our staple and principall commodities, it is not vented, but at tenne for one profite;[4] and admitte the

---

[1] Jobson continues to maintain this significant claim (but see p. 143, n. 2 above).
[2] The long confrontation between the English and the conjoint Spanish-Portuguese empire had been halted, at least in Europe, by a 1604 peace treaty, followed by a 1609 truce for twelve years between Spain-Portugal and England's Protestant ally, the United Provinces; hence, even when Jobson wrote, between 1621 and 1623, England remained at peace. However, overseas the position was more unsettled. While the English government sought amity with Spain, as reflected in Jobson's eirenic language, the Portuguese were inclined to distrust the dominant partner in the conjoint empire, to wish to maintain the confrontation with 'heretics' and 'pirates' overseas, and thus in West Africa to regard the English as illegal intruders who should be fought off.
[3] Jobson had, of course, an interest in depicting the local Africans as friendly and welcoming.
[4] Such claims of high profit, often made by promoters of overseas enterprises, were regularly proved to be misleading. One reason was that a ten-fold mark-up earned much less when transport, labour, and risk costs were taken into account. Jobson's Adventurers had already lost one ship and its goods.

discovery should not proove; yet there will be found places of trade sufficient, and that within the limit of faire recourse, to vent and put of such a reasonable proportion as shall bring a returne of that advantage, as shall be able to beare the charge of a further search, and likewise answer the expectation here at home of any reasonable minded adventurer, provided they doe arme themselves by knowledge, of what those things are which are vendible,[1] and likewise how to attaine unto those places, and order their occasions, where those returnes, are to be made; whereunto is added that the expectation is not long, in respect of other voyages, when as the returne is such, that within the compasse of tenne monethes, the whole voyage is to be performed, both out and home, allowing the ship to bee set foorth from *London*, and to to make their [there] her returne againe.[2] /163 = 155/

Moreover by the last discovery, so many hundred miles up the River, all which way is perfectly known, and from part to part observed, and every reach in order by me set downe, and carefully kept,[3] which may not onely cleare any doubts and difficulties in that already knowne way, but likewise enable the judgement for passing further, and especially order and give directions, what boates or vessels are most apt and proper to follow the discovery withall, as well for speedier passage, as also for the most advantage, to a more profitable returne.

And further we may take into consideration, how the times and seasons of the yeare, are unto us discovered, that the turbulent and infectious seasons may be provided for, and men advised the better to beare them, and provide for themselves, whereby (as it shall please God to give a blessing) those inconveniences may bee avoyded, which formerly have beene fallen into, and things more necessary carryed along, which through ignorance heeretofore have beene neglected, together with diverse other abuses, that by experience, no doubt, may be amended.

And lastly, in taking leave of you the noble gentlemen Adventurers in this hopefull Discoverie, let mee (under correction) say unto you, Be not discouraged, let not the jangling dispositions of any, whom your owne wisedomes leades you to see aime onely to make up their owne ends, dishearten you. And if it please you examine the condition of what is past, which, if I mistake not, may bee this [? thus] set downe. The first adventure was lost, and miscarried through want of Care and Judgement of /164 = 156/ those Sea-men and Merchants who had the managing, by over-much trust of supposed friends, who should at the very best have beene no otherwise thought

The vagrant Portingall.

---

[1] Jobson is no doubt thinking of the unexpected demand for salt.
[2] Jobson makes a reasonable point. The necessarily slow return on capital reduced the profit in overseas ventures relative to that from domestic investment; nevertheless, compared with voyages to destinations in the Americas or Asia, the passage to and from River Gambia was relatively brisk, making for a quicker return on the capital.
[3] Presumably this statement refers to his 'Journall'. But unless the original was much more detailed than the form in which Purchas printed it, perhaps Jobson also kept a notebook recording navigational detail and even sketch maps. Perhaps this was his 'Small Journall'.

and conceited of, then suspicious enimies, who have now discovered themselves, which will ever stand for a warning to avoyde the like, and trusting them any further: And that is all you have for that mony.

The second may some-wayes be laid uppon the Sea-men, whose understanding should have avoyded unseasonable times, and especially Discretion should have led them to have shunned watering in the very height of unseasonablenesse;[1] but it may be excused for want of experience, insomuch as there had never beene any triall made, so high in the River before to any effect, to discover the unholsomnesse, with the operation thereof, whereby so many of them lost their lives, and brought again another losse upon you, wherein the power of God was manifested, by whose onely hand they fell, and those few that returned, were sent to testifie, what they had felt and fallen into, whereby you have gained a perpetuall knowledge, for observing seasonable times, for your better proceeding heereafter, the valuation whereof being truly understood, may advance the imployment, which onely remaineth in that losse, to make you satisfaction.

And for this third and last, wherein mine eyes have beene a witnesse, how accounts are brought in, and perfected with you, I am ignorant, but I presume, as bad as it was, what with the returne that was made, and the remaynes brought home, of the *Cargazon* that was sent, you cannot (being /165 = 157/ justly dealt withall) receive any losse, but for gaine, it was never intended towards you, the whole businesse being carried by those you gave credit, and countenance unto, with an absolute hand, to abate and discourage your desires, for wading further uppon these Adventures, as by the manner thereof appeareth, which I have already particularly acquainted you withall, and unnecessary to be remembred heere; Onely this remaines, to make good, what I in carefull duety desired to lay open unto you, from whence that first intent, of giving that blow of discouragement unto you, did arise: you have beene since subject to divers other incounters, and all occasions are still earnestly pursued, to imbrace that oportunitie, that will give leave or way to strike you; And whereas they seeke to discourage you, yet by all publique and secret meanes can be devised, they both have and doe still addresse themselves, to proceed and goe on in the same adventure, as you both know, and have had just cause to except against. And apparent it is, that notwithstanding you in your generous dispositions have sate downe by the losse, yet there is that [you] have gained. But allow (if it please you) all had beene lost, if you shall againe consider, what charges and expences have beene layd foorth, and disbursed in Discoveries of this nature, nay in those of farre lesse expectation; with the recoveries and satisfaction, that afterwards they have made, even to this our native Countrey, whereof I forbeare

---

[1] Jobson repeats his view that taking drinking water from the river at particular times of the year caused the sickness. In practice, alternative water supplies may well have been more dangerously polluted as a result of human contacts. It is likely that the presence of mosquitoes amid a local African population spread malaria among the English, although they may well have also acquired, as well as fevers, dysenteric conditions through eating and sharing local food and drink.

examples, in regard they are not hidden from your true and ingenious knowledge: only in regard of some great resemblance, that may be to this intended busines, /166 = 158/ I may commend to your considerations, the voiage into *Muscovie*,[1] wherein the Marchants have that long passage of so many hundred leagues up a River, and by a customary trade, is brought to bee held as an ordinary passage, the Countrey, being fitted accordingly, by which use, it is now no other wayes unto them, then (as wee may terme it heere) our Westerne passages up the River of *Thames*; wherein were more probability for the attaining of this we ayme at, in regard our River is at all times open, and not subject to cold, nor those extreame frosts, which to the *Muscovy* trade, are so great hinderers: So that if you would conclude amongest yourselves, of a sufficient stocke, and be armed with a bancke, the ground of merchandising, to follow resolutely your undertaken enterprize, For so much as to mee belongeth, I dare affirme, you are upon the most promisingst occasion, that ever in our little Iland was undertaken, most especiall[y] considering by how small a charge it maybe perfected wherein as experience hath made me the Writer, to acquaint you with each particular, So likewise I offer my selfe up, both with my life and fortunes, even with my uttermost indevours, in your behalfes, (by Gods especiall blessing) to bring to perfection, what I have heere related, which is left with my selfe, to your worthy considerations.

FINIS.

---

[1] The earliest English voyages to Russia were made in the 1550s.

JOBSON'S LARGE JOURNALL

(as 'Extracted' in *Purchas his Pilgrimes*, 1/7/1, 921–6)

NOTE. The italicizing of names in the text and side notes is here omitted. Although most of the side notes merely repeat information found in the text or in Jobson's book, a few provide information not found elsewhere, and a number give spellings of River Gambia toponyms different from those in the text, suggesting possible miscopying of Journall items by Purchas. All are therefore included below.

A true Relation of Master RICHARD JOBSONS Voyage, employed by Sir WILLIAM Saint JOHN, Knight, and others; for the Discoverie of Gambra, in the **Sion**, *a ship of two hundred tuns, Admirall; and the Saint* **John** *fiftie,*[1] *Vice-Admirall. In which they passed nine hundred and sixtie miles up the River into the Continent. Extracted out of his large Journall.*

We set sayle from Gravesend, on Saturday the fift of October, 1620. On the five and twentieth, we departed from Dartmouth, we sayled from Dartmouth to the Canaries.[2]

The fourteenth of February,[3] we came to an anchor in Travisco Road,[4] where we found three Frenchmen, and one Flemming. Francisco a Portugall here dwelling was busie to enquire if we went to Gambra, having a letter as he said from M. Cramp, who had lately departed thence for Sierra Liona, set forth by the Company. This Portugall fearing just revenge for the ship taken and men betrayed and murthered by them in Gambra, had procured a Letter in behalfe of some of his friends.[5] In the River of Borsall we entred, where we tooke a small Boat belonging in part to Hector Numez, the principall in that Treachery and Murther aforesaid and detayned some of his goods therein for satisfaction,.taking thereof a publike Inventorie, that if any other could lay just clayme they might be restored. This was done by punishing Numez, and to terrifie others from like treacherous attempts, not without effect. The Portugals were glad they so escaped, knowing and cursing Numez his villanie.[6] The Portugalls which trade here, and inhabit the River

The Katherine betrayed.

Gambra Portugals which trade.

---

[1] The Glasgow reprint has 'sixtie', wrongly.
[2] Cf. *GT* 7–8, each text having additional information.
[3] The month is an error: for February read November (cf. *GT* 8).
[4] For Travisco, see Itinerary (Appendix A), p. 208, n. 2.
[5] It appears that Cramp, on his way to Sierra Leone on behalf of the Guinea Company, had delivered a letter at Rufisque exonerating named Portuguese from the attack on the *Catherine* in 1619, presumably for them to show to the 1620 English, and seemingly in response to a request from them. Perhaps they had been of some assistance to the English who escaped overland in 1619, possibly by helping them to find ships on which to return home. Not mentioned in *GT*.
[6] The River Saalum episode is not in *GT*.

are banished men, Renegadoes, and baser people, and behave themselves accordingly.[1]

The generall winds. We built a Shallop, and lanched it the two and twentieth.[2] The next day we set sayle up the River, and the tyde spent, anchored against a litle Iland on the South-side some foure leagues up.[3] From October till May, the winds are generally Easterly, and downe the River which much hindred our course up the same.[4] We past up by tydes, intending to stay at a *Tankorovalle* Towne called Taukorovalle,[5] but over-shot it in the night, and the next morning were against another Towne foure /922/ leagues higher, called *Tindobauge.* Tindobaug. Our ship with her Ordnance might here come both sides the River.[6]

Here dwelt Emanuel Corseen a Portugall, which told us that Master Tomson was killed by one of his Company,[7] and that the rest were in health. It was intended the Sion should stay here,[8] and therefore the Kings Customers [*sc.* customs officers] were paid, who dwelt some six miles from the River, but had his drunken Officers to receive them.[9]

The Voyage up the River. Leaving her with five and twentie men and boyes, on Wednesday, the nine and twentieth, the Saint John and two shallops, we set sayle up the River twelve men in the bigger, with Henry Lowe, and thirteene in the lesser with my selfe, which with the Boat towed her up in calmes.[10]

Pudding Iland. On the first of December, we came to Pudding Iland, sixteene leagues from the ship.[11] The second, we anchored against a little Creeke which leads *Maugegar.* into a Towne Maugegar [*sic*: Mangegar].[12] We went to this Towne, meeting by the way a Portugall, called Bastian Roderigo, who gave mee an Ounces skinne.[13] On Munday the fourth, the King with his Alcade came aboord, and drunke himselfe with his Consorts so drunke, that the Customes were

---

[1] Repeated *GT* 5, and enlarged 28–32. Jobson's hostile generalization was too sweeping, and not all the Portuguese traders in Senegambia were in any sense 'banished', renegade, or overall 'base'. See the section of the Introduction on 'The African Setting/Polities and Peoples/The Afro-Portuguese'.
[2] See Itinerary (Appendix A), p. 209, n. 2.
[3] Ibid., p. 209, n. 5.
[4] Hence, 'from May to October, it blowes up the River' (*PJ* 926).
[5] See Itinerary (Appendix A), p. 210, n. 1.
[6] Ibid., p. 210, n. 2. The river had narrowed to the width of a cannon-shot from each side of the ship in mid channel. Tendabaa is not mentioned in *GT*.
[7] Emanuel Corseen = Manuel ??Cozinha. For the death of Thompson, cf. *GT* 7.
[8] Itinerary (Appendix A), p. 211, n. 1. The *Sion*, at an unstated date, sailed up to Kasang (*PJ* 925).
[9] The information in this sentence is not in *GT*. It is noted hereafter that customs duties are paid at several other localities: whether they were paid earlier is uncertain. They were usually a payment to African rulers for permission to collect water and wood. Here, as probably elsewhere, they were paid at least partly in goods, as demanded, especially alcoholic drink, some instantly consumed. The drinking indicates that the rulers, or in this case, the officers, were not Muslims, or at least strict Muslims.
[10] Itinerary (Appendix A), p. 211, n. 2.
[11] Ibid., p. 211, n. 3.
[12] Ibid., p. 211, n. 4.
[13] Ounce = leopard.

deferred till next day. Henry Lowe agreed for a house, and left there Humfrey Davis, John Blithe, and one Nicholas a prettie youth, which two last dyed there.[1] On the seventh, we passed thence by a Towne on the Northside, called Wolley, Wolley, bigger then any wee had yet seene,[2] and in the after-noone came to an anchor at Cassan (where the Katharine was betrayed) where no Portugall would now be seene. This King is under the great King of Bursall. The Alcade shewed us friendship, and told us that the Portugals had hired men of that Towne to kill us as we went up, in some narrower Streights of the River, for feare whereof we could not get any Blackman to goe with us to be our Pilot and Linguist. This Towne is populous and after their manner warlike.[3] We here had intelligence, that Salt is a good Commoditie above in the River, and that within eight dayes there would come a Caravan from Tynda for Salt before this place.[4]

[margin: Wolley, Wolley. Cassan. Portugals perfidie.]

On the fourteenth, we came to a Towne on the South-side, called Pompeton, above which dwels no Portugall in this River.[5] Next morning we came to the Port of Jeraconde, two miles from which dwelt Farran a perpetuall Drunkard, but which held his Countrey in greatest awe. Hence Henry Lowe sent a slave with a Letter to Oranto sixteene miles off, where the English dwelt.[6] On the seventeenth, Matthew Broad[7] and Henry Bridges came to us by Land, which were exceeding glad after so long space to see their Countrimen, as we also to heare them report their securitie amongst those wild

[margin: Pometon. Jeraconde. English at Oranto.]

---

[1] None of these occurrences at Mangegar is noted in *GT*. Jobson makes the first of several references to Gambian rulers drinking to excess.

[2] Itinerary (Appendix A), p. 212, n. 1.

[3] For Kasang, see Itinerary (Appendix A), p. 212, n. 2. Jobson describes Kasang and its ruler more fully in *GT* 44–6, 59–61, but does not allude to the Portuguese threat, the failure to obtain a pilot/interpreter, the demand for salt and the Tinda caravan.

[4] The salt was brought up-river in canoes and mostly originated in the salines of River Saalum (see *GT*, note 101). Tynda/Tinda was the interior district visited by Thompson.

[5] Itinerary (Appendix A), p. 213, n. 1. That no Portuguese lived higher (repeated, with the Portuguese man named: *GT* 118) may or may not have been correct; certainly Portuguese traded higher up-river, but Jobson does not report meeting any.

[6] *GT* does not name Jeraconde or Oranto, for whose possible locations see Appendix B. In fact, *GT* does not allude other than vaguely to the English survivors whom Jobson now rescues, or to the mid-river base he establishes, 'Whiteman's Town' (*GT* 68–9). Apart from a reference to the blind king, *GT* 58, none of the information in this paragraph is in *GT*. Lowe was 'principall factor' (*GT* 40), and is mentioned by name again later (*PJ* 922). The reference to a slave is of interest, not least given Jobson's subsequent denial of English slave trading (*GT* 89), and it seems unlikely that Purchas introduced the term. Yet neither text indicates where and how this man was acquired, or indeed that any Africans other than the two taken aboard at Batto were with the party. The English may, however, have been given or lent him, as an interpreter and messenger, at one of the halts in their progress up-river, and if so most likely either by the Portuguese man at Pompeton or the headman at Jeraconde. At Kasang they had been unable to obtain 'any Blackman to goe with us to be our Pilot and Linguist' (*PJ* 922). The wording does not completely rule out being given a slave at Kasang but makes it unlikely. If the original Journall really contained the term, the stress laid on the man's slave status, yet his qualification as a messenger being apparently that he knew the neighbourhood and could be trusted to carry a message 16 miles without decamping, may lend some support to the suggestion above, that either the Portuguese man or the Jeraconde headman provided him. If the former, he was most probably a domestic slave.

[7] Broad went up-river with Jobson (*GT* 924).

Oranto.

Ferumbas faith.

Batto.

people. Broad said, much good might bee done up the River, but that it must be done without delay, the River falling daily. Comming within six miles of Oranto we landed and went thither, where Brewer which had beene at Tinda with Tomson, filled us with golden hopes. But the neglect of bringing Salt thorough ignorance or emulation was a hinderance. The King of Oranto abode on the other side of the River; his name Summa Tumba, a blind man and subject to the great King of Cantore.[1] We went to him and had a speech made to him of thankfulnesse, for our Countreymens kind usage; His answere was repeated by the mouth of another, after the fashion of the Countrey: which Ceremonie done, he made hast to drowne his wits in the Aqua-vitæ and good liquor we brought him. His Custome paid, we departed.

The one and twentieth, I sent away my Boat, & the next day came abundance of people; some to sell; all to begge; the King sometimes by his Wife, sometimes by his Daughter, but every day his Sonnes were there, and likewise divers others of the better sort, but Count, from many great persons: which word they use for commendations. You must returne something againe, or it will be ill taken.[2]

On Christmas day, Ferambra sent us as much Elephants flesh as one could well carrie, new killed. This Ferambra went foure miles off, and was a friend of our people, and when the Portugals had dealt with the King of Naoy, to kill them all, who sent his forces to performe it, he put himselfe and his people in Armes for their defence, and conveyed them over the River to his Brother, called Bo John, and saved their goods.[3] On the one and thirtieth, came the Shallop backe.[4]

We being ten white men, went the second of January from Oranto for Tinda:[5] the first tyde we went to Batto, Bo Johns Towne, and there agreed with a young Mary-bucke to goe with us. Lowes emulation hindred us with delayes, both now and before.[6] On the sixth, Sumaway, King of Bereck under the great King of Cantore, came aboord with his Wife, and begged our courtesie.[7] We tooke in Sangully, a blacke Boy, who had lived with Master

---

[1] Itinerary (Appendix A), p. 213, n. 5. The title *suma* could indicate a petty ruler who was also heir to the ruler of a larger polity, but it seems to have been used at times more loosely, to indicate the governor of a district.

[2] The information in this paragraph is not in *GT*. The reference to 'Count' is puzzling and the sentences may be garbled. But perhaps the 'many great persons' were addressed in Portuguese as *Conde* 'Count', the term certainly being used at a later date in Senegal to describe leading individuals (Hair, *Barbot on Guinea*, p. 137, n. 17, 'Condy'). The last sentence refers to the African expectation that gifts will be reciprocated.

[3] See Itinerary (Appendix A), p. 214, n. 1. The margin note gives Feramba (misprinted Ferumba), which may be more correct, but *GT* also gives Ferambra. Naoy is an error for Nany (*GT* 100), that is, Niani. While *GT* 100 repeats Ferambra's defence of the English, it omits the passage across the river. The Christmas gift of elephant meat is described at greater length in *GT* 149=141.

[4] Itinerary (Appendix A), p. 214, n. 2. This reconnaissance up-river is not mentioned in *GT*.

[5] Jobson now follows Thompson in exploring the upper river. See the Itinerary (Appendix A), p. 214, n. 3.

[6] *GT* refers to Boo John's town, but not by the name Batto; however, it gives a name to Jobson's marabout, Fodee Careere/Kereere (*GT* 63, 109). For the probable locality of Batto, see Appendix B. For Bo John and Lowe, see the Itinerary (Appendix A), p. 214, n. 4; p. 215, n. 1.

[7] The visit of the king is not in *GT*. See Itinerary (Appendix A), p. 215, n. 2.

Tomson, and spake prettie English.¹ On the ninth, we anchored in a vast place both at noone and night, where was a world of Sea-horses, whose pathes where they went on shore to feed, were beaten with tracts [*sic:* tracks] as great as London high-way.² Next morning we anchored at Massama-coadum, fifteene leagues from Pereck. On the eleventh, at Benanko.³ The twelfth, after rockie passages to Baraconda. The tyde went no further.⁴ Beyond were no Townes, neere the River, nor Boates nor people to be seene.⁵

Sea-horses, high-wayes.

On the fourteenth, Bacay Tombo, the chiefe man of the Towne, came a board with his wife, and brought us a Beefe. We hired another Marybuck, because they are people which may travell /923/ freely: & now were ten white and foure blacke.⁶ Having now the streame against us, we durst not for feare of Rockes in the night, nor could for immoderate heat in the Suns height proceed, but were forced to chuse our houres in the morning till nine, and after three in the afternoone.⁸ We past by Wolley a small River,⁹ and found above shallow waters, wherein were many Sea Horses curvetting and snorting hard by us, one came swimming by us dead and stinking, yet the Negros were displeased they might not eate him.¹⁰

Marybuckes, sacred persons, by the superstition of those parts, and are their Priests and Merchants.⁷

On the seventeenth, on both sides the River we saw thousands of Baboones and Monkies. A Sea Horse gave the Boat a shrewd blow, but did no harme. We had still our Canoe before us to sound the depth.¹¹

Sea-horses abounding in the fresh water, both in the water and on shoare. They are like a horse, but with clawes on their feet, and short legs, tuskes, manes, etc. Monkies and Baboones.

On the eighteeneth, we were forced to enter the River naked, very fearefull of the Bumbos, (so they call the Crocodiles) and carry the Boat against the current, and over or thorow the sand, heaving and shoving till we come in deepe water.¹²

The nineteenth, we met with a violent current, that all the strength of sixe Oares could prevaile, but a mile in an houre.¹³ The twentieth, on the Starboard side, we had Cantore River, which hath a faire entrance, where Ferran Cabo is the great King.¹⁴ On the one and twentieth, we sent a shoare to the Mountaine tops, whence might be perceived onely Desarts, replenished with

---

¹ Cf. *GT* 83.

² See Itinerary (Appendix A), p. 215, n. 4.

³ *GT* begins its account of Jobson's journey up-river from Baarakunda, so omits the information on the passage from Batto. See the Itinerary (Appendix A), p. 215, nn. 4, 5.

⁴ On the 'rockie passages' and the tide, see ibid., p. 215, n. 7; p. 216, n. 1.

⁵ Ibid., p. 216, n. 1.

⁶ Cf. *GT* 82–3 (where the headman brings 'two beeves'); and see Itinerary (Appendix A), p. 216, n. 2.

⁷ *GT* includes a lengthy account of the 'Marybuckes' or maraboults: *GT* 61–2.

⁸ Enlarged in *GT* 11, 15 (but 'until 9. or 10.', with the marginal note '7. houres in 24.'), 83 (4 hours in the morning and four in the evening).

⁹ Itinerary (Appendix A), p. 216, n. 4.

¹⁰ See p. 216, n. 5 above. Purchas' marginal note has 'clawes' instead of 'paws'.

¹¹ *GT* does not specify this sighting of monkeys and baboons but describes the animals at length: *GT* 143–5. For the canoe, cf. *GT* 83, and for three blows from a hippopotamus, *GT* 21.

¹² For more than one undated instance of this action, *GT* 18–19.

¹³ Not specified in *GT*.

¹⁴ Itinerary (Appendix A), p. 217, n. 1. The river and the ruler are not noted in *GT*.

terrible wild Beasts, whose roaring we heard every night.¹ The Blackes are so afraid of the Bumbos, that they dare not put their hands into the water, divers of them being by them devoured. Yet did they avoid from us, whether it were our noise or multitude which caused it. Some we saw thirty foot long, yet would not come neere us.² On the two and twentieth, walking on the banke, I espied sixteene great Elephants together hard by me. A Blacke with me fell a trembling. The sedge in the place was almost as high againe as our heads; so that we could not be seene till we were within Pistoll shot of them. We saw divers little ones by their sides. We made an offer to shoot, but the Peece would not off, which they perceiving began to run, in a miles space not so much as turning nor looking behind them; making speed to the Mountaines, like a Deare in the Forrest. The Moores wondred at our adventure.³ On the three & twentieth, we were faine to enter the water, & by strength of hand, to carry the boat a mile & a halfe into deeper water. On the foure and twentieth, we towed her, sometimes adding ha[u]ling by the Boats side, as sholds and trees permitted: and met with one vehement current, overthwart broken rockes, so that we were forced to hold her by force, till one taking the Anchor on his neck, waded above that quicke fall, and letting it fall, we ha[u]led by our hasor [hawser], and escaped that gut.⁴

The five and twenty, troubled with sholds, we heard as we passed, a gush of water, hidden by the greene trees, with which water we stored our selves; that of the River being so ranke with a muskie sent of the Crocodiles, as we supposed that it was dista[s]tefull; whereas this was pleasant.⁵ One of our Moores was taken, and like to be lost in a Whirlepoole; notwithstanding, he could swimme well, had not one of our men laid hold on him as he rose the third time, almost spent, from under water.⁶ On the sixe and twentie, we were comforted with the sight of the hill of Tinda, being high rockie land.⁷ We sent three Moores thither with a present to the King, and to Buckor Sano, a Merchant of Tinda,⁸ intreating him to come downe to us with provision, for we had no flesh. Deare and Fowle were plentifull on both sides the River, had we beene provided of a good Peece.⁹ And the River fish did so taste of Muske, that (like the water) we could not endure the shoare [*sic:* taste]. I went

---

¹ 'Desarts' = wildernesses, unpopulated areas. Cf. *GT* 85.
² Jobson discusses crocodiles at length in *GT* 16–19.
³ The elephants observed the men, not, as Jobson's wording suggests, the gun (they were, in any case, presumably unacquainted with guns). The episode is not specified in *GT*, but for elephants fleeing from shots, see *GT* 142.
⁴ These episodes are not specified in *GT*, but there are general references to hauling the boat in *GT* 18.
⁵ 'Muskie' water noted three days before reaching the highest point up-river, but not the fresh spring (*GT* 19).
⁶ Itinerary (Appendix A), p. 217, n. 2.
⁷ Ibid., p. 217, n. 4. Not mentioned in *GT*.
⁸ Ibid., p. 218, n. 1. Buckor Sano had been sought at Tinda by Thompson but missed (*GT* 84–5). Jobson fails to specify that a message had been left at Tinda by Thompson to inform Buckor Sano of the English interest, and that the English would return, but it may be presumed that this happened.
⁹ The shortage of food is only casually indicated in *GT* 85. The statement that they lacked a 'good Peece' is puzzling, since the party carried a number of guns and Jobson four days later

ashoare to view the River, & might see sometimes twentie Crocodiles one by another: and in the night, specially towards breake of day, they would call one to another, much resembling the sound of a deepe Well, and might be easily heard a League.¹ We past the sholds, and against Tinda River, recovered steepe water, and saw many Sea Horses, which love deepe waters.² On the thirtieth, we killed an Anthelope bigger then any Windsore Stagge, the blood of him drew a world of Eagles, and other Fowle; amongst which came one Stalker, a Fowle higher then a man, which we likewise killed.³ Presently after, came our men backe with Buckor Sanos brother, and a servant of the Kings, with Hens. Our Deare was killed in good season for their entertainment: the report passing among them current, that with our thunder (so they called our Guns) we could kill whatsoever we would. They much fearing the same, as having never seene or heard it, whereof we made good use.⁴ [*Antelope. Gun-thunder.*]

On Thursday, the first of February, came Buckor Sano with a troupe of forty people, amongst which his wife and daughter. Having tasted of our strong Waters, hee lay drunke aboard that night (he was never so after) and was sicke the next day. He gave us a Beafe, and many of the people brought Goats, Cocks, and Hens, which we bought easily.⁵ On Saturday, we began to trade our Salt, which is the chiefe thing they desired; other things they asked for, which we had not provided: slaves (he told us) were the things they held dearest; for any thing else we should have, if we would maintain our comming thither, he would provide it.⁶ We had some Elephants teeth, Negros Clothes,⁷ Cotten Yearne, and some gold of them. We refused to buy Hides, because we would not lade our Boat downe the River, the water falling every day, which wee kept note of by the shoare. The people came daily more and [*Salt, chiefe trade.*]

---

shoots an antelope and a bird. However, the reference is perhaps to the 'Peece' which misfired four days earlier and which may have had some special quality.

¹ Cf. *GT* 19.
² Tinda River is the river Jobson 'went ashoare to view', presumably by walking along the bank of River Gambia towards the junction. See Itinerary (Appendix A), p. 218, nn. 2, 3.
³ Ibid., p. 218, n. 4. On antelope and deer generally, and on the Stalker bird, a Crowned Crane (*GT* 143, 155/147–156/148).
⁴ Enlarged in *GT* 86–7. It is highly unlikely that the merchant and his entourage, who apparently travelled down-river regularly, apparently as far as the coast, were ignorant concerning guns and their firing, since the Portuguese traders had used guns for shooting animals, and at times enemies, for more than a century. But there is no evidence that guns had reached the interior peoples.
⁵ Much enlarged in *GT* 86–8, which, however, does not mention the trader's wife and daughter.
⁶ Enlarged in *GT* 89–90, which includes the famous refusal of Jobson to buy slaves. But here the statement about slaves being 'the thing they held dearest' is unclear. It may mean that, among commodities on offer to Europeans, the Africans charged most for slaves. It is perhaps less likely that it means that the merchant is especially interested in buying slaves from the English – his expectation not entirely implausible if the Portuguese sometimes conveyed slaves up-river (although this appears to be unevidenced). However, the wording may be Purchas's and a misunderstanding of the original text.
⁷ 'Negros Clothes' = locally-made cotton cloths. Probably these had been purchased at the mid-river base, and they provide an instance of a common feature of the period, Europeans acting as commercial intermediaries in West Africa, by trading African commodities bought from African merchants for other African commodities sold by African merchants.

more to us, and upon the shoare they built houses, we also had a house open to trade under, so as it seemed like a pretty Towne.¹ Our Blackes went over the River, and three dayes after brought other people, which built a Siege /924/ Towne on the other side the River.² And within three dayes there were five hundred, which were a more Savage people; having breeches of beasts skins, neither had they ever seene any white people before. The women would run and hide themselves when we came neere them at their first comming; but after grew bold to buy and sell with us. These people likewise were all for Salt, and had Teeth and Hides store.³ Our Salt was almost gone before they came; for we had but forty bushels at first.⁴

<small>Bajay Dinko usko was the chiefe man, & called by the name of his Countrey, under the great King of Cantor.</small>

Bajay Dinko the chiefe was aboard, very desirous we should come againe.⁵ On that side wee saw likewise there was Gold, and those people had familiarity with each other, whereby it seemed they had trade and commerce, by some higher part of the River.

<small>Juddies or Fidlers.</small>

On the seventh, the King of Jelicot on Tinda side, under the great King of Wolley,⁶ came downe with his Juddies or Fidlers, which plaid before him and his wives, such being the fashion of the great ones. These Juddies are as the Irish Rimers: all the time he eats, they play and sing songs in his prayse, and his ancestors: When they die, they are put in an hollow tree upright, and not buried,⁷ we gave him a Present, and he a Beefe to us.

<small>He bought and sold etc. for us.</small>

On the eighth, Buckor Sano would needs be stiled the white mans Alcaid; I tooke it kindly, and put about his necke a string of Christall, and a double string of Currall. Broad gave him a silver chaine, and with drinking a cup of Rosa-solis, and shooting off five Muskets, a solemne cry, Alcaide, Alcaide, was proclaimed: he adding his fidlers musicke, the people also ready with their bowes and arrowes, his wife with matts on shoare to attend the solemnity. So soone as he came on shoare, he frankely gave his nuts to the people, rejoycing in this new honor. These nuts are of great account through all the River,

<small>Nuts of precious esteeme. He seemeth to be the Cola.</small>

and are a great favour from the King: five hundred of them will buy a wife of a great house. Their taste is very bitter, but causeth the water presently after to taste very pleasant.⁸ This done, he went to the Kings house, who sate without doores, their fashion being assoone as it is darke, to make a fire of Reed

---

¹ Cf. *GT* 87, 89.

² Enlarged in *GT* 94–5, which does not use or suggest an explanation of the term 'Siege Town' – perhaps it means that the houses were collectively enclosed with some form of a wall or fence. For the people, see Itinerary (Appendix A), p. 219, n. 3.

³ The sentence seems garbled and should probably conclude 'Teeth and Hides in good store', that is, in quantity.

⁴ The amount of salt carried is not in *GT*.

⁵ Cf. *GT* 95, where the name is given as Biage Dinggo, and it is not said that he is 'called by the name of his Countrey' or that he is under the King of Cantor (as stated in the margin note but not in the text), although the latter point follows from his coming from the south side of the river. For the name, see *GT*, p. 144, n. 3.

⁶ Itinerary (Appendix A), p. 219, n. 4. Cf. *GT* 96, where the king is described as the 'immediate King of that land whereon wee kept our trade' and there are no references to 'Tinda side' (that is, east of the confluence) or his dependence on Wuli.

⁷ Jobson described the '*Juddies*' or praise-singers at length in *GT* 105–8.

⁸ Kola is discussed in *GT* 134, 134bis.

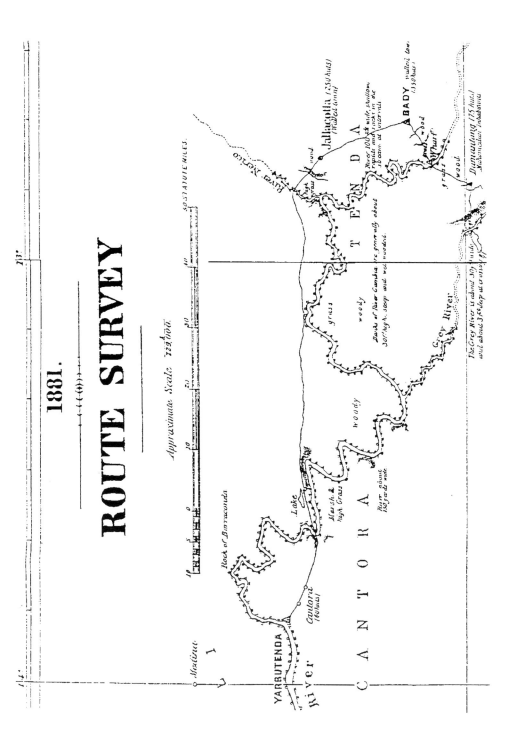

without doores, and the best sort to have matts, on which to sit downe, and use their Ceremonies. He placed me by the King, and went himselfe sixe paces off, right before him, and made a speech, which one repeats after him as he speakes, to intreat his kind usage to the white men. The King answered with a like speech, giving us liberty to shoot any thing on the land, and none should offend us. Buckor Sano kneeled downe, and gave him thankes, and sent unto him in our behalfe the Currall and Christall. Whereupon the King made a long speech, concluding, that he would give him that land whereon we were, for us, for ever. Upon which words, Buckor Sano pulled off his shirt in token of thankefulnesse, and kneeled downe naked, untill divers Marybucks with their hands raked up a heape off the ground, upon which he lay flat with his belly, and covered him with the earth lightly from head to heele. Then with his hands he threw the earth round about amongst us all: after which, the Marybuckes gathered a round heape againe together, and compassing it with a round ring of the same earth, [a Marybucke] wrote with his finger as much as the round would containe: which done, Buckor Sano tooke of that earth into his mouth, and put it forth againe, and then taking both his hands full of the earth, and our two Marybuckes following him upon their hands and knees, they came to me where I sate, and threw it into my lap. This done, he rose up, and two women were ready with clothes to wipe him, and a third woman with a cloth to fan him, and stepping a little off, he had his best clothes brought him, which he put on, and his sheafe of arrowes about his necke, a bow and an arrow in his hand. He came in againe, and twenty more, with bow and arrowes, after he had gone twice or thrice about, presenting himselfe by drawing his arrow up to the head, as if he were to shoot, he delivered them, and sate downe by me. The rest with their bowes and arrowes came one after another, and kneeling at his foot with their faces from him presented their bowes, as hee did. Then began others to dance after their fashion, at the end whereof they began to make severall speeches, (for every one of the better sort will have his speech) wherewith we were weary, and left them for that night.[1] Our manner was to set our watch with a Psalme,[2] which they hearing, would be still, and after a shot would leave us quiet till morning.

I shewed this Bucker Sano a small Globe, and our Compasse, whereupon he told us that he had seene with his eyes a Countrey Southward, whose houses were all covered with gold, the people wearing iron in rings through their lips and eares, and other places, to which place hee told us it was foure moones travell.[3] Hee told us likewise of a people which hee called Arabecke,

The Countrey given to the English.

In this manner the Kings take possession of the lands they came [? come] to.

Iron preferred before Gold.

---

[1] This whole episode is repeated at even greater length in *GT* 96–100, but with 'three guns' not five muskets and forty not twenty men with bows and arrows; however, it omits mention of Broad, the wife, the 500 nuts, the great house, the third woman, and the additional speeches.

[2] Not mentioned in *GT*.

[3] 'Southward' can hardly be correct, unless the 'four moones travell' represents an extremely vague reference to distant goldfields in southern Africa – but their existence would be more likely to be known to Jobson than to the merchant. It is more likely that this is either a Purchas misreading of, or a misprint for, 'northward'.

who came unto this Countrey, and would be at a Towne, called Mombarr, but six dayes journey from Tinda, the second moone after, which was in March. And there was a Town called Jaye, from whence much gold came, but three dayes journey from Mombarr, whither these Arabeckes went not. More I might have knowne, had not the emulations of my companie hindred, who would not suffer the blacke boy to let me know what he speake.[1] Much Gold.

Some people which came to us, were of Combaconda, a Towne foure dayes journey thence, which we thinke is Tombuto.[2] A Marybucke was here of Master Tomsons acquaintance, borne in Jaye, which would not company with the people of Tinda, but came to us, and told us that many people were comming, but were sent backe by some that returned, and reported our Salt was gone. He offered, if we were past these people, he would undertake to bring us to Mombarr /925/ and Gago.[3] We made haste to be gone: for by our marke the water was sunke above six inches, promising to returne in May, when the water increased. We called this place Saints [*recte*: Sainte] Johns Mart.[4] Combaconda. Tombuto.

On Saturday the tenth of February, we came away, the wind and streame served, but wee durst not sayle for the sholds, nor row by night. On Wednesday at night, having but three miles to Baraconde, by the Moores intreatie wee went thither over Land, and passed easily in six dayes downe, what had cost us twelve dayes labour and trouble.[5] We had a great chase at an Elephant, wounded and made him flie but lost him in the high sedge, and after found him in the River, where being shot in the eare he turned head on us, and made us row off, and leave him thrice wounded, our Peece failing in the discharge. Barraconde.

Munday the nineteenth, we came to Butto, Bo Johns Towne, [where] wee had our first Marybucke and the blacke Boy which spake English, whose age of sixteene yeares was now ripe for their Circumcision. Hither we came in season for that Solemnitie, hearing before we came, shoutes, Drummes and Countrey Musicke. The Boy knew the meaning, and told us it was for cutting of prickes, a world of people being gathered for that purpose, like an English Faire. Under everie great tree, and among all their houses at night were fires without doores, and in especiall places dancing, the Musicall Instruments made with Keyes like unto Virginals, whereupon one playes with two stickes which have round Balls of leather at the end, about their wrists Iron Bracelets. They are called Ballardes, and contayne some seventeene Keyes.[6] The women for the most part dance with strange bending of their bodies, and cringing of their knees, their legges crooked, the standers by keeping a Circumcision. Daunces.

---

[1] Cf. *GT* 90–1, 92–3, without mention of the globe and compass, the iron rings, or Jaye, this toponym being mentioned instead by a marabout (*GT* 100).
[2] Itinerary (Appendix A), p. 219, n. 2.
[3] Cf. *GT* 100–101.
[4] Cf. *GT* 104–5, without the six inches.
[5] No further details about the down-stream journey to Batto appear in either text. See Itinerary (Appendix A), p. 219, n. 6.
[6] For this instrument, cf. *GT* 106–7.

time in clapping their hands together to grace the dance. If the men dance, it is one alone with such Swords as they weare, naked in his hand, with which he acteth.

About two furlongs from their houses under a great tree were many fires, and much drumming with great noyse: here they said were those which were cut, but would not suffer mee to goe see. Some distance beyond we might heare a great roaring noyse, which they fearfully said was the voyce of Hore, that is, after their imposture a Spirit, which approacheth at great Feasts, for whom they provide store of Rice, Corne, Beefe, and other flesh readie drest, which is instantly devoured. And if he be not satisfied, he carries some of their Sonnes (the uncircumcised Females he regards not) and keepes nine dayes or more in his belly, then to bee redeemed with a Beefe, or other belly-timber: and so many dayes after must they be mute, and cannot be enforced to speake. This seemes an illusion of their Priests to exact Circumcision, and the hoarsenesse of some shewed, they had lost their throats in that roaring. This roaring, shouting and dancing continued all night. We saw our blacke Boy circumcised, not by a Marybucke, but an ordinary fellow hackling off with a Knife at three cuts his præpuce [*sc.* foreskin], holding his member in his hand, the Boy neyther holden nor bound the while. He was carryed to the rest, nor would they suffer our Surgeon to heale him.[1] The people in twentie miles space came in to this Feast with their provision.

I made haste backe to Setico,[2] to meete the Tinda Merchants, and on the sixe and twentieth, being within two miles of the place, I received a great and dangerous blow by a Sea-horse which indangered our sinking, but we made shift to stop it with some losse. We came to Setico foure miles from the water side, the greatest Towne we saw in the Countrey, higher then which the Portugall Trade not, and from hence carry much Gold; the most of the Inhabitants Marybuckes, and the Towne governed by one of them, called Fodea Brani.[3] They are stored with Asses and Slaves, their Merchandize Salt. The chiefe Marybucke dying, there came multitudes of people to his Funerall. Of the Grave-Earth digged for him every principall Marybucke made a Ball mingled with water out of one pot, which they esteemed as a Relike. They lay all sweet smels they can get into the ground with him, and tooke it kindly that I bestowed some. Much Gold is buried with them, or before by themselves in a private place, for their use in another World. Much singing, or howling, and crying is used many dayes about the Grave.[4] This recourse was also to establish his eldest Sonne in his dignitie, to which many Presents are sent. I saw among other beasts one Ramme of a hayrie Wooll like Goats. Sonnes succeed their Fathers, but the Kings Brethren take place before the Sonnes. The sicknesse of our men in the Saint John, hastened my departure.[5]

---

[1] For a fuller version of the circumcision ceremony, see *GT* 109–18.
[2] Itinerary (Appendix A), p. 220, nn. 1, 2.
[3] Fodee Bram, *GT* 63–4. For the visit to Sutuko, cf. *GT* 63–6.
[4] Cf. *GT* 70–2.
[5] Not in *GT*.

Sunday the eleventh of March, I returned [to Butto], and on Wednesday came to the Saint John. The next day, I set forward to the Sion, and on Saturday came to Pompetan, where the Portugall made us good cheere. Hee told us of the Devils giving notice of our beeing in the River, and comming up, which the circumstances made probable.[1] On Munday, we came to Cassan, a hill where the Sion did ride: the Master and many others dead, and not above foure able men in the Company.[2]

Devils oracles.

Here we lay from the nineteenth of March, to the eighteenth of Aprill,[3] wee weighed and came the next morning to anchor against Wolley, Wolley, under the King of Cassan. Whiles wee were there, came a new King from the King of Bursall to take possession of the Countrey, the old King being ejected as the Sonne of a Captive woman, whereas this was right Heire by both /926/ Parents, and now comne [sic] of age, who now transported himselfe and his over the River, to give place to this new King, which promised us all kindnesse.[4]

The twentieth, we came to Mangegar, within a mile of which, every Munday is a great concourse and market, but miserable Merchandize. The last of Aprill, the Saint John came to us, and the fourth of May we sayled downe the River together.[5] From May to October, it blowes up the River except in the Ternado, which comes for the most part South-east. On the eighteenth, we prepared our Shallop. On the nineteenth, we set up Tents on the shoare.[6] The King of the Countrey called Cumbo, came to us, and was very kind and familiar, promising all favour, labours of calking and other businesse, watching and Musketos, which here exceedingly abounded, did much molest us.[7] On the ninth, wee turned out of the River. Next morning before day, we had a violent storme, or Ternado, with Thunder, Lightning, and exceeding store of raine. This weather is frequent from May to September. Wee put in at Travisco for Workmen, our Carpenters being dead. Thence we hasted home.[8]

---

[1] Cf. *GT* 118–19.
[2] None of the information about the return journey given in this paragraph is in *GT*.
[3] Jobson gives no explanation as to why the ships stayed at Kasang for a month, despite the mortality and sickness.
[4] Cf. *GT* 59–60.
[5] Itinerary (Appendix A), p. 221, nn. 2, 3.
[6] Ibid., p. 221, n. 5.
[7] Unless the king provided guards for the English camp, 'watching' seems out of place. Perhaps there is a copying error and the passage should read – '… other business, watering and wooding. Musketos, which …'.
[8] The references in the last two paragraphs to the passage down-river, other than the reference to the prevailing wind, are lacking in *GT*. For the passage home, see Itinerary (Appendix A), p. 221, n. 6.

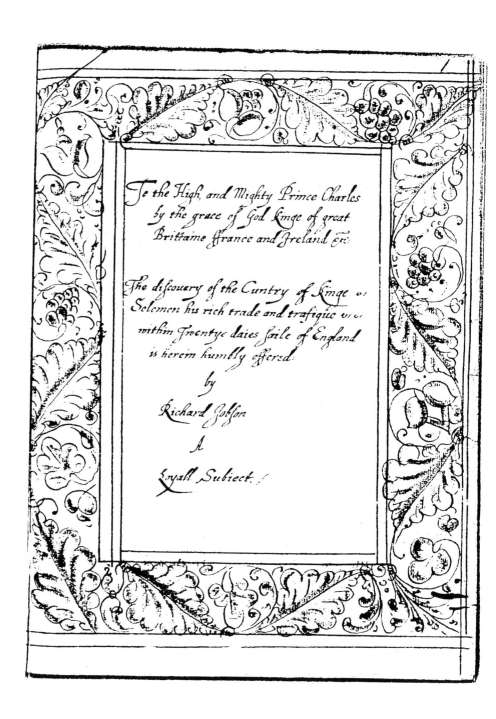

To the High and Mighty Prince Charles
by the grace of God Kinge of great
Brittaine ffrance and Ireland &c:

The discouery of the Cuntry of Kinge
Solomon his rich trade and trafique
within Twentye daies saile of England
is herein humbly offered:

by

Richard Jobson

A

Loyall Subiect

JOBSON'S PETITION (?1626)

British Library, Royal MS L8 A LVIII, 275, ff.1–5v

*To the High and Mighty Prince Charles by the grace of
God Kinge of great Brittaine ffrance and Ireland etc.*

*The discovery of the Cuntry of Kinge Solomon his rich
trade and trafique within Twentye daies saile of England
is herein humbly offered*

by

Richard Jobson

A

Loyall Subject.

/2/
Most Gracious Soveraigne

Maie it please your *Ma:^{tie}* to pardon my attempte who am ledd in zeale to your *Ma:^{ts}* honour & my countryes service to manifest my duties. That whereas some two yeares since I presumed in a fewe leaves to write this discovery of kinge Solomons traffique, unto your princelie father of happie memory, which some reasons leade mee to conceave might or did miscarry, the rather, in regard I was never called to any attempt of the same, as my humble desire mentioned, It maie please your *Highnes* to take into your consideraton the greate importance of soo weightie a busines, which offerd it selfe unto your *Ma:^{tie}* being but 20 dayes saile from you the people being allready made & fitted with a longing expectaton that this our nation should bee the people with whom they intirelv desire a Commerce. Myselfe was the first white man made tryall of their conditon & brought them such commodities [as] were very acceptable. I speake of those higher parts of the River I traded in, being some 320. leagues, or 960. myles up the river, from whence is an undoubted possibilitie of the returne of great quantitie of rich gould, within the compasse of ten monthes, both out & home and to continue from yeare to yeare, but the hinderaunces have bin such since my Accompt of the discovery, by reason of the differences betweene the Gentlemen and the Marchaunts, who should avoyd it, as hath spoild the proceeding, insomuch as it hath not onlie bin disputed at the moste hon[ora]^{ble} counsell, but likewise brought & argued in the high courte of *Parliament*, and remaynes, as yet undetermined, notwithstanding the gentlemen presumeing on the greater libertie; resolv'd the last yeare to put on upon it. but by the secret practise of the Marchaunts they were hindered of such convenient shipping as they desired & was fitting for them; and in conclusion forc't to set forward in a Spanish bottome; bought here in the River of Thames, which proved rotten, & so omynous to their proceeding, hazarding the lives of us all that were ingaged in her, albeit wee went not farre, only bearing it in the channell, [so] were inforcte to put in no farther then *Dover*.

Again that the truth maie appeare unto your *Ma:^{tie}* it was much to bee doubted amoungst some of the zealous mynded gentlemen that were adventurers, as also even amoungst some of ourselves, that were to goe the Voyage whether our designe could prosper, in regard many of the gentrye that made upp the Adventurers were Romish Catholiks, by meanes whereof they had shipt amoungst us /2v/ divers of their owne faction whereby wee were devided, even here at home before wee went, whereas in so great an action, wherein the cheifest place the glory of God is pretended, & consequentlie the honor & wealth of king & country no care can be too greate even to avoid the least inconvenience, and haveing considered some accidente befell us many of us resolve the stoppe this last voyage was God's blessed hinderaunce, And most gracious sovereigne the fullnes of tyme is sure appointed for every great action. Therefore maie it please your *Highnes* to call this busines to an attempt, wherby there maie be some direct course ordered for the proceeding in it, there will appeare more & better hopes for the gayning of this gowlden expectaton then the king of *Spaine* could have in his first attempte of *America*, wherein experience makes mee bolde, & I should be moste willing to expresse my reasons more largely. Only here breifly as I maye for feare of offending I presume to present some part of my proceedinge.

It maie therfore please your *Matie* to understand that in the North latitude of 13 degree & ½ there is a River called by the name of *Gambra*, into which from the harbour of *Dartmouth* in the West part of England I sayled in 20 dayes. This river as appeareth comes very farre from the Inland parts of *Africa*, for proofe whereof myselfe with only nyne men more of my owne countrymen went upp in a small shallop so neare as I could with carefull dilligence observe about some 320. leagues or 960. myles, haveing in that passage had conference with six of their pettie kings besides other principalls & commaunders of Countries, and from them all found faire and freindlie acceptaunce: More especially in the highest parts where I was amoungst them, for as they are a black people & undoubtedly the very *Ethiopian*, so these in those higher parts wee were at had never seene white people before, yet notwithstanding such freindly commerce was betwixt us, & so acceptable unto them was those poore commodities wee brought that a faire exchange was return'd, and a most earnest desire came from them to encourage us to continue our recourse for better animation whereof the king of the highest place commynge where our boate did /3/ ride did there, after a moste solemne manner deliver unto mee a most ample possession of that part wee were one [on], being done in their fashion with prittie & curious ceremonies. From these people wee received perfecte intelligence that wee were come within the confines of those countries where the Moores of *Barbary* fetch their gould, & what commodities they brought to fetch the same, likewise of the trade they had with an unknown rich people, such as divers Authors doe write of, but more especially a principall Marchaunt who had recourse unto us, constantly affirmed that high with in the country there is infinite store of Gold, and of soe meane esteeme as the people preferring Iron doe weare in their ears and lipps rings of Iron, using gowld (as hee confidently affirmed) to lay upon their houses for covering, of which himselfe was an eye witnesse haveing bin foure severall tymes at the place; & being by us demaunded, if hee would carrie us thither, hee made answer they had many enemyes in the wildernes & wast places they past alonge before they came thence with whom they had divers encounters, & received many wounds shewing for a witnesse severall hurts upon his owne bodie; but then wee tould him wee would carry alonge some of our musketts or small peeces and with them wee would defende our selves & the company wee went withall, & destroye such as offended us, hee was very glad of our motion inasmuch as our gunnes unto them are very fearefull, haveing never seene nor heard the like before, calling the reports of our peeces in their languag' thunder, & certifying from one place to another, how a white people was come upp amoungst them, that with thunder kild both the wild beast of the feild and the fowles of the aire. And all such of the people as used to travaile, with whom wee had any familiar commerce & discourse did constantly reporte of the aboundaunce of gould was farther up in the country.

And that it maie appeare how in this land, & neere these confynes King *Solomon* had his trade; If it maie please your Matie to observe what in our sacred bible is set downe, where in the first booke of Kings, & second of the Chronicles, it is specified that *Solomons* navye was three yeares on their voyage, for which tyme it maie bee well maintained how so much might be spent to gaine this place, and make returne unto their home againe, but the very mayne and pryncipall poynte of all, is in the examinaton what commodities Kinge /3v/ *Solomons* navye return'd & that will make it moste plaine, it might bee here; for the words say once in three yeares the navye returnd & brought with them gould & silver, Ivory, Apes & peacocks; nowe in desyneing of that some of treasure; gowld is sett downe by certayne number, silver never mentoned, which I conceave avoydes the doubtfullest question, because silver is scarce in this country, & that little they have is much esteemed amoungst them: But for the aboundaunce of gowld, if there were none other trade knowne to come from them, but what the Moores of *Barbarie* have with Trade supplies the greater parte of *Europe* with gowld, it were sufficient to testifie the plentye; whereof the marchaunts of countreys who trade in those parts can give good accompt, & in their conferrences speake exceeding largely, howbeit it is latelie known that in the Moores returne with their gowlden ladening they are compelld to discharge part of their burthen to pay a tribute for their passage through the countrye of *Sydan Alley*, who maie restrayne them if he please and hath a castle on the mountaines where he commaunds them to laye down their peeces, which *Sydan Alley* is but lately found and in his country wee have no commerce. So likewise from all parts of *Affrica* bounding upon the mayne Ocean, what place or people some ever they are that trade, they allwaies finde & bring from thence rich gowld, which is a principall likewise to bee noted, for it is said of *Solomons* travailes, that they brought so much of fyne gowld, nowe in all other parts of the world where gould is found, eyther in the King of *Spaynes* socalled *America*, where is such plenty or any other place or country, whatsoever, their gowld compared to this of *Africa* is much more baser, & of lesse esteeme. And for this River wee can justifie that what was brought us answered the reporte.

And next for Ivory it proves much this was the place of trade, this might satisfie such as suppos'd *Solomons Ophyr* was in *America*, for in all voyages made from thence, wee have not seene one *Olaphants* tooth, neither have wee heard of any such beast there. In the East Indies there is moste plentye, but yet our marchaunts shippes that goe into those parts carry from /4/ hence thither many of those *Oliphants* teeth which come from *Ginney*, and in no River in all those parts of *Africa* is there more *Oliphants* then in this River I write of, and of the greatest stature, which are all wild & walke togither in Companyes, for the people are very fearefull of them but wee were us'd to sett upon them and found them as other beasts, & have brought home many of their teeth both great and large.

Againe for Apes no place affords more, for wee see of these Baboones, and Monkyes, thousandes togeither, and wee observe in the Baboone a kinde of government.

And lastly for peacocks this place affords all over the country infinite store of beautifull birds, but not the same [as] our peacock is, somewhat lesse, about the bigness of our peahenne, his feathers lye all over him like eyes, verye pleasingly and the flesh is much better then our peacocks flesh, their feeding is upon the ground just like a peacock, and wee meete them with greate flocks of them togeither, most commonly neere the Townes, or the peoples howsinge. with our speeres wee kill many of these beautifull byrdes, and likewise the people when they take them either younge or otherwise alive will present them unto us, who for his fayre coulour & manner of breedeing maie bee called a peacock in our translation, where that byrde is not knowne. Therefore I contend that even this place affords the commodities *Solomons* navy returnd home, and no place else that is knowne, can so directly bringe them alltogether.

Againe maie it please your Matie vouchsafe to heare what as a poore seaman I observed within this River concerning the Queene of *Sheba*, who as our scripture testifieth came from the farthest part of the earth to visitt *Solomon*, and brought with her greate store of gowld, some pretious stones, and such sweete odours as is set downe not to have byn there since; & that shee return'd from *Solomon* with all things her hart desired, wherein first concerning her country, Josephus makes it plaine shee was Queene of the *Ethiopians*, and there is no question can bee made, but these people amounge whome I was are the true & perfect *Ethiopians*. for better approbaton it is said, that after shee sawe *Solomons* state & heard his wisdome, being therat astonished shee promised that *Solomons* god should be her /4v/ god, and that shee & her people would serve the god whome *Solomon* and his people did, and whether the time of her staye there, was to have her people shee brought with her perfectly instructed, therby to informe themselves, & be able to teach others at home, or whether Solomon in his prynceley care did send of his religious people alonge with her, wee doe finde amoungst these the true and absolute forme of the Leviticall lawe and those which are the *Bissareas* or *Marybucks*, as wee call them, being they that teach & instruct them in their religion, first have their Townes & dwellings set out by themselves, & they doe match & marry togeither within their owne tribe, & bring upp their children all to write & read, which none of the temporall people are taught or understand; likewise amoungst them only remains those large bookes, & great volumes of manuscripts contayning their religion, the character whereof is like the *Hebrew*, of which writing I brought home with mee to see if our schollers could understand it. And these *Marybucks* are greate travellers going in companyes carrying their youth with them, who are most of them laden with greate bookes & in all things wee sawe concerning their religion they did not differ from the Leviticall lawe except in the number of their wives & women, wherein they followe the custome of the kings & temporall people; that for necessitie have many women, as may be better explained. Likewise in conferrences they were pleased to heare, wee understood the names of Adam & Eve, of Moses, of Noah & of the fludd, [in] which thinges they were perfect. And this most likely to be received & brought unto them by that travelling Queene.

Nowe for remotenes it maie well be said, these parts to be the uttermost of the earth if wee observe them from Jerusalem. And for the things shee brought to *Solomon*, concerning gould sufficient is said before; and it is not unlikely where these rich mynes are but that some stones of valewe wilbe found, – [*illegible words*] parts wee never yett sawe any but by attayning farther upp wee maie.

And for the odours which is said shee brought, there is no country better then this, for all the Sea coast yeilds plentie of Amber Greece & the woods and mountaines are full of muske & civitt catts & very many codds wee bring home, which wee have of the people for poore commodities./5/

And one thing more I did observe amoungst them, they much affect sweet odours & when any pryncipall person dies, hee is buried with all sweete things, which argues their esteeme of odours, & that shee might carry her sweete present from home; and it is like the intelligence shee had of *Solomons* state might come unto her by the trade his people had in the country, which trade is at this daye found in this River, if it maie please your Matie to commaund that it might be effectually [*overwritten*: ? lookte] after.

And in regard it hath pleased God to give mee leave to see the country & in my reason am confident of the riches wilbe present gaine from thence, besides a farther expectaton, that it may please god to make your *Highnes* Lord of that greate countrye and weake armed people forasmuch as they have a prophesy amounge themselves, they shall bee subject unto a white people. These things in my dutie makes me bold to present this paper unto your *Matie*. And although it bee presumption in mee to motion it, your *Highnes* would bee pleasd to call to mynde *Columbus* hys offer to your famous great grandfather King *Henry* the seventh which hee refused in regard of much moneye was to bee layde out in it; whereas this voyage is to be performed with a small charge and that charge in tenne monthes space by *Gods* blessing, without all doubt will bring that gaine as will warraunt the business.

Therefore maie it please your *Matie* to take this designe into your owne *Princlie* consideraton. And in case your high and weightie affaires otherwise will not permitt yourselfe to take an accompt hereof in particular from mee, to vouchsafe to appoint some honorable and judiciall persons whome you shall think fitt to examine my propositione herein, to the ende the same being well understood, my labour & endeavor for your *Ma^{s}* honour & proffit & the good of my country: the onlie marks I ayme at maie be spedely supported least delay (*the ruyne of all good actions*) occasion the losse of this noble & hopefull enterprize, wherin expediton is the rather to bee usd, for the deferring of tyme maie worke mee out of the love & remembraunce of these black people; two maine helpes of businesse of this nature; and will much further advaunce your *Maties* service /5v/ herein, Before other Nations, by our neglect maie stepp in before us, & reape the fruite of this discovery, though I your *Maties* subject to your *Highnes* honour, had the first footing there.

This I humbly submitt to your *Ma^{s}* royall wisdome and my life to your service humbly remayning

                        Your *Ma^{ts}* moste humble
                             devoted subject

                            Richard Jobson

APPENDICES

APPENDIX A

Jobson's itinerary

Jobson's topographical references are to be mainly found, not in his book, but in the material 'Extracted out of his large Journall' printed by Purchas. The Itinerary below, based on both sources, includes all topographical references, and the notes supply identifications and details about places visited. However, before setting out a detailed itinerary of Jobson's voyage, it will assist the reader to present the material in brief summary.

Leaving England in early October 1620, the two ships of the expedition reached the 'Little Coast' of Senegal, immediately beyond Cape Verde, in mid November, and the mouth of River Gambia on **17 November**. After halting for some days while a boat was built and probably while provisions were bought from villages on the river bank, and halting again for one day at an island, on **24 or 25 November** the ships reached **Tindobauge** (modern form of name, Tendabaa), about 70 miles up-river. The larger ship, the *Sion*, was left here, although later it proceeded to Kasang. A week later, at **Mangegar**, a further 60 miles up-river, three Englishmen were left behind, presumably to trade. On **7 December** the *Saint John* reached **Cassan** (Kasang), some 165 miles up-river. Staying here for only a day or two, the English pressed on, and by **14 December** they had reached **Pompeton**, claimed to be the furthest limit of Portuguese residence. A few days later the English established a middle river base, the remaining survivors of the previous voyage having been contacted at an interior town, **Oranto**. The *Saint John* was left at this base.

Over the next fortnight, while a boat apparently explored further up-river, Jobson contacted a nearby ruler and some trade was done. The boat having returned, on **2 January 1621** Jobson took a small party up-river in it, first reaching **Batto**, where he engaged two Africans. During the rest of the month the English made their way up-stream, through many difficult passages, particularly near and after **Baraconda** (Baarakunda), 330 miles up-river, until they approached **Tinda River**. At this furthest point, some 250 miles from the coast and 460 miles up-river, on **1 February** they were met by an important African trader. He obtained for them, from a local ruler, a grant of land and trading rights. But the English soon ran out of goods for sale, and could only promise to return in the future, although some information about the gold trade was collected.

On **10 February** the party began the journey down-river. On **19 February** it was again at **Batto**, where a circumcision ceremony was witnessed. Jobson returned a short distance up-river to make a visit to the important trading town and Islamic centre of **Setico**. After a stay of perhaps a fortnight, he left there on **11 March** and returned to the middle river base. The *Saint John* now sailed down-river and on **19 March** rejoined the *Sion* at **Cassan**. The ships did not leave Kasang for several weeks, perhaps to allow Jobson to collect local information. They reached **Mangegar** on **20 April** and **30 April** respectively, and left the river on **9 May**, having spent six months up-river.

ITINERARY

Note that the distances up-river subsequently cited have been measured from Banjul using a 1980 map, and these often disagree with earlier estimates, partly because different starting-points and different channels around islands in the river were used in calculations. Note also that the Itinerary uses Jobson's version of toponyms, whereas the notes use, when possible, present-day versions.

1620

5 October	The *Sion* (200 tons), and a pinnace, the *Saint John* (50 tons), sail from Gravesend: *PJ* 921.[1]
25	Leave Dartmouth: *PJ* 921.
4 November	'Up with the Iland of Launcerot' (Lanzarote): *GT* 8.
5	Pass 'Canary Island': *GT* 8.
14	Reach **Travisco** Road (Rufisque on the 'Little Coast' of Senegal): *PJ* 922.[2]
?	Enter the **River of Borsall** (River Saalum) : *PJ* 922.[3]

---

[1] 'Saturday the fift of October, 1620', but 5 October (O.S.) was in fact a Thursday. This may be a miscopying by Purchas. Jobson supplies few dates in his book, and although on his boat journey he claimed to have noted and respected Sundays (*GT* 83), Purchas's extracts from the Journall tend to supply dates without days of the week. However, on the few occasions when both are supplied the day of the week is correct.

[2] In *PJ* the date is given as 14 February, but this is clearly an error, and a date of 14 November makes sense. The name 'Rufisque' is a French corruption of Portuguese 'Rio Fresco' ('River of Fresh Water'); the Dutch called the locality 'Vischers Dorf' ('Fishermen's Village'). Jobson's version, Travisco, appears to be a garbled mixture of the Portuguese and Dutch, with the additional first syllable perhaps an abbreviation, found on a map or chart, of 'Terra', wrongly linked. As 'Trefisco(e)', the name appeared in other English 17th-century sources (e.g., PRO, HCA 13.36, 6.1.1603/4; 13.60, 5.5.1646). Rufisque, on the mainland SE of Cape Verde and originally a fishing village, because of its sheltered anchorage became popular with European vessels as a place at which to trade and obtain fresh water and provisions. In time it became the main port for the Wolof polity of Kayor. Some Portuguese settled there but in the later 16th century it came under French influence, while the Dutch, who acquired the nearby island of Gorée in 1617, also established a trading post at Rufisque (R. Mauny, 'Notes d'histoire sur Rufisque d'après quelques textes anciens', *Notes africaines*, 46, April 1950, pp. 47–9). For a drawing of Rufisque in the 1680s, see Hair, *Barbot on Guinea*, opp. 50. Its importance faded in the 19th century when nearby Dakar became the main port of call. On the return journey to England the ships called again at Travisco – see the final entry of this Itinerary.

[3] 'Borsall' represents Wolof *buur Saalum* 'king of Saalum', Saalum being a Wolof (or at least Wolof-controlled) polity which exercised some control over lesser polities along parts of the north bank of River Gambia. At this point, Jobson seized a boat belonging to Heitor Nunes, a Portuguese allegedly responsible for the massacre of some of the crew of Thompson's ship, the *Catherine*, at Cassan (Kasang) in 1619, and the theft of the ship (*PJ* 921).

APPENDIX A

17	Reach **River Gambia**;[1] begin to build a shallop[2] and (probably) buy provisions.[3]
22	Launch the shallop: *PJ* 922.
23	Sail up River Gambia; 'anchored some 4 leagues within the mouth of the river' (*GT* 8), or 'anchored against a little Island on the south-side some four leagues up' (*PJ* 922),[4] probably James Island.[5]

[1] Whereas *PJ* fails to provide the date of arrival at River Gambia, *GT* states that, after passing the Canaries, the ships had 'layd all that land[;] the 17. of November, we came to an anckor in the River of *Gambra*, having had some occasion of stay by the way, to the losse of neere three dayes, so as our whole travaile from *Dart-mouth* thither was in 20. dayes, we anckored some foure leagues within the mouth of the river' (*GT* 8). The stay of three days must have been spread over the only two halts mentioned, Rufisque and River Saalum. Counting twenty days from 25 October, the date the ships left Dartmouth, and allowing for three days lost, makes the date of arrival at River Gambia 17 November. With the editorially-added punctuation suggested above, *GT* confirms the calculation.

[2] A small shallop could be conveyed on deck, and probably its parts, or many of them, could be carried from home, to be assembled later. Possibly some of the boat-building was done aboard the ship at sea but completed after the English vessels had reached the mouth of River Gambia. Although it seems likely that the boat was assembled or built ashore between 17 and 22 November, it is curious that Jobson says so little about the estuary region – see the next note. Very soon we hear of two shallops (*PJ* 922; cf. *GT* 40). Was one carried out aboard ship, or did each ship build a shallop?

[3] Did the English buy provisions after entering the river? This is not specifically recorded, perhaps because of a cut in the Journall by Purchas. But Jobson states that the Mandinka near the mouth of River Gambia would, under certain conditions, supply provisions (*GT* 27–8), and his knowledge of the possibility may indicate that the English did in fact obtain provisions in the estuary, presumably while building their shallop. (It is perhaps less likely that Jobson was referring to the down-river journey when the ships halted for three weeks near the mouth of the river, although it is plausible that provisions were also then obtained for the halt and the journey home.) The ships spent several days in the estuary before moving up-river, giving them time to make contacts and buy provisions, as well as to work on a shallop. Yet, alternatively, Jobson may have obtained the information about provisions from any colleagues on the *Saint John* who had travelled on the ship when it previously visited the river. A related point is that, later in *GT*, Jobson refers vaguely to customs paid 'in the mouth of the River' (*GT* 47), without making it clear whether this is a general observation or one specific to his voyage; *PJ* only notes customs paid higher up the river, at Tindobauge (Tendabaa): *PJ* 922. It seems unlikely that Jobson reckoned Tendabaa to be in the mouth of the river, and it is not implausible that customs had in fact been paid when entering the river, probably to the ruler of Barra (local name Niumi/Nyoomi), particularly if the English landed in his territory in order to work on their shallop. (Again, it is less likely that Jobson was referring to any customs that were paid during the halt before leaving the river.) Barra could provide European vessels with up-river interpreters (for the engagement of interpreters whom 'we call pilots' by a Portuguese vessel, apparently in the 1580s, see Part II below: Donelha, f. 22). However, if the ships collected a pilot when in River Saalum (and there was certainly local traffic between the two rivers), it is conceivable that dues covering River Gambia were paid there, the ruler of Saalum at times considering the River Gambia estuary to be within his authority, as overlord of Niumi.

[4] The wording in both texts makes it seem as if the first halt within the river was at the island, but the shallop cannot have been built there since *PJ* states that it was launched the day before 'we set sayle up the River' and the island was reached.

[5] On the course of the river in general, see the section on 'The River' in the Introduction. Jobson described the mouth of the river as being 4 leagues broad, that is, some 12 miles (*GT* 10). This is close to the actual distance of about 10 miles at the widest opening, that between Bunyadu (Buniada) Point on the north and Cape Point on the south, although the narrower opening between Barra (Baara) and Banjul Island is only 2¼ miles across. 4 leagues within the mouth of the river would have brought the English only to the first island, Dog Island. This is so close to the north bank that from the middle of the river it appears to be only a locality on the north bank, and from there people can walk across to it at low tide, through the mud. Since ships could run aground on the mud flats near the island, they preferred not to anchor there, except when waiting for the tide to turn. Thus it seems more likely that the English vessels anchored on the south side of the next island, then known as St Andrew's Island but later in the century renamed by the English

Pass **Taukorovalle** (Tankular) – 'over-shot it in the night': *GT* 10.¹

24 or 25   Reach **Tindobauge** (Tendabaa), 69 miles up-river; customs paid to the officers of an inland king.² The *Sion* remains here, but later, at an

---

James Island. This lies another 10 miles up-river, or some 22 miles from Banjul, and is visible from lower mid-stream. It was named St Andrew's Island in circumstances described by Cadamosto, who, after estimating the width at the mouth as 6 to 8 (Italian) miles, and the width within the river as 3 to 4 miles at the narrowest, placed the island 10 miles (about 9 English miles) up-river (see Part II: Cadamosto, [2], [17]), either an under-estimate – as Jobson's most likely was (if not a miscopying by Purchas) – or else indicating that the name first applied to Dog Island and was only later transferred to James Island. In the 1650s the Courlanders clearly distinguished between Dog Island ('Hondes Eylandt') and St Andrew's Island ('Sant Andres') (1651 map and a voyage journal in Otto Heinz Mattiesen, *Die Kolonial-und-Überseepolitik der kurlandischen Herzöge im 17. und 18. Jahrhundert*, Stuttgart, 1940, p. 272, Karte 1; Edgars Andersons, *Tur Plivoja Kurzemes Karogi* [New York], 1970, map). Furthermore, in 1646 a ship carrying Spanish missionaries anchored 'in a port called Gelufer [Jufure], where they found a Dutch vessel, anchored alongside a neighbouring island called San Andrés', Jufure being opposite James Island (Mateo de Anguiano, *Misiones Capuchinas en Africa*, ed. Buenaventura de Carrocera, 2 vols, Madrid, 1957, II, pp. 64, 267). Sources later than Jobson indicate that dues on behalf of the king of Nyoomi were at times collected at Jufure; that the drinking water there was of better quality than that at Barra; and that pilots and interpreters were often hired there for an up-river journey. But Jobson makes no mention of the English landing on the mainland when anchored off the island, or taking on pilots or interpreters at this stage (or at any other in the lower river). Possibly some crew members who had been on the previous English voyages felt sufficiently acquainted with the river as far as Kasang, and additionally the English may have hired a pilot while at Rufisque or, as suggested above (p. 209, n. 3), in River Saalum. Although it cannot be totally ruled out that contact was made with Jufure while the ships anchored at James Island, since the halt was apparently one of only a single day any significant contact seems unlikely.

¹ Taukorovalle, Tauckro Valley (*GT* 10), Tangeevalley (Huntington map), Tencrouualle (Gerbier map), Tamkevalle/Tankikevalle (Fitzhugh map), Tankerval (Olfert Dapper, *Naukeurige Beschijvinge der Afrikaensche Gewesten*, Amsterdam, [1660] 1676, p. 415), Tancoroale (Part II below: Lemos Coelho, 18), represents Tankular (Tankural) on the south bank of the river, in Kiyang district and about 30 miles from James Island and 52 miles up-river (not 'some 30 leagues' = 90 miles, a gross exaggeration: *GT* 10). The name may have been originally Tankorowal, in the Bainunk language (Stephan Bühnen, 'Place names as an historical source ...', *History in Africa*, 19, 1992, p. 92). Built along the river bank, Tankular was an important trading centre, with resident Portuguese. The tide, currents, and winds made crossing to Nyoomi and Badibu districts on the north bank fast and easy, while paths to Manduar and Jali to the south led to Fonyi district, inhabited by the Jola and Bainunka peoples. In daylight the town would have been clearly identified from mid river by the larger trees, silk cottons and baobabs contrasting with the mangrove fringe along the bank, but at night it would have been very easy to miss. The river here is about 2 miles wide but so shallow near the north and south banks that only canoes can cross it, with larger vessels having to remain in midstream. Jobson comments on the river up to Tankural that 'it doth spread it selfe, into so many rivers, bayes and creekes', too 'intricate' to examine into, and he names none (*GT* 10). Before reaching Tankular a series of creeks is indeed passed, as shown on the Gerbier map, where the creeks, today with local names, are given European names.

² Tindobauge, Tendaba (Gerbier map), Tendena (Huntington map), Tondeba (Dapper, *Beschrijvinge*, p. 415), Tagamdaba (Part II below: Lemos Coelho, 20 – wrongly located), Tendeba (1732 map, Moore, *Inland Parts*) represents Tendabaa, on the south bank 17 miles up-river from Tankular, a major canoe port for Kiyang district. An old trade route ran south from the river, fish, salt and other commodities being taken inland. The Gerbier map shows, inland from Tendaba, 'Aldeas Del Rey Phaian', meaning 'towns of King Phaian', the last word being apparently a corruption of 'Pharan', that is, the Mandinka title *farang* 'lesser ruler'. Jobson's king lived inland 'some six miles'; the king's town was probably at or near presentday Kwinella, 4 miles inland from the river, where today the district chief of Central Kiyang resides. (This *farang* may have been the one who in 1616 befriended the survivors from a French vessel making their way to River Gambia from further south – see Introduction, p. 12, n. 2.) At Tendabaa Jobson made his first recorded contact with a Portuguese in River Gambia (who told him of Thompson's death) (*PJ* 922). The stay for several days was partly to pay dues to the local ruler, probably watering dues but also perhaps dues because the *Sion* was to lie there for a period, with some of the English party.

APPENDIX A

29	unstated date, sails up-river to Cassan (Kasang): *PJ* 922, 925.[1]
	The *Saint John*, with two shallops, sails up-river: *PJ* 922.[2]
1 December	Passes **Pudding Island** (probably Elephant Island): *PJ* 922.[3]
2	'anchored against a little Creeke which leads into a Towne Maugegar' or **Mangegar**, about 133 miles up-river: *PJ* 921; *GT* 122–3.[4]

[1] Perhaps the depths of the river were to be tested by the smaller vessel before the larger proceeded, although there is no mention of a message being sent back to the *Sion*, and it is not stated when this ship moved up-river to Kasang. Or perhaps the *Saint John* represented an advance party moving up-river more quickly, by means of occasional towing (see next note), than the *Sion* could achieve, being probably too large to tow as easily or manoeuvre as speedily.

[2] Jobson has a very clumsy sentence. 'Leaving her [the *Sion*] with five and twentie men and boyes, on Wednesday, the nine and twentieth, the Saint John and two shallops, we set sayle up the River twelve men in the bigger, with Henry Lowe, and thirteen in the lesser with my selfe, which with the Boat towed her up in calmes' (*PJ* 922). This appears to mean that the shallops (Lowe and 12 men in the larger, Jobson and 13 men in the smaller, oddly), together with the ship's boat, towed the *Saint John* (with an uncertain number of men left aboard) when it was impossible to progress otherwise. Jobson had explained that the winds blew down-river at this time of year, hence the need to tow the ship at times. Generally, however, ships and boats went up-river with the tide, anchoring when the tide turned, hence Jobson remarked that the ships 'past up by tydes' (*PJ* 921). The number of men left in the *Sion* is not indicated, but it is likely that some of the 25 with Lowe and Jobson were crew from that ship, although it cannot have been seriously depleted of men since it subsequently passed up-river.

[3] The name 'Pudding Island' is not found on any map, or in any other text, and the reason for its so naming defies speculation. The first major island encountered on this stretch of the river would be Elephant Island, some 40 miles up-river from Tendabaa (perhaps the distance estimated by Jobson as 'sixteene leagues from the ship', i.e. the *Sion*). Elephant Island was named as such on the Gerbier map ('Ille Oliphante'), and the name is still used. The earliest reference appears to be in 1552; 'about halfway to Cantor is an island which our people, because of the many elephants there, call the Isle of Elephants' (see Part II below: Barros). A low marshy island, Elephant Island is unlikely to have had any number of elephants on it, and the 'many' were most probably seen on the nearby banks of the river, although a few may have swum across. If 'Pudding Island' was indeed Elephant Island, it is curious that the English should have renamed an island which had a long-established name (appearing for instance in 1600 in Pory, *Historie*, in Brown, *The History*, I, p. 18), and it might indicate that they lacked contact with earlier knowledge of the river. Elephant Island represents the point at which the mangrove fringe on the river begins to thin out, salt water not being consistently present throughout the year in the reach up-stream as far as Niani Maru, although the exact position at any time in this reach is complex and varying, depending on such factors as the seasons, the extent of rainfall, the tides, and the output of fresh-water tributaries.

[4] Jobson provides the earliest reference in print to Maugegar/Mangegar, Mongagar (Gerbier map), Mansegaer (Mattiesen, *Überseepolitik*, pp. 156, 158), Mansiguer/Mansibaer (Dapper, *Beschrijvinge*, pp. 415, 420), Macugár/Majagar/Manjagar/Manjaguar (Part II below: Lemos Coelho, 1669, 15; 1684, 20). The location of this town was described in a later source as follows:

> From Degumasamsam [Duma-sansang] to the port of Majagar is eight leagues, the latter being situated on the north side. It is also very difficult to recognize because its entrance is by a very small stream. But there is no other stream in the land, and by this point the river does not have as many mangroves. You make your way in by boat, and the port is close by, as is the village too, but not in sight. ... In the same land, at a distance of one league, is the port of Caur [Kau-ur]. (Part II below: Lemos Coelho, 1669, 15; 1684, 22–3)

The town of Mangegar has disappeared. But about a mile west of the later important centre of Kau-ur is a stream (now called Simbara Bolong), leading from open country down to the main river, and Mangegar apparently lay close to this stream. A latterday map (1948, 1:50,000) shows, NW of Kau-ur, and 3 miles from the river and 2 miles from the stream, a small village, Manjaharr (known to the Wolof as Manjaxar), whose name seems to reflect a memory of the past. The 'town' 1 league inland from the 'port of Caur' mentioned by another Portuguese source (Part II below: Donelha, f. 24), may have been the same as Mangegar. Jobson met a Portuguese 'by the way', apparently a resident (*PJ* 923); the Gerbier map marked near 'Mongagar' a 'Casa de Diego Gratie' (probably the first appearance of this odd toponym, which was repeated in corrupt forms on later printed maps), and a later source spoke of 'some Portuguese and mulattos' living

4–5	The king and his 'Alcade' visit the ship; Lowe, the chief factor, rents a house for three of the English to stay (and presumably trade): *PJ* 922.
7	The *Saint John* passes **Wolley, Wolley**, a town on the north side, 'bigger then any wee had yet seene' (*PJ* 922);[1] and anchors at **Cassan** (Kasang), 166 miles up-river, which Jobson describes at length: *PJ* 922; *GT* 44–6.[2]
?	After perhaps two days, the *Saint John* with the two shallops proceeds, on a difficult course,[3] and via uncertain locations.[4]

there, although this may have been mere deduction from the name, Dapper, *Beschrijvinge*, p. 415. Having travelled some 64 miles from Tendabaa to Mangegar in three days, Jobson's ship was averaging 20 miles per day up-river. For Jobson's return to Mangegar, see 20 April 1621 below.

[1] The locality of 'Wolley, Wolley', that is, Oulaoula (Part II below: Almada, 6/11 [English version only]), Wolouuole/Wolowolo (Gerbier map), Ola Ola (Part II below: Donelha, f. 24ᵛ), seems to be represented in all later sources by a different toponym, first given as 'Nanhimargo' (Part II below: Lemos Coelho, 1669, 16 – probably in error for Nhanimargo, 1684, 26). This is today the important port of Niani Maru. In earlier centuries it could accommodate small ocean-going vessels, and the adjoining area was rich in the timber suitable for repairing ships (Part II below: Donelha, f. 24ᵛ–25). Described as the 'first port of Niani' (Part II below: Lemos Coelho, 23), in 1724 Stibbs found the king of Kasang residing there (Moore, *Inland Parts*, p. 254).

[2] As 'Cassan', Kasang was shown on the Gerbier, Huntington and Fitzhugh maps, as it already had been on a Portuguese-derived printed map of 1602 ('Guiné' by Luís Teixeira, reproduced in Donelha, *Descrição*, fig. 3), and it was extensively described in the later Portuguese sources (see *GT* 44–6 and p. 108, n. 2). When a Portuguese visited Kasang, apparently in the 1580s, there were nine Portuguese ships in the port (Part II below: Donelha, f. 25). Jobson, who seems to have spent only a day or two at Kasang when going up-river but a month on his return down-river, described the town's fortifications and the king's house (*GT* 44–6). Kasang was where the massacre of the crew of the *Catherine* had taken place, but the Portuguese who had been responsible were not encountered by Jobson, having removed themselves, although probably only temporarily. Jobson was unable to obtain interpreters, allegedly because the local people feared that the Portuguese were going to attack the English when further up-river (*PJ* 922). The town has long since been abandoned but the site is marked on a river chart of 1942, as is the 'Red Hill of Kassang' down-river from the site. An archaeological survey of the area was carried out in 1974 (by Professor Matthew Hill), but no major archaeological work has yet been undertaken there. For Jobson's return to Kasang, see 19 March 1621 below.

[3] Kasang was as high as the *Sion* went (so stated, *GT* 23) when it came up from Tendabaa at an uncertain date, and the *Saint John* must have struggled to proceed. The up-river course turns sharply to the south immediately after Kasang, and then makes a series of sweeping curves eastwards. Navigation is difficult for sailing vessels, because of the frequent sandbanks and because the main channel is too narrow to permit tacking against contrary winds. See the section on 'The River' in the Introduction. That both shallops accompanied the pinnace has to be deduced. Lowe as well as Jobson certainly went up-river to the middle river base but only one shallop is specifically mentioned (*PJ* 922). However, Jobson's remark that he went 140 leagues further than 'the other shallop' (*GT* 40), probably means that the shallops separated after reaching the base. For Lowe's subsequent movements, see p. 213, n. 6 below.

[4] The identification of place names now becomes difficult. Leach's 1732 map (in Moore, *Inland Parts*, 1738), the first printed map of River Gambia to detail the river beyond Kasang, shows the following localities. On the north side, in the interior of Wuli district – Settiko (Sutuko), Caunkade/Kings Town, Pholeykunda (Fula Kunda, a Fula town); and on or near the river – Fatatenda Factory, Barracunda. On the south side, in the interior of Kantora district – Kings Town, Malo; and on or near the river – Pry (Pirai), Sama Port (Sami), Jabo (Yabu-tenda), another Pholeykunda. Until the 1890s survey of the boundary between the colonies of Gambia and Senegal, all later maps simply copied the 1732 names. Later travellers to the interior (such as Mungo Park in 1795–7) did not follow the windings of the river, hence in their writings the names of places higher up were rarely recorded. The sketch map made as a result of Governor MacDonnell's 1848 journey to the upper river, for 105 miles above Baarakunda Falls (see Appendix D), has no place names, apart from Grey River, a name given by the Governor on this occasion (*Reports exhibiting the past and present state of Her Majesty's Colonial Posessions for 1848*, London, 1849, pp. 325–7 and map – the original map in the PRO). Similarly, Lt. Dumbleton's 1881 map of the 'Upper Gambia River' from Yarbutenda to

APPENDIX A

14	'Reached **Pompeton** on the south side, above which dwels no Portugall'; a Portuguese entertains the English: *PJ* 922; *GT* 118.¹
15	Reaches the port of **Jeraconde**, and a message sent sixteen miles to **Oranto**, where survivors of the previous voyage are living: *PJ* 922; *GT* 128.²
17	Two of the survivors arrive overland: *PJ* 922.³
?	The party moves up-river to within six miles of Oranto, forming a temporary middle river residence on the riverside called 'Whiteman's Town', where the *Saint John* will be left: *PJ* 922; *GT* 68–9, 100.⁴
?	Jobson visits Oranto, then the king on the south side: *PJ* 922.⁵
21	One shallop 'sent away'.⁶
25	Ferambra, the local governor and a friend of the English, living four

---

Bady Wharf (see p. 193) is almost devoid of names. The countryside was then allegedly deserted as a result of local wars, and the only names shown between the two localities are Rock of Barraconda, Grey River, Nerico River, and Jallacotta (H. N. Dumbleton, 'Expedition to the Upper Gambia, 1881 ...', 'Map of the Upper Gambia ...', PRO). 20th-century maps, however, indicate numerous ports on the middle river (Kasang-Baarakunda), many of these being crossing places between Wuli and Kantora districts, and since almost all of the localities noted on the 1732 map are still shown, they presumably existed in the intervening centuries. This degree of continuity suggests that they may well have already existed in Jobson's day, a century before 1732 – indeed some are named in later 17th-century Portuguese and English sources. But Jobson fails to name most of them, perhaps because the English deliberately failed to halt at those visited and associated with by the Portuguese, lest there was danger.

¹ Pompeton is not shown on modern maps, and Jobson does not indicate how many days' journey from Kasang was involved. But 'Pompetongg' was shown on the 1661 Vermuyden map, on the south bank in the loop of the river at whose base is now Kerewan; and a 19th-century traveller, listing places passed on a down-river journey, gave 'Pooom-patory' as the name of a locality between Mamayungebi Hill (otherwise 'Arse Hill') and Kunting (Thomas Eyre Poole, *Life, Scenery and Customs in Sierra Leone and the Gambia*, 2 vols, London, 1850, II, p. 171). This fits the 1661 map location and places Pompeton more or less opposite presentday Karantabaa. The ship had therefore travelled about 55 miles up-river from Kasang, probably representing a journey of three days, hence Jobson's stay in Kasang was short. When moving down-river the journey from Pompeton to Kasang took two days (*PJ* 925). *GT* states that, when arriving at Pompeton on his way up-river, it was explained to Jobson that diabolical means had led to his being expected; but in *PJ* the explanation is given to him when calling at Pompeton/Pompetan a second time, on his way down-river, on 18 March (*PJ* 925), a call not mentioned in *GT*. He probably did call twice, but one of the references to the explanation of the diabolical means would seem to be in error.

² Jeraconde and Oranto are difficult to identify and locate, and hence it is impossible to be sure about the exact location of the middle river base Jobson is about to establish. See Appendix B.

³ Jobson's reference to eight survivors of the 1618 voyage living on the river, all of whom except Thompson returned to Britain (*GT* 6, 128), presumably indicates that altogether seven Englishmen were rescued at Oranto.

⁴ For discussion of the location of Oranto, see Appendix B.

⁵ Oranto was said to be 6 miles back from the river. On one view of its location (see Appendix B) this would seem to place it in the middle section of presentday Wuli district. However, the boundary of the Wuli polity may have been further to the east, which might place Oranto nearer the boundary, and perhaps in a no man's land between Niani and Wuli polities. Apparently the Oranto domain straddled the river, since the king, Summa Tumba, lived on the south side (*GT* 58). According to Jobson he was subject to the 'King of Cantore', which is difficult to explain if Oranto was in Wuli – perhaps the area was in dispute between the two polities. Two of the rescued Englishmen living at Oranto had previously travelled up-river with Thompson, and at least one of the three rescued Englishmen named by Jobson, Matthew Broad, who was perhaps one of the two, now accompanies Jobson up-river (*PJ* 922, 924).

⁶ This was presumably 'sent away' to investigate conditions up-river, but since Jobson calls the shallop 'my Boate' (it is later referred to as a shallop) it was presumably not the the shallop captained by Lowe (see p. 211, n. 2 above), whose activities at and from this this point are obscure (see p. 214, n. 4 below).

	miles away, sends elephant meat for Christmas dinner: *PJ* 922; *GT* 100, 149/141.¹
31	The shallop returns: *PJ* 922.²

1621

1 January	An aged marabout warns Jobson of the difficulties and dangers up-river: *GT* 79.
2	Jobson 'went from Oranto for Tinda', in a shallop with nine other whites and towing a canoe³ – the second shallop apparently concentrating on trading⁴ – and in one tide reaches **Batto**, where a young

---

¹ Ferambra is presumably *farang-baa*, 'great/superior royal agent or representative'. Ferambra's town was 'within three miles of Tobabo Conda [Whiteman's Town], the place where our habitation upon the land was' (*GT* 100 – *PJ* 922 says 4 miles). Ferambra had defended Thompson and the other English men from the horsemen of the king of Niani (*GT* 100), and this almost certainly indicates that his residence was north of the river, in Wuli territory (see also Appendix B). Ferambra's overlord, the *mansa* of Wuli, most probably lived much further inland – in 1724 the *mansa* lived at 'Cussana' (Moore, *Inland Parts*, p. 266), perhaps the 'Caucade, King's Town' shown on the 1732 map, although the capital was later moved to Medina in the west of Wuli.

² The boat had been gone some ten days, but Jobson does not divulge where it had been, hence, if its journey was up-river, as seems most likely, how far it had penetrated or what contacts had been made. Judging by the time taken on Jobson's subsequent journey, it may have reached at least half way to Baarakunda. Perhaps Jobson remained silent about this journey by others in order to make his narrative more dramatic, the first part of his own up-river journey being made to seem more pioneering (for Englishmen) than in fact it was.

³ Jobson generally refers to the 'boat', but once to 'going up in the other shallop' (i.e., other than Lowe's) (*GT* 41). It eventually carried fourteen men, and appears to have been the same smaller shallop which Jobson brought up from Tendabaa with the same number of men. He refers to travelling with 'the strength of six Oares' (*PJ* 923), which indicates that those aboard took turns at rowing (and that it cannot have been rowed at any stage by only the Africans). He notes that, below Baarakunda, the English brought provisions from Fulbe women, 'in most places … where we did lie for Trade' (*GT* 36). This seems to refer to the up-river exploration and perhaps indicates that some general trading was done en route, although *PJ* does not mention this. The canoe, which was perhaps only acquired after Africans were hired, was used to go ashore and bring back provisions and guests, 'reserving our boate in the middest of the River, as a castle and refuge for us' (*GT* 83). Stibbs also halted in mid-stream (Moore, *Inland Parts*, 284).

⁴ The manning of the two shallops at Tendabaa was in order 'to goe up the river', the larger boat being for 'the principall Factor to follow his trade in' (*GT* 40). This wording does not make it clear whether the division of men was to allow only Jobson's party to make the final exploration up-river after the *Saint John* halted, or whether it was left open for both boats to go. A reference to 'Lowe's emulation … both now and before', inserted in the Journall after a reference to Jobson reaching Batto (*PJ* 922), can certainly be read as indicating that Lowe accompanied Jobson up-stream from the base at least as far as Batto. Although it is possible that Purchas inserted this comment at the wrong point, the possibility that the two shallops separated at Batto rather than at the base cannot be ruled out. If this was the case, then it was perhaps Lowe, rather than Jobson, who engaged the Africans at Batto. Nevertheless this is not the impression that Jobson gives, admittedly by his failure to specify Lowe's later movements. What were these? A stray reference to undated activities of the 'principall factor' (*GT* 40–1) probably explains what Lowe did after separating from Jobson, whether at the base or at Batto. While trading in his shallop, part of his crew died 'before he came backe to the shippe'. This could mean either that he returned to the *Saint John* at the base after trading probably both up and down the river, or that he traded down-river from the base, or from Batto, until he met the *Sion* at Kasang. The latter course is perhaps the more likely. Jobson claimed that he had ascended the river further than the other shallop by 140 leagues (420 miles) (*GT* 41), but even if he was counting from where the *Saint John* lay it was only about 200 miles. He was perhaps inferring that the second boat did not travel far up-river from the base – if any distance at all. Jobson trades on the spot (*PJ* 922).

APPENDIX A

	marabout, Fodee Careere, agrees to accompany the English: *PJ* 922; *GT* 63.[1]
6	Visit of Sumaway, king of Bereck/**Pereck**, 'under the great King of Cantore'; Jobson hires a youth, Samgully: *PJ* 922.[2]
8	Samgully's mother complains that he may miss the circumcision festival: *GT* 109.[3]
9	Anchorage at an unnamed place with many hippopotamus: *PJ* 922.[4]
10	Anchored at **Massamacoadum**, 15 leagues from Pereck: *PJ* 922.[5]
11	Anchored at **Benanko**: *PJ* 922.[6]
12	'after rockie passages to **Baraconda**' (Baarakunda),[7] 330 miles

[1] Oranto was in the interior, so despite the wording Jobson must have returned to the river, in order to travel one tide to Batto. Unfortunately *baa-to* means 'at the river' in Mandinka and the place is therefore unidentifiable by name – for a possible location, see Appendix B. Batto is described as the town of Bo John. Jobson thought that Bo John was a title in the local political hierarchy (*GT* 58), which appears to be wrong. According to traditions, the Bojang were a Mandinka clan living at one time in Wuli, before moving eventually to Combo. The town may well have been named Bojang-kunda (Bojang's town), but no place of that name exists today. From Jobson's account, the town of Batto was a mile distant from its port on the river, but sounds from the town could be heard on the river bank (*GT* 110). Jobson states that Fodee Careere was 'with mee, both up into the highest part I went, as likewise all the time I followed any trade in the river' (*GT* 63), but this was not strictly correct, Jobson having traded some days before reaching Batto.

[2] The king is 'of Bereck' but a few lines later the place is 'Pereck'. This is almost certainly the 'Paraki-cunda' of the Vermuyden map, shown up-stream from 'Color' (i.e. Kulari') and down-stream from 'Zama' (i.e. Sami), locating it in approximately the position of modern Perai. Although this is up-stream from the probable site of Batto, the sequence in the text, 'King of Bereck ... came aboord... We tooke in Sangully', seems to indicate that the king travelled to, and was met by Jobson at, Batto, since Jobson later notes that the youth came from there. The term *suma* appears to have been a title of varying significance, denoting in some districts an heir apparent, often in charge of a frontier district while awaiting his turn to rule, in others a high official, either at the court or in charge of a district (Ba Tarawale, *Mandinka-English Dictionary*, Banjul, 1980; Bühnen, 'Place names', 90, note 70; Donald R. Wright, *The World a Very Small Place in Africa*, New York/London, 1997, p. 111).

[3] Samgully/Sangully, aged sixteen or about seventeen (*PJ* 925; *GT* 108), who had lived with Thompson and later with the English survivors (*GT* 108), and spoke good English, had missed the annual circumcision festival in the previous year by being away from his home town; and if he missed it again there were penalties. Jobson supplies the date, and also states that the mother overtook the boat after two tides, which suggests that the boat left Batto on 6 January. This was the date of the king's visit, but perhaps the boat left in the evening, after the ceremony.

[4] Jobson's account of these animals (*GT* 20–22) probably derived from observations at this point. On this stretch of the river Stibbs complained about 'the hideous Noises of Elephants, River-Horses and Allegators, all Night' (Moore, *Inland Parts*, p. 269).

[5] Massamacoadum is unidentified. For the stated distance of Massamacoadum from Bereck/Pereck, 15 leagues or 45 miles, see Appendix B. Massamacoadum being only one day's journey from Bananko, it probably lay between Passamassi and Yarbutenda.

[6] The village of Bananko, just across the present Gambia/Senegal boundary, is now abandoned.

[7] There are 'rockie passages' both before and after Baarakunda, the latter more severe and known to later sources as the 'Barracunda Falls'. Possibly Purchas made an error in summarizing Jobson's Journall. In 1724 Stibbs described

> an Overfall ... not above three Leagues above Barracunda ... From the North side of the River runs a solid Bed of Rocks one third over ... near 10 Feet above the Water of the River ... It was close to this Side that I found a Passage, and a very strait one, for our Canoas rubbed the Rocks on both Sides. From the South Side, for above one third across the River, was another plain and equal Bed of Rocks ... under Water about ten Inches ... Between ... the rest of the River was choak'd up with large single Rocks ... (Moore, *Inland Parts*, pp. 274–5).

In 1912 the Falls were again described, as

> only a rapid running over a bar of submerged rocks which stretch across the river from the left

	up-river, where the tidal assistance stops: *PJ* 922.¹
14	Visit by the head man of Baraconda town, and another marabout and a fourth African are hired: *PJ* 922–3.²
15	Leaves Baraconda and 'going against the streame': *GT* 83.³
?16	Passes by Wolley, a small river: *PJ* 923.⁴
17	Sees 'thousands' of baboons and monkeys; the boat struck by an hippopotamus; the canoe ahead, sounding: *PJ* 923.⁵

---

bank continuously until about twenty feet from the right bank, where a channel capable of taking about four feet draught at high water in the dry season, when the influence of the tide slackens the sluice which runs through the pass. The north bank rises from the deep water as a cliff ... to the height of ten to twelve feet' (Reeve, *The Gambia*, p. 124 (photograph of the 'Falls'), p. 127).

¹ 'The tyde went no further ... beyond were no towns, neere the river, nor boates nor people to be seen'; and again, 'they have not the use of any boates, above where it ebbes and flowes, so farre as we have hetherto beene' (*GT* 15). Jobson regularly refers to the river below Baarakunda as that of 'the flowing and ebbing'. Since the immediate course 'beyond' was difficult, it is likely that boat traffic was indeed negligible, but unlikely that further up there were no canoes, and Jobson later acknowledged this (*GT* 95). Elsewhere Jobson stated that the tide flowed 'unto a Towne called Baraconda, or some little above' – a distance he reckoned as 'near upon 200 leagues', that is, 600 miles, fitting his statement that the total journey was one of 320 leagues or 960 miles, both distances gross exaggerations (*GT* 10, 11). Later in the century, Hodges justly remarked that his own miles were 'not such miles as Capt. Jobson made when he makes 200 leagues from James Island to Barrowcundy, the which I could never make to be more than 558 miles' (Stone, 'Journey of Cornelius Hodge', p. 89). However, Jobson's 200 leagues was actually from the river mouth, not James Island, and, more important, Hodge's 558 miles was also an exaggeration, although not quite as extreme as Jobson's. According to the present-day navigational authority, in the dry season the tide ascends to Yarbutenda, that is, about 15 miles short of Baarakunda; in the rainy season, not as far (*Admiralty Pilot* Vol. I, pp. 294, 302 and tidal diagram). From the middle river base, and allowing for two days spent at Batto, the boat had travelled in eight days some 60–75 miles, thus averaging 7–10 miles daily.

² The marabout was Selyman (Suleman, i.e. Solomon, a Biblical hence Koranic name), and a relative, Bocan Tombo, was also hired (*GT* 83). (The chief man of the town was called Bocay Tombo – perhaps so, but otherwise a copying slip.) Jobson spoke of the party as now comprising 'ten white and foure blacke'; and elsewhere as '10 Whites on a shallop and 4 Blacks that I hired to carry up a Canoe' (*GT* 12). Perhaps this last reference means that only the Africans were competent to manage the canoe. Again, 'I did hire Blackemen, as I had occasion to use them, to serve as Interpreters, likewise to send abroade, and to helpe to row, so that when I came to passe the flowing, and to goe all against the currant, I did furnish my selfe, of four able Black-men' (*GT* 18). This sentence can be read as implying that the four were hired at Baarakunda, which was not the case, only two being hired there.

³ 'Eleven dayes travel', or '12. dayes against the currant, which wee travailed in the moneth of January, when the water was at the lowest of his nourishment' (*GT* 12, 83). Jobson claims that there is no longer a flood tide to assist up-stream passage, although in 1724 Stibbs found a small tide for the first 50 or so miles after Baarakunda. January was certainly in the dry season, but the season continued for at least two more months, during which period the river levels declined further. Jobson now encounters a difficult section of the river, with frequent sandbanks and large shoals, dangerous rocks, and strong currents in narrow channels. The party row the boat only in the cooler parts of the day – 'onely 7 houres in 24' – and not at night 'for avoidance of trees sunke, rockes, and sholes'; and Jobson claims that note was taken of the obstacles, perhaps implying the drawing of a river chart (*GT* 15).

⁴ Probably River Niaoulé, near Gouloumbou, a short distance above Baarakunda Falls. In 1724 Stibbs met shoals only 20 miles beyond Baarakunda and struggled on for another 40 miles (by his own reckoning), but up to and for some distance beyond his final halt he reported only dry rivers. Unlike Jobson, however, he noted an occasional inland town and made occasional contact with individual Africans along the river (Moore, *Inland Parts*, pp. 276–86).

⁵ Probably supplying much of the information about baboons in *GT* 144–5. Jobson states that his boat was thrice attacked by hippopotamus, 'one of the blowes very daungerous' (*GT* 21). In 1724 Stibbs had similar anxiety about hippopotamus attack (Moore, *Inland Parts*, p. 287). The fake account of the 1661 voyage included a reference to such an attack (ibid., Appendix III, p. 16), but this was based on the Jobson reference.

APPENDIX A

18	The Englishmen 'forced to enter the river ... and carry the boat against the current': *PJ* 923.
19	The boat meets a 'violent current ... but a mile in an houre': *PJ* 923.
20	Reaches '**Cantore River**' on the south side, with 'a faire entrance': *PJ* 923.[1]
21	'sent ashoare to the Mountaine tops', a view of 'onely Desarts': *PJ* 923; *GT* 85.[2]
22	Sixteen elephants seen on the bank: *PJ* 923.
23	The boat carried through low water for a mile and a half: *PJ* 923.
24	The men tow the boat, 'as sholds and trees permitted', and haul it through a narrow passage against a 'vehement current': *PJ* 923.
25	Watering at a fresh-water stream; many crocodiles in the river; one black saved from a whirlpool: *PJ* 923.[3]
26	Sight of 'the hill of **Tinda**';[4] three of the blacks sent to a king and the

---

[1] 'Cantore River' is Grey River, known in French and Senegalese sources as the Kouloutou. It meets River Gambia some 400 miles up-stream of the latter. 'The Grey River is 15 yards wide at its mouth and 6 feet deep in mid stream, current running about 1½ miles per hour, banks generally about 30 ft. high and wooded' (Dumbleton map, 1881). The first person to take a sailing vessel up this river was Reeve in 1908 (Reeve, *The Gambia*, p. 72; for the date, see the listing of a map, 'Levé de la Haute-Gambie de Yarbouténda à Bady Wharf', as charted by Reeve and Staub in March–April 1908, in E. Joucla, *Bibliographie de l'Afrique occidentale française* [Paris, 1937]). For air photographs showing the confluence between rivers Kouloutou and Gambia, see *Report to the Government of Gambia and Senegal: Integrated Agricultural Development in the Gambia River Basin*, United Nations Food and Agricultural Organization, Rome, 1964, page e. The English party are still in the district known to the Portuguese as 'Cantor', the Mandinka polity of Kantora, hence the name of the river. Jobson states, however, that the area was under 'Ferran Cabo', that is, *farang Kaabu*, the ruler of Kaabu, a Mandinka polity to the SE, and, if Jobson's information was correct, the overlord of the ruler of Kantora.

[2] That is, viewed from hill tops (but it is not stated on which side of the river), the area appeared deserted. Jobson wrote that 'from Baraconda ... we never hard, nor saw of any Towne, or plantation, nor recourse of any people unto us, but what we sent for' (*GT* 15–16). For lack of habitations in 1848, see Appendix D, para. 78. The area was similarly described, from much the same spot, in 1881. 'From the summit of some hills, whose headlands abutted at the river, an extensive view was obtained of the surrounding country, which, as a rule, is flat and uninteresting and appeared to be quite uninhabited. Indeed the eye searched in vain for homestead or hamlet, for clearing or cultivation, or other sign of human occupation of the land ... Not a single canoe or other vessel was seen throughout the length of the river from Yarbutenda to the Wharf of Bady, a distance of some 180 miles' (V. S. Gouldsbury, 'The Upper Reaches of the River Gambia beyond Barrakunda Falls', in *Correspondence relating to the recent expedition to the Upper Gambia under Administrator V. S. Gouldsbury* [London, 1881]). The day of the week in which Jobson's party stopped to view the district was a Sunday, but it may be doubted whether this was relevant.

[3] Jobson discourses on crocodiles, their smell, and the Africans' fear of them (*GT* 16–19). The black saved from the whirlpool was almost certainly Fodee Careere, 'my Alcaide' (*GT* 63, 76).

[4] For a photograph of 'Tenda Hill' near River Nérico, see Reeve, *The Gambia*, opp. 124. What 'Tinda' represented in Jobson's day is unclear. Today 'Tenda' denotes a large interior region to the SE of the confluence of the rivers Gambia and Nieriko, thus including the town of Jalakoto, but extending far to the east and apparently bounded on the south by River Niokolo Koba (1980 map of Sénégal, Institut Géographique National, 1:1,000,000). The term has also been used by linguists to denote a single language (which perhaps gave the region its name), or a group of related languages, or as a general term for several contiguous groups possibly related, the former alternatives located within, and the latter covering, an even larger region, stretching eastwards from at least River Kouloutou (Grey River). Jobson only reached the western fringes of the narrower region, in the Jalakoto district, hence his failure to delimit or discuss 'Tinda'. Whether Tinda at this date also included an area to the NW of the confluence is unclear, but that this was so might be indicated by Jobson' statement, if correct, that Tinda was under Wuli (*PJ* 924). A century later, the statement that 'the country on the Cantore side is populous

	merchant Buckor Sano: *PJ* 923; *GT* 84.¹
?	Having passed some shoals but stopped by another one,² the English approach **River Tinda**, 460 miles up-river and the highest point reached: *PJ* 923; *GT* 12, 84.³
30	Jobson sees animals and birds and hunts on land (*PJ* 923);⁴ when the messengers fail to return, the other Englishmen complain and threaten to leave: *GT* 85.
31	One messenger returns, with representatives of the king and the merchant: *PJ* 923; *GT* 86.⁵

... but on the other side are no Towns ... till you come to Tinda' (Moore, *Inland Parts*, p. 286), may or may not mean that Tinda extended some distance west of River Niériko.

¹ The messengers apparently left on 26 January, a Friday, and expected to return by 'Sunday night' (*GT* 84). Jobson later supplies both a date and its day of the week, Thursday, 1 February (*PJ* 923), and the day is correct (see p. 208, n. 1 above). Thus Sunday was 28 January, and the following references to Monday, Tuesday and Wednesday can be dated. Tinda/Tenda is usually considered the name of a district (previous note), but Jobson understood it as the name of 'the Towne' (*GT* 83). Very much later Park stated that 'several towns within sight of each other [are] collectively called Tenda, but each is distinguished by its particular name' (Park, *Travels into the Interior*, entry for 28.5.1797). The king who is contacted by the messengers was 'King of Tinda' (*GT* 86), but the king who subsequently arrives is the king of Jalokoto. Could it be that the two were one and the same?

² Whereas *PJ* seems to indicate that the boat proceeded further after the messengers were sent away, *GT* states that the messengers left when the final shoal was reached and the boat could go no further. If the latter is correct, then this entry is of the same date as the previous one.

³ Although Jobson does not indicate on which side of River Gambia was the mouth of River Tinda, there can be little doubt that this was River Niériko (Mandinka -*ko* 'river'), the next large tributary after Grey River, joining River Gambia on its north bank. While a lesser stream than the main river, it is curious that Jobson refers to this fairly lengthy tributary as a 'little river'. Further, it was 'said to runne neare unto the place', that is, Tinda Town. If indeed Tinda Town was Jalokoto (n. 1 above), this town lies about 6 miles in the interior, equidistant from the two rivers. Jobson must have been misinformed. At the confluence, River Gambia approaches from more or less due south, which may have been discouraging to Jobson who had so far travelled eastwards. As regards the halting point, Jobson says – 'we recovered within halfe a league of the place or Rivers mouth', which might mean either that the tributary river's mouth was half a league from the inhabited place they were seeking, or else that they halted half a league downstream from the tributary river's mouth. The latter seems more decisively indicated by another statement that the party was stopped by a shoal 'which stayed us we could passe no higher ... [having] laboured to get neare to this Tinda' (*GT* 84). The statement in *PJ*, possibly a summary, does not clarify matters. 'We past the sholds, and against Tinda River, recovered steepe water, and saw many Sea Horses, which love deepe waters' (*PJ* 923). Elsewhere, Jobson says that, from where they were stopped by a shoal, 'being past that place, the river shewed himselfe againe, with faire promising, so farre as wee had occasion to looke, neare a league' (*GT* 12). Since the boat was accompanied by a canoe, the latter could have been carried over the obstacle and relaunched. Although Jobson does not specify that this occurred, it would explain how he knew that the boat halted a mile and a half before the river mouth and that the later water was deep and harboured hippopotamuses. An alternative or supplementary explanation is that Jobson states that he 'went ashoare', and he may therefore have walked along the bank towards the confluence to investigate the further scene. However, since the boat contained the goods for trading, where it had halted could be considered the effective end of the voyage. The boat had travelled from Baarakunda about 130 miles in twelve or thirteen days, thus averaging – despite the shoals – some 10 miles daily.

⁴ Jobson killed an antelope 'bigger then any Windsore Stagge', and a large bird. But again, on Wednesday, 31 January, Jobson and his companions hunted and killed a beast 'as bigge in body as a great stagge, and had wreathed hornes' (*GT* 86). The description makes it sound as if this was the same antelope – although no doubt many such animals were killed – and if so, Jobson supplied different dates for the same occurrence.

⁵ If the boat did move up-river after the messengers left (see n. 2 above), it is not explained how the party from 'Tinda' knew where to find the English.

APPENDIX A

1 February	Buckor Sano and an escort arrive, followed by 200 men and women; he 'lay drunke aboard that night': *PJ* 923; *GT* 86–7.
2	Trade begins; female slaves offered and declined: *GT* 88–9.[1]
3–10	Salt especially in demand; more people arrive; 'market house' built (*PJ* 923–4; *GT* 89); Jobson eventually inquires about gold, from Buckor Sano and from a visitor, the latter conveying news to 'our Marchants below': *PJ* 924; *GT* 92–3, 101–2).[2]
5–8	Messenger sent to the interior; the message brings, after two or three days, large numbers of 'a more Savage people' from the 'other side'; Jobson discusses with their 'principall man': *PJ* 924; *GT* 94–5.[3]
7	Visit of 'The King of **Jelicot** on Tinda side, under the great king of Wolley': *PJ* 924; *GT* 96, 98.[4]
8	With Jobson's agreement, Buckor Sano proclaims himself 'the white mans Alcaid'; he approaches the king, and on behalf of the English obtains for them trading concessions: *PJ* 924; *GT* 97–100.[5]
10	The English party leave, proceeding down-river in falling water levels: *PJ* 925.
14	Within three miles of Baraconda, they are persuaded to continue overland: *PJ* 925.[6]
19–?	**Batto** reached, where Jobson witnesses a circumcision ceremony: *PJ* 925; *GT* 109–18.[7]

[1] Jobson writes at considerable length about events between 2 and 10 February (*GT* 86–104). It seems most likely that the English conducted their trading, when not directly from the boat, on the SE bank of the confluence, assuming that Buckor Sano had arrived from Jalakoto or at least from the direction of that town.

[2] The town of 'Combaconda' (probably a misprint for 'Tombaconda'), some four days journey away, which Jobson thought might be Tombutto (Timbuktu), was almost certainly Tambakunda, some 30 miles north of River Gambia, a town on an important overland trade route. The 'Merchants below' may be either Lowe's party or the traders left at Kasang.

[3] The 'other side' was presumably the district or region to the SW and south, between Grey River (River Koulountou) and River Gambia. The people who were not Mandinka or Jaxanke, and wore animal skins for clothing, were perhaps Bassari; but the 'principal man' was seemingly a Badyara/Pajadinka (*GT*, p. 144, n. 3). These contiguous ethnicities are found today on, or near, the uppermost stretches of rivers Koulountou and Gambia, some considerable distance from Jobson's halt (D. Westermann and M. A. Bryan, *Languages of West Africa*, Oxford, 1952, pp. 16–17); but perhaps in earlier times they were nearer.

[4] This was the ruler of Jalakoto, a town and polity south of River Niériko and west of River Gambia (now running from due south), the town half a dozen miles from the confluence, near which Jobson was now installed to conduct trading. How much of the confluence district was part of the Jalakoto domain is unclear, but the land granted to the English was probably close to and SE of the confluence and therefore undoubtedly in 'Tinda'. See the reference to the king in p. 218, n. 1 above. Whether Jalakoto actually had Wuli as its overlord is otherwise unevidenced.

[5] Jobson named the site formally granted the English by the king, Saint John's Mart (*PJ* 925; *GT* 104), presumably after the name of the ship. The site was abandoned when the English retreated and the name left no trace.

[6] It is not stated what happened to the boat but presumably the Africans brought it, once lightened, down-river and through the Falls and other 'rockie passages'. Jobson must then have rejoined the boat, since he states that 'we ... came to an anker at the port [of Batto]' (*GT* 109).

[7] The date is given in *GT* 109 as 9 February, an error. It was probably during this stay at Batto that Boo John quarrelled with the son of a king (apparently the king of Oranto) and Jobson had to intervene (*GT* 57).

?	Jobson 'made haste backe to Setico', to meet merchants from Tinda; on the way the boat struck by a hippopotamus: *PJ* 925.¹
?	Reaches 'with my boate as neere as I could come' to Setico [Sutuko].²
?–?	Travels to nearby **Setico** with two other Englishmen; learns about Islam and inquires about the gold trade: *PJ* 925; *GT* 63–70, 80–2.
?	Witnesses the funeral of the senior marabout: *PJ* 925; *GT* 70–2.
11 March	Leaves, having heard of sickness on the *Saint John* down-river: *PJ* 925.³
14	Reaches the *Saint John*: *PJ* 925.
15	The pinnace sets off down-stream to join its consort: *PJ* 925.
18	Reaches **Pompetan**/Pompeton, where Jobson is entertained by a hospitable Portuguese: *PJ* 925; *GT* 118–19.⁴
19	The *Saint John* joins the *Sion* at **Cassan** (Kasang): *PJ* 925.⁵
18 April	Jobson leaves Cassan in the *Sion*: *PJ* 925.⁶

Jobson's accounts read as if the boat stopped at Batto on its course down-river, after which it carried on to the base where the *Saint John* lay. But since the base and Batto were fairly close, if the texts have chronological omissions the boat may in fact have gone directly to the base and then returned to Batto. Even if the former was the case, Jobson may have communicated with the base from Batto, and conceivably there may have followed some exchange of men. The 'Chirurgeon' was present at Batto with Jobson, but on balance it is perhaps more likely that he had accompanied Jobson up-river than that he had stayed with the ship.

¹ By 'haste back to Setico' Jobson does not mean that he was returning to Sutuko, which he had not previously visited, but that he was travelling back up-river some distance – elsewhere described, also clumsily, as 'a second returne up some part of the River' (*GT* 41). Jobson does not detail the journey up-river, which was presumably in the same shallop which had ascended earlier, although not necessarily with the same English crew and probably with only one African, Fodee Careere. A canoe was also taken (*GT* 70). Sutuko/Sutukobaa was a traditional trading centre – see *GT*, p. 125, n. 1. Although Jobson refers to Sutuko as the principal town of Cantor (Kantora), it is in fact on the north side of the river, although not far inland, and therefore in the territory and polity evidenced later as Wuli. It may confirm Jobson's statement that a generation later the port of Sutuko was still referred to as 'the port of Cantor' – see Part II below: Lemos Coelho, 1669, 26. The history of the area is obscure, being known only from oral tradition, but perhaps before the Wali established their rule on the north side, this riverside area of the district was also considered part of Kantora, the polity to the SE.

² Jobson refers to 'the place or port whereat my boat did ride' (*GT* 70), without naming it. Fatatenda is generally considered to have been the main port of Sutuko. A generation after Jobson's journey it was termed 'the first port of the kingdom of Oli [i.e. Wuli]' (Part II below: Lemos Coelho, 48), and it was certainly the first port with a short route to Sutuko to be encountered when travelling up-river. But the town could in fact be approached rather nearer, by sailing further up-river to the most northern part of the next turn of the course, and then landing on the north bank, preferably at the crossing from Sami (Samitenda) (see Moore, *Inland Parts*, 269), later maps as well as traditions indicating a road to Sutuko from this point. The distance by this road would be slightly less than the distance by the route from Fatatenda, so this may have been where Jobson landed. He states that he was within 3 miles (*GT* 113), or 4 miles (*PJ* 925), of the town, and he describes a canoe ferry on the river at this point, confirming an important road to Sutuko from the south (*GT* 70).

³ Jobson gives the day of 11 March as Sunday, which is correct.

⁴ For Jobson at Pompetan, see p. 213, n. 1 above.

⁵ Only four able men were left on the *Sion*, the others being sick or dead. Among those dead were the master and the master's mate (see Introduction, p. 26, nn. 2, 3). While at Kasang much fishing from the ship was done (*GT* 23).

⁶ The *Sion* had been left at Tendabaa in late November and it is not stated when it moved up to Kasang. But it had been lying off both places for over three months. Considering the mortality the crew of the *Sion* had suffered, it is surprising that another month was spent at Kasang. The passage down-river from Kasang was in fact leisurely, the ships not leaving the river until two months after Jobson reached Kasang. Possibly part of the crew of the smaller ship transferred to the larger, and perhaps the ships delayed in the hope of

APPENDIX A

19	The *Sion* anchors at **Wolley, Wolley** (*PJ* 925–6; *GT* 59–60);[1] the *Saint John*, 'being ready to come away' from Kasang, is contacted by the new king: *GT* 60.[2]
20	The *Sion* lies off **Mangegar**, where Jobson observes a market: *PJ* 926; *GT* 122–3.[3]
30	The *Saint John* joins the *Sion*: *PJ* 926.
4 May	Both ships sail down-river: *PJ* 926.
18	A shallop is 'prepared': *PJ* 926.[4]
19	Crews go ashore, in **Cumbo** (Kombo) on the south side of the lower estuary, for caulking 'and other businesse'; visited by the king: *PJ* 926.[5]
9 June	The ships leave River Gambia: *PJ* 926.
?	Put in at **Travisco** (Rufisque) for repairs, 'our Carpenters being dead', before sailing for England: *PJ* 926.[6]

---

some of the sick recovering. Presumably trading continued at Kasang, and it was probably during this period that Jobson collected most of his information on the town. The 'second day after I was gone from the Towne', the new king of Kasang arrived and he spoke to the 'Factor' in the pinnace (*GT* 59–60), therefore Jobson transferred to the *Sion* at Kasang. Down-river at Mangegar, the *Saint John* 'came to us' (*PJ* 926).

[1] While at Wolley Wolley Jobson heard of the deposition of the existing king of Kasang by his overlord, in favour of a younger man (*GT* 59–60; *PJ* 925–6). Wolley Wolley seems to have been part of the dominion of the king of Kasang; in 1723 Stibbs found the king living at Yanimary (Niani Maru), which was probably a renaming of Wolley Wolley (Moore, *Inland Parts*, p. 254).

[2] The new king entered Kasang on this date (*PJ* 925). Since the *Saint John* reached Mangegar ten days after the *Sion*, it may have been held back at Kasang by the need to establish friendly relations with the new king.

[3] Presumably the single survivor of the three men left at Mangegar was picked up, although Jobson does not mention this (*PJ* 922).

[4] 'Prepared' perhaps to carry tents and tools ashore.

[5] The ships were being made ready to go to sea. There are a number of places where the crews could have gone ashore (and perhaps beached the ship for caulking): Kembuje or Bafuloto near Brikama (one of the old royal towns), Bereto near the mouth of the Lamin Creek, or the sheltered side of the island of Banjul (the presentday Half-Die area of Banjul town). It is uncertain whether the English had spent much time ashore in the lower river previously, but during the fortnight they now spent on land in Combo they were troubled by 'Musketos, which here exceedingly abounded' (*PJ* 926).

[6] A voyage of 'tenne monethes ... both out and home' (*GT* 162), would bring the ships to England in August 1621. While the outward voyage took only twenty days, the return, being against the prevailing direct winds and hence on a less direct and much longer course, would take more time, probably at least four weeks. But arrival in later July, rather than arrival in August, cannot be counted out and might be reckoned to be within ten months. The *Sion* had been hired from its owners for a period of eleven months and a subsequent law suit revealed that it had exceeded that period by an unstated amount (PRO, HCA 3/95, 6.2.1621). However, the total period doubtless included some weeks at Gravesend (*PJ* 921), when it was being prepared to sail, and possibly some days on its return when it was being unloaded and tidied up for handback.

The 'Middle River'

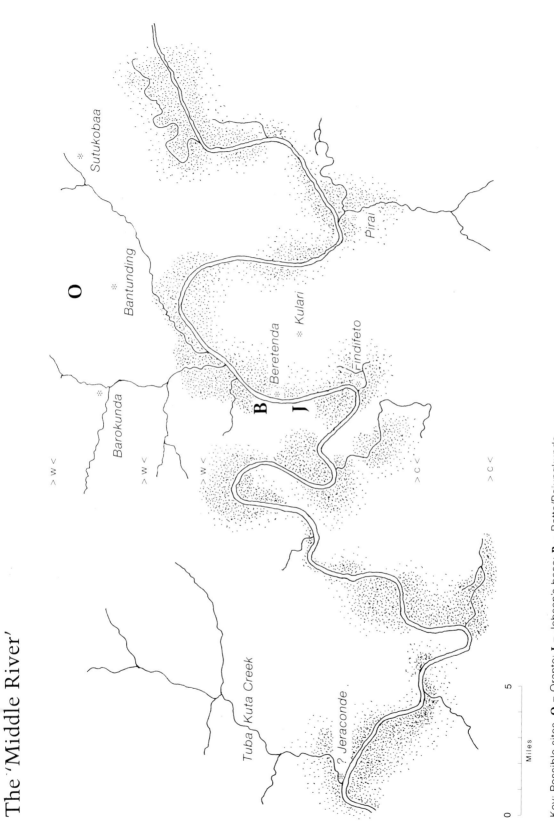

Key: Possible sites, **O** = Oranto; **J** = Jobson's base; **B** = Batto/Bojungkunda.
> w < = ? c. 1620 western boundary of Wuli.
> c < = ? c. 1620 western boundary of 'Cantor'/Kantora.

APPENDIX B

The middle river base

1. Both Thompson and Jobson operated from a middle river base in order to explore up-river. Where Thompson's base was located is nowhere indicated, although it is likely that it was at or near the inland town of Oranto, where his comrades were later found.[1] Jobson's base, whose name is uncertain, was nowhere given an exact topographical description, but it was certainly in the vicinity of Oranto. Since the Jobson texts relate several episodes to the base, it is worthwhile to attempt to locate it. Because of the difficulties in relating the two sources and interpreting their vagaries the discussion has to be long, tortuous, and in the end only tentative.

2. No place named Oranto appears in later sources or on maps, with one possible exception.[2] But all that can be firmly deduced from this possible exception is that Oranto lay between Kasang and Sutuko, which is patent from the texts. A later description of this stretch of River Gambia is fairly detailed, but the toponyms supplied are not those in Jobson's texts and cannot be related to them, regrettably.[3]

---

[1] A 1912 assertion that 'Thompson made a settlement at Fata Tenda' (Reeve, *The Gambia*, p. 48), although subsequently followed by a number of historians, has no evidential basis.

[2] A name difficult to decipher but seemingly 'Orantie' or 'Orantis', which may well be the same as Oranto, is given on an anonymous English map of the eastern Atlantic seaboard which shows River Gambia (Henry E. Huntington Library, San Marino, California, HM 2098: we are indebted to William B. Frank and Mary Robertson of the Library for studying the original and producing the alternative reading). It occurs within the following sequence on the north bank of the river: cassan, quila cundie [or cundis], orantie/s, sette-coe. The Huntington map is undated but undoubtedly of the period 1600–1650. Its toponymy is basically Portuguese, with occasional anglicisms, of which 'sette-coe', by its spelling, is patently one. The dozen names along the river are some in Jobson, although spelled differently, some not. It may therefore be either of a date earlier than the English voyages, or later. But it would seem most likely that the relevant names derived from the voyages of 1618–1621, yet if so the source was apparently not Jobson. The map has only a crude representation of the course of the river, but the identifiable names, 'cassan' (Kasang) and 'sette-coe' (Setico/Sutuko), are in the right order. Moreover, 'orantie/s' occurs between 'cassan' and 'sette-coe', and such a location of Oranto would fit Jobson's account, inasmuch as from his middle river base, beyond Kasang, he went up-river to Sutuko. The name 'quila cundie' contains *kunda*, the Mandinka ending for the name of a town, e.g. *Jula-kunda* 'trader's town'. The English pronunciation of 'quila' might well be /kwila/, rather than /kila/, but neither Kwilakunda nor Kilakunda are known or acceptable Mandinka toponyms. Possibly some miscopying has occurred, the correct name being, perhaps, Julakunda, but this is a name found today in several locations. As for 'oranto/orantie/s', *oran-* is per se an impossible beginning for a Gambian placename, since local placenames invariably begin with a consonant. 'Oranto' must therefore involve a considerable degree of corruption. All therefore that can be said of 'Orantie/s' is that it tends to weaken any supposition that, although the name Oranto appears many times in Jobson's writings, he miscopied a note and consistently reproduced an erroneous name. It is more likely that he wrote Oranto because this was the name he had heard from the English survivors, and that it was they who had produced the corrupt form.

[3] The Portuguese trader, Lemos Coelho, gave a detailed account of 'ports' up River Gambia, including those on the stretch indicated (see Part II below). Between Kasang and the ports of Sutuko, Lemos Coelho listed the following ports – those not identifiable from later sources italicized – with the distances between them and from Kasang as noted: Cação (Kasang), Conicomco (+ 6 leagues), Jurume (+4 = 10), Lamé (+8 = 18), [Fulos' Pass], *Bunhacú* (+8 = 26), Peripho (+1 = 27), Maresamsam (+7 =

3. On his way from Kasang to Oranto, an unstated distance, Jobson passed 'Pompeton' and 'Jeracunde', neither places identifiable from modern maps.[1] But the suggested location of the former near Karantabaa (see Itinerary, p. 213, n. 1) is reasonably secure. Jobson's writings thereafter evidence the following stages up-river (all *PJ* 922, except where otherwise stated).

(a) After arriving at Pompeton on 14 December, the 'next morning' the ship reached the Port of Jeraconde.

(b) From Jeraconde, apparently on the north bank, a message was sent 16 miles overland to Oranto, and in response, two Englishmen from Oranto reach the ship on 17 December.

(c) Up-river an unstated distance, and after an unstated time of travel, the ship reached a point six miles from Oranto overland.

(d) Apparently it was at this point, on the riverside, that the middle river base was established.

(e) Near the base was the town of Ferambra, seemingly a representative of the king of Wuli, as well as a Fulbe settlement.

(f) From 'the place where our men lived', presumably the base, up-river to Sutuko was 'some 16 leagues' or 48 miles (*GT* 32).

(g) Up-river from the base 'one tide' was 'Batto', the town of Boo John.

(h) Jobson's party was visited by the king of Bereck/Pereck, identified as Pirai, an important locality between the ports of Sutuko.

(i) Up-river from Pirai 15 leagues (45 miles) was an unidentified locality (Massamacoadum) and two days' journey from there Baarakunda, an identified locality 160 miles up-river from Kasang.

4. It is impossible to produce a solution to the problem of the location of the base which exactly fits all these clues. This may be because Jobson exaggerates distances and because the extracts from his journal published by Purchas are incomplete and omit essential stages in the journey. However, it is clear that the middle river base was located somewhere in the central section of the stretch of 65 or 75 miles

---

34), ['Arse Hill'], Rio de Alea [Sandugu Creek] (+8 = 42), *Ponor* (+6 = 48), *Jalacuna* (+6 = 54), Nhamenacunda (+8 = 62), Perifo (+7 = 69), Sumacunda (+1 = 70), Nhacoi (+4 + 74), Findefeto (+6 = 80), Fatatenda [port of Sukuko] (+4 = 84), Pirai (+4 = 88), Same [port of Sutuko] (+4 = 92). But Lemos Coelho's distances, like those of Jobson, are exaggerated. The 42 Portuguese leagues or 155 English miles between Kasang and Sandugu Creek stated by Lemos Coelho represent, in actuality. only some 65 miles; and the 46 leagues or 169 miles between Sandugu Creek and Same represent only some 75 miles. Since Lemos Coelho regularly estimates the distance between ports in round figures, mostly 6 or 8 leagues, probably each distance is over-estimated. A curious difference between Lemos Coelho's account and that of the Englishman, is that three physical features emphasized in the former (and inserted above in square brackets), a famous magic hill (later termed 'Arse Hill') and two notable narrowings of the river awkward for navigation (Fulos' Pass and Fundo Feito/Findefeto) are not mentioned in the English account, thus making it more difficult to estimate distances. Several of the names in the Portuguese account (e.g. Ponor and Jalacuna) do not appear in any later sources, and in fact none of the names throw direct light on the placing of the middle river base. Finally, it may be noted that whereas the Portuguese call at ports between Kasang and 'Cantor' approximately every 10–15 miles, many on the south side of the river, the English ignore all of these – or at least Jobson does not mention any.

[1] The name Jeracunde appears to be derived from *jeere*, a Mandinka personal name (a nickname for Omar), and a place so named is unlikely to have retained the name for more than a generation or two. It is therefore not surprising that Jeracunde is not recorded elsewhere.

between Pompeton and the two ports of Sutuko. The distances stated by Jobson will now be examined.

5. If from Pompeton to Jeraconde was only some twelve hours' travel ('the next morning'), hence probably only 10–12 miles, then given that Oranto was 16 miles overland from Jeraconde, the point on the river within 6 miles of Oranto must have been down-river from the centre of the indicated stretch. This would fit the distance between the base and either of the ports of Sutuko being 48 miles. But Jobson greatly exaggerated all the river distances that can be checked, and it may be that Purchas omitted a day entry between Pompeton and Jeraconde.[1] It will now be argued that the base was more probably higher up-stream.

6. From Bereck/Pereck/Pirai to Baarakunda was stated by Jobson to be 45 miles plus two days' journey, say, 65 miles. But the true distance is some 33 miles. Accepting that this is an instance of Jobson's exaggeration of river distances allows the base to be further up-stream.

7. A different approach is to consider the topography of the river. As shown below, both Oranto and the base were opposite the territory of 'Cantor', and therefore they were on the north side of the river, as seemingly was Jeraconde. Upstream from Pompeton, and on the north bank, two creeks exit. The first, Sangedugu (or Sami) Creek, 12 miles upstream from Karantabaa, has only low-lying or marshy land around its mouth so that vessels had to go up it several miles to find a landing place and any settlement. The mouth of Tuba Kuta Creek, 38 miles from Karantabaa, is more promising as a calling place for the English, since dry land comes down to the river at this point.[2] Although it to some extent fits the requirements for the base area, it appears to be ruled out by the consideration stated below. Instead, this may have been the site of Jeracunde – supposing Purchas to have telescoped the Journall entries between Pompeton and Jeraconde.[3]

---

[1] Given that Pompeton was reached on 14 December and the English survivors arrived at the river on 17 December, Purchas may have cut entries for 15 and 16 December which showed that in fact the ship only reached Jeraconde on 16 December, the messenger leaving for Oranto that day. In a single day the ship could not have sailed 38 miles from Pompeton to the site of Jeraconde proposed below.

[2] In the present century there was a small wharf here called Dasilame Tenda, although census returns suggest that it was occupied by only one family, no doubt owners of a canoe used to ferry goods and persons across the creek and the main river. Dasilame town, 3–4 miles inland, now Muslim, is said to have been previously Soninke or non-Muslim (Nancy Ann Sheehan, *Tenure and Resource Management in the Gambia: a case study of the Sandu district*, Madison, 1994, p. 44). A nearby place, Sumakunda, although recently occupied by a Fula community, was presumably once the residence of a Mandinka *suma*, or local ruler. This area therefore to some extent fits the location of Oranto, the base, and neighbouring settlements.

[3] In his book, Jobson referred to 'Ferran', who 'had the government of a countrey, where we had much and often trade, and for the most part, kept a Factor' (*GT* 58). Although *farang* was a title for a lesser ruler and was therefore probably applicable to several notables along the river, Jobson's only specific reference to an individual *farang* was as follows: 'we came to the Port of Jeraconde, two miles from which dwelt Farran a perpetuall Drunkard, but which held his Countrey in greatest awe' (*PJ* 922). Assuming that the 'Farran' near Jeraconde was the same individual as the 'Ferran' of the country with much trade, it would seem that a factor was stationed at Jeraconde, presumably as long as the middle river base was occupied, although the existence of this arrangement is not mentioned elsewhere. If Jeracunde was indeed at the mouth of Tuba Kuta Creek, the *farang* perhaps lived at Sumakunda (see previous note).

8. A third locality some 20 miles further upstream from Tuba Kuta Creek is the preferred site for the base, on the grounds that it most probably lay within contemporary Wuli territory. Ferambra (3(e) above) appears to have been a frontier representative of Wuli, inasmuch as he saw off an attack on the English survivors by the king of down-river Niani (*PJ* 922). If the 1732 map of the river (in Moore, *Inland Parts*) is correct, and if we may suppose that the boundaries between polities there shown represented those in existence a century earlier – admittedly, a slightly incautious supposition – then Wuli began higher up-river than the Tuba Kuta Creek.[1]

9. If Jeraconde was in fact at the mouth of the Tuba Kota Creek, the stated overland distances of 16 and 6 miles (3(b) and 3(c) above) would seem to place Oranto in the district of the present town of Bantunding – supposing Jobson's estimates of land distances to have been more accurate than his estimates of river distances (which is perhaps reasonable). As for the base, in order to be within six miles of Oranto, it was most likely located somewhere on the north-tending course of the river up-stream between Findefeto and Fatatenda, perhaps near Limbambulu (a recent town). This would be far less than 48 miles from even the further of the two ports of Sutuko (3(f) above), the distance being only about 25 miles. From Pompeton it would be some 55 miles upstream.

10. The existing capital of Wuli, Medina, was destroyed during the Soninke–Marabout wars of the nineteenth century, and Bantunding, not shown on earlier maps, is locally stated to have been only subsequently established, by refugees from the troubles. Indeed, some present-day Wuli traditions assert that the only town in southern Wuli before colonial days was Sutuko, elsewhere only temporary farming villages being in use. While Bantunding represents a recent settlement, it has been claimed that a small village of griots existed earlier on the spot (p.c. P. Weill, 1997). Water supply is a difficulty in Wuli, and settlements away from streams depend on wells, often very deep ones, and therefore continuity of settlement in one locality, to make use of the wells, is plausible. Oranto may therefore have been close to the site of Bantundung, if not actually at it.

11. Tests involving the few dates supplied by Jobson do not rule out the suggested location for the base. After landing and going to Oranto, Jobson returned to the river and crossed it to visit the king of Oranto. On 21 December he began trading, apparently at the riverside, and it may be assumed that the middle river base was being set up where he traded, at the point first arrived at. The English had reached Pompeton on 14 December. Allowing for not fewer than two days for the visit to

---

[1] The same boundary is shown on recent maps of electoral districts as well as on a map purportedly representing the position at the beginning of the present century (in Galloway, 'History of Wuli'). A degree of confirmation of the 1732 boundary is afforded by a Portuguese statement of 1669 that the first port of Wuli was Fatatenda (Part II: Lemos Coelho, 1669, 26). Village traditions claim that the the zone east from Tuba Kuta Creek to opposite Basse, often called Nyakoi, was settled from Bundu and not from Wuli (interviews with elders, Kerewaan, 1957); and in the present century the inhabitants have successfully complained of being wrongly included, for administrative purposes, in Wuli (instead of Sandu). Although it thus seems that, in Jobson's day, Wuli did not extend to Tuba Kuta Creek, it is possible, however, that the Niani/Wuli boundary was not a clearly defined one, and that there was a no man's land between between the two polities, where herders could go and individuals gather products from the bush, but which required the nearby presence of a Ferambra to safeguard Wuli interests.

Oranto and the king, the journey from Pompeton to the middle river base would seem to have taken not fewer than three and not more than five days. If the ship travelled upstream at 12–15 miles per day this would make a journey of between 36 and 85 miles. More convincingly, on its return down-river, from the base to Pompeton the ship took from a Thursday to a Saturday, say three days, and from the base to Kasang from Thursday to Monday, say five days (*PJ* 925). From the suggested site of the base to Kasang is some 110 miles, so the speed of the ship would be 22 miles per day; at this speed, the base would lie 66 miles up-river from Pompeton, an estimate produced by this rough calculation not far from the 55 miles indicated above.

12. To fill out the picture, it will be useful to consider other points about the middle river base. It was certainly in the vicinity of Oranto, but Jobson's references to the precise relationship between the two are typically vague and unhelpful. The king of Oranto lived inland 'on the other side of the River ... subject to the great King of Cantore' (*PJ* 922), and the Kantora domain being to the south, the reference places Oranto on the north side. Jobson does not explain why the king lived where he did, but there could be many reasons.[1] If a reference to hunting with horses 'especially on the north side' having been witnessed by 'divers Englishmen', and the English were Thompson's party at Oranto, this would also seem to place Oranto on that side (*GT* 48).[2] After visiting both the town and the king, Jobson 'sent away my Boat', began to trade, and was visited by the king's relatives; this suggests that the English were setting up the base at the point where the *Saint John*, having halted, was to lie off for the next few weeks. In his book, Jobson referred to the middle river base as 'our habitation', 'our habitation upon the land', 'our dwelling', 'our Land-dwelling', 'the place our men dwelled at', 'houses seated by the River side', and 'where we had houses built'; and it was given the Mandinka name of Tubabo Conde, meaning 'Whitemen's Town' (*GT* 32, 56, 58, 68–9, 100, 134, 149/141).[3]

13. A possible source of confusion is that Jobson's account of his visit to Sutuko, 'above the place where our men dwelled at', which he reached by sailing to the 'port of Setico' (*GT* 63), includes a paragraph which begins – 'The place where we had houses built, and walled with straw for our owne uses, was seated by the River side, upon the top of the banke; and by the people of the Country, called *Tobabo Condo*, the whiteman's towne' (*GT* 68–9). This might seem to describe a temporary residence at the port of Sutuko. However, the paragraph continues by mentioning 'an ancient Mary-bucke called *Mahome*' and another Mary-bucke called Hammet, the first very friendly, the second less so, both of whom lived nearby. Elsewhere Jobson

---

[1] Conceivably he had quarrelled with neighbours – the king drank alcohol in an an area noted for staunch Muslims. Or perhaps, as he was blind, he had been forced to hand over authority at Oranto town to a relative.

[2] Horses appear to have been associated only with the king of Niani and the *buur Saalum*, rulers on the north side (*GT* 100). However, it is conceivable that the hunting with horses was seen by the English who crossed overland to Cape Verde, and if so not necessarily around Oranto.

[3] When a small Portuguese vessel passed, five Englishmen were occupying 'the place where our Land-dwelling was' – the number of Englishmen may have been that left to guard the *Saint John* after both shallops had left (*GT* 32). But alternatively the episode may relate to Thompson's men a year earlier, after he had gone upstream, supposing that his base was also at the riverside.

refers to the two, 'in the towne where our housing stood', and how Mahome warned him about the exploration of the upper river, which was to begin from the middle river base (*GT* 79). Thus, the description above must be, not of a shore base at the port of Setico, but of the middle river base. When Jobson reached the Tinda region, the English traded from their shallop, although their African visitors build temporary housing; and presumably in the port of Sutuko the English similarly operated from their vessel, and not from temporary housing ashore.[1]

14. The English base, although formed apparently by temporary houses on the riverside, seems to have been close by an existing African town site. Jobson refers to the local Africans 'amongst whom we have settled ourselves' (*GT* 32), and of the Africans as 'our neighbours, in the towne where our housing stood' (*GT* 70). Distant 'some hundred paces', presumably from the English houses but probably still part of the same settlement, was 'a small towne of religious people', that is of 'Marybuckes' (*GT* 69). A Fula village was some little distance away, perhaps on the road to Oranto (*GT* 116). The 1732 map and later maps name several places called Fula Kunda (Fula Town) at various points close to the river, but since it is generally considered that Fula settlements were migratory this is dubiously helpful. Jobson throws in another obscure element when he casually remarks that once he was 'comming from Bo Johns house to the place at Faye' (*GT* 117). The name 'Faye' is not mentioned elsewhere and no such name appears on maps. Was it was the name of the town where the English had their base, or a local name for the exact spot by the river where the English houses stood? Or was it the name of the Fula settlement? A Fula village in present-day Kantora is called Sincu Faye.

15. Three or four miles from the middle river base lived the friendly Ferambra, whom Jobson visited (*PJ* 922; *GT* 100, 149/141). Ferambra had earlier protected Thompson and his English comrades from attack by the horsemen of the king of Niani, sending them to safety 'over the river', and this again would seem to place both Ferambra's house and Thompson's place of refuge, perhaps Oranto, north of the river (*GT* 100; *PJ* 922). Jobson once walked 'from our dwelling to *Ferambras* house, in the way we were to passe through a towne of the Fulbies' (*GT* 116) – unless he omitted to say that he crossed the river first, this would indicate that the middle river base was on the same side as Ferambra's town, that is, on the north side. Jobson described the town as a 'good spatious plantation' (*GT* 148–9), which presumably means that it was surrounded with trees, but otherwise tells us little about it. It is not made clear whether Ferambra's town was inland or further along the river, and we cannot suggest a location.[2] A marabout whom Jobson met up-river had a wife in Ferambra's town, which indicates that it was a Muslim settlement (*GT* 100).

16. It is, however, possible to speculate on the location of the residence of the 'King of Oranto'. More or less opposite Limbambulu, but inland from the south bank of the river, is Kulari, a settlement shown on the 1661 and 1732 maps as

---

[1] In 1724, Stibbs found no town at either of the ports of Sutuko: Moore, *Inland Parts*, pp. 266, 269.
[2] The 1661 map shows two places along River Gambia called 'Farambacunda' or 'Frambacunda', but neither anywhere near the district of this Ferambra's town; however, the office of Ferambra was clearly not uncommon.

'Color'. On the 1732 map it is placed in the polity of 'Tomany' with the border of 'Cantore' slightly to the east, but earlier Portuguese sources, like Jobson, describe the whole south bank of the river as being in 'Cantor' (*GT* 47). The ruler of Kulari was therefore perhaps the same as the 'King of Oranto ... under Cantor'.[1]

17. From the middle river base, Jobson sailed 'first tyde' to 'Batto, Boo Johns Towne' (*PJ* 922), and later, returning down-river, came to 'Butto' [*sic*] (*PJ* 925) or 'the port of Boo Johns town', the town standing 'some mile from the waterside' (*GT* 109). (On return down-river to 'the port', Jobson describes having to climb a bank 'high from the river' to approach the town, suggesting a very limited 'port': *GT* 110). The reference to 'first tyde' may indicate that the distance sailed was that of a whole tide, many miles – or it may simply indicate that the party left on the earliest tide and may therefore have travelled only a short distance. Unfortunately, the toponym 'Batto' is unidentifiable, Mandinka *ba-to* meaning 'at the river'.[2] The town was perhaps actually called Bojangkunda, but no such name is found in other sources or on maps. A conjectural location is the former crossing place of Beretenda, on the north side, about nine miles from the suggested location of the base. But one counter-indication must be noted. When Ferambra protected Thompson and the English, he 'conveyed them over the River to his Brother, called Boo John' (*PJ* 922). From this it might be inferred that Batto was on the south side. This is puzzling; no solution can be offered and it may be unfair to back up the suggestion above by the argument that Purchas must have made another error.[3]

18. Immediately following the reference to Batto, Purchas's Extracts state that, after arriving at Batto on 2 or perhaps 3 January, on 6 January 'Sumaway, King of Bereck under the great King of Cantore, came aboord', and that later the shallop proceeded 15 leagues beyond 'Pereck' (*PJ* 922). In his book Jobson referred to 'Summaway, King of the next place, and he came downe and was aboard' (*GT* 58). Possibly 'came downe' means, not 'came down the river', but 'came down country to this port'. Despite the juxtaposition in Purchas, it is therefore possible that the encounter with the ruler was not at Batto but after the English had moved further up-river to Perai, a documented port between the two ports of Sutuko (Fatatenda and Same). We have seen above that Perai is not Jobson's '15 leagues' from Baarakunda, but it would be a suitable stopping point when travelling towards Baarakunda. From the conjectural site of Batto to Perai is only 3–4 miles, so that although a meeting at Perai makes sense, that the king 'came down' the river to Batto cannot be ruled out.

---

[1] The Mandinka inhabitants of Kulari, having been replaced in the present century by Serahuli, have moved to the north bank.
[2] In 1738 it was stated that the Mandinka call the river 'Bato' (Moore, *Inland Parts*, p. 298). For the meaning of the term, see Appendix A (Itinerary), p. 215, n. 1.
[3] According to Jobson, a quarrel broke out between the king of Oranto's son and Boo John 'in my house'. Since the king of Oranto lived on the south side of the river and Jobson's house was on the north bank, the king's son must have been visiting Jobson, but the episode throws no clear light on where Boo John lived (*GT* 56–7).

19. All of the specific identifications proposed or suggested above are a good deal less than certain. But there can be no doubt that the area of important English activity, at the middle river base and Oranto, was within a fairly narrow district north of the river and between the Tuba Kuta Creek and Fatatenda.

APPENDIX C

Text of the Gerbier map, in translation[1]

River Gambia and the City of Cassan in Africa
River Gamby, called Camboya in Portuguese, lies between the Kingdom of Guinea and Malagueta. 'Camboye' means 'fishery'.[2] It is situated at eight degrees north of the equator. At its mouth is a small island separating River Mosquitoes and River Palms.[3] On the north side of the island there is much gold in the sand which can be collected from the streams and the sand, and it is only necessary to wash this because the gold in nugget form is pure. Secondly, all along the banks of River Gamby there is much of this gold, especially on the east side, where there are many streams about a league and half up-river, and at 40–50 [? leagues] where the hills

---

[1] The original text is as follows (words in *italics* are in code in the original).

La Rivière Gambia et La Cité de Cassan en Afrique / La Rivière Gamby en portugais appellée Camboya, est entre le Royaume de Guinée, et Malagette, Camboye veut dire Pescherie, elle est située a huit degrez de la ligne Equinoctiale du costé du Nort, Il y a a l'embouchure de la dite Rivière une petite Isle laquelle separe les Rivieres Mosquitos et Palmos: il y a du costé du *nort* de la dite Isle quantité d'or en Sable que l'on peut ramasser dans les ruisseaux et dans le Sable, qu'il faut seullement laver, d'autant que l'or est pur en gravier. Secondement, Il y á tout du long des bords de la Riviere Gamby quantité de cest Or, mais principallement du coste O *oriental*, ou il y a plusieurs *ruisseaus*, environ *une lieu et demie* montant la Riviere, et á *quarante* ou *cinquante* [? lieues] ou commencent *les montagnes* sur les [word missing]. L'or demeure apres les pluyes, il y á la *montagne* sur laquelle le *soleil* donne ses *raions* sur *midi* dans le *creus* de laquelle les pluies ont fait couler une Masse tres grande d'or en gravier qu'il faut seulement laver avec de l'eau, celuy qui a donné ceste relation dit avoir remarqué que les habitans n'avoyent pas de ce temps la prins guarde á ceste grande Masse d'or, mais qu'ils s'ocupent seullement a ramasser celuy qui se trouve dans le Sable au bords du Rivage, qu'ils vendent pour du Linge de petite Valeur, et des bandelettes d'escarlatte, des Couteaus, haches, Coraux de verre de toute couleur, et des Espingles, ce sont des Mores fort suptils et adroits en ouvrages d'or, de bagues, chaisnes, et cordons de fil d'or, que les Orphevres en Europe ne scauiroint mieux faire. ils sont grands trompeurs, ils meslent l'or avec l'argent quand ils en peuvent attraper. il ne faut pas traicter avec eux en Or, que pour celuy qui est en grain. Le pais est plat au commencement. Ces Mores trafiquent en dents d'Elefants, corne de Rinocere, Gomme Elemie. le Pais abonde en chair, poisson, vollaille, vin de Palme, et en Esclaves, qu'il faudroit transporter en Amerique, ou sont les mines d'or, pour les y faire travailler. Est a notter qu'il y a grande quantité de Sel en la Riviere Gamby, que l'on peut avoir pour rien, pour s'en servir á saller le poisson, ce qui donneroit grandissime profit, et c'est un advantage á quoy on n'a pas pris guarde; il y a aussi grande Abondance de Cire, et comme celuy qui donna ceste relation au Sr Forbicher (Theologien Escosois) avoit visité les endroits plus secrets dans ceste partie du Monde qu'il asseuroit que les endroits nottez cidessus en chifre n'estoyent pas connus aux Mores, et par consequent pas auz autres Nations. Forbicher donna l'original de ceste relation au Roy Jaques 6me, qui la recommenda au Duc de Buckingham, qui faisoit la charge de grand Admiral d'Angleterre, des mains duquel ie l'ay receue.
  B. Gerbier Douuilly

[2] No known Portuguese source refers to Gambia as 'Camboya', i.e. Portuguese *camboa* 'seashore fish weir'.

[3] If this really relates to River Gambia and the island is therefore Banjul, these river names are untraced in any other source.

begin on the [missing word]. The gold remains after the rains.[1] A hill is found there on which the sun shines its rays towards the south, in the hollow of which the rains have washed a very large heap of gold nuggets that only need to be washed in water. The man who supplied this account said that he had noticed that the inhabitants had to date paid no attention to this large heap of gold, but only busied themselves with collecting the gold found in the sand along the river, which they sold for cloths of little value, scarlet cummerbunds [?], knives, axes, multi-coloured glass corals, and muskets. These Moors are very clever and skilled in gold working, making rings, chains and girdles of gold thread which the goldsmiths of Europe could not better. They are great fraudsters, mixing gold with silver when they care to cheat. It is best not to trade with them for gold other than gold dust.[2] The land starts off flat. These Moors trade in ivory, rhinoceros horn, and *gomme elemie* [? gum arabic?]. The land abounds in meat, fish, fowl, and palm wine; and in slaves which they have to transport to America, where the gold mines are, to make them work there. Note that that there is much salt in River Gambia which can be obtained free, to be used in salting fish, which brings great gain, and this is a benefit to which no attention has been paid. There is also a great amount of wax.[3] And since the man who supplied this account to Sieur Forbicher, a Scottish theologian, had visited the most secret places in this part of the world, he was sure that the places noted in the figure above were unknown to the Moors and hence to other Nations.[4] Forbicher gave the original of the account to King James the Sixth, who commended it to the Duke of Buckingham, who holds the office of High Admiral of England, from whose hands I received it.

B. Gerbier Douuily[5]

[1] Alluvial gold is not found in River Gambia, not near the mouth or even up-stream.
[2] The references to gold-working and gold-trading are probably from an earlier source on Gold Coast, possibly the 1605 French translation of [Pieter de Marees], *Beschryvinge ende Historischen verhael vant Gout Koninckrijck van Gunea ...*, Amsterdam, 1602, chs 43–44.
[3] Salt and wax relate to River Gambia, but the salt was sold there, not obtained for nothing.
[4] Obscure. If he means the gold places, they are not marked on the map; and the places marked were of course well-known to both the local 'Mores' and the Portuguese. The unknown man who supplied the account was perhaps not the same person who drew the map (or supplied the original details for the map), since the account is largely fictional whereas the map has original accurate details.
[5] James I was king 1603–March 1625. George Villiers, Duke of Buckingham (1592–1628), was Lord High Admiral 1619–28. Balthazar Gerbier (1591?–1667), born Dovilly, near Middelburg, later Sir Balthazar and terming himself Baron Dovilly, came to England and was in Villiers's service from 1616 as an artist and art-adviser. 'Forbicher' was probably John Forbes (Forbes being at the time pronounced 'Forbis', and 'Forbicher' being a confusion with 'Frobisher') 1568?–1634, a Scottish cleric and author of theological treatises denying royal supremacy in the church, who after 1606 first lived in France, then served as a pastor in Holland, notably 1611–21 at Middelburg. However, while contact between a Scottish cleric and a Scottish king is understandable, Forbes's views raise some doubts about the identification. (We are indebted for the 'Forbicher' information to Roy Bridges and Grant Simpson of the University of Aberdeen.) Thus, the account reached the king in or before 1625, and Villiers – assuming that he received it when Lord High Admiral and not when earlier in royal service – between 1619 and 1628, therefore Gerbier could have copied it any year between 1619 and 1628. It is hence problematical whether it was copied before or after the 1621–22 English voyage. Moreover, none of the above throws light on the author of the account and, if a separate person, the original composer of the map; or, unfortunately, on the circumstances of the collection of the map information, other than that the date was earlier than 1626. But there seems a distinct possibility that the original information came, not from English sources, but from French or Dutch, and that it reached England because of the impending or current English voyages. A tiny clue hinting at Continental provenance may be afforded by the re-appearance of the distinctive toponymy of the Gerbier map in the 'Pas

APPENDIX C

caarte van Rio Gambia' in Jan van Keulen, *De nieuwe groote lightende zee-fakkel*, pt. 5, Amsterdam, 1683, although apparently not in any intermediate or later map. The view of Gray, *History of the Gambia*, p. 20, that the account was based on Jobson's petition as 'perused by a Scotch chaplain named Forbicher' is without foundation. Apart from mention of gold from River Gambia, the two documents have nothing in common.

APPENDIX D

Account of the Upper River by Governor MacDonnell, 1848[1]

76. The Upper River, as the part of the Gambia above MacCarthy's Island is generally called, is perhaps in one respect most interesting, because the least known. The Gambia is navigable at any season of the year to vessels drawing nearly 9 feet of water to more than 360 miles from its source,[2] and may be regarded as navigable for a smaller class of vessels at any season of the year as far as Barraconda, which is supposed to be 450 miles from Bathurst. A rock at that place nearly crosses the river, and impedes further navigation in the dry season; though during the rains the waters rise nearly 30 feet over the rock. Very little is even at this day known of the river above Barraconda, and till the last few years not much was known of the river even so far.

In 1844, I proceeded from Fattatenda by myself in an open boat to Barraconda, and there discovered a passage through the rock at the right bank, which tallied exactly with a description given in 1724 of the rock by an old traveller. This passage, however, was nearly useless, as above the rock the shallows are so numerous as to render the navigation a matter of impossibility in several places during the dry season for anything larger than a row-boat.

77. I had been requested by some of the leading merchants last year to allow the colonial steamer, "Dover", to proceed as high as possible up the river in the rainy season, with a view to ascertaining the navigable character of the river and the possibility of approaching Bondou by the waters of the Gambia. Having occasion to visit the Upper Gambia in the "Dover" last January, I availed myself of the opportunity to ascertain the propriety of complying with the wishes of the merchants, being myself most anxious to give effect to them. Having visited all our various trading stations in the Upper River, I brought the "Dover" about 5 miles beyond Yabbatenda and 65 beyond Fattatenda, and there I was obliged to leave her, owing to the shallowness of the river. I then proceeded with two boats of the steamer, and accompanied by Mr. Brown, Lieutenant Child, 2nd West India Regiment, and the present Acting Secretary, as far as 105 miles above the rock of Barraconda. I and my party were obliged to sleep five nights in the woods whilst engaged in this undertaking, and none of us suffered any inconvenience therefrom.

78. I cannot but acknowledge that the results of the above expedition on the whole disappointed me. It is true, that the Gambia contains generally an astonishing body of water considering the great distance from the sea, at which we had arrived, no less than 505 miles; but on the other hand there occur at several intervals insuperable impediments to navigation during the dry season, whilst the banks

---

[1] From the report accompanying the Blue Book for the year 1848 (Despatch No. 41 of 16 June 1849).
[2] Probably 'source' is a slip for 'mouth', since the source was not then known.

of the river, though here and there diversified with table hills, and in general about 30 feet high, bear evident marks of being overflowed in the rainy season to a considerable distance from each bank. The consequence is, that there are no towns or inhabitants on the banks. We met one or two parties of the natives engaged in elephant hunting, but their own descriptions of the inhabitants and towns further in the interior led me to the conclusion that they were of the poorest class, and that there was nothing to induce a merchant to undertake the risk and expense of opening a communication with them.

79. I was also of opinion that it would be extremely hazardous to send the "Dover" beyond Barraconda in the rainy season; she could, as far as depth of water is concerned, easily pass any obstacle presented in the dry season, but in the rains what are now islands and cliffs, and even the present banks of the river, would become dangerous shallows and rocks, with a current running over them at the rate probably of seven or eight knots per hour. I therefore consider that any further discoveries ought to be prosecuted by much the same sort of expedition as that conducted by myself, and not with a vessel like the "Dover".'

80. I cannot also but consider the stories told by ancient navigators of the quantity of gold found along the banks of the Upper Gambia as fabulous, or nearly so. There is gold to be found, but it appears to come from the direction of Bondou, or from countries much further in the interior than I penetrated, and which have a shorter road to a more ready market on the Senegal, and in the Nunez, or even at Sierra Leone. One traveller in modern times proceeded further up the river than myself, I mean Mr. John Grant, a merchant of Bathurst, who went as far as Tenda, the name given to the district the confines of which I reached; but I find nothing in his journal militating against the opinions which I have formed. It is doubtful whether he proceeded so far by water as myself, but he left his boat and made his way by a short cut overland to a town near a small river running from the northward into the Gambia.[1]

This river, called by the natives the Nylarico, has been the source of much speculation: but even supposing it navigable, which is highly improbable, it would be difficult to get to it. It is possible, and only just possible, that there may be a communication between the upper streams of the Gambia and those of the Senegal. And though such a question is one highly interesting, and which might I think, be settled without much difficulty, it is still a disputed matter, and, I believe, the most authentic information on the subject at present in addition to that now transmitted to your Lordship, is that contained in the 26th Chapter of "Mungo Park's Travels". At about 80 miles from Barraconda I discovered a small river flowing apparently from the south into the Gambia, and I took the liberty of calling it by your Lordship's name, "Grey River". From all that I can learn from the natives its course is of considerable length, but the countries through which it flows are unproductive and thinly peopled. I enclose a chart made from a sketch of the Gambia, by myself, above Barraconda. As I had nothing but a pocket-compass to assist me in compiling such a chart, it can have but small pretensions to accuracy. Nevertheless as it is the

---

[1] Presumably Jalakoto near River Niérico (eds).

only one in existence of that portion of an important river, and might be a help to future travellers, I judge it best to forward it, The tortuous windings of the river which it depicts, prove the Upper Gambia to be a very circuitous road into the interior.

[81–4 dealing with a visit to Bondou in 1844 omitted.]

85. It appears to me, however, that the main difficulty to the extension of trade in the Upper River is the fact that it does not pay; and if that be the case no merchant is likely to do much towards its extension. The expenses attending it are very great, and must continue to till it be possible to employ a thoroughly honest class of men as traders, or discover some means of recovering debts due by the natives, and till the expenses of the river craft now employed in it can be diminished. With reference to this latter point, I have hopes that the colonial steamer may prove of great use hereafter; but at present, there is much discouragement prevailing on the subject, and the feeling of the merchants, already in part carried out, is to endeavour to concentrate at MacCarthy's Island the trade which for some years back they endeavoured to catch, as it were by planting numerous agents here and there in the Upper River.

86. The immediate cause of this feeling on the part of the merchants is, that the Joulahs or native factors have latterly raised the per centage charged by them on the value of such goods as they convey to the British factories. The Joulahs are an extraordinary people, belonging to the Mandingo race. I had much opportunity of studying their habits and mode of trade, having visited numbers of their towns, and I lived with them, and under their protection, for a considerable time in the interior. They appear to me to fill the same place among the natives here, as far as trade is concerned, as that which I have seen occupied by the Jews in Poland, only that they are of a higher caste in the estimation of those amongst whom they dwell. In other words, all the commerce of the country is under their protection and passes [through] their hands. The various coffles or caravans from remote districts in the interior are conducted by one or more Joulahs attached to each coffle, and are passed from one Joulah town to another. They feed the travellers and protect them when possible against aggression, whilst they exercise generally a paramount influence in the councils of the native chiefs. For all these services on conducting a caravan to its destination they expect a recompense; and for years they have received from the trader to whom they conduct the travelling merchant and his goods, a premium of 10 per cent. on the value of the entire amount bartered.

87. Latterly, however, or rather, to my own knowledge, for a period much longer than the merchants had supposed, the Joulahs rose in their demands, being encouraged thereto by the rivalry amongst the agents of the merchants, who might be considered as bidding against one another for the traffic furnished by the Joulahs; and the presents now demanded by the latter may altogether be looked on as amounting to little short of 25 per cent. on the amount paid by the merchant to the owner of the produce conducted by them to the British traders. The consequence of these demands has been an attempt to resist them, &c. and effort by the merchants to draw their trade down to MacCarthy's Island, where it would not be attended by those expenses. That attempt is met by an effort, or at least by a threat

of the Joulahs, to divert the trade altogether from the banks of the Gambia, and in that situation matters rest at present; but by firmness and union amongst the merchants, I am convinced the Joulahs could be compelled to yield.

88. I regret, under the above circumstances, that I have no means of ascertaining the real value of the produce obtained in the Upper River; but from the conduct of the merchants, who on such a point may safely be supposed the best judges of their own interests; the value of that produce does not equal the expense of obtaining it, an expense made up rather of losses by bad debts, dishonest agents, and large presents to the Joulahs, than by the actual cost of the produce.

[added in 1850][1]

32. We did not, however, find near the banks of the river any, or at least few signs of cultivation or inhabitants. Nevetheless, apart from the possibility of extending our commerce, the mere geographical question of the direction and extent of the course of the Gambia is one replete with interest, and which I hope may ere long be set at rest.

33. It does not appear that much expense or danger would attend such an expedition if undertaken at the proper season, viz., the end of December or beginning of January. I and my party bivouacked 15 nights in the woods after leaving the "Dover", and returned in perfect health. The abundance of game to be found in the country would ensure provisions, the carriage of which is in all such undertakings a great difficulty. In proof of the abundance of game, I may mention that I and my party shot several elephants and numbers of deer, river-horses, and guinea fowl, though we did not seek particularly for game, or leave the banks of the river for that purpose.

34. Although I cannot but consider it as very extraordinary that this colony should have existed so long, and so little as yet be known of the course of the Gambia. When I arrived here first the most vague and strange reports on the subject obtained credence; but I think that having made three expeditions beyond Baraconda, and returned without incurring any considerable danger or inconvenience, I have been the means of dispelling such fanciful theories, and awakening in their place a more rational curiosity ...

---

[1] In the Annual Report on Gambia for 1850, also by Governor MacDonnell.

PART II

OTHER EARLY SOURCES ON RIVER GAMBIA

Cadamosto (1455 and 1456 voyages)

Usodimare (1455 voyage)

Diogo Gomes (1455/1456 and ?1458 voyages)

Duarte Pacheco Pereira (*c.*1508)

Valentim Fernandes (*c.*1508)

João de Barros (1552)

Francisco de Andrade (1582)

André Álvares de Almada (*c.*1594)

André Donelha (1625)

Francisco de Lemos Coelho (1669/1684)

OTHER EARLY SOURCES

The Portuguese reached River Senegal and Cape Verde in 1444. In the next few years they explored the 120 miles of coast between Cape Verde and River Gambia and the mouths of the intermediate rivers, and they attempted to make contact with the local Africans, in order to trade. Very little is known about these voyages, but on several of them episodes occurred in which Africans and Portuguese fought and some of the latter were killed, including Nuno Tristão and (perhaps) his whole crew in (probably) 1446.[1] That even before 1450 River Gambia was reached and a preliminary exploration of its estuary carried out is possible, perhaps likely. Indeed, it is conceivable that Portuguese vessels penetrated up the river some distance. However, the earliest accounts of visits to the river relate to the mid 1450s, these being the accounts of **Cadamosto** (in Italian) and **Diogo Gomes** (in Latin), to which may be added the brief and obscure statements of Cadamosto's companion, **Usodimare**, in a letter (in Latin). Which party, that of Cadamosto or that of Gomes, reached the river first is uncertain. While neither account specifically claims initial exploration of the estuary and lower river, each can be read as inferring this, and neither appears to allude to any predecessors. However, it appears virtually certain that Diogo Gomes was the earliest European visitor – and in that sense explorer – up-river to beyond Kasang and as far as 'Cantor', this being the stretch of the whole river described in the present work as the middle river. What is absolutely certain is that the two accounts jointly supply, in fair detail, the earliest references to significant and peaceful Afro-European contacts on River Gambia, extending the peaceful trading that was under way on the Senegal coast to the west, and replacing the earlier incidents of conflict.

In the 1500s, two accounts of the West African coast were written, in Portuguese, by **Duarte Pacheco Pereira** and **Valentim Fernandes**, and each included substantial passages on River Gambia. Both sets of passages drew on Portuguese experience of the river in the later fifteenth century; Fernandes, dealing solely with the western coast of Guinea, also drew slightly on Cadamosto's account, which reached print in 1507; but otherwise both presented original material. Like the Gomes account,

---

[1] In a lengthy article, first written in 1946 and enlarged and much revised in 1971, Avelino Teixeira da Mota examined the references to the Nuno Tristão disaster. In the course of concluding that it occurred on one of the branches of River Saalum, west of River Gambia, he discussed, in considerable detail, topographical and chronological aspects of the visits to River Gambia of Cadamosto and Diogo Gomes (A. Teixeira da Mota, *Mar, além Mar: estudos e ensaios de história e geografia*, Lisbon, 1972, pp. 99–249; see also 253–73). In 1446 a group of caravels apparently led by Estevão Afonso reached a 'quite large river' with a sandbank on which one caravel stuck and had to be abandoned, and then 'going some way' came to a land with great fields and many cotton trees, and many estates growing rice, also other trees of various kinds', and 'everywhere there seemed to be marshlands'; finally, 'Guineans' with bows and poisoned arrows appeared. The description, to be found in Zurara's cronicle composed *c.* 1448, probably relates to River Saalum, but if this is River Gambia, as Teixeira da Mota argued, it is the earliest description (ibid., pp. 148–53, 221–3).

these two accounts remained in manuscript for several centuries and hence, in general, did not influence later writers. Also concealed in the archives for centuries was a later original account of the Cape Verde Islands and the mainland trading places prepared, in 1582, by **Francisco de Andrade**, which included a short passsage on River Gambia. An exception to this concealed evidence was the reference in print to River Gambia, in the mid-sixteenth century saga of Portuguese out-thrust by **João de Barros**, who seems to have had a glimpse of the Pacheco Pereira account, but elsewhere presents original material, probably drawn from oral information. However, Jobson does not seem to have encountered this printed source.

In the later sixteenth century, River Gambia was visited regularly by traders from the Cape Verde Islands, as witness two lengthy Portuguese descriptions, parts of accounts of the whole western Guinea coast, finally written in around 1594 and 1625, by **André Álvares de Almada** and **André Donelha**. This trading continued in the seventeenth century and produced another description, again part of a wider account of the coast, by another Cape Verde Islands trader, **Francisco de Lemos Coelho**. Written first in 1669 and enlarged in 1684, it described the river from personal experience in the 1650s and early 1660s, a generation after the first English experience and the account by Jobson.[1] Like the three earlier accounts by Diogo Gomes, Duarte Pacheco Pereira and Valentim Fernandes, none of these three later Portuguese accounts appeared in print for several centuries. Jobson was therefore acquainted only with Cadamosto – or, more likely, only with those pieces of information from Cadamosto's account which found their way into several general works during the sixteenth and early seventeenth centuries.[2]

The interest of these other early writings for the present edition is two-fold. In many respects, they confirm Jobson's account. But they also provide a historical perspective on its content, particularly in respect of the earlier European encounter

---

[1] Other pre-1650 references to River Gambia are disappointing. Guides to the Guinea coast, Portuguese and Dutch, merely refer to the mouth of the river; and those Dutch, Courlander and French accounts which appeared after bases were obtained in the lower river either limit themselves to descriptions of that district – as it happened, a district not discussed by Jobson in any detail – or discuss Senegambia broadly, apparently drawing mainly on Senegal experience. The references to River Gambia in one other Portuguese account however deserve mention. Father Baltasar Barreira, S.J., included these in a paper he wrote in 1606, discussing the coast and its peoples in terms of the likely success of a Jesuit mission (António Brásio, ed., *Monumenta missionaria africana*, 20 vols, Lisbon, 1952–88, 2nd ser., IV: item 45; English translation, P. E. H. Hair, ed. and trans., *Jesuit Documents on the Guinea of Cape Verde and the Cape Verde Islands 1585–1617*, Liverpool, 1989, item 13). But Barreira's single paragraph on River Gambia refers mainly to the Mandinka and Islam, and since it is based on information deriving from the community of Cape Verde Islands traders, it echoes the same information in Almada and Donelha. Finally, a much-quoted early (brief) reference to River Gambia is dubious. A Spanish chronicle written in the 1480s described a 1476 episode when Castilian vessels by trickery seized a 'king of Guinea' and his entourage, and took them to Spain as slaves (J. W. Blake, *European Beginnings in West Africa 1545–1578*, London, 1937, 44; translated text in J. W. Blake, ed., *Europeans in West Africa, 1450–1560*, London, 1942, pp. 212–13, 216–17). The heading to the passage refers to the 'king of Gambia'. The text does not confirm this and aspects of the story suggest a wrong attribution, perhaps editorial – e.g., previous regular trade with the Portuguese, and the king riding on a horse, points which suggest a Wolof ruler of the Cape Verde district of Senegal, rather than a 'king of Gambia'. (Nevertheless, the reason for the insertion of 'Gambia' eludes and puzzles us.)

[2] The extent to which Jobson was acquainted with Cadamosto's account is unclear. The two accounts contain a number of similar references to the same items, but these tend to be items likely to be observed and noted by every European visitor in earlier centuries, e.g. the exotic animals and plants.

OTHER EARLY SOURCES

with the river and its peoples, which was effectively an Afro-Portuguese encounter. They are annotated accordingly, but are not given full editorial treatment.[1] Cross-referencing to Jobson's material is limited to a few major passages; for other cross-references, see the Index.

---

[1] In particular, the three recently-published accounts are lightly annotated. For Donelha, see the printed edition. The section on River Gambia in Almada, and the whole of Lemos Coelho, as yet lack published annotation. No attempt is here made to annotate the section of the latter, the whole work being still under study.

TEXTS

CADAMOSTO (1455 and 1456 voyages)

Alvise Cadamosto, a young Venetian trader working for the Portuguese during two voyages to Guinea, in 1455 and 1456, which included visits to River Gambia, wrote (or concluded) his account of the voyages after returning to Venice in 1463, probably immediately.[1]

1455

[1] … we set sail, still holding our southerly course within sight of the shore, which appeared to us continually more beautiful, more thickly covered with green trees, and always low, until at length we reached the mouth of the river of Gambra.[2] Seeing that it was very wide, not less than three or four miles[3] in the narrowest part, so that our [three] ships could enter in safety, we decided to lie there with the

---

[1] The translation is based on that of G. R. Crone, ed., *The Voyages of Cadamosto and Other Documents of Western Africa in the Second Half of the Fifteenth Century*, Hakluyt Society, London, 1937, pp. 56–61, 66–75, revised, partly in the light of the variorum text in Tulia Gasparrini Leporace, ed., *Le navigazioni atlantiche del veneziano Alvise da Mosto*, Venice, 1966, pp. 78–86, 96–109, a work also containing useful annotation. Occasionally Crone's mock-medieval English has been modernized. The Portuguese translation and the annotation in Damião Peres, ed., *Viagens de Luís de Cadamosto e de Pedro de Sintra*, Academia Portuguesa da História, Lisbon, 1948, are worth notice. A French translation (Frédérique Verrier, ed., *Voyages en Afrique Noire d'Alvise Ca' da Mosto (1455 & 1456)*, Paris, 1994, is disappointing in its scholarly apparatus. For an authoritative examination of the Cadamosto voyages in the context of the Portuguese exploration of Guinea, see Teixeira da Mota, *Mar*, 99–249, 253–73. Four versions of the text exist, two in pre-1500 manuscripts (but neither in Cadamosto's hand) and two in print, the first print of 1507 and the version in Ramusio ([G. B. Ramusio], *Delle navigationi et viaggi*, 3 vols, Venice, 1550–59, I, 1550, pp. 105–21), the last undoubtedly containing editorial additions. A critical comparison of the texts has yet to be made, as have a completely reliable English translation and an edition with well-informed Africanist annotation. There being no standard chapter division or pagination of the text, the extract here is given separate paragraph numbering, in square brackets [ ].

[2] Cadamosto's reference to River Gambia is the second earliest recorded, and the earliest to appear in print. One of the manuscript texts refers to 'Gambra' throughout, the other to 'Cambra' (Leporace, *Navigazioni*, p. 79, etc); however, the Benincasa maps, almost certainly based on Cadamosto's information, use the form 'Gambia', as do later maps and the early Portuguese accounts (Teixeira da Mota, *Mar*, table between pp. 122–123, 237, with variant 'Gambya'). But 'Gambra', derived (via Eden) from the printed Cadamosto, occurs in Hakluyt, *Principall Navigations*, pp. 84, 94; and in the 1588 Guinea patent in Hakluyt, *Principal Navigations*, II/2, p. 123 (although, curiously, in the earlier edition it appears as 'Gambia': *Principall Navigations*, pp. 240, 242), as well as at II/2, pp. 188, 192. Jobson's use of 'Gambra' was either derived from Hakluyt, or as he may be claiming at one point (*GT* 10), from maps which also repeated Cadamosto's variant. Later occasional use of 'Gambra' most probably derived from Jobson.

[3] Cadamosto uses the Italian mile of 1480m, while Portuguese sources use a league of 5920m (Leporace, *Navigazioni*, p. 282). In English (land) miles, these represent 0.92 and 3.68 miles. The distances supplied for the entrance to River Gambia are reasonably correct.

intention of finding out on the following day whether this was the country of Gambra, which we were so keen to find.[1]

[2] On entering the great river, which at its mouth is no less than six to eight miles wide, we concluded that the country must be Gambra, so keenly sought by us, and that along the river we would most likely find some suitable solid ground where we might easily carry out profitable trade in gold, spices (*spetie*) or other valuable commodities.[2] The next day, the wind being very favourable, we sent the small caravel ahead, well manned and carrying bombards,[3] together with one of our boats, with orders that, as the caravel was small and drew little water, they were to advance as far as possible, and if they encountered any sandbanks in the estuary, take soundings. If they found sufficient depth of water for our [larger] vessels, they were to withdraw and, after signalling to us, anchor.

[3] These instructions were followed by the caravel. Having discovered about four (spans) of water at the mouth, it [withdrew and] anchored according to orders. When the caravel came out, it was decided to send our boat, armed, in company with that of the caravel, farther up the river, since both were small. They were instructed that, should the blacks (*negri*) of the land come in their canoes (*almadie*)[4] to attack them, they were to return at once to the ship without attempting to fight. This was because we had come there to trade in the country peacefully, and with the approval of the blacks, which would be better accomplished by tact than by force. The boats having gone forward, they took soundings in many places and finding nowhere less than 16 (spans) of water, continued for about two miles. The banks of

---

[1] Like several later references, these remarks are ambiguous – the name of the river is known but Cadamosto speaks of finding it, perhaps as if it had not been visited before. Since it is likely that Cadamosto had been everywhere preceded by Portuguese vessels, he is the target of modern Portuguese historians for claiming to be an explorer when he was not. The claim, however, is to be largely blamed on the patriotic exaggeration of certain modern Italian historians. Cadamosto makes his text interesting by awarding features new names and noting the astonished reaction of Africans to his arrival, which can be read as indicating that no white had preceded him in a particular locality, but he never outrightly claims first arrival. That his account was wholly written to deceive, by deliberately presenting himself as everywhere an explorer, is a modern Portuguese misunderstanding and misrepresentation of an account essentially that of a business trip. In any case, the ships of the two Italians were accompanied by, perhaps escorted by, perhaps even guided by, a Portuguese caravel, so that Valentim Fernandes later wrote that 'River Gambia was discovered by [blank], a servant of the Infante, accompanied by two merchants, each in his own ship' (Th. Monod, A. Teixeira da Mota, and R. Mauny, eds, *Description de la côte occidentale d'Afrique (Sénégal au Cape de Monte, Archipels) par Valentim Fernandes (1506–1510)*, Bissau, 1951, f.104). Moreover, Cadamosto earlier stated that, before he left Lisbon, Prince Henry had told him that 'not very far beyond this first kingdom of Senega was another called Gambra', where there was much gold, although the latter information was not said to come from Portuguese explorers but from captured Africans (Crone, *Cadamosto*, ch. xxxiv; Leporace, *Navigazioni*, p. 73). Furthermore, there may be an alternative explanation of the reference to finding the river. The land to the west of the river entrance is low-lying, with sandbanks and rocks off shore, and the entrance can be difficult to spot from the ocean (particularly in the season when dust storms reduce visibility). However, it cannot be ruled out that the Portuguese had already made use of African pilots familiar with the coast and entrance and that the ships with Cadamosto were able to draw on the knowledge gained.

[2] 'solid ground': the lower Gambia has mangrove swamps on both sides, with the exits of tributary streams often difficult to detect, and only occasional points where solid ground comes down to the water's edge.

[3] Crude cannon, firing like mortars.

[4] The term 'canoe' only reached Europe after the discovery of the Americas, hence Cadamosto uses the term employed by the Portuguese (from Arabic *al-ma'-dīya*) to denote light African craft or canoes.

the river proved to be very beautiful, being bordered with high green trees. [But] since the river made many turns higher up it appeared unnecessary to them to proceed farther.[1]

[4] On the return, there issued from the mouth of a stream which flows into this great river, three canoes (we call them *zopeli*)[2] which, from what I observed later, are all made of a single portion of a large tree hollowed out, fashioned like the little boats which are towed behind our ships. When our boats saw these canoes, being doubtful whether they might not have come to do them injury, and having been warned by other blacks that in this country of Gambra all the bowmen used poisoned arrows,[3] they took to their oars in obedience to their instructions, although they were sufficiently numerous to defend themselves, and returned with all possible speed to the ship. They did not return so rapidly however but that the canoes were close behind, within a bowshot of them, when they reached the ship, for they are very swift. When our men had boarded their ship, they began to call out to the canoes to draw near. These slowed down, and approached no nearer. There were about 25–30 blacks in each;[4] these remained for a while gazing upon a thing which neither they nor their fathers had ever seen before, that is, ships and white men in that river,[5] without showing any wish to parley, despite all that was done and was said to them, and then went about their own affairs. And so that day passed without further incident.

[5] The following morning, at about the third hour, we on the other two ships made sail with a favourable wind and tide to seek our consort and in God's name to enter the river, hoping that in the country farther upstream we might find more civilized (*humane*) people than those we had seen and met on the water. Having joined our consort, she made sail in company and we began to enter the river, the small caravel leading the way over the shallows, we following one behind the other.

---

[1] The 'many bends' begin about 80 miles up-river but the first curve, after a more or less straight course, occurs about 39 miles up-river, before Tankular.

[2] A Venetian term for a light vessel.

[3] The Portuguese had encountered poisoned arrows among the Serer, the people to the west of River Gambia, and immediately before reaching River Gambia, probably in River Saalum, one of Cadamosto's slave interpreters was killed when sent ashore (Crone, *Cadamosto*, ch. xxxiv; Leporace, *Navigazioni*, pp. 77–8). For poisoned weapons at Kasang, see Almada, 5/6, below.

[4] If the figures are correct, these can hardly have been ordinary fishing canoes. Yet Cadamosto does not describe them as war canoes with armed men (although a later source referred to 'very large canoes in which they sometimes go to war': Almada, 6/10, below), and the war canoes subsequently met are smaller. Perhaps they were goods canoes, now packed with men anxious to see the strange sight of the ships. However, they probably came from a creek on the north bank, in the territory of Niumi, and having reconnoitred the new arrivals, reported back, resulting in the subsequent attack from that quarter.

[5] This can be read as Cadamosto claiming that no white had preceded him in the river. But he may have meant that these particular Africans, at this particular locality, seemed never to have encountered a European vessel and whites, which could have been the case. However, as Cadamosto later states, information about the whites had already reached River Gambia from Senegal (and the ships and perhaps their white sailors had been sighted at River Saalum), hence the Africans cannot have been altogether overcome with surprise by the sight. In any case, Cadamosto may have been wrong about their never seeing whites before in the river, if, as is likely, Portuguese vessels had already visited there.

[6] Having passed the sandbanks and sailed about four miles upstream, we suddenly perceived several canoes coming up behind us (I do not know from whence they came) as fast as they were able. Seeing this, we turned upon them, and being anxious in respect their poisoned arrows, which we had been informed were mortally wounding, we protected our ships as best we could, and stood to arms at our stations, although we were poorly equipped. In a short time, when we were almost at a standstill, they reached us. I, being in the leading ship, split the canoes into two sections, and thrust into the midst of them: on counting the canoes, we found they numbered seventeen, and were of the size of large boats. Checking their course and lifting up their oars, their crews lay gazing as upon a marvel and novelty, never before, we judged, having seen ships with sails.[1] We estimated on examination that there were about 150 men at the most.[2] They appeared very well-built, exceedingly black, and all clothed in white cotton smocks (*camise*): some of them wore small white caps on their heads, very like the German style, except that on each side they had a white wing with a feather in the middle of the cap, as though to indicate they were warriors.[3] A black stood in the prow of each canoe, with a round shield, apparently of leather, on his arm. They made no movement towards us, and we made no hostile response; then they perceived the other two vessels coming up behind me and advancing towards them. When they reached them, without any other salute, they all threw down their oars and began to shoot off their arrows.

[7] Our ships, seeing this development, at once discharged four bombards: hearing these, amazed and confounded by the loud bangs, they threw down their bows, and gazing some here, some there, stood in astonishment at the sight of the

---

[1] Since elsewhere on the Guinea coast there is later evidence that fishermen off-shore sometimes used mats as sails to help them along, it is perhaps unlikely that the Gambian Africans were totally ignorant concerning wind power as a propellant on water, even although Cadamosto later asserts this ([18]). Apart from knowledge communicated from Senegal, Muslims who travelled long distances would have seen wind-powered vessels, if not on River Niger, certainly when on pilgrimage to Mecca. But no doubt the canoe-men were impressed by the size of the sails, the elaborate rigging, and the dependence on wind power for such – as seemed to them – enormous vessels. For canoes with sails, see Pieter de Marees, *Description and Historical Account of the Gold Kingdom of Guinea (1602)*, eds Albert van Dantzig and Adam Jones, Oxford, 1987, p. 116 (a tree-bark sail), 118 (sails of rushes or mats); K. Ratelband, ed., *Reizen naar West-Afrika van Pieter van den Broecke 1605–1614*, 's-Gravenhage, 1950, p. 14 (Senegal); Guy Thilmans, 'La Relation de François de Paris (1682–1683)' *Bulletin de l'Institut fondamental d'Afrique noire*, sér. B, 38, 1967, p. 25 (River Gambia, mats). In a general description of Senegambia, Cadamosto went into more details about African 'surprise' at European vessels, their admiration for the 'construction of our ship', notably the mast, sails, rigging and anchors, and for the strangers' capacity to navigate out of sight of land – but no mention is made of previous total ignorance of wind power. In another passage, however, a party of Africans did not know 'the use' of mast and cross-yards, which may or may not indicate total ignorance (Crone, *Cadamosto*, chs xxxiii, xlvi; Leporace, *Navigazioni*, pp. 71, 113).

[2] If these counts were correct, the canoes were crewed by fewer than ten men each, on average.

[3] The caps and white smocks suggest Mandinka and Islamic influence. For feathers in caps, see Almada, 5/16, below. The uniform dress suggests that these were official and military vessels of a local Mandinka ruler, presumably the ruler of Niumi. Since Niumi appears to have been a dependency of the polity on River Saalum to the west, possibly Niumi had been alerted to the forthcoming arrival of Cadamosto's vessels. Cotton was grown and cloth woven up-river and Cadamosto has later references to such cloth.

shots falling into the river about them.¹ After watching thus for a considerable while, and seeing no more they overcame their fear of the thunder claps after many shots had been fired, and taking up their bows, began afresh to shoot with much ardour, approaching to within a stone's throw of the ships. The sailors began to discharge their crossbows at them: the first to do so was a bastard son of the Genoese, who hit a black in the breast so that he immediately fell dead in the canoe. His companions, perceiving this, pulled out the arrow and examined it closely, in astonishment at such a weapon: but this did not restrain them from shooting vigorously at the ships, the crews of which replied in like fashion so that in a short space a great number of blacks were wounded. By the grace of God, however, not one of the Christians was hit.

[8] When they saw the wounded and dead, all the canoes with one accord made for the stern of the small caravel, where a stiff fight was waged, for her crew were few and ill-armed. Seeing this, I made sail for the small vessel and towed her between our two larger ships amidst a discharge of bombards and crossbows. At this, the blacks drew off: we, lashing our three ships together by chains, dropped anchor, which, as the water was calm, held all three.

[9] We then attempted to parley with the blacks. After much calling and shouting by our interpreters, one of the canoes returned within bowshot.² We asked of those in it the reason for their attack upon us, notwithstanding that we were men of peace, and traders in merchandise, saying that we had peaceful and friendly relations with the blacks of the Kingdom of Senega, and that we wished to be on similar terms with them, if they were willing. Further, that we had come from a distant land to offer fitting gifts to their king and lord on behalf of our king of Portugal, who desired peace and friendship with them. We besought them to tell us in what country we were, what lord ruled over it, and what river this was and its name; and we told them they might come in peace and confidence to take our wares, for we were content that they should have as much or as little as we pleased.

[10] They replied that they had had news of our coming and of our intercourse with the blacks of Senega,³ who, if they sought our friendship, could not but be

---

¹ This is the earliest recorded use of gunpowder on the Guinea coast, probably in West Africa, and perhaps throughout Black Africa. Its use in North Africa by the Portuguese, and conceivably by North Africans, appears not yet to have influenced warfare on the southern side of the Sahara. Bombards were crude mortars, impossible to aim at exact targets, and Cadamosto's account suggests that not even near-hits on the canoes were achieved, the effect of the shots being entirely in the surprise and fear they created. Hand-guns do not seem to have been used in Guinea until a later date.

² Cadamosto earlier explained that immediately after the arrival of the Portuguese in Guinea some slaves from Senegal were sent to Portugal specifically to learn Portuguese and to return as interpreters (Crone, *Cadamosto*, ch. xxxv). The Portuguese developed a regular and efficient system for dealing with the multifarious languages of Black Africa: see Sousa Viterbo, 'Noticia de alguns arabistas e interpretes de linguas africanas e orientaes', *O Instituto*, 52, 1905; 53, 1906, (many instalments); P. E. H. Hair, 'The Use of African Languages in Afro-European Contacts in Guinea: 1440–1560', 5, 1966, also in *Africa Encountered*, item VI, 5–26. At this point the language of communication was most probably Mandinka, but the interpreter was more likely a Wolof (given as 'Jolof' in all early sources) who also spoke some Mandinka.

³ The Portuguese had been in contact with polities in Senegal for about ten years. Apart from trading contacts between Senegal and the Gambia River, the Wolof ethnicity extended from Senegal to points on River Gambia, and parts of the north bank of the river were still under the control of Wolof polities. It is therefore not surprising that knowledge of the whiteman had reached River Gambia, even perhaps before any whites were seen.

wicked men, for they firmly believed that we Christians ate human flesh, and that we only bought blacks to eat them;[1] that for their part they did not want our friendship on any terms, but sought to slaughter us all, and to make a gift of our possessions to their lord, who they said was three days distant.[2] Theirs was the country of Gambra, and to the river, which was very large, they gave a name which I do not recall.[3]

[11] At this moment the wind freshened; realizing the ill will they bore us, we made sail towards them. They, anticipating this move, scattered in all directions for the land, and thus ended our engagement with them.

[12] Thereupon, we discussed among ourselves, particularly taking the advice of those who directed our ships,[4] whether we should proceed farther up the river, if possible for at least one hundred miles, in the hope of finding a better disposed people. But our sailors, who wished to return home and not to essay further dangers, began with one accord to shout out, declaring that they would not consent to such a course, and that what had been done was sufficient for that voyage.[5] When we saw that this was their general desire we agreed to give way in order to avoid dissention, for they were pig-headed and obstinate men. Accordingly on the following day, we departed thence, shaping our course for Cape Verde, to return, in God's name, to Spain [*sc.* Iberia].

[omission of a passage on the Southern Cross]

[13] In this place we found the night to be 13 hours, and the day 11 hours, that is, in the first days of July, or rather on the first of the month.

---

[1] While the Portuguese on arrival in Guinea no doubt often described themselves to Africans as 'Christians', partly because of their earlier hostile contact with Saharan Moslems, and while it is just possible that the Gambian Africans had heard the term from contacts with Senegal it is doubtful whether they had any idea what it meant. It is therefore more likely that 'we Christians' is Cadamosto's term and that the Africans did not actually refer to 'Christians'. Yet, on the other hand, the curious belief that whites ate black slaves (repeated in later centuries by Africans), while it may have derived from earlier African beliefs about the alleged cannibalism of rival ethnicities, when stated in Cadamosto's form, that 'Christians ate human flesh', does have a ring about it of Islamic derogation of the Eucharist. And there were certainly some Moslems in the River Gambia region.

[2] Cadamosto did not know from which bank of the river the canoes had come, but if the ruler was 'three days distant' this probably referred to the ruler of River Saalum, the overlord of Niumi.

[3] Duarte Pacheco Pereira later asserted that the river was called 'in the Mandinka language Guabuu' (text below). Despite a degree of resemblance between 'Gambra/Gambia' and 'Guabuu', the latter term to be pronounced 'Kabuu', it may be doubted whether the two terms are related. (For a possible explanation of the name 'Gambia', see *GT*, p. 84, n. 3.) Kaabu was an interior Mandinka polity stretching from the upper river for a considerable distance to the SE, and the name may only have become known to the Portuguese when they sailed upstream. The name of the river which Cadamosto could not recall, if it was not Kaabu, may have been *ji*, Mandinka for 'water' (as later stated by Jobson, *GT* 10), perhaps with an additional term meaning 'great water'.

[4] Presumably those directing the ships were Portuguese, and their advice may indicate that one or more of them had previous knowledge of River Gambia, at least of the estuary and lower river.

[5] Resistance to further exploration or continuance of trade by the seamen was a not uncommon feature of the later history of European marine enterprise in Guinea. In particular, it operated in the River Gambia voyages of Thompson and then Jobson. High mortality, perhaps higher among the lower-deck members than among the officers, no doubt contributed to the demand to leave Guinea. But traders like Cadamosto had an interest in prolonging the Guinea experience – hence his angry reference to 'pig-headed' sailors.

[14] This country is hot at all times of the year. It is true that there is some variation, and what we call a winter: thus, beginning in the aforesaid month [of July] until the end of October it rains continuously almost every day from noon, in the following way: clouds rise continually over the land from ENE or ESE, with very heavy thunder, lightning and thunderbolts. So an excessive quantity of rain falls, and at this season the blacks begin to sow in the same manner as those of the kingdom of Senega.[1] Their sustenance is entirely millet [*melio*] and vegetables, meat and milk.[2] I understand that in the interior of this country, on account of the great heat of the air, the rain which falls is warm.[3]

[15] In the morning, when day breaks, there is no dawn at the rising of the sun, as in our parts, where between dawn and sunrise there is a short interval before the shadows of night disperse: the sun appears suddenly, though it is not light for the space of half an hour, as the sun is dull and, as it were, smoky on first rising. The cause of this appearance of the sun early in the morning, contrary to what happens in our country, cannot, I think, arise from any other circumstance than the extreme lowness of the land, devoid of mountains, and all my companions were of this opinion.

[16] Little or nothing can be said of the features of this country of Gambra from what we were able to see and to learn on my first voyage - particularly from our own observations - for, as will have been understood, the people of the coast were so rude and savage that we were unable to have speech with them on land, or to treat about anything.

1456

[17] ... We sailed on until we arrived, another time, at the aforementioned river of Gambra, which we entered at once, and without coming into further contact with the blacks or their canoes, we sailed up-river by day, with the lead always ready. The canoes of the blacks, a few of which we came across, kept along the river banks, not daring to accost us.[4] About ten miles upstream we came upon an islet, similar to a *polesine*,[5] formed by the river. When we had cast anchor by it, on a Sunday, one of our sailors, who had been prostrated many days by fever, departed this life, and though his death depressed us all, nevertheless, wishing to do that which would be pleasing to God, we buried him on this island. His name was Andrea; for which reason we decided that the island should in future be called 'St Andrew's Islet', and thus it will always be known.[6]

---

[1] Cf. *GT* 126–7.

[2] Rice is not mentioned here, but is later.

[3] Much of this climatic information must have been obtained, not during his brief visits to River Gambia, but when Cadamosto was in Senegal. Nevertheless it indicates that by the time Cadamosto wrote, in the early 1460s, the Portuguese had correctly assessed the climatic seasonality of Senegambia and, *pari passu*, Guinea. But the references to Gambian agriculture and food, while correct, are slight and vague. For *melio*, see Leporace, *Navigazioni*, p. 280. For the seasons, cf. Fernandes, f. 116^v; Almada, 6/13; both below.

[4] These were no doubt simple fishing or ferrying canoes, and it is curious that Jobson never mentions the traffic in these, or the less frequent traffic in long-distance trading canoes.

[5] '*polesine*' = a low island in rivers of North Italy, Leporace, *Navigazioni*, p. 289.

[6] The name did, in fact, persist for about two centuries.

[18] Leaving the island and sailing always upstream, we were followed at a distance by some canoes of the blacks. We and our interpreters having displayed a friendly attitude to them, the interpreters called to the blacks, displaying silken stuffs and other articles, and gave them to understand that they might with safety draw near, that we would give them these garments, and that they need have no fear, for we were humane and well disposed men. Little by little the blacks drew nearer, gaining confidence in us, until at last they drew alongside my caravel, and one of them who could understand my interpreter boarded the ship. He marvelled greatly at her, and at the method of navigating by means of sails, for they knew no method except by rowing their canoes with oars, and considered no other way possible.[1] He was overcome with astonishment at the sight of us white men, and marvelled no less at our clothing, so different to his – especially since most of them went naked, or, if clothed, in a white cotton smock. We made much of the black, giving him many small items of little value, with which he was exceedingly pleased.

[19] By asking him many questions, at last I ascertained that this was the land of Gambra, and that the principal lord was Farosangoli,[2] who, he said, dwelt far from the river inland towards the midday sirocco [*sc.* SSE], as he pointed, at nine or ten days' journey. This Farosangoli was subject to the Emperor of Melli, the great Emperor of the Blacks, but nevertheless, there were many lesser lords who dwelt near the river, some on one bank, some on the other: he offered, if I were willing, to bring me to one of them, called Batimaussa,[3] and to treat with him to enter into friendship with us, since it seemed to him that we were well disposed persons. This

---

[1] See p. 247, n. 1 above.

[2] Farosangoli (cf. Gomes, ff. 276, 277ᵛ, below): the first part is a mis-hearing of *farang*, a Mandinka title for a lesser ruler (cf. Almada, 5/2; Donelha, f. 15; both below), with possibly some confusion with Mandinka *foro* 'individual of high rank'. For attempts to identify 'Sangoli', probably the name of an interior polity, see Teixeira da Mota, *Mar*, pp. 245–7; and Sekeny-Mody Cissoko, 'Le royauté (*mansaya*) chez les Mandingues occidentaux, d'après leurs traditions orales', *Bulletin de l'Institut fondamental de l'Afrique noire*, sér. B, 21, 1969, pp. 325–38, which relates it to a semi-legendary 'Sankola', an entity accepted as historical in Dibril Tamsir Niane, *Histoire des Mandingues de l'Ouest: le royaume du Gabou*, Paris, 1989. As Cadamosto states, the polity in question lay well to the south of River Gambia, and 'terra farosangali' was marked on a 1468 Benincasa map in the interior of what appears to be the Jeba estuary (Teixeira da Mota, *Mar*, p. 119).

[3] Ramusio printed the more correct term, Battimansa (Leporace, *Navigazioni*, p. 98), and perhaps Batimaussa involved a misreading of notes at one stage, although it appears in both manuscripts and the first print. 'Battimansa' (cf. Gomes, f. 278, below) was the *mansa*, Manding for 'king, ruler', of 'Bati'. Cadamosto does not state on which side of the river 'Bati' lay, but Diogo Gomes can be read as indicating that it lay on the opposite side to Niumi, that is, on the south side, though higher up. If so, 'Bati' may be related to Bateling, a town in Kiang district, and oral evidence collected in 1993 referred to an episode of 1900 in this locality involving a 'Bati Mansa' (Mododu Jammeh with Nancy Ann Sheehan, *My Own Beef*, Land Tenure Centre, University of Wisconsin-Madison, 1993, p. 17). But there is no reference to a polity termed 'Bati' in early texts or on maps, other than the Cadamosto and Gomes references. More or less opposite Kiang is the longer-evidenced polity of Baddibu (not mentioned by Jobson but appearing on the 1732 map in Moore, *Inland Parts*); and on an alternative reading of Gomes 'Bati' may have lain on the north side and hence might be thought to be an earlier representation of 'Baddibu', as suggested in Teixeira da Mota, *Mar*, p. 248. But the stated linguistic basis for this equation of the names is unacceptable (DPG). Finally, since at the distance stated up-river the south bank is more open than the mangrove-lined north bank, and since Cadamosto speaks as if Batimansa was contacted at a settlement on or near the river, it is arguable that the topography favours the south side.

offer was very acceptable to me: so, taking him on board and treating him hospitably, we sailed up the river until we reached the place of the said lord Batimaussa, which according to our estimate was about sixty miles and more from the river mouth.[1] Note that while ascending this stream we were going eastwards and that at the spot where we dropped anchor the stream was much narrower than at the mouth, being in our judgement no more than a mile in width. This river has many branches which join together.

[20] Arrived at this place, we decided to send one of our interpreters with the black into the presence of this lord Batimaussa, with a present for him, an *alzimba* of Moorish silk - as we should say a surcoat (*camisa*) - which was quite fine and made in the land of the Moors.[2] He was to say that we had come by command of our Lord King of Portugal, a Christian, to establish firm friendship with him, and to inform him that if had need of the products of our country, our king would send them to him each year, and many other messages.

[21] The interpreter went with the black, and, in brief, we treated so with the lord that when we parted from him we had not only secured his friendship, but had sold many articles, for which we received in exchange black slaves and a certain quantity of gold, but not of much account with respect to what we had anticipated, because the rumour of it had been much greater in the reports of blacks of Senega: indeed we found very little by our standards, but by those of these very poor people it was considerable.[3] Gold is much prized among them, in my opinion, more than by us, for they regard it as very precious: nevertheless they traded it cheaply, taking in exchange articles of little value in our eyes.[4]

[22] We remained here about fifteen days,[5] and in this time many blacks dwelling on one side or the other of this river came to our ships – some to gaze upon a sight so strange to them: others to sell some small items of theirs, or little rings of gold. The articles they brought were cotton cloth and thread, cotton cloths woven in their fashion, some white, others variegated, white and blue striped, or red, blue, and white, excellently made. They also brought many apes (*gatto maimoni*), and baboons of various species large and small, of which there are very large numbers in these parts.[6] These they bartered for objects of little worth, giving ten *marchetti* for the

---

[1] The estimate of 60 miles and more, say 55 English miles, would bring Cadamosto to a stretch near to Bateling (see previous note), where the river is about one English mile wide. It appears that the ruler had a residence on the river but the exact locality is not stated (cf. Gomes, f. 278, below).

[2] The Portuguese often expressed the continuity of their outthrust, and minimized the novelty of their arrival in West Africa, by presenting African rulers with commodities obtained in North Africa, commodities with which the rulers were already to some extent familiar, thanks to the Saharan trade route.

[3] Cadamosto, who had traded in the eastern Mediterranean, is the first European on record to refer to comparative poverty as a significant feature of Black Africa.

[4] Possibly the apparent contradiction about the value of gold is explained by the fact that the Europeans first offered to buy gold objects which had ritual or sentimental, rather than economic value. Later observers did not support Cadamosto's view that gold was regarded as a 'precious' commodity.

[5] Another text says two days and Ramusio about eleven. Given the amount of information obtained, two days is implausible, fifteen plausible.

[6] The speed with which the Africans brought 'apes' (or perhaps monkeys) and baboons for sale to the Europeans suggests that there was already local trade in these animals. Jobson has much to say about baboons.

value of one. They also brought for sale civet [*sc.* musk], and the skins of cats from which civet is obtained. They gave an ounce of civet in exchange for an article not worth forty or fifty *marchetti*, not that they sold by weight, but as I say by estimation. Others brought fruits of various kinds, among them many small wild dates, not very good ones. According to them they are good to eat and many of our sailors ate them, finding them of a different flavour to ours, but I had no desire to eat them, for fear of diarrhoea.[1] In this way we had each day fresh peoples of various tongues down at the ships. They are constantly journeying from place to place up and down the river in their canoes, carrying women and men, as with us our boats during the floods.

[23] But all their navigation is by means of oars. They row standing up, so many on each side. They always have one extra rower at the stern, who rows now on one side, now on the other, to keep the boat straight. They do not use rowlocks for their oars, but hold them steady with their hands. The oars are fashioned thus: they have a shaft, like a short lance, a yard and a half in length, and at the head of this shaft they nail, or rather bind after their fashion, a round disc. With this style of oar they row their boats along the sea coast from point to point.[2] There are many mouths of streams which they enter and leave in safety, but they do not commonly venture far outside their own country, for they are not safe between one district and the next from being taken by blacks and sold into slavery.[3]

[24] At the end of a few days we resolved to depart from there, and to proceed to the mouth of the river, because many of our men began to suffer from a high fever, sharp and continuous: so we left suddenly.

[25] Of the things to be recounted of this country, from our own observations, and from the information we gathered during the short time we were there, we shall deal first with their faith. It is, in general, idolatry in various forms: great credence is placed in spells, and other diabolical methods with which they are acquainted, yet all recognize one God, and some of them are of the sect of Muhammad. The latter are men who wander through the world and the lands of the Moors, not remaining tied to their homeland, for the yokels know nothing of such things.[4]

[26] In their way of life they conduct themselves in almost all respects similarly to the blacks of the kingdom of Senega. They eat the same foods, except that they have more varieties of rice than grow in the kingdom of the first blacks; also they eat

---

[1] Cadamosto's caution probably reflected his experience in the eastern Mediterranean. The tree *Phoenix reclinata* (Mandinka *koroso*) produces very small wild dates, as well as a form of palm wine when tapped near the base.

[2] The reference at this point to the 'sea coast' (*per la costa del mar*) is puzzling. Trading canoes did move by sea between River Saalum and River Gambia, and it is conceivable that Gambian war canoes did occasionally enter the sea to raid further south. But the context relates to canoes within River Gambia. It must be a more general reference, either to canoes seen along the coast of Senegal, or to 'sea' in the sense of any waterway.

[3] The enslavement on the river before the Europeans arrived was most probably in order to send slaves in two directions: first, from south of the lower river northwards to the Wolof polities (routes still in use for trade, and travelled by DPG); and secondly, perhaps mainly from the middle stretches, up-river and to some extent further into the Islamic interior.

[4] Cadamosto introduces Jobson's 'Marybucks' (*GT* 61 ff.). Cf. Fernandes, f. 109; Almada, 5/10; Lemos Coelho, 17; all below.

the flesh of dogs, which I have never heard is eaten elsewhere. Their garments are of cotton, whereas almost all the blacks of Senega go naked. They are clothed because of the great abundance of cotton. The women are also clothed in a similar style, except that they delight, when young, to work designs upon their flesh with the point of a needle, either on their breasts, arms, or necks. These appear like those designs of silk that are often made on handkerchiefs: they are made with fire, so that they never disappear.[1]

[27] This region is very hot, and the more one advances towards the south, the hotter these countries become. On this river particularly it is very much hotter than on the sea, because it is covered with numerous and very large trees which are everywhere throughout the country. Concerning the size of these, I may say that, at a spring near the river bank from which we drew water, there was a very great and broad tree; its height, however, was not in proportion to its size, for while we judged it to be about 20 paces high we found the girth by measurement to be about 17 paces round the foot. It was hollow in many places, and its branches were very large and threw a deep shade around.[2] Even larger trees are to be found, so that from such trees it may be concluded that the nature of the country is good, and fertile, being bathed with many waters.

[28] There are large numbers of elephants in this country.[3] I have seen three wild ones, for they do not know how to domesticate them as in other parts of the world. It was either on this voyage or another, after arriving in a caravel and entering this River Gambia,[4] when the ship was in the middle of the river we caught sight of these three elephants emerging from the forest and going towards the river bank. Some of us jumped into the boat to go to them for they were some distance off, but when the animals saw us approaching, they returned to the forest. Later I saw a small one, dead: for, to satisfy me, another black lord named Gnumimenssa,[5] dwelling near the mouth of this river of Gambra, set out to hunt it with many blacks, following for two days before they killed it. They go hunting on foot, carrying no other weapons for the attack save the *azagaie* [*sc*. spear] which I have described above, and bows, all the weapons they carry being poisoned. You should know that they seek elephants when they are in the woods where they normally live, for they prefer muddy places, because, like swine, they like to be in mud.[6] So where there are many trees the blacks place themselves behind them, and wound the elephant with arrows or poisoned spears. They advance scrambling and jumping from tree to tree, so that before the elephant, which is an unwieldy animal, can escape, it is wounded in many places

---

[1] Cf. Jobson, *GT* 55.

[2] The tree was a baobab (*Adamsonia digitata* L.): see Leporace, *Navigazioni*, p. 289. 3 miles west of Tendabaa is Tubab-kolong 'Whiteman's Well' (the Mandinka toponym evidenced on the 1661 Vermuyden map, and cf. Lemos Coelho, 18, below) – was this Cadamosto's 'spring'?

[3] Most later observers discuss elephants, elephant hunting, and ivory: cf. Fernandes, f. 111; Almada, 5/4; Lemos Coelho, 65; all below. As does Jobson (*GT* 139–42).

[4] This seems a curious wording, but presumably 'another' means the first voyage.

[5] The *mansa* or ruler of Niumi, the polity on the north bank at the entrance to the river, not mentioned specifically by Jobson.

[6] Woods are not necessarily muddy places, but presumably the thought behind the reference was to elephants wallowing in muddy pools, and to the woods of the district reaching the river in the form of mangrove swamps.

without being able to defend itself.[1] I may say here that in the open, with no trees near, no man would dare to face one, for no man can run so fast that the elephant, without breaking out of its ordinary pace, cannot overtake him; for considering his size his pace is very rapid: if it happens by a mischance that an elephant pursues a man in the open and overtakes him he attacks him with nothing but the great trunk of his muzzle, which is somewhat like that of a pig, except that the pig's snout is not mobile, as is the elephant's. It resembles a large tough lip, which, unlike the pig, he is able to twist, extend and shorten at will; winding this trunk around the man he hurls him so far into the air that often he is dead before he falls to the ground. This was told me by many blacks. But the elephant is not however so ferocious an animal that it will attack men without first being annoyed.

[29] I saw this small elephant lying dead on the ground. Its tusks were no more than three spans in length. Of these three spans, one was embedded in the jaw, so that the tusks were actually only two spans in length. For this reason they said it was a young animal, since some have tusks 10–12 spans long: but small though it was, we judged that its carcass equalled those of five or six of our bulls.

[30] I was given this elephant by the lord, that is, I was allowed to take whatever portions of it I wished, and the remainder was given to the hunters for food. Gathering from this that its flesh is eaten by the blacks, I had a portion cut off, which, roasted and broiled, I ate on board ship, to establish this thoroughly, and to be able to say that I had eaten of the flesh of an animal which had never been previously eaten by any of my countrymen, Actually the flesh is not very good, seeming to me tough and insipid.[2] I also brought away one of its feet and a portion of its trunk to the ship, and many of the hairs from its body too, a span and a half in length or more, and very thick. These, with a portion of the salted flesh, I presented later in Spain to my lord the Infante Dom Henrique of Portugal, who received them as a handsome gift, being the first that he had had from that country, and he much liked to receive exotic objects coming from far parts and from the lands discovered by his enterprise.

[31] I wish it to be understood that the foot of the elephant is round, almost like the foot of the horse, but not with a hoof like the horse, being entirely a black, thick callosity, around which are five claws, level with the ground, round, and little larger than a *grossone* [coin]. The foot of this little elephant was not so small but that it was a span and a half in width, across the sole in every direction, for, as I have said, it is quite round, like a platter. I was also given by the same chief another elephant's foot which I measured several times across the sole in the presence of many people, and found it to be three spans and an inch in breadth in either direction. I also presented this to the Lord Infante, with a tusk twelve spans long, which with the said foot he ordered to be given to the Lady Duchess of Burgundy as a worthy present.[3]

---

[1] Contrary to popular beliefs, the reduction in the elephant population of West Africa was effected, not by Europeans, but by a growing population of Africans, in part defending their new farms from a predatory animal, and using, not guns, but traditional weapons and skills.

[2] Diogo Gomes was also presented with elephant meat (below, f. 278ᵛ), as was Jobson (*PJ* 922; *GT* 149/141).

[3] Isabel, Duchess of Burgundy, was a sister of the Infante Henrique.

Do not believe that the elephant cannot bend its knees, as I have heard said at times; on the contrary it moves, bends, and rises like any other animal.[1]

[32] Also in this river of Gambra, and in many other rivers in the country, in addition to the cockatrice [*sc.* crocodile] and divers other animals, there is found an animal called a 'fish horse' (*pesse cavallo*).[2] This animal is in nature something like the 'old man of the sea' (*vechio marin*), which lives now on the land, now in the water, and maintains itself in both elements.[3] It is formed thus - its body, the size of a cow's, with short limbs, has cleft feet, and its head is shaped like a horse's, with two large tusks, as a wild boar has. These tusks are very large, for I have seen some two spans in length, and at times longer. This animal frequently comes out of the water, and walks about on the banks like a quadruped: the like has not been found in any other parts to which we Christians have sailed, save these countries of the blacks.[4]

[33] There are also to be found bats, or as we say *nottole*, of three spans and more, and many other birds of different kinds to ours, and especially innumerable parrots. This river is rich in fish, unlike ours in form and flavour, but none the less very good to eat.

[34] As I have said above, we left the country of the chief Batimansa, on account of illness among our men, and in a few days cleared the river.

ANTONIOTTO USODIMARE (1455 voyage)

Usodimare, a Genoese merchant mentioned by Cadamosto, accompanied the latter into River Gambia in 1455 and 1456. On 12 December 1455 Usodimare wrote from Portugal, in bad and often ambiguous Latin, a letter to his creditors, seeking finance for another voyage to Guinea.[5] Usodimare is an obscure character, details of his role in Guinea are lacking, and the brief references in his letter to his 1455 voyage and River Gambia include a measure of fanciful speculation and wishful thinking. Nevertheless, through the haze the Cadamosto voyage can be detected. Despite its extreme limitations, the letter represents the very earliest known reference to River Gambia, and to the name (in the form 'Gamba').

… my good fortune has been to make my way in a caravel to the parts of Guinea. I penetrated where no Christian had been,[6] more than 1300 miles, and having come

---

[1] On the elephant, cf. *GT* 129–32.
[2] The crocodiles and hippopotamuses of the river are invariably discussed at length by later observers (below, Pacheco Pereira; Fernandes, f. 112; Barros; Almada, 5/7, 6/7), including Jobson, *GT* 16–22.
[3] Probably the walrus.
[4] More knowledgably, Ramusio added 'except, perchance, in the Nile' (Leporace, *Navigazioni*, p. 108).
[5] The letter (or perhaps a poor contemporary copy) was discovered and first published in 1802. Since then it has been printed several times, either in its original Latin, or in Italian and Portuguese translation. The Latin text presented here in original English translation (apparently the first) can be found in *Monumenta Henricana XII*, Coimbra, 1971, pp. 191–3, which has bibliographical annotation; in Italian translation, see Rinaldo Caddeo, *Le navigazioni atlantiche di Alvise da Cà da Mosto*, Milan, 1929, pp. 153–5; in Portuguese translation, Luís de Albuquerque and Maria Emília Madeira Santos, eds, *Portugaliae monumenta africana*, I, Lisbon, 1993, pp. 77–8. For comment on the contents, see Leporace, *Navigazioni*, pp. xi, 73, 282, etc; Teixeira da Mota, *Mar*, pp. 173–4, 234.
[6] Or, perhaps, 'was'.

upon River Gamba, of very great extent [?at its mouth], I entered it, knowing that in this district gold and pepper are available.[1] The fishermen there attacked me with bows or rather poisoned arrows, reckoning that we were enemies.[2] Seeing that they did not wish us here, I felt obliged to withdraw. About 70 leagues from there a noble black lord gave me 40 head [of slaves] and some tusks, parrots and a little musk, in exchange for some cloths presented to him. Mindful of what I had intended, I conveyed with me to the lord king of Portugal an agent/counsellor (*secretario*) of his [*sc.* the black lord's], with some slaves. This agent undertook to treat for peace with that king of Gamba.[3] And the lord king noting this [? the hostility], wished to exclude me from the enterprise, but, on the prayers of the agent, was content that I should go along with him to those parts.

… Truly, from that land to the country of Prester John is less than 300 leagues, not, I mean, to Prester John himself, but to the boundary of his territory. And if it prove possible I may see the captain of the king of Melli,[4] who was six days' journey from us, with 100 men, and with him were five Christians belonging to Prester John, and I spoke with some of his army.[5] I met there one of our own nation, I think from the Vivaldi galleys lost 170 years ago, who spoke to me, and the agent confirmed that none of that group [*sc.* ? no one descended from that group] remained save him.[6] He [? the agent] also spoke to me of elephants, unicorns, civet cats, and other strange things, and of men with tails who ate children.[7] Although this may seem to you incredible, you should believe me that one day further on from there you lose the pole star.[8] The reason why I could not stay was because provisions were lacking,

---

[1] For Cadamosto's similar pre-knowledge about gold in River Gambia, see p. 245, n. 1 above.

[2] Note that Usodimare fails to mention Cadamosto. According to Cadamosto, Usodimare had his own ship which had met Cadamosto's ship while both were separately sailing to Guinea. The two Italian merchants then agreed, in alliance with a third ship, an official Portuguese vessel (and the Italians may have been travelling on licenced Portuguese vessels), to find and investigate River Gambia. The reference to the attack fits Cadamosto's account, but the attackers were not mere fishermen, although no doubt fishermen were seen.

[3] Assuming that the distance is a gross exaggeration (or a copying error), the black lord was probably in River Saalum, since it is likely that the ruler there would be able to influence directly or indirectly those in River Gambia who had attacked the Cadamosto/Usodimare vessels.

[4] This appears to be the earliest reference in European sources to 'Melli', i.e. Mali.

[5] At the time the Portuguese were anxious to reach 'Prester John', that is. Christian Ethiopia, which was believed to extend across Africa and hence closer to Guinea than in actuality. Hence Diogo Gomes, who was also in River Gambia in the same year, carried an 'Indian', probably an Ethiopian, who had been ordered by Prince Henry to explore overland eastwards (see Gomes, f. 278^v, below). In Cadamosto's account, information about interior 'Melli' is only recorded as being obtained on the second, 1456 voyage, when a subject of the 'Emperor of Melli', Farosangoli, was stated to live 8–9 days away from the point on the river where the information was obtained (see Cadamosto, [19], above). The reference to five Ethiopian Christians is impossible to explain and may indicate a garbled passage.

[6] It is usually assumed that, if indeed a Genoese was met, he was a survivor from a post-1444 trading ship which had been lost on the coast. A report reached Lisbon that four survivors of the Nuno Tristão disaster were captives in the interior (C. R. Beazley and E. Prestage, eds, *The Chronicle of the Discovery and Conquest of Guinea. Written by Gomes Eannes de Azurara*, 2 vols, Hakluyt Society, London, 1896, 1898, ch. 94, p. 287). The Vivaldi brothers set out in 1291, purportedly to sail around Africa, and patently their initiative was still remembered in Genoa nearly '170 years' later.

[7] Baboons or monkeys seizing babies? Stories to this effect are found in the folklore of many West African peoples.

[8] Arguably, the one day takes the traveller back from River Saalum to River Gambia, where Cadamosto noted the emergence of the Southern Cross (noted above, Cadamosto, between [12] and [13]).

white men not being able to eat their foods without falling sick and dying, these foods only being edible to blacks born there. However, the air is excellent, and the land more beautiful than [any] under heaven, and almost equinoctial. In July, the day is 12½ hours long, the night 11½. ...

DIOGO GOMES (1455/1456 and (?)1458 voyages)

Diogo Gomes, an officer and agent of the Infante Dom Henrique, was in Senegambia during part of the 1450s. He seems to have reached River Gambia shortly before or after Cadamosto, but was apparently the first Portuguese to ascend the river as far as Cantor, where he made inquiries about the interior gold trade. He was still alive in the early 1480s when he related an account of his voyages to a German visitor, just possibly from notes but probably in greater part from memory, so that elements have to be received with some caution. Nevertheless the account depends on no known anterior sources.[1]

/f.275ᵛ/ Some time after this the Lord Infante equipped at Lagos a caravel, named *Piconso*, and appointed Diogo Gomes captain. And he equipped two other caravels to go as well, of which he appointed Diogo Gomes captain-in-chief. ... [the ships reach Guinea and explore the coast] /f.276/ ... We returned to the ships, and on the next day made our way from Cape Verde, and we saw the broad mouth of a river, three leagues in width, which we entered, and from its size correctly concluded that it was River Gambia.[2] We entered it with the wind and tide in our favour, as far as a small island in the middle of the river, and there remained that night. In the morning, however, we went farther in, and saw many canoes (*almadias*) full of men, who fled at sight of us, for it seems they were the same who killed the Christians whom I mentioned earlier, and their captain.[3] The next day, however, we saw, beyond the beginning of the river,[4] some people on the right-hand bank, to whom we went, and made a peaceful contact with them. Their lord was called Frangazick, a nephew of Farisangul, the great prince of the blacks.[5] There I received from the blacks 180

---

[1] The narration of Diogo Gomes was written down in colloquial Latin and perhaps edited more than once, and the text was only printed in the mid nineteenth century. The translation is based on that in Crone, *Cadamosto*, pp. 91–101, but has been considerably revised in the light of T. Monod, R. Mauny, and G. Duval, trans and eds, *De la première découverte de la Guinée récit par Diogo Gomes (fin XVe siècle)*, Bissau, 1959, pp. 32–53, which includes Africanist annotation.

[2] The wording implies previous knowledge of at least the mouth of River Gambia.

[3] Gomes undoubtedly refers to the killing of Nuno Tristão and his men in 1446, but this was not in River Gambia. (It most likely occurred in River Saalum, see Teixeira da Mota, *Mar*, p. 120). It is possible that Gomes, in old age, confused the attack on Nuno Tristão with the attack on Cadamosto's ships, although no whites were then killed, always assuming that Gomes followed Cadamosto into River Gambia. Most probably the earlier attack involved the polity of Saalum, and the later attack its subordinate, Niumi.

[4] *ultra caput fluminis*, literally 'beyond the head of the river' – this perhaps means that after passing the wider estuary (the 'small island' being probably James Island) they now penetrated the somewhat narrower stretch, towards Tankular.

[5] Cf. Cadamosto, [19], and p. 251, n. 2 above.

arrateis weight of gold, in exchange for our merchandise, cloths, necklaces, &c.[1] They told us that the blacks on the left bank would not speak with us because they had killed the Christians. The lord of that country had a certain negro, named Bucker, who was acquainted with the whole country of the blacks, and finding him /f.276ᵛ/ perfectly truthful, I asked him to go with me to Cantor, and promised to give him a mantle and smocks, and everything necessary. I made also a similar promise to his lord, which I kept.[2]

We ascended the river, and I sent a captain with his caravel into a certain harbour, named Ulimaijs.[3] Another caravel remained in Animaijs,[4] and I went up the river as far as I could and reached Cantor, which is a large town near that river.[5] On account of the thick growth of the trees on both sides of the river, the sails could not be of use, and I sent out the black whom we had brought with us, to make it known to the people of the country that I had come there to trade merchandise.[6] In consequence,

---

[1] Gold was not generally offered to Europeans in the lower river – perhaps Gomes' memory was at fault. But a merchant may have brought some down-river to trade.

[2] The name 'Bucker' represents Bokari, a common Islamic name – and this local form of the name was also the name of the African trader whom Jobson met up-river, Buckor Sano. That a guide for Cantor high up the river could be obtained on the lower river indicates the range of traditional riverain trade and communication. Since there was no contact with 'the blacks on the left bank', presumably those of Niumi, although the text is not altogether clear about where Bucker was recruited, 'the lord of that country' must refer to the previous lord on the right bank, since a promise is made to him. This is important to establish because of a future episode (discussed in p. 261, n. 6).

[3] 'Ulimaijs', later 'Ulimansa' and 'Ollimansa', are assumed to be a corrupt forms of Wulimansa (i.e. *Wuli mansa* 'king of Wuli'), Wuli being the polity on the north side of the middle river (Jobson's 'Wolley': *GT* 48). This is the earliest record of the name Wuli. Two centuries later Fatatenda was described as the 'first port of Wuli' (Lemos Coelho, 48, below), and from there a road led overland to Sutuko, which most probably was Gomes' 'Cantor', referred to in the next sentence. If Gomes remembered correctly and if, as the wording seems to be saying, he went on to Sutuko from the port of Wuli where he left a captain, either (a) he went from Fatatenda overland, in which case 'up the river' is misleading; or (b) from Fatatenda he travelled up-river, apparently in the third ship (or, if the third ship was left down-river, without this being mentioned, in a boat), to a point on the river nearer Sutuko, perhaps Same. Alternatively, in the 1450s Wuli had a port further down-river than Fatatenda, and Gomes travelled up-river from that port to Fatatenda, before going overland to Sutuko. However, the fact that the text back-tracks from the port of Wuli to a part of Niani, which should have been mentioned first, does not increase confidence in the details of Gomes's journey.

[4] ''Animaijs' is assumed to be a corrupt form of Nianimansa, Niani (Jobson's 'Nany': *GT* 100) being the polity on the north side of the river between Baddibu and Wuli. The port was perhaps Jobson's Cassan (Kasang).

[5] It is uncertain whether the Portuguese had heard of 'Cantor' before this voyage, or even by the 1480s when Gomes's account was collected. Hence this appears to be the earliest recorded reference to the name. Assuming that no Portuguese had preceded Gomes to Cantor, his penetration of the interior up some 300 miles of river, most of whose course was previously unknown to European sailors, was an extraordinary feat. If local African pilots were taken, which is conceivable, it would also indicate considerable confidence in their skill. Yet the journey is noted so casually that it may be that the river had been already penetrated for some distance by earlier Portuguese. At later dates Cantor (Kantora) was normally the name of a district and polity on the south side of the upper river. But since it seems likely that the the town visited was Sutuko (Jobson's 'Setico'), it must be assumed that in the 1450s Cantor included part of the north side. According to tradition, the capital of Cantor was, at one time, Sonkunda, a fortified town much higher up-river and near the port of Yarbutenda, but later Portuguese sources do not suggest any contact with this area, the contact being instead with Sutuko, as Jobson confirms.

[6] It is not clear whether the ship actually reached the port of Cantor or whether it was halted by the difficulties later encountered by Jobson, the final advance being by boat.

the blacks came in very great numbers. Peace having been established with them,¹ the report that the Christians were in Cantor spread throughout the whole land, and they assembled from all parts, viz., those from Tambucutu (Timbuktu)² in the north, and men living to the south towards the Serra Geley,³ and also some came from Quioquum [Kukia],⁴ which is a great city, surrounded by a wall of baked tiles. I learned from these latter that in this city there was a great amount of gold, and that caravans of camels and dromedaries crossed to there carrying merchandise from Carthage or Tunis, from Fez, from Cairo, and from all the land of the Saracens, to carry back the gold, which is abundant there, having been brought from the mines of Mount Gelu. The opposite side of this mountain (range) is the range of hills (*serra*) called Serra Lyoa (Sierra Leone).⁵ They told me that this range of mountains begins at /f.277/ Albafur, and runs southwards, which pleased me much, because all the rivers [of western Guinea], large and small, descend from these mountains and run westwards. And they told me that other rivers run eastwards from the mountains and are large rivers, and that near that city was a great river, named Emin.⁶ They said that there was also an extensive stretch of water (*mare*), not very broad, on which were large canoes like ships, and that those in one part [or: on one side] of the water made war on those in another part [or: on the other side], and that those in/on the eastern part/side were whites. On my enquiring what lords ruled in these districts, they answered that in the part inhabited by blacks was a certain lord named Sambegeny, and in the eastern part was a lord called Semanagu, and they were always at war.⁷ A short time before, they had a great battle, won by Semanagu. Admedi, a Saracen of Termezen, who overland went there and who had traversed the whole land, informed me that he had been present at a battle on land and water. When I afterwards related all these things to the Lord Infante, he told me that a merchant in Oran had written to him two months before about the warfare and battle between Semanagu and Sambegeny, and he therefore believed all [I said]. /f.277ᵛ/

These are the things told me by the blacks who were with me at Cantor. I questioned them as to the road which led to the countries where there was gold, and

---

¹ Probably this wording was intended to indicate a contrast between the hostility of some Africans on the lower river and the friendliness of those higher up. In his later years the Infante had told his captains to act peacefully towards the peoples they encountered.

² While Gomes believed this was Timbuktu, more likely it was the much nearer town of Tambakunda. Jobson was similarly confused (*GT* 102).

³ Perhaps the Fuuta Jalon highlands.

⁴ Later 'Quioquia'. Kukiya was the ritual capital of the Songhay polity which had in the previous century conquered the Mandinka nuclear polity of Mali and now ruled at least part of its territory. Kukiya was situated on River Niger, a little below Timbuktu. Whether Mali retained influence over the River Gambia region by this date is uncertain and controversial, but local informants from the time of Gomes even up to Jobson's day certainly liked to assure Europeans that it still did.

⁵ Sierra Leone was reached and named by the Portuguese in 1460 or 1461, some years after the ascent of River Gambia by Gomes, an indication that he is relating his account at a much later date – or that the reference is an editorial insertion, by the collector of the old man's oral information.

⁶ It is usually assumed that these vague statements indicate a belief in a linked system of rivers deriving from an inland lake, the Senegal, Gambia, and others running westwards, the Niger running eastwards.

⁷ 'Semenagu' may be the semi-mythical warrior Sumanguru, who appears in the Mandinka oral saga of Sunjata. As for 'Sambegeny', Samba is a Fulbe forename.

asked who were the lords of that land. They said that the king was Bormelli,[1] and that the whole land of the blacks on the right side of the river was under his dominion and the blacks were his subjects,[2] and that he lived in the city Kukia. They said that he was lord of all the mines, and that he had before the door of his house a lump of gold, just as it was created in the earth and not yet exposed to fire, so large that twenty men could scarcely move it, to which the king always tied his horse.[3] He kept this lump of gold not because of its value but because of its splendour and size when found. And (they said) the nobles of his court wore gold in their nostrils and ears. They said that this eastern region was full of gold mines, and that the men who went into the pits to get the gold sands, carried them out and gave them to women to wash and to extract the gold from the sand. And that these men did not live long, on account of the air which escaped from these pits.

I inquired the road from Cantor to Kukia, and was told that to Morbomelli from Cantor the road is eastward to Somanda, and from Somanda to Commuberta and to Cereculle and other places, the names of which I have forgotten.[4] And in these aforenamed places is great abundance of gold, as I can well believe, for at the time I saw the blacks /f.278/ who travelled by these roads come laden with gold. And they said that Forisongul was a subject of Mormelli, who is lord of the right bank of River Gambia. While thus holding peaceful intercourse with those of Cantor, my men became worn out with the heat, and so we returned in search of the other two caravels. In the caravel which had remained in Ollimansa, I found nine men had died, the captain, Gonçalo Affonso, very ill, and all the rest of his men sick, except three. I found the other caravel 50 leagues lower down towards the ocean, and in it five men had died.[5] We immediately withdrew, and made for the sea, and I came to the place where I had hired the black traveller, and gave him what I had promised him. They then informed me that on the other, that is, the left side of the river, there was a certain great lord called Batimansa,[6] and I desired to make peace with him, and

---

[1] Bormelli = *buur Mali*, 'king of Mali' (later 'Mormelli' and 'Morbomelli'). Since the term *buur* is Wolof, Gomes must have picked up the name on the coast or in the lower River Gambia and forgotten this, or else he was using at Cantor an interpreter who spoke both Wolof and Mandinka and passed on the title in its Wolof form or a mixed form. Manding merchants at Cantor would have spoken of *Malimansa*. The ethnonym and/or language name 'Manding/Mandinka' does not seem to have been known to the Portuguese in this early period.(its earliest recording appears to be in a slave document of 1480: Albuquerque and Madeira Santos, *Portugaliae monumenta africana*, I, 1993, p. 257, correcting P. E. H. Hair, 'Black African Slaves at Valencia, 1482–1516: an Onomastic Inquiry', *History in Africa*, 7, 1980, p. 132).

[2] This implies that the other side of the river was not under Mandinka control, presumably being subject to Wolof rulers.

[3] For an equally fictitious story about the abundance of gold in the interior, as told to Jobson, see *GT* 91.

[4] The placenames are unidentified but 'Cereculle' appears to be the ethnonym Serahuli/Sarakole/Sarawule, which represents an interior ethnicity whose members speak a cluster of dialects related to the Mandinka language.

[5] Cf. the later experience of the Jobson voyage in respect of its retreat because of sickness. The distance on the river between 'Ollimansa' and 'Animaijs', 50 leagues or 180 miles, if between Fatatenda and Kasang is reasonably correct.

[6] Going up-river, Diogo Gomes described 'Farisangul', where apparently he hired his guide as being on the right bank, and he now describes 'Batimansa' as being on the other side, the left. If 'Batimansa' represents 'Bati' (see p. 251, n. 3 above), coming down-river it would lie to the left, but not 'on the other side'; if it represents Baddibu, it would be on the other side but not to the left. However early

sent to him that black who had been with me at Cantor. The lord of that land desiring to speak with me in a great forest on the bank of the river, brought with him innumerable men armed with poisoned arrows, azagays, and swords and shields (*adagas*). I went to him, carrying him my presents, and some biscuit, and some of our wine, for they have no wine except that from palms, that is, date trees.[1] He gave me three blacks, two females and one male, and he was pleasant and extremely gracious, /f.278ʳ/ making merry with me and swearing to me by the one true God that he would never again make war on the Christians, but that they might travel safely through his land to trade their merchandise.[2]

Being desirous of putting this to the proof, I despatched a certain Indian named Jacob, whom the Lord Infante had sent with us, in order that, in the event of our reaching India, he might be able to hold speech with the natives.[3] I ordered him to go to the place which is called Alcuzet, with [? to] the lord of that country, where previously he and a certain a knight had gone, through the country of Geloffa (Jolof) to find the country of Gelu and Tambucutu [Timbuktu]. This Jacob the Indian related to me that Alcuzet is a very fertile[4] land, having a river of sweet water and abundance of lemons, some of which he brought with him to me. And the lord of that country sent me elephants' teeth, one of them very large, and four blacks, who carried the tooth to the ship. They came in peace as far as our ships, and thus I was safe from them. Afterwards I went to his house, where there were many habitations of the blacks. Their houses are made of water-reeds covered with straw, and I remained with them for three days. Here were many parrots and many leopards, and he gave me six leopard skins, and ordered that an elephant should be killed and its flesh carried on board the caravels.[5]

---

travellers usually described the right and left banks of African rivers in terms of up-river journeys. Although the text as we have it does not specifically say so, Gomes had to turn his ship around after returning his guide, and he then ascended the river again some distance in order to reach 'Batimansa', in which case the left side would be the north side and now the opposite side to 'Farisangul'. This would favour the identity of 'Batimansa' with Baddibu. But it seems as likely that the text is simply muddled, at one or several points.

[1] Although common date palms are not the same as the oil palms from which palm wine is generally tapped, a small date palm (see p. 253, n. 1 above) does provide wine and is to be found in localities up-river from the estuary, where in fact the oil palm is rare.

[2] A mystery is created by the fact that Cadamosto and Gomes are alleged to have visited Batimansa within a short time of each other, but neither account refers to the other visitor and each account makes the ruler appear as if he is being for the first time contacted by Europeans. Moreover, it is odd that the ruler promises not to war on Europeans again, since as far as we know there had been no hostility between the Portuguese and 'Batimansa'. The ruler guilty of hostility was that of Niumi, lower down the river. Perhaps Gomes was forgetful or his editor confused, and the text mixes up a meeting with 'Batimansa' and the meeting about to be described with the ruler of Niumi. For the ruler to swear 'by the one true God', if not words put in his mouth by Gomes, indicates a Muslim.

[3] It is generally assumed that, by 'India' Prince Henry and Gomes meant Ethiopia, which at this time was often thought to extend most of the way across Africa, and that Jacob was an Ethiopian. Cf. the reference to Prester John in Usodimare, above. Other than what Gomes states, nothing else is known about this mission.

[4] The text has *viciosa* 'vicious', but this seems to be a mistranslation of Portuguese *viçosa* 'luxuriant, rich'.

[5] This apparently relates to an earlier attempt to contact the interior, by travelling from somewhere on the Senegal coast, an attempt otherwise unevidenced. Alcuzet is an unidentified locality (if not, in fact, a personal name or title). But since Gomes was to visit there (unless the account mistakenly conflated two separate episodes), it was apparently not far from River Gambia; indeed, since elephant meat was presented by the ruler to the caravels, it was probably fairly close to the river. It stood along a stream, and if it had been previously reached from the Senegal coast, it most likely lay between River

It was here that I learned the fact that all the wrongs done to the Christians had been done by a certain king called Nomymans, who possesses the land which lies on that promontory.[1] I took great pains to make peace with him, and sent him many presents by his own men in /f.279/ his own canoes, which were going for salt to his own country, where salt is plentiful but of a red colour. He greatly feared the Christians because of the injuries he had done them. I went by the river towards the ocean, as far as the harbour near the mouth of the river, and he sent to me many times men and women to test me, whether I would do them any harm, but, on the contrary, I received them with a friendly face.[2] When the king heard this, he came to the river-side with a great force, and sitting down on the bank, sent for me to come to him, which I did, with due ceremony on my part, as best I could. A certain bishop of their church was there, a native of Mali,[3] who asked me about the God of the Christians, and I answered him according to the intelligence which God had given me. Finally I questioned him respecting Muhammad, in whom they believe. What I said pleased his lordship the king so much that he ordered the bishop within three days to leave his kingdom. Springing to his feet, he declared that no one, on pain of death, should dare any more to utter the name of Muhammad, for he believed in the one and only God, and there was no other God but the one in whom his brother, the Infante Henry, said that he believed.[4] Calling the Infante his brother, he desired that I should baptize him, and so said all the lords of his household and his wives. The king himself said that he would have /f.279ᵛ/ no other name than Henry, but his nobles took our names, such as Jacob, Nuno, and so on, as Christian names.[5] I remained that night on shore with the king and his barons, but I did not dare to baptize them, because I was a layman.

On the next day, however, I begged the king, with his twelve principal men and eight of his wives, to come to dine with me on board the caravel, which they all did

---

Saalum and River Gambia, the intermediate land being intersected by many creeks. If the sequence of the account is correct, it presumably was close to 'Batimansa'. The reference to fresh water at Alcuzet was perhaps made with the thought that part of the course of River Gambia was salt water.

[1] By 'promontory', Gomes means the land between the lower River Gambia and the river to the north, River Saalum. 'Nomymans' is Niumimansa, the ruler of Niumi, the polity on the north side of the lower River Gambia. Niumi was subject to a polity on River Saalum which had almost certainly been involved in the earlier episode which led to the deaths of Nuno Tristão and his crew, and Niumi was most probably responsible for the attack on Cadamosto's party in River Gambia.

[2] No doubt the women were sent as 'comfort women', partly to test whether the white men were fully human, partly to attempt to ascertain their general attitudes and intentions.

[3] As was common at the time, Gomes considers Islam a heretical Christian sect, and an itinerant Moslem teacher (who was probably also a merchant) therefore as a 'bishop'. Cf. Almada, 5/10, below ('Abbot or Provincial'); also a Jesuit writing c. 1615 who stated that the highest Islamic clerics 'correspond to the rank of bishop or archbishop among us' (Manuel Álvares, *Ethiopia Minor and a Geographical Account of the Province of Sierra Leone*, trans. P. E. H. Hair, Department of History, University of Liverpool, 1990, pt 1, ch. 3).

[4] If the king actually behaved like this and uttered the sentiment recorded, the wording suggests that he was already under Islamic influence. However, the later drinking of wine by the king and his party suggests that the influence was limited – it may have been the alcoholic liberation that led them to express enthusiasm for Christians.

[5] Taking European – in this case Portuguese – names was normal practice in earlier times when Africans acceded to Christian beliefs. But given that the exchange of views was conducted through interpreters, it must be doubtful whether the names were meaningful to the Africans – and indeed how much of the reported discourse was actually said by the Africans.

unarmed. I gave them fowls and meat cooked after our own fashion, and wine, both white and red, as much as they pleased to drink; and they said to each other that no nation was better than the Christians. Afterwards, when we were on shore, he desired that I would baptize him; but I answered that I had not received authority from the supreme pontiff. But if he so wished, I would speak to the Lord Infante, who would send a priest to baptize them. He immediately wrote to the Lord Infante to send him a priest, and someone noble to inform him about the faith; and (also) send him a falcon for hunting, for he wondered greatly when I told him that the Christians carried a bird on the hand which caught other birds. He wished him also to send two rams, and sheep, and ganders and geese, and a pig, as well as two men who would know how to build houses and wall his city with beaten earth.[1] All these things I promised that the Lord Infante would fulfil. At my departure he and all his people wept, so great was the friendship which had sprung up between him and me. /f.280/

It so happened that for two years no one went back to Guinea, because King Affonso, with a fleet of 352 ships, sailed to Africa and took the powerful city of Alcacer dalquivi [Al Ksar el-Kebir, *recte* Al Ksar es-Seghir], for which reason the Lord Infante, being fully occupied, gave no attention to Guinea. I then reminded the Lord Infante of the matters about which this king had written, and he ordered everything to be done as I had promised.

After leaving that king of Gambia I pursued my way to Portugal, and sent one caravel with those who were in the best health straight home. The other caravel remained with me because the people on board were sick. ... /f.280ᵛ/ ... /f.281/

After the Lord Infante returned in the fleet with King Affonso, I reminded him of what King Nomimans had said to me, so that he should send to him all those things which had been promised. This the Lord Infante did, and sent thither a certain priest, a relation of the cardinal's, the Abbot of Soto de Cassa, to remain with that king and instruct him in the faith. He also sent with him a young man of his household, named João Delgado.[2] This was in the year 1458. ... /f.281ᵛ/

Two years afterwards the Lord King Affonso equipped a large caravel, in which he sent me out as captain, and I took with me ten horses and went to the land of Barbaçins, which is between the Serreos (Serer) and King Nomimans. ... And the King gave me authority over the shores of that sea, that whatever caravels I might find off the land of Guinea should be under my command and rule, for he knew that there were caravels there which carried arms and swords to the Moors, and he ordered me to capture such caravels and bring them to him to Portugal. ... /f.282/

---

[1] Possibly Gomes or another Portuguese wrote the letter for the king. Otherwise it can only have been written, in Arabic, by a Moslem literate. But the request for a pig is non-Islamic. The request for masons, if true, may mean that Gomes had shown the ruler an illustration of a European town, although at this date it cannot have been a printed book. However, the information about superior housing and city walls may alternatively have reached him from a Muslim activist boasting about physical features in the Islamic world.

[2] Nothing appears to be known about this mission, which was almost certainly the earliest Christian mission to Guinea. It is just possible that the 'young man' was the same 'young man' who in 1447 had been captured on River Senegal and then ordered by Prince Henry 'to be instructed as a priest so that he could be sent there [to Guinea] to teach the faith of Christ', Monod, *Description*, f. 90.

This happened in the port of Zaza. ... While we were there, a caravel arrived from Gambia ...[1]

DUARTE PACHECO PEREIRA (c.1508)

Duarte Pacheco Pereira, a royal officer and mariner, served in West Africa in periods between the 1480s and 1520s. His account, basically a *roteiro* or navigational guide to the coast of Africa beyond the Mediterranean, incorporated much general information, derived apparently from personal observation and unnamed oral informants. It is unclear whether his personal experience included visits to River Gambia. His account, completed c. 1508, was not published until the nineteenth century. It seems to have been seen by Barros but otherwise did not influence the early sources that follow.[2]

Item. By standing four leagues out to sea and sailing 15 leagues SE from Rio dos Barbaciis [River Saluum], you will come to the mouth of Rio de Guambea [River Gambia].[3] ... This river is 13°5′ north of the equator. High tide flows NW and SE. Half a league to the N of this palm forest is the mouth (*ramal*) of the river at the present time,[4] and he who wishes to enter it must make his way E-by-S and in the deepest part he will find 2½ spans at low tide and 3½ at full. It is noteworthy that the tide flows with such force in this river that it runs up it 180 leagues and more. At 150 leagues from its mouth is a district called Cantor, where there are four towns, the principal of which, called Sutucoo [Sutuko], has some four thousand inhabitants. The other three are Jalancoo, Dobancoo and Jamnamsura, and they are all enclosed, with wooden [defences] and are distant from the river half a league, a league and a league and a half.[5] At Sutuko is held a great fair, to which the Mandinguas (Mandinka) bring many asses. When the country is at peace and there are no wars, these same Mandinka come to our ships which at the bidding of our prince

---

[1] Zaza was apparently a port on River Saalum, the name now not known. The reference implies that, by 1460, the Portuguese had begun regular trade with River Gambia.

[2] Duarte Pacheco Pereira, *Esmeraldo de Situ Orbis: Côte occidentale d'Afrique du Sud Marocain au Gabon*, ed. Raymond Mauny, Bissau, 1956, liv. 1, cap. 29. The present translation revises that in G. H. T. Kimble, *Esmeraldo de Situ Orbis*, Hakluyt Society, London, 1937. Exceptionally, the manuscript seems to have been the source of navigational material in Manuel de Figueiredo, *Hydrografia ... com os Roteiros de ... Guine* ..., Lisbon, 1614.

[3] By analogy with instances of toponyms and ethnonyms beginning gu- elsewhere in this text, 'Guambea' is to be pronounced Kambia. However, almost all earlier sources (texts and maps, but see p. 244, n. 2 above) write 'Gambia' (Teixeira da Mota, *Mar*, table between pp. 122–123, 237), so this spelling must be idiosyncratic. It may, however, be relevant that certain dialects of Mandinka sound a normal 'g' as 'k'.

[4] *ramal* has the meaning of an assemblage of branches, and several streams enter River Gambia near its mouth, hence 'estuary'. In his text Pacheco Pereira has previously stated that the mouth of River Senegal shifts from time to time, hence, presumably, his 'at the present time', although in fact River Gambia does not have a shifting mouth.

[5] The punctuation in the original makes it uncertain whether the wooden defences and the ½–1½ leagues distance from the river apply only to the three towns or include Sutuko, which is in fact at the largest stated distance inland. The names of the three towns are not in use today. But Jamnasura may well represent Jagrançura, at one time a port of Sutuko (below, Almada, 5/12) and Jalancoo has a resemblance to Jalacuna (Lemos Coelho, 44, below), as noted in Teixeira da Mota, *Mar*, pp. 159, 224 (although the suggested Mandinka etymologies are inaccurate). DPG collected in Bantundung (Wuli) a tradition that Dobanko was one of the former ports of Wuli and was close to present-day Limbambulu; and today Dobankoo is the name of a locality on a small estuary between Limbambulu and Fatatenda (p.c. Sidia Jata).

visit these places, and they buy common red, blue and green cloth, kerchiefs, thin coloured silk, brass bracelets (*manilhas*), caps, hats, the stones called '*alaquequas*' (cornalines) and much more merchandise, so that in time of peace, as we have said, 5,000–6,000 doubloons of good gold are brought from there to Portugal. Sutuko and two of the other towns belong to the kingdom of Jalofo (Jolof), but being on the frontier of Mandinka they speak the language of Mandinka. This River Gambia divides the kingdom of Jolof from the great kingdom of Mandinka, the former being called in the language 'Emcalhor' as I have said above;[1] the river itself is also called in the Mandinka tongue Guabuu [Kaabu].[2] When ascending the Guabuu the kingdom of Jolof is on the north and that of Mandinka on the south, extending nearly 200 leagues in length and eighty in breadth. The king of Mandinka can put into the field 20,000 horsemen, and infantry without number, for they take as many wives as they choose.[3] When their king is very old and cannot govern or when he is afflicted with a prolonged illness, they kill him and make one of his sons or near relatives king. 200 leagues from this kingdom of Mandinka is a region where there is abundance of gold. ... [the 'silent trade' in gold in an interior region]

Returning to River Gambia, it contains water-horses (*cavallos marinhos*) larger than oxen, of every colour that ordinary horses have. In shape they are like oxen, with cloven feet like oxen, but in their neck, face, hair, ears and flanks they are like horses; and on their neck they have two small horns or tusks of two spans in length and thick as a man's upper arm. They live in the river, usually in the shallow parts with the water up to their belly, but also in the depths when they choose; they also come ashore to graze and sleep in the sun, the majesty of great Nature thus providing for them both on land and in water.[4] There are also in this river many large lizards, some of them 23–24 feet from head to tail; they live in the water and come ashore for reproduction, when they lay eggs under the sand much larger than ducks' eggs, and when these are hatched they are a span in length and at once enter the water and grow up there. They are dangerous beasts and eat men, oxen and cows.[5] Many other things concerning River Gambia I omit because I am no friend of prolixity, although it is no bad thing when satisfactorily employed.

The people of this country all speak the language of Mandinka and are Moslems, following the sect of Mahomet. They wear blue cotton smocks (*camisas*) and pantaloons (*seroulas*) of the same material.[6] They have many vices and take as many wives as they like; lust is universal among them. They are very great thieves, drunkards, liars and ingrates; all wickednesses that exist they have.[7]

---

[1] 'Emcalhor' represents Kayor, a Wolof kingdom on the Senegal coast, whereas 'the kingdom of Jolof' is usually taken to refer to an interior Wolof kingdom.

[2] It is not clear whether an alternative name for the river is being indicated, or whether it is being suggested that Gambia/Guambea/Kambia and Guabu/Kaabu are etymologically related. For a possible explanation of the name 'Gambia', see *GT*, p. 84, n. 3.

[3] The inference being that polygamy multiplies population, a widely-believed supposition where it is practised but not necessarily correct.

[4] Cf. *GT* 20–22.

[5] Cf. *GT* 16–17.

[6] Short smocks and baggy trousers were regularly described by later writers as Mandinka male dress.

[7] The reason for this condemnation is not clear. Other Portuguese sources tend to take a markedly favourable view of the Mandinka.

VALENTIM FERNANDES (*c.*1508)

Valentim Fernandes, a native of Bohemia but a printer and publisher in Lisbon, in the later 1500s assembled a corpus of material on the Portuguese out-thrust, including an account of the western coast of Africa from Morocco to Sierra Leone. He appears to have composed the account himself (although its Portuguese was perhaps finalized by a native speaker), and based it largely on information, most probably oral information, obtained from individual Portuguese who had worked on the coast. His account does, however, include extracts from Cadamosto and was therefore written after 1507, the date of the first printing of Cadamosto's account. The corpus remained in manuscript and soon found its way to a library in Germany where it remained hidden until discovered and published by German scholars in the mid nineteenth century. His account of western Guinea therefore did not influence the later writings on that region detailed below. He says relatively little about River Gambia but a great deal about the Mandinka. This people occupied a large district stretching far into the interior, and away from River Gambia. Nevertheless, since the Portuguese almost certainly obtained most of their information about the Mandinka from their contacts with those along River Gambia, much of what Valentim Fernandes wrote about this people would seem to apply to the Gambian Mandinka and is therefore comparable with what Jobson wrote about the Mandinka in the next century.[1]

/f. 104/ 1455. River Gambia was discovered by [blank], a servant of the Infante, two merchants being with him, each with his own caravel, they being a Genoese named Antoniotto and a Venetian named Luys de Mosto. ...

/f.105/ River Gambia is otherwise called River Cantor. Here the kingdom of Mandinga (Mandinka) begins. Many ships enter this river to trade horses as well as all other kinds of merchandise, because they trade on the north side with the Gyloffos (Jolof/Wolof) and on the south side with the Mandinka.[2] Ships pass up the river for 300 leagues, that is, ships of 50–60 tons. These ships trade cloths for gold, of which there is much here. Merchants from far-away lands come to the ships to trade; however, all are black and [only] a few light tawny. The river is unhealthy for Christians who come here.

The Mandinka people differ from the Jolof people in language, but in religion and customs they are all one.

/f.105ᵛ/ The king of Mandinka is called Mandimansa because those of this land in their language call the province of Mandinka, *Mandi*, and *mansa* means 'king' in

---

[1] The standard text of the Valentim Fernandes corpus is Joaquim Bensaude and António Baião, eds, *O Manuscrito «Valentim Fernandes»*, Academia Portuguesa da História, Lisbon, 1940. (For checking of the transcription of another section of the account of western Guinea against the original manuscript, see P. E. H. Hair, 'The Text of Valentim Fernandes's Account of Upper Guinea', *Bulletin de de l'Institut fondamental d'Afrique noire*, sér. B, 31, 1969, pp. 1030–8. No English translation of the account of the African coast has yet appeared, but French translations exist, the relevant one being a bilingual text, with Africanist annotation (Monod, *Description*). The account of the Mandinka includes a little material from Cadamosto, Fernandes' corpus including a Portuguese translation of an abbreviated version of the printed text of Cadamosto. The borrowings are the whole of f. 104–104ᵛ, with traces in ff. 111–13: the former passage is therefore omitted below. Those sections of the remainder of the account most relevant to Jobson's account, about three quarters, have been extracted and translated.

[2] The extent to which at this date the Wolof controlled all the ports on the north side of the river is uncertain, but it is likely that the petty rulers on both sides were already Mandinka.

their language, so they call their king Mandimansa. This king is lord of many vassals and they pay much tribute. He lives 700 leagues in the interior in a town walled with beaten clay, called Jaga. ... [description of the king of Mandinka]

/f.106ᵛ/ ... The king and any other man can have as many wives as he can buy. And as the number of wives, so the same number of dishes at dinner and other meals, because each woman brings him her dish, since he lives by himself and has no woman in his house. /f.107/ The first night he sleeps with the first wife in his house, the second night with the second wife, and so on to the last, then he begins again with the first. The women support themselves, their husbands, and their slaves.

If any of our whites arrive at a house of a black, even the king's, and ask for a wife or a daughter, he sends anyone the man chooses to sleep with, and he does this for good friendship and not from force. A father himself hands over any girl the man chooses and if he wants her sisters, these too. If anyone sleeps with his own sister or daughter, as often happens, this is no crime. ... [theft and adultery]

/f.107ᵛ/ ... What these lands produce are green parrots, gold but not much, male and female slaves, cotton cloths, mats, hides, a little musk, and monkeys with tails. The Portuguese take there brass bracelets, beads of *maçanûngu*, red cloth, mantles from Alentejo, cotton carried from the Cape Verde Islands, and horses - one horse here for seven blacks. There is no money in this land or in all Guinea, goods being exchanged for each other.

/f.108/ There are great lords in this land and great merchants. Throughout Mandinka land there are great lords called *foroes*, an office or dignity among them like that of royal magistrate (*corregidor*) or royal governor among us, and one such is much respected by them.[1]

The custom among them, both great and small, is that when individuals meet who have not seen each other for some considerable time, just as we embrace, they kneel, their elbows on the ground and their hands over their eyes. They beat their elbows on the ground many times, then beat the ground with one elbow and with the other hand snatch up earth and throw it behind or over themselves. When an individual goes to see a lord he wears nothing except breeches, the rest of him being naked.[2] Also when one lord goes to see another lord they do the same. ... [other lordly ceremonial]

/f.108ᵛ/ ... It is the custom among the Mandinka that they speak frequently through intermediaries. For instance, when 3, 4 or 10 are assembled and one of them wishes to say something and says it, another of them rises, though the first one said everything so loudly that the others heard and does not hesitate to say the same thing even more loudly, and this is the custom in or their land.

/f.109/ This race of Mandinka is the largest tribe of one language in all Guinea, larger than any other.

In this land many hold to the sect of Mohammed, and hence many *bisserijs* travel

---

[1] The term is probably Mandinka *foro* 'person of high rank, noble', nowadays applied to an individual or individuals in any major village. But there may be some confusion with the more commonly evidenced term *farang* '(regional) governor', Jobson's 'Ferran'.

[2] Cf. *GT* 48, 98–9.

through the land, these being Moorish priests who teach their faith to the people. The others are all idolaters in the same way as the Jolofs, as has been said.

Mandinka and Jolof males are all initiated as stated, in this way. Once a youth is 14 or 15, he assembles with many others, 15–20 of them, and an old man, their *bisserij*, takes them together into a wood outside the village at the time of the new moon. After the youths have been initiated, the old man returns, but the youths stay in the wood until they see the next new moon. So for a month they remain away from their homes, and their relatives bring them food to the edge of the wood, while they hide for shame and do not care to eat until the relatives are gone.[1] When the new moon arrives, and the youths have to return to the village, their fathers kill cows and goats and make a great feast for them, with dancing and singing.

/f.109ᵛ/ The Mandinka are well-built men, like the Jolofs, but the Mandinka are more truthful, likeable, amiable, and of a better condition;[2] and they trade more. They are very good at working with their hands, in needlework and weaving, and many other crafts. They show great respect to strangers in their lands, for which reason they are respected by all, the people of Guinea as well as the Christians. They trade very far into the hinterland, further than any other people of this land, and, passing through the interior, they go as far as the Castle of Mina.

The women of this land and all Guinea grub the land, work it, and sow it, to feed their husbands, and they spin cotton, making many cotton cloths to dress themselves as well as to sell.

Men and women are dressed like the Jolofs, the men in breeches called *magalhões*, their women in a narrow cloth of four-fingers width. Unmarried girls, and women for their glory and honour still virgins,[3] go about naked, not covering their shameful part, and they wear a string belt above their loins covered with little white shells, and in front over their shameful part they wear three little pieces of wood hanging down, with apertures a finger-length long, but once deflowered they cover their shameful part. Boys go quite naked until they are twenty or married.

/f.110/ Their feeding is like that of the Jolofs except that they have more rice, so much that they bring it to sell and exchange, also wine, oil, meat and other foodstuffs, because this land of Mandinka is very adequately supplied with rice, millet (*milho*), etc. It has a great stock of cows, that is, in the interior, also asses, and very many sheep without wool.

They feed on rice, millet, *milho zaburro*, and yams - boiled and roasted - the *coco* [cocoyam] plant, and beans. The poor people who lack yams or rice feed on wild *norças*, boiled and seasoned like *tremoços* (lupins) here because they are always bitter.

Their houses are made of fired or sun-dried bricks, and are round like tents and roofed with straw from from the fields, or dry hay. Some are of clay and some of wattle and daub, roofed with thatch.

---

[1] Cf. *GT* 110–17.

[2] 'of a better condition' – obscure: does it mean 'more comfortably off'?

[3] 'for their glory and honour': presumably this curious phrase is inserted because the writer is thinking of celibate female religious in Christian Europe – irrelevantly. The literal translation of the Portuguese phrase for 'still virgins' is, significantly, 'not corrupted', that is, not corrupted by entering into sexual activity.

Men in this land do not marry but buy their wives in the following way. They go to where they want to find some suitable young women, then the man goes to her father, and makes a bid to him of as much as he is able to afford. According to the standing of the father and the daughter, the man pays the father in gold, or in rice or millet, or in cows, etc, until the father is satisfied. Then he takes the girl to his house and uses her as his wife. But if he is not satisfied with her, he returns the girl to her father and the father must repay him what he gave for her.[1] They can marry anyone except their mother and sister; no blame attaches to any other.

/f.110ᵛ/ ... [weapons] ... They have no iron, except the iron that the merchants who come from beyond River Gambia bring, supplying the larger part of Gambia, this iron, coming from the hinterland of Sierra Leone, being conveyed by this river.[2] ... [weapons] ...

/f.111/ In Mandinka are many elephants, hence there are great hunters there, who kill the elephants with arrow-heads on a lance-shaft which they hurl. There are many wild buffalo; many leopards; many cats with long tails in a variety of patterns and colours; many antelopes; great troops of russet-coloured gazelles; many hares – [but] no rabbits; few cows, [only] small ones; many dogs – the fat ones they eat; wild pigs; a few civet cats. Goats, raised in planked huts for fear of leopards, give birth twice a year, two kids each time. No horses are bred in Guinea and none can live here. ... [birds, snakes] ...

/f.112/ ... The crocodile are very large, 30 feet long, and when men, women, or cows go into the river the crocodiles kill them and eat them. ... [hunting crocodiles] ... In this river are many fish-horses [hippopotamuses]. ... [trees, fruits]

/f.113ᵛ/ In these lands there is much wax and honey, and a large number of bee-hives in trees, made of straw and covered over with clay.

There are many places with 5,000–10,000 or more inhabitants, each place with its king. The kings fight each other, and capture persons whom they sell to Portuguese ships or to the Arabs.[3] Many places in these lands are surrounded by a wooden defence, with gates and gate-keepers, these places being guarded as here [in Europe], so no-one can enter until the king knows who he is.

/f.114/ Cape St Mary is the headland of River Cantor to the southwards. In these lands they make palm wine thus. The palm at its uppermost height throws out very large bunches [of leaves] like the heads of the pine, and if from any bunch they wish to draw wine, or oil, for the former they cut each bunch below at its base, and place there a gourd morning and night. As much wine runs out in daytime as at night,

---

[1] Latterly at least, the return of a wife would only be if a virgin had been promised and she proved otherwise.

[2] The 'hinterland of Sierra Leone' means Susu country in the southern parts of the Futa Jalon highlands, but the iron could not have been transported from this area all the way by River Gambia. Nor is it likely that the iron was brought to the coast at Sierra Leone, possibly via the Scarcies Rivers, and then transported up-coast by sea, since this would have involved Portuguese vessels. However, once it reached the upper River Gambia overland, it is plausible that it was distributed down-river, by boat as well as overland. Lacking iron in Portugal, and also because the use of iron for weapons was thought by the papacy a reason for forbidding iron imports to infidels, the Portuguese never imported large quantities, a disadvantage which enabled the later Europeans to reduce Portuguese dominance in the marine import/export trade of Guinea.

[3] That is, slaves are sold to the interior slave trade as well as to the marine export trade.

and it lasts 15 days. This wine is as sweet and tasty as malmsey, and is white like milk. It is alcoholic like ours, but if kept from one day to the next becomes sour like vinegar.[1]

If oil is wanted, they let the bunch mature and do not cut it, but when ripe they cut it and take away the nuts and boil them very full. Then they break them and boil them again, and then press them out in a cloth, and the result is the oil. When the bunches begin to grow they are black but after they are ripe they turn red, and so is the oil.

The palms of this land that give dates produce such small ones that they are worthless.

/f.114ᵛ/ ... From Cape Verde to here there are two winters[2] and two summers each year. They sow twice and harvest twice, rice and millet, etc; harvesting once in April and again in September, and when the rice is harvested then they sow yams, and so cultivate throughout the year.

All that has been said up to here applies generally to the Barbacijs, the Jolofs, the Mandinkas and the Tucuraes.[3] These four regions have each their own language, whether on the coast or in the interior.

JOÃO DE BARROS (1552)

Barros, who held a post at Lisbon in the official Guinea trade, although he had briefly visited West Africa in the 1520s, as far as is known never visited Senegambia, and his account of River Gambia – a small section of his extensive and influential printed account of the early Portuguese outthrust – is therefore a compilation, albeit probably in the main from oral information.[4]

This land, commonly called Jalofe by the natives, lies between these two remarkable rivers, Sanagá (Senegal) and Gâmbea, which, as their courses are long, receive several names from the different peoples who live on their banks. ... [River Senegal]

The other river, the Gambia, [at least] from the trading-place (*resgate*) of Cantor, has not such a variety of names. Almost the entire river, as far as the Trading-place for Gold to which our ships go – about 180 leagues from the bar, or, on account of its twists and turns, 80 as the crow flies – is called by the blacks Gambu, and by us Gambia. The greater part of its course is tortuous, with small turns, especially from the trading-place downstream, until it enters the sea in 13½°, SE of the Cape we call Verde (Green). It brings a greater quantity of water than the Senegal, and is much deeper, because it receives many fierce tributaries with large quantities of water. These rise in the interior of the country called Mandinga (Mandinka), and their principal sources are those of Ptolemy's River Niguer and Lake Libya.[5]

[1] Cf. *GT* 131.
[2] Text corrected.
[3] The Barbaçijs (Wolof *buur ba Sine* 'ruler of Sine', but in modern Wolof *buur Sine* (Jean Boulègue, *Le Grand Jolof*, Paris, 1987, p. 16), were the Serer people SW of the Wolof; the Tucuraes or people of Tekrur were the Fulbe.
[4] João de Barros, *Ásia, Primeira Década*, 1552, eds H. Cidade and M. Murias, Lisbon, 1954, liv.2, cap.8. The translation is revised from that in Crone, *Cadamosto*, pp. 135–41.
[5] Barros believed, with others of his time, that River Senegal and River Gambia both linked up with an interior river, River Niger, which possibly turned eventually into the Nile.

As the river descends tortuously, its waters are broken up so that they do not strike violently against our ships proceeding up the river. About half-way to the Trading-place, it forms a small island, which our men call the 'Island of Elephants' on account of the many elephants there.[1] Above the Trading-place for Gold there is a rock completely blocking the passage; the King João of whom we have spoken sent agents to remove it, but this proved to be too expensive and difficult.[2]

In general, both rivers, the Gambia and the Senegal, produce a great variety of fish and aquatic animals, such as the horses we call sea-horses [hippopotamus]; very large lizards which in shape and nature resemble the Nile crocodiles celebrated by so many authors; and also snakes, which are small and not as monstrous as often painted and fabled.[3] The animals drinking the waters of these rivers are so numerous, and of so many varieties, that even elephants go in herds, as our cattle do here. Gazelles, pigs, leopards, and all kinds of game, their names unknown to us, are found here in as great numbers and varieties. ... [Senegal River and the Jolof]

Because of this river [Gambia], the more populated land is that which lies along it, where there are certain cities, the chief one, Timbuktu, lying three leagues to the north of the river.[4] Because of the gold carried there from the great Province of Mandinga, many merchants from Cairo, Tunis, Oran, Tlemcen, Fez, Morocco, and other kingdoms and dominions of the Moors go there ... Other peoples from the interior of Mandinga come to the Trading-place of Cantor, which our ships reach by River Gambia. And though the sands of these two remarkable rivers, Senegal and Gambia, do not carry as much gold as those of our Tagus and Mondego [in Portugal], the opinion of men is so unreasonable that they do not appreciate so much what they have near them, as what they expect to gain through much danger and toil, such as those endure who go in search of gold to these two barbarous rivers.[5] ... [an attempted fort on River Senegal]

Moreover, as by the Castle of Arguim, the Trading-place of Cantor, Sierra Leone, and the fortress of Mina, a great part of the land of Guinea was bled of its gold, so this fortress on River Senegal would tap the gold coming to the said two markets, which lay close to its banks, and it would not fall into the hands of the Moors, who go to seek it by camel caravans across many deserts, in which sandy plains of Libya many of them are often buried on their journeys.[6]

---

[1] The earliest reference to this toponym. The island is actually only one third of the way to 'Cantor'.

[2] While there are certainly rocky barriers across a large part of the course of River Gambia at two points, Barros appears to have confused River Gambia with River Senegal, where the attempted demolition of 'a rock' was earlier recorded (P. E. H. Hair, 'The Falls of Félou: a Bibliographical Exploration', *History in Africa*, 11, 1984, p. 115).

[3] Perhaps a reference to Cadamosto's account, by now in print, which spoke of monstrous snakes in Senegambia (Crone, *Cadamosto*, ch. xxviii). River Gambia does, however, have fair-sized pythons.

[4] Barros means that Timbuktu is located near River Niger, the interior river he assumes to be linked to River Gambia.

[5] In fact, very little gold passed down River Senegal. The reference to the sands of the rivers of Portugal must be figurative, indicating the riches and greater economic opportunities of the home country.

[6] A commonplace justification of Portuguese intrusion into the gold trade of West Africa, that it passed into Christian hands what had previously enriched North African Moslems.

FRANCISCO DE ANDRADE (1582)

A senior local-militia and administrative officer (*Sargento-mor*) on Santiago Island, Francisco de Andrade wrote a report on the Cape Verde Islands and the trading districts of the mainland coast, specifically to inform the government of the new joint monarchy in Madrid about the military and economic condition of the Portuguese dominion in western Guinea, over which it had just gained control.[1]

The fifth trading-place is River Gambia. From the mouth of its bar to the port and town of Yambor is 30 leagues of salt water, and from here to the port of Cantor, 90 leagues of fresh water. The whole 120 leagues can be sailed on the tides, and the river has a depth permitting vessels of 80 tons to come and go on the flood and ebb tides. Throughout the river there is trade in white cotton cloth, and some black, and in *teadas* (woven cloths), these being carried to other rivers; and in slaves, ivory and wax carried to this island.

The final port, that of Cantor, cannot be gained by any ship from this island or other [Guinea] district, because it is reserved to His Majesty's Treasury or to his contractors.[2] At one time there was traded in this port 10–12,000 *cruzados* worth of gold dust, in exchange for copper bracelets, and some ivory and wax. Now less is traded, since the last ship that went there brought only some 3,200 *cruzados* of gold, but the excuse was given that they carried few trade goods and that much gold remained to be traded.

The blacks of this trading-place state that twenty days' journey into the interior is a very large capital city called Tunbuqutum, whose lord is the Grand-Fulo, and near it is a very large lake, which empties on one side as River Gambia and on the other as River Senegal.[3] The mouth of the latter is in 15½° [North], 125 leagues from this island, going east or east by NE. Further, the blacks say that the gold is brought from the distance of Çofala, to which they also carry the bracelets with which they buy the gold.[4] In this river [Gambia] there are great plains and forests, rice in quantity, and [other] foodstuffs, the blacks watering these, in lack of rain, with the fresh water from the river, there always being plenty.

---

[1] The report of Francisco de Andrade, dated 26 January 1582, is reprinted in Brásio, *Monumenta missionaria Africana*, 2nd ser., III, pp. 97–107. The report was prepared on the orders of Pedro Sarmiento de Gamboa, in command of a Spanish fleet heading for the Straits of Magellan which had halted at the Cape Verde Islands to land a Spanish governor and troops. For a French translation of the whole report, with a useful introduction, see Jean Boulègue, 'Relation de Francisco d'Andrade sur les îles du Cap-Vert et la côte occidentale d'Afrique, 1582', *Bulletin de l'Institut fondamental d'Afrique noire*, sér. B, 29, 1967, pp. 67–87. The present English translation of a section (pp. 103–4, 106) is original.

[2] Reporting to the king, Andrade refers to the royal monopoly in the gold trade – whether in fact this inhibited ships belonging to Cape Verde Islands traders from visiting Cantor may be doubted, and if any did visit there it may be doubted whether they limited what they obtained to commodities other than gold.

[3] This misinformation appears to confuse Tambakunda and Timbuktu, as well as wrongly linking the waterways, the rivers Niger (the lake), Senegal and Gambia.

[4] The usual explanation of this error is that Arabic-speaking informants told the Portuguese, correctly, that the gold came from the lands of the heathen, in Arabic *gaffir*, written in Portuguese *cafre* (English, kaffir); by which the Portuguese understood the region they themselves termed 'lands of the Kaffir', the coast of East Africa, hence Sofala. Cf. Almada, 5/14, below.

THE DISCOVERY OF RIVER GAMBRA

... The tenth trading-place is the rivers of Sierra Leone, where trade is in many slaves and in wax, ivory, some gold, and a fruit like a chestnut called cola, cargoes of which are shipped, it being valued throughout Guinea, especially in River Gambia, the main trading-place for this. ...

ANDRÉ ÁLVARES DE ALMADA (c. 1594)

Almada, a Cape Verde Islands trader, finalized his account of western Guinea in the early 1590s. But his trading experience of the region, especially of River Gambia, went back to the 1560s; and he also probably drew on the experience of the Cape Verde Islands trading community, some of the traders having lived for periods on the river as well as regularly trading there.[1]

Chapter 5. Which discusses the Kingdom of Gambia, otherwise called the Kingdom of Cantor, which is the kingdom of the Mandingas, and of great extent.

1. The Kingdom of Gambia begins at the entrance to its very famous river, five leagues from the bar of the River of the Barbacins. The river can be entered very easily and without risk, because the entrance is like a bay. It has to the leeward Cape St. Mary – which is in Mandinga territory – and to the windward a number of islands, some swampy, some not, lying between the River of the Barbacins and the Gambia River, (all of them) covered with forests of mangrove and other trees. Some of the islands are settled, some are not; and they are called the islands of Jubander. Between the islands lies a little river called the Rio de Lagos: it leads into the River of the Barbacins, near the palace of the King of Broçalo which is called Ganjal.
2. The Gambia River is settled throughout its length, on both banks, by Mandinga blacks. Each twenty leagues they have a king, who is subject to other rulers called *Farons*, this being a title among them which counts higher than that of king. Thus, the whole of this river is extensively settled with blacks and has many kings. The King of Broçalo, whom we discussed in Chapter 4, is the ruler of the north bank of the river for many leagues, and he has kings under him who obey

---

[1] Almada's account, surviving in several copies and versions, was first published in 1733. The English text is from André Álvares de Almada, *Tratado breve dos Rios de Guiné (c. 1594)*, trans. and ed., P. E. H. Hair, *Brief Treatise on the Rivers of Guinea (c. 1594)*, additional notes by Jean Boulègue, 2 vols, Department of History, University of Liverpool, 1984. The translation is of a unpublished variorum text organized by the late Avelino Teixeira da Mota: for published texts, see André Álvares d'Almada, *Tratado breve dos Rios de Guiné*, ed. Luís Silveira, Lisbon, 1946; André Álvares de Almada, *Tratado breve dos Rios de Guiné do Cabo Verde*, ed. António Brásio, Lisbon, 1964. A small part of Almada's account was summarized in Fernão Guerreiro, *Relação anual das coisas que fizeram os Padres da Companhia de Jesus nas partes da Índia Oriental, e no Brasil, Angola, Cabo-Verde, Guiné, nos anos de seiscentos e dois e seiscentos e três...* Evora/Lisbon, 1605, reprinted in modernized orthography, ed. A. Viegas, Coimbra, 1930, liv. 4, chap. 9, ff. 131^v–137^v. For Almada, see A. Teixeira da Mota, *Dois escritores quinhentistas de Cabo Verde: André Álvares de Almada e André Dornelas*, série separatas 61, Agrupamento de Estudos de Cartografia Antiga, Junta de Investigações do Ultramar, Lisbon, 1971 (reprinted from *Liga dos Amigos de Cabo Verde - Boletim cultural*, suplemento, November 1970, pp. 10–44). The chapter headings below are from the original text, but the numbering of paragraphs is from the Liverpool translation.

him and pay tribute. It is true that sometimes they revolt when one king dies and another succeeds by forcefully imposing himself on the kingdom, but as the monarchy is powerful the king soon reduces them to subjection again.

3. This river, as well as being itself very beautiful and very large,[1] contains many beautiful tree-covered islands, some of them two leagues long and more than one league across. On these islands there are many game birds, that is, royal herons and common herons, doves and pigeons, and especially large numbers of cranes, flamingoes – a bird the size of a crane – ,teal, and large black ducks – the ones that have spurs at the tips of their wings. And there are many animals to hunt, gazelle, deer, and many other sorts.

4. The river brings down fresh water. In summer, fresh water is found to within thirty leagues from the sea, and in winter when the river is in spate to within six or seven leagues. It is a river which possesses a large trade in slaves, in black and white cotton cloths, in raw cotton, and in wax – although no hives are built,[2] the bees are so numerous, and the forest so great, that honey and wax are plentiful. Much ivory is obtained there, more than in any other river of Guinea. When travelling by boat on the river it often happens that one sees elephant on the land in herds, like cattle, and ships often meet herds crossing the river from one side to the other. From the river one also sees herds of buffalo, gazelle, and another animal called in the language of the blacks *dacoi*, the size of a buffalo, which they say is the true *anta*.

5. The whole land furnishes foodstuffs in abundance, rice, the *milho* called 'maçaroca', and other ground crops. Most of the settlements are near the river, because of the trade with our people, and there are many well-built villages along it, containing large numbers of houses and inhabitants. Some of the houses are of clay and wattle, round in shape, and covered with straw or palm-thatch; others are of straw, not clay, but are built the same round shape.

6. These blacks are very war-like, and in this land there are more weapons than in any other land in Guinea. The reason for this is that, as they have iron here which they smelt, they make spears, darts, knives and arrows in quantity. The poison used by the Mandingas is more venomous than any other (known in Guinea). (This we know from what) we saw at the port of Caçan. At nine or ten in the morning, the blacks and our men had a serious affray which left dead on each side. At vespers, when the blacks had withdrawn and our people sought to bury the dead, those who had been struck by poisoned arrows could not be carried away from where they lay dead, because the poison was so effective that already their bodies were decaying, to the extent that if they were lifted up by an arm it fell off from the body, and the same thing happened if a leg was lifted. All that could be done was to dig graves where the corpses lay and push them in. Such is (the power of) the poison used by the blacks. They are mostly a treacherous lot. All those on the south side of the river are bad: they take a delight in killing whites and seizing ships, which they have done on several occasions. One can only go there in a strong ship carrying a stout crew

---

[1] In an earlier detached passage, Almada spoke of the river as follows: 'River Gambia has very heavy winter rains, with great storms and hail showers, and (therefore) is very beautiful, (its banks) being covered with many trees ...' (Almada, *Tratado breve*, ch. 1/24.

[2] Incorrect. Hives were constructed from basketwork and hung in trees; see Fernandes, f.113^v, above.

well-armed, and it is necessary at all times to keep a careful watch on the blacks, since they never behave other than treacherously. Along the river and its creeks are certain military fortifications which the blacks call *sāosans*.¹ These are made of very strong wooden stakes, their pointed ends embedded (in the ground), and a rampart of earth behind. Each has its guard towers, bastions, and parades, from which they fight by shooting arrows. They also make a kind of pitch with tar which they heat up in vessels. And when the enemy attack they hurl these vessels at them to make them withdraw. As stated, they make their fortresses along the river and its creeks; (this is) because of (the supply) of water and because they have boats in which they attack other places. Hence, when they are at war, they rob those who pass by.

7. Along the river on each bank there are many villages of Fulos, who live in these parts after having left their own lands in search of the pasturage and water which they need for their animals. Hence, the district has large numbers of cattle. Along the river are very beautiful meadows, which they call *lalas*, in which many kinds of game are to be found at all times, both beasts and birds. The river has excellent fishing, and very fine plaice can sometimes be taken. It contains large crocodiles, which often seize men and cows, and carry them off to eat in their lairs. But crocodiles are so made that in the deep of the river they are unable to seize or harm any creature. (They are dangerous) only in places where they can strike their tail into the ground, for if they cannot do this they lack the power to do anything. There is no risk except along the shore, where the river has little depth. So many crocodiles live in this river and do so much damage, that the blacks in the settlements they inhabit have the practice of building within the river a fenced enclosure, which acts as a rampart. Within this, they can water their stock in safety and wash and draw water, where otherwise they would run great risk (of being seized).

8. Up this river, on one side or the other, are many kings, twenty leagues or less separating one from another; but there are other kings with large territories, and (even) emperors among them called *Farin*.

9. The clothes they wear, the arms they carry, and the oaths they take are like those of the Jalofos and Barbacins. The slaves which they own and sell are enslaved either in war, or by the courts, or else by being kidnapped, for they go about robbing one place or another, being great thieves. (As a result) they sell large numbers of slaves. (But Christians are) forbidden to buy from them black slaves which have been kidnapped. It has been known here in Guinea, (especially) in this river and in the Rio Grande, for the blacks to bring certain slaves to sell to our people, and when our people refused to buy them, because this is forbidden, the blacks who brought them and offered them for sale killed them on the spot, so that (their kidnapping of them) would not be discovered. I am not sure that it would not have been better to have bought them, since this would have meant that they received baptism and became Christians. (However) I do not meddle further in this business, for it involves points (of moral law) which I am not competent to determine.²

¹ The variant *cāosans* in one MS. But the Mandinka is *sansang* 'fortification, fortified place'.
² This passage probably inspired Father Barreira S.J. to write a report on local enslavement in 1606 (Hair, *Jesuit Documents*, item 16; P. E. H. Hair, 'A Jesuit Document on African Enslavement', *Slavery and Abolition*, 19, 1998, pp. 118–27).

10. To return to the river. In this district there are more of the devout *bixirins* than in all the rest of Guinea, because there are many establishments of this religion throughout the district, and many pilgrims who go from kingdom to kingdom. On the north side of the river, there are three large major establishments, corresponding to monasteries with us, which arouse great religious feeling and devotion in the blacks, and in which these 'monks' live, and (also) those who are studying to become *bixirins*. The first establishment is at the mouth of the river and is greatly venerated by them, because from it the open sea can be seen, which they say is an extraordinary sight. The second is located 70 leagues further up the river, in a stretch where it narrows and forms three channels, which (later) meet again, making some of the land at this point an island. The stretch of river where the monastery is found is called Malor. The third establishment – which is also the object of great religious devotion – is located 50 leagues away from the second and 120 from the first, in a town one league inland (from the river), called Sutuco. The chief of these 'monks', a rank like that of Abbot or Provincial with us, they call *Alemame*, and he wears a ring like a bishop. All three establishments are on the north side of the river. The *bixirins* write in bound books, which they make themselves, as already stated. In these they tell many lies, and the devil gives other blacks ears to hear and believe the lies. These heathen priests go about looking thin and worn out by their abstinences, their fasts and their dieting, since they will not eat flesh of a creature killed by any person who is not one of them. They wear long clothes, and over these capes and tippets of baize or leather, with large black and white hats, which are brought them by our traders. They make their ritual prayers with their faces turned towards the east, and before doing this, first wash their nether parts and then their face.[1] They recite their prayers all together, in a high voice noisily, like a group of clerics in choir, and at the end they finish with '*Ala Arabi*' and '*Ala Mimi*'. Both those who live in the establishments and those who live outside have wives whom they keep with them.

11. The imported goods which the people of this river value most are as follows: wine – they would die for it, and they call it *dolo* – ,horses, white cloth from India, Indian beads as on the coast, Venetian beads, pearls large or small, small Venetian beads, red thread, red cloth, 24–weave cloth, scarlet fabric, cowries, paper, nails, copper bracelets, barber's basins, copper cauldrons weighing one or two pounds, and copper scrap. But of all the imported goods the most esteemed is cola, a fruit produced in Serra Leoa and the neighbouring district, and worth so much in this river that they would give anything in exchange for it, foodstuffs, cloth, slaves or gold. And it is so valuable that the blacks carry it as far as the Kingdom of the Grand Fulo, where it is worth a great deal, and also into the other rivers of our Guinea.

12. In this river, up-stream 120 leagues, on the north side, in the port called Jagrançura of the town called Sutuco, there is trading in gold, which is brought here in caravans by certain Mandinga merchants, who are also *bixirins* and make their prayers as others do. The gold they bring here comes mostly in the form of powder, with some in coins, and is very fine quality. The merchants are very expert with

---

[1] Cf. *GT* 69–70.

weights as they are in other points (of their trade). They carry accurate scales, the arms of which have silver inlay and the cords are of twisted silk. They carry little writing cases of unpolished leather without fasteners, and in the drawers they carry the weights, which are of brass, and are shaped like dice. The scales carry a larger brass weight of one pound, shaped like the pommel of a sword. The gold they transport in laces, in scraps of cloth, in the quills of large birds, and in the hollow bones of cats, which they hide in their clothing. They carry it this way because they go through many kingdoms and spend many days on the road, and are often robbed, despite the fact that the caravans take with them officers and guards. Depending on its worth, a caravan may have one thousand archers, or more, or less. Copper bracelets are the merchandise they chiefly buy with the gold. It seems to me that (our) trading in these bracelets brings (us) no profit, or if there is any profit it cannot be much, since one pound of gold buys 1440 bracelets. (However,) there is much profit with other goods, *cano de pata* – an elongated precious stone which comes from India – *,brandil*, paper, and all the other goods mentioned earlier, except horses and wine – for these merchants do not drink wine. And also barber's basins, and small kettles of one pound in weight and more.

13. I myself took part in this trade in 1578.[1] Because some people wondered whether the merchants had come by order of the Turk, to obtain copper to be made into guns,[2] I carefully inquired of the gold merchants where they were going to get the gold and why the people there wanted bracelets.

14. Thus I learned with certainty that the bracelets are used only as ornaments and adornment by the people (of the gold region). They wear them on their arms and legs, and value them as much as, and even more than, we value bracelets of gold. They do not use gold because they value it little, having so much of it in their lands. Without exception, (all) this gold and the gold which comes to Tumbocutum comes from the hills of Sofala.[3] For when I spoke to Anhadalen, the leader of the caravan, and asked him exactly where he was going and where he was taking the bracelets, he told me it was to the Cafres (Kaffirs), using the actual term. When I asked him why they wanted them, he told me it was to wear on their arms and legs. When I asked them [*sic*] how much they gave him for each bracelet, he replied that he would not tell me that, since they were not such dull-witted merchants as to fail to make high profits on goods which they carried so far, for they spend many days on the road and pass through many lands, at great risk to their persons. And they bring the same gold to the Kingdom of Galalho, called by us Gago, and to the Grand Fulo. What makes me testify more strongly that they want the bracelets only as ornaments to wear is this. About a thousand of the bracelets I took were broken into pieces, and I asked the captain of the caravan guard if he would buy them, and he told me that they were useless. When I said that I would give him two broken ones instead of one good one, he replied that even if I gave him ten for one he would not take them, because they would be of no use, they would only take whole ones capable of being worn. Hence the suspicions I had entertained disappeared.

[1] One MS has 1579.
[2] Presumably brass cannon.
[3] Cf. Andrade, above.

15. These merchants take over six months on their journey. But as they are blacks and lacking in energy it is surprising that they do not take much more time. They follow a route which fringes (the lands of) all the blacks of our Guinea, on the interior side, and they go (this way) by order of a black emperor whom all the Guinea blacks we have discussed are subject to, called Mandimansa, whom none of our people has ever seen. As soon as his name is mentioned, all the blacks who hear it immediately uncover their heads, such is his authority. The Mina people call this king the Great Elephant, and he is so well known that all the blacks respect his name for more than 300 leagues around.

16. On the occasion mentioned I had to leave the trading place (without obtaining) five quarters and eight pounds of gold which had come in the caravan, because I had no (more) goods to exchange. Today this trade is lost, because no ship has gone there for eight years; the merchants, seeing that there is no trade for them, must have joined those (trading) at Tumbocutum. Some Moors come to this trading place and bring gold, exchanging it for glazed earthenware bowls, red cloth, and a few coins, if they find any of these things there. The clothes of these merchants are the same kind as those of the Mandingas; the clothes of the guards who come with them are different, being large tunics and baggy trousers whose width continues to more than a hand-span below the knees, then they narrow like boots and cover the whole leg. They fix many feathers on their tunics, and in the caps they wear. They carry short swords like the other blacks, and two knives, one in the belt and the other attached to the upper left arm. The arrows they carry are short and the bows small. They say that they prefer these because (the arrows) are of no use to their enemies who have large bows, while the arrows of their enemies are of use to them. Although their bows are small, they shoot arrows accurately. They also carry spears and very strong shields made of poles and reeds.[1]

Chapter 6. Which discusses the other features of the Gambia River.

1. At the entrance to this river, the land on each side is flat but completely covered with a thick forest of mangroves, trees so tall and so thick that if their wood were not so heavy they could provide masts for ships of large tonnage. And there are other trees which have a very good and hard wood, in colour apricot or red, called there charcoal-wood.[2] The mangroves extend (inland) to the tidal limit of salt water, and there they stop. Meadows called *lalas* are then revealed. The most beautiful are on the north side, where fine fields of sugar-cane could be established, and these might be watered by the river itself rising and falling, although (in fact) there

---

[1] In one MS the concluding passage reads – 'the clothes of these merchants are long shirts which go below the knee and are cut away at the neck, with sleeves to the elbows, and long trousers, the bottoms hanging about a hand-span below the knee. And they wear on their heads very tall red caps called *turquimas*, seemingly because they obtain them from the Turks, but some wear black hats. They drink no wine, and they eat raisins, dried figs, and all kinds of candied fruit. The soldiers who come with these merchants as their guards wear the same clothes, with many feathers on their shirts and caps. They carry bows and quivers of arrows, the arrows short and the bows small, but the bows shoot well.'

[2] Mandinka *kembo*.

is no shortage of water here since it rains a great deal. At the point where the meadows are revealed, at a place called Balangar, some rising ground emerges which continues up-stream alongside the meadows and acts as a wall around them. This higher ground extends more than 100 leagues up-stream, and the further it goes the higher it gets. It stands less than one quarter of a league from the river. All this is on the north side. On the south side there are some rounded hills, but they are not continuous, nor do they border the river as those on the north.

2. The river is navigable for more than 160 leagues, and what stops navigation further is a narrow stretch with a rock over which the water falls from a height. The blacks say that if a boat was built beyond this, it might well be able to proceed up-stream many leagues further.[1] The rising and falling tide reaches as far as the foot of the rock over which the water falls, and when it is high tide at the bar of the river it is low tide in all the upper part. From the land, the flowing of the tide in and out can barely be observed; it can only be detected during the period of time when the ship turns (on its anchor). The tide rises so high with the rains and the water coming down from the hills that ships cannot stay at the Trading Place for Gold between the middle of June and December.

3. Iron can be obtained in this river: (the ore) is dug locally. The blacks smelt it, and make bars one hand-span long, and three fingers wide at one end and two fingers wide at the other. Our people trade for this iron and bring it to the Rio Grande and the Rio de São Domingos. There is (also) silver here, seemingly of good quality, and the blacks make bracelets and rings. But our silversmiths cannot make good articles out of it, because they say that it breaks, as if it contained some other metal.

4. I must not fail to report something I saw in this river, at a place called Fulos' Pass. The river is a very large one, very fast flowing, very deep and very wide. There (once) came here, in the course of war, an army of Fulos which had already reduced to subjection the Mandingas in every district it came through. It was so great that it covered all the *lalas*. The army decided to cross to the other bank of the river, but had no boats to do this. Though the river is a league or more across, the Fulos flung in stones (and made a ford), so that the whole army could pass over. Many assert that so numerous was the army that it was only necessary for each soldier to bring one stone. Be that as it may, they filled up the river, and the whole army went across with its baggage, which was very considerable, since they brought with them many horsemen, and (also) donkeys, and herds of cows, which marched with them. (In battle,) the archers took up their position among the cows, and shot arrows from there. When the Fulos wanted the cows to halt, they spoke to them in their language and they stopped, and when they wanted them to resume the march, they spoke (again) and they set off. The Fulos carried swarms of bees, which they launched against their enemies when the wind was blowing towards them. The army was a terrifying one. Never had a military force on such a scale been seen by these nations. Destroying and ravaging all, it passed through the lands of the Mandingas, the Cassangas, the Banhuns and the Buramos, for more than 100 leagues. It

---

[1] One MS has – 'From the port of the trading place for gold a ship can go up-stream for four tides, but cannot pass further because of a rock...'. Probably the 'rock' represents the Baarakunda Falls.

crossed everything in its path until it reached the Rio Grande, the country of the Beafares, where the Fulos were defeated and put to rout. This must have been eighty or ninety years ago.[1]

5. Later on, the flood waters came down from the hills, and the river broke through (the ford) on the south side near the land, and made a channel there, along which ships can go if they keep close to the land, so close that they touch the trees with their yards. But the rest (of the ford), though covered with water, remains a shallow. The place is called Fulos' Pass; it is twelve leagues above Lame, and I have been through there twice.

6. I saw another (strange) thing in this river, which again I must not fail to repeat. (Indeed,) if necessary, it could be sworn to on affidavit, by those who were with me on the voyage. Going up-stream from Fulos' Pass, there came to our notice troops of monkeys the size of hares, of a reddish colour, more red than orange; and in each troop there was one monkey who rode on the back of another, like a man on a horse, and those who rode were neither the largest nor the smallest of the monkeys. The blacks of this country said that the one who rode was the king or captain of that group. The blacks spoke to them in the language of the country, and they replied loudly in a grotesque voice, as if they were speaking words. In all the troops we met, there was always one monkey riding in the manner described.[2]

7. In this river there are many hippopotami, who emerge from the river to graze on the land. These 'water horses', like land horses, are of all colours. Their colours, their neighing, the shape of their ears, these make us speak of them as 'horses'. But the shape of their body is that of an ox, and the body is bigger than that of a horse. Their legs are short, so short that the blacks make very low fences in their rice fields, to stop them eating and destroying (the rice), and the animals cannot go over them because of their short legs. They have hooves which are split and divided into two like those of oxen, and a short head with long teeth, a hand-span in length, more or less, and bent back. They say that these teeth are a cure for piles. (But) many say that the hooves of the animal are more effective than the teeth in curing this disease, and that it must be the hooves on the left side. The blacks kill many of these 'horses' in their rice fields, and they eat the flesh. They kill them for two reasons, because they eat their rice, and as food. The meat is white: I have seen it eaten, but never eaten it myself. The animals give birth in the river, under the water, and small boats, such as launches and canoes, run a risk when near a female with young, because she will attack them and sometimes break open the boat. As soon as the young can walk, they come out (of the river) to graze on land with their mothers.[3]

8. This Kingdom of the Mandingas is very large, since it extends up the Gambia River, which is navigable one hundred and fifty leagues; and then goes on further and penetrates into the interior until it meets the Jalofos on the north, the Fulos on the NE, the Beafares on the ENE, inland, and the Cassangas and Banhuns on the east. At the head of the river can be found two *Farins*, who are emperors among the

---

[1] One MS says 120 years.
[2] These baboons were probably seen at the locality now known as Monkey Court, on the north bank, just short of Karantabaa. Cf. *GT* 144–5.
[3] Cf. *GT* 20–22.

blacks; one on the north side where gold is traded called the Farim of Olimansa [*sc.* Wuli *mansa*], to whom we give gifts; and Farim Cabo on the east side, to whom we also give certain gifts. The present Farim Cabo is part Beafar and part Mandinga. The Mandingas stretch so far that they surround the Beafares on the interior side, as we shall explain later. The weighing officers, whose job it is to weigh the gold when people come there to trade, have to live on one the north side and one on the south, and to both of them we give gifts.

9.   The chief trade among the blacks, one we have not previously discussed, is the trade in cola, a fruit which grows on a tree; and there are only cola trees within the territory of Serra Leoa. The cola grows in a bristly pod like a chestnut, and is so highly valued among the blacks that all those in other parts want it and buy it, and it is carried as far as to the Moors. For cola, the blacks give all the kinds of goods they have in the Gambia, that is, slaves, black and white cotton cloth, gold, foodstuffs, and anything else that is asked for it. Cola is worth more in River Gambia than in any other river of Guinea. The blacks make use of it in the same way as betel leaves are used in our Indies. A black will travel around all day chewing a cola nut, which is like a chestnut. They suck the juice and believe that it serves as a medicine for the liver and the bladder. We also employ it for the same purpose, but the blacks make more use of it than we do. If they have a headache, they chew it and rub their forehead with the paste. Cola keeps from one year to another, and even longer if required. It is wrapped in the large leaves of a tree called *cabopa*, and placed in long baskets called *colecas*, which carry two thousand nuts each or a little less. It has pleased God that there should be none of this fruit in Guinea except in the territory of Serra Leoa, as said above, in order that it should gain the value that it has for the benefit of many. If sown in other parts, though it grows into trees they never flourish. In the Rio Grande there is one cola tree, and in the Rio de São Domingos another, and these produce cola, but in the rest of Guinea there are none, as we have said; and these trees do not even produce enough cola for the villages where they grow, because the blacks are always eating cola.[1]

10.   For 70 leagues up-stream from the entrance of this river, the inhabitants have very large canoes in which they sometimes go to war, such large ones that they have attacked French launches and captured them. At the prow they have thick wooden screens which keep off musket balls; hence they can assault boats, and they have captured some of our ships. After 70 leagues up-stream, canoes are less in number, and the higher up one goes the less there are. This is for two reasons, because the blacks are not good sailors, and because they lack beaching-places for their canoes, since the land on each side of the river is high here and there are no beaches to land the canoes on. If they had canoes they would do much harm to (passing) boats, since all the blacks on the south and east side are evil and treacherous, as already mentioned. The river has three fords or 'passes'. Going up-stream, the first is at Malor, the second is Fulos' Pass – described above – ,the third is at Janguemangue, near the trading place for gold.

11.   The Mandingas make large quantities of salt, which they bring to the upper part of this river to sell. Here it is very valuable, since there is no (locally made) salt

---

[1] Cf. *GT* 134–5.

above 60 leagues up-river: salt can be made only as far up the river as the sea water reaches. The Mandingas carry the salt up-stream in canoes. The depot for salt is at a village called Oulaoula, one league from the port of Casan. This is where the blacks store it, before sending it through the interior as far as the Grand Fulo, and by sea as far as it can go.

12. The tide goes 150 leagues up the river, and a ship (at anchor) swings round with the rising and the falling tide. When the rains come, at the end of June, ships cannot stay in the port of the gold trade: the water rises so high that they are unable to lie there, because their ropes cannot hold them. Sixty leagues on [down-stream], at the port of Casan, such is the force of the flood water that in August ships no longer turn with the incoming tide.

13. 'Winter' begins in these parts at the end of April or beginning of May. The blacks work in their rice fields from May on. The rice fields remain under water for more than three months, since the rising of the river floods all the *lalas* between June and November. From the flooded *lalas* the blacks recover their rice plants, and transplant them into drier *lalas*, where they soon give their crop.

14. There is much timber in this river and many boats could be built, oared vessels as well as decked ships. On Cabopa Island there is much timber, (especially) some very large spars which are used for masts of ships of large tonnage, or are sawed into planks. There are many wild fruits, senna and tamarinds being plentiful: the pulp of the latter is sold in large balls.

15. All these Mandinga blacks, together with the Jalofos, Barbacins and Fulos, wear white striped caps, shaped like crowns; and when they speak to each other they take them off as we do.

16. Above the port of a *Faran* called Jaroale, on the south side, about ten leagues up-stream from the bar, ships of up to sixty tons can go along a creek called Bambaro. This penetrates to the land of the Banhuns, where much cotton, wax and ivory is traded.[1]

17. The river we are discussing used to be the best in Guinea, with more trade than all the others. With five or six different types of goods one could buy a slave who could not have been bought for five *cruzados* of good money. Today all is changed and prices are high, due to whites who have spoilt and corrupted everything. There is no village on the coast or for many leagues into the interior which lacks white adventurers (resident there), and acting as trading agents for the English and the French. This has reached the point where the blacks no longer respect the whites, saying that they are as persistent as flies in milk: even if one falls in and dies, this does not stop more coming. For it has happened that the blacks have killed whites in this river, yet it has not scared off the other whites. Before these (intruders) came, the river was more peaceful than it is today, and the blacks used to come aboard ship to sell cloths and food.

18. There is good senna in this river, and with the root they make a medicine. But they have the superstitious belief that if they eat the senna itself, their mothers will die.

---

[1] One MS has – 'Eight or ten leagues from the bar, above the port of a *Faran* called Jaroale, an arm of the river called Heretics' River penetrates to the land of the Banhuns...'.

19. This river is somewhat unhealthy, because it is completely surrounded by high trees: these prevent it being swept by winds and therefore it has many gnats [sc. mosquitoes] and flies. The healthiest place is the port of Casan, up-stream 60 leagues, because the land is open around the river: the wind is free to blow and the port has pleasant breezes, and for this reason it is healthy. This town and port is the chief trading place on the river.

Chapter 7: Which discusses ... the blacks who live to the South of Cape St Mary

1. Cape St Mary is encountered as one comes out of River Gambia ...
2. To the south of this cape the Blacks are still Mandingas. The district is called Combomansa, and there is trade in rice and wax. ...

ANDRÉ DONELHA (1625)

André Donelha, a Cape Verde Islands trader, is known only through the account of western Guinea he finalized in 1625, an account which is, therefore, in this respect, almost contemporary with Jobson's writings. However, Donelha's long and interesting account of River Gambia centres around a visit he made to 'Casan'/Kasang, which he dates to 1585. Within his general account he draws on information from fellow members of the Cape Verde Islands trading community, and possibly this influences the section on River Gambia, but his personal observations of the coast appear to fall within the period between the 1570s and the 1600s. It would seem, therefore, that he describes Portuguese activities in the river a generation before Jobson's visit. Donelha's account was published only in 1977.[1]

/f.15/ ... In our Guinea there are four chief Farims, for though the kings who are kings over kings take the title of Faran, they are not as powerful (as Farims), and they each obey one of the Farims, as follows.[2] Farim Cabo, who is the most northern of them all, (rules) in the interior of the Gambia River one hundred and twenty leagues from the sea, where the port of Cantor is, and where there is great trade in gold, wax and ivory. Up to there and for twenty leagues further, large ships may ascend the river. Further on, certain shoals and rocks hinder navigation, and this place is called Fundo Feito (Bottom Struck). The Gambia River is very full-flowing, broad and roomy, and has great trade and riches. It is peopled on both banks, from

---

[1] André Donelha, *Descrição da Serra Leoa e dos Rios de Guine do Cabo Verde (1625)/An Account of Sierra Leone and the Rivers of Guinea of Cape Verde (1625)*, ed. Avelino Teixeira da Mota and P. E. H. Hair (Lisbon, 1977 – also available in a Portuguese/French version). The foliation supplied below is that of the original manuscript as given in this edition. The edition is particularly relevant to the present study because of its detailed annotation on the River Gambia, annotation supplied by the general editor, the late Avelino Teixeira da Mota, an authority on the earlier textual and cartographic sources of the region. This annotation is not repeated in the present edition.

[2] DPG doubts whether the distinction between *farim* and *faran* is other than a dialect difference. A supreme ruler or overlord would have been more likely to be entitled *mansa*.

the bar upwards, with Mandingas. The Farim is also a Mandinga, with his residence above Cantor; he is the lord of all the kings of the /f.16/ Mandingas, who are many, and of the Jalofos, (and lord of) Berbecim and various other kings to the north.

CHAPTER 10: About the great and beautiful Gambia River and the trading ports in this river.

The Gambia River is one of the three great rivers in our area, and I believe it to be the largest. In winter fresh water reaches to the bar, in summer salt water penetrates about six leagues. On both banks it is entirely peopled with Mandingas, under various kings, all themselves Mandinga; some kings of the north bank are subject to the kings of the Berbecins. /f.22/

To enter the Gambia when lying opposite to it, you must have sight of Cape St. Mary, which lies to the south. Coming to within little more than a league and a half of it, carry on and come in sight of the trees on the south side, since the land on the north does not show up. Find on the same south side two higher trees, called the Two Sisters, and steering by them you can reach a point a little more than half a league from the land. Going further in, the bar of the river will appear and the land on the north side. If you want to go further in, you must keep away from the land to the north, because a reef of rocks sticks out, and must stay in mid-stream or move more towards the south side, which is on the right hand entering, because it is clear and deep. Once past this reef, there is no further danger. The river is very wide and deep, and suitable for large ships.

On the north side, soon after entering comes the port of Barra and the kingdom of the first Mandinga king. Ships anchor one fifth of a league from land. Those going to Cantor pick up here two blacks as linguists or Interpreters, called here *chalonas*, and we call them pilots, to encourage them. These men behave very correctly in matters that concern them, since, if any accident occurs, the other Mandingas can imprison them; and even if they got away, the journey back (home) by land is very difficult, because of the various kings who are generally at war and because there are many creeks, forests and swamps which have to be passed. We always take two, so that if one happens to die, the other serves as witness and no blame falls on the captain of the ship. But in fact we treat such people very well and give them all they need until they return home. And if we care to do so, we leave one person with goods at the port of Barra to conduct trade (there).

Opposite this port, towards the south, is an island called Combo, /f.22ᵛ/ lying along the river and of white sand throughout. From one land to the other is little more than one league. This island produces much rice and is very beautiful.

Going on from this port of Barra, which is on the north side, up-stream on the same side is the port of Lagoa. Going up-stream we come upon a tree almost in the middle of the river, called the Ship Tree, because it looks like a ship under sail; it stands on a small island called St. Andrew's Island. Having passed this small island, there is a creek on the south side; about two leagues from its mouth is a town of a king called Jaroale. The creek and the town are called Berefete. This king Jaroale, because he is wicked, we call Red Cat. Ships anchor in the river, at the mouth of the creek, and fire off a cannon or musket; soon they come out from the land in canoes, and if they wish to carry out any trade, they stay two days.

Going higher up the river from this port of Berefete, on the south side is found a point of land called Julacote, where I have seen a group of nearly twenty elephants as we went by quite close to the land, for the river here is very deep. When Julacote Point is passed, going up-stream towards the east we come upon a rock almost in the middle of the river but closer towards the south side, which is why we always keep to the north bank; at high-tide the rock does not appear, at low-tide it can be seen above water, the size of a cask. Soon after passing this rock, up-stream on the south side is a town called Guian. This place is mostly on level ground, and on a slope leading to the upper part of a hill, and it is completely walled around with timber stakes, the war-canoes being kept afloat within the *tabanca* (*sc.* fortification), because of the attacks made upon this place by Jaroale or Red Cat, King of Berefete.

After passing this port of Guian, which also supports a king and a kingdom, up-stream the river bends imperceptibly to the right, towards the /f.23/ NE, up to the port of Jambor, thirty leagues from the bar. Jambor also has a king, and being on the north side it is subject to the kings of the Berbesins. In this port is a beautiful gourd-tree which is more than five arm-spans around at base. We anchor near the land, the river here being very deep, and when we jump ashore, to the right, up-stream, at little more than a stone's throw, is a creek covered with trees, straight and deep. If ships think that enemies are coming here, (the crews) guide them into the creek, holding back the branches of the trees with their hands, and they go up the creek almost as far the town, which is removed from the river half a league.

Going up-stream from this port of Jambor, towards the east, on one side and the other we come upon very high mangroves, high and thick and straight, which, if the wood was not so heavy, would make masts for large ships. ...[1] We travel on the south side, quite close to the mangroves, not because there is danger on the north, but in order to come to the /f.23ᵛ/ port of Jagra. Hence we make our way along the mangroves to locate a hippopotamus skull, placed on a mangrove root in earlier times, and as soon as we reach the skull we moor in the shade of the trees, which are well out in the river, beyond the sight of land. We fire off a musket, and presently canoes from the town of Jagra come out from between the roots of the mangroves; and *tangomaos* always come in the canoes to buy and sell, as happens in all the ports previously mentioned. After staying there two days, we leave behind someone from the ships with goods to trade, having paid the king the gift which is the normal due, a flagon of wine or its value.

We continue the voyage up-stream, (travelling) always on the flood tides, although the river runs fresh water, as already mentioned. Presently we come to a marshy island covered with very tall mangroves; here one can anchor at a hammer's throw from the island and pass to the north or south side. And here the river is very deep, more than twenty fathoms. This island is called the Island of Elephants. As soon as we pass this we come to the Devil's Turn, so called by the sailors because,

---

[1] Omission of a description of the mangrove.

going up-stream with the tide and the wind at the stern, in an instant the wind switches to the front of the ship, because the river twists here to the west. It is necessary to strike sails, and with the launch ahead, to be taken in tow, the prow to the west, for a little more than half a league; then the river turns to the east, as far as the port of Caur, which stands on the north side on the right bank of the river.

The town is in the interior about one league. It is also called the Port of Palm-trees, because there are two or three near the landing-place. This port is not covered over (with trees), the trees being small. It has a range of hills opposite, the first to be seen from the river, called the Red Range. When landing at this port, you jump from the launch on to dry land, because here the river is about twenty /f.24/ fathoms deep, but after jumping ashore if you want to go to the town, clothes are taken off and carried on the shoulders. Twenty steps from the river you begin to enter a swamp, called here *bulanha*,[1] an off-shoot from the same river, into which I have sunk up to the waist, but at each step it gets shallower. After travelling a sixth of a league we reach mud, and when we have travelled about a third of a league - this is the case in winter - we reach dry land, and here we put our shoes on and dress and make our way towards the town, which is a league from the river, as I have said. It is true that down-stream on the left side of the landing-place is a creek which goes in for half a league, but it is a bad place to land, being morass and mud. I have written all this above so that it may be seen how much trouble for what they gain it costs those who travel in Guinea. There is good trade here in cloths, cotton and wax.

Past the port of Caur, which lies some forty leagues from the sea, the river up-stream turns imperceptibly to the SE, as far as the port of Culenho, which lies on the south side. The ship anchors at the mouth of a creek which goes towards the south, and a musket or cannon is fired off; soon *tangomaos* in canoes come along the creek to trade. To go ashore, you get in the launch and go up the creek a quarter of a league, to land on dry land, but there is (still) a pool of two or three pikes' length to go through, with water up to the knee; from there to the town, which is another quarter of a league or more, the ground is flat and dry. In this town of Culenho lives the *falfa* who is the captain; two leagues from there lives the king. In (all) my life I have not seen so many crocodiles as I saw in this creek, very large and fat ones, some coming up out of the creek, others coming from the mud to put themselves in the creek since the land here is very marshy, a thing I have never seen elsewhere. /f.24ᵛ/ I found this town in alarm because a little time before they had seized a ship, but I calmed them with friendly greetings and gifts.

After this port of Culenho has been passed, the river widens for more than a league and a half. On the other side, to the north, is the port of Ola Ola. Soon we reach a beautiful and cheering island, called Pinto's Island; it seems that someone of this name was buried there. Ships pass on either side. The river is deep, more than fifteen fathoms deep. The island is flat; it has trees, but only in some parts. A ship going by on the north side will see the sails of another on the south, above the land. Its width may be half a league or a little more, its length two leagues. On this island there are some trees which throw their branches over the river, and ships

---

[1] A *bulanha* is actually a bund or water-terrace, here marking off rice fields and holding back water.

which wish to fit themselves with a mast attach their tackle to a branch and make a mast, for near the island the river (-bank) slopes down sharply, and ships can approach the island without danger and it is possible to jump to land from on board.

After Pinto's Island, which is uninhabited, the river up-stream is less wide, and a depth of more than twenty fathoms is found in the channel. On the north side the land is somewhat higher. On the same side we go by the foot of a high hill, standing alone; it is called Red Hill. On its slopes, though these are steep, it has small trees. At the top of the hill the Mandingas have an idol of Mohammed, where they go on pilgrimages to make '*imshallah*' to the cursed Mohammed, since there is no nation of blacks which has taken with more fervour to the religion of Mohammed than have the Mandingas.

Passing this Red Hill, we find an island, called Cabopa Island, because it has on it many *cabopa* trees, very straight ones, /f.25/ which are cut for masts of ships, and the leaves are gathered to wrap up cola. It also is uninhabited. Passing this island we reach the port of Casan, the chief port for our Gambia trade.

CHAPTER 11: About the Port of Casan and the Gambia River again.

The port of Casan is sixty leagues from the bar and the sea. It stands on the north side. I found seven ships trading in the port, and later there were nine of us. I found there many well-known *tangomaos* [sc. resident Afro-Portuguese]. I met a black Mandinga youth, by name Gaspar Vaz ...[1] It was no small advantage to me to meet him in River Gambia, because he was of service to me in everything, and what I bought was at the price current among the people themselves, very different from the price they charged the *tangomaos*. And he served me as interpreter and linguist, called there *chalona*.

The town of Casan stands a pistol-shot away from the port. The port is handy, we disembark on dry land, and there is a certain amount of sand. Near the port are some high trees, under which the ground is as bare as if it had been swept, and here the black women hold a market when ships are in port; they bring for sale rice, *milho* (millet), *cuscus* (sc. gruel of steamed millet), hens, eggs, milk, butter, country fruit and other things. Below these trees, boats are caulked, masts are shaped which are cut on Cabopa Island, and rigging and ropes for the ships are repaired.

The town is small, and built in a circle, with round houses of baked brick, white-washed with a white clay that resembles lime. There are some houses with an upper floor, such as the houses of the duke or *sandeguil* and most of them have benches of brick within, to sit on. All have doors, the wooden locks and keys being the first I had seen. The town is entirely surrounded with high timber stakes, called *tabanca*; outside this, a deep and wide ditch which runs around the whole town is full of water in winter-time. There are four bridges and four gates; the bridges being of

---

[1] The youth had been a slave with on Santiago Island and had trained as a tailor. Donelha describes a conversation.

palm-trees. At the east gate is an open space, with some high trees. In their shade is a paved /f.26/ square covered with thick mats. They make their '*imshallah*' in this paved area. Nearby, on the west side, there are bowls of water, in which they wash their feet and hands when they come to pray. One day, when the sandeguil or duke was in this area with many *bixirins*, who are priests, and many people, to pray and make '*imshallah*' to their cursed 'Mafoma', whom they do not agree is called this but rather Mohammed, I passed near them, without speaking to the duke or making any show of doing this. ...[1] /f.27/

Here at Casan excellent round shields are made, and if we order them are made in our style. They are covered with leather and painted. And the *tangomaos* have sheaths made for their swords and daggers, and have them decorated, since there are good craftsmen here. There is great trade in cloths, cotton, wax, ivory, gold and hides of /f.27ᵛ/ various animals, but all the gold, wax, ivory and hides they take to the Jalofo Coast, to the French, English and other nations. (These foreigners) even come up the Gambia to undertake this trade with the blacks, and they draw immense profit from this river. The king lives two leagues from here.

CHAPTER 12: Continuing the Gambia River up to Bottom Struck, and other matters.

When I was in this port of Casan, I had news that Christovão Caiado, who was at Cantor, sixty leagues up the river, one hundred and twenty leagues from the bar, in a fine trading ship out of Lisbon to trade for gold and other goods in that port, was dead. Since I had been sent from the trading station on this island as captain of another trading vessel, I took the largest launch there was at hand, (and set off) with my clerk and some good native assistants, together with muskets, powder and ammunition, and in the company of four canoes which had come from up-river to exchange gold, wax and other goods, for cola. We went up the river, which is very large and deep. I took a sounding-lead, and because we had to moor on the ebb-tides, kept towards the land to avoid the current in the river, which is strong, like (that of) water coming off a hill, but never found less depth than four fathoms. From Casan up-stream, the river turns imperceptibly to the SE, and from there to Cantor makes several twists, now to the east, now to the SE, always tending (however) to the SE.

We passed by Codichar and Lame, good trading ports. We reached Cantor in three days, since we sailed or rowed day and night with the tides. I found /f.28/ Christovão Caiado's ship, with its yards raised in order to go downstream. I made an inventory of the goods he had bought, gold, ivory and wax; and of the goods which remained to sell. I handed all of them over to the clerk of the ship, Izidro da Cruz. We came down-stream in five days, as we had good winds. The river is very beautiful, large and deep; and both banks are covered with trees, which go right down into the river. The ship had two decks, and behind its boarding-net were many fine bronze cannon. From Casan we travelled with four ships belonging to

---

[1] Omission of a description of a visit to the 'duke'.

tangomaos, which had already finished trading and were going to the Rio Grande; in their company the ship went to the Rio Grande to be handed over to Diogo Anriques, the factor, as ordered. ...[1]

The trade of Cantor is of great profit. The king has the title of Farim, which among these blacks means 'emperor' and 'overlord', but he is obedient to the great Mandimansa. At the period that I was there, which was in 1585, ships sailed up beyond Cantor twenty leagues, almost as far as Bottom Struck, but did not pass beyond because of shallows and rocks, which stopped passage further. But today, since the last years of governor Francisco Correa da Silva, whom God keep in His kingdom, ships sail more than one hundred leagues (further) up-stream.[2] And it is said that the river up-stream is very large, spacious and deep, and the further up-stream (one goes) /f.28ᵛ/ the more water (there is) and the deeper (it becomes) ; and I will presently explain why. It is said to be fertile and pleasing (country), with a large amount of trade in all things, so that today in Guinea they call this river the New Peru, as I have heard from several persons, because of the riches that they are drawing from it. The reason why this river has been closed for so many years I will now state. ...[3]

As I said, the Gambia River remained blocked from then on, and that place was named Bottom Struck. But in the last years of governor Francisco Correia da Silva, certain enemy boats entered the Gambia and made their way up-stream. The ships of the *tangomaos* which were at Casan, hearing news of this, took refuge up-stream at Cantor; there they heard that the bandits were coming up from Casan. The *tangomaos* ordered Bottom Struck to be sounded, and they found an open channel, and passed up-stream, until they received news that the pirates had gone. Since then the *tangomaos* make their way (through Bottom Struck). But they say that this voyage takes many months, however the profit is great. /f.29ᵛ/

CHAPTER 13: Giving the reason why the Gambia River draws more water up-stream, and about its source.

... It is well known that the best traders in Guinea are the Mandingas, especially the *bixirins*, who are the priests. These people, as much for the profit they draw as (because they desire) to spread the cursed sect of Mohammed among the uncivilised, make their way through all the hinterland of Guinea and to all the seaports. And hence at any port, from (those of) the Jalofos or of Rio São Domingos or of Rio Grande, up to Sierra Leone, one finds Mandinga *bixirins*. And what they bring to sell are fetishes in the form of ram's horns and amulets and sheets of paper with writing on them, which they sell as (religious) relics, and while they are selling all this stuff they are spreading the sect of Mohammed in many districts. And they

---

[1] Donelha describes the handing-over of the ship and goods.
[2] A doubtful, and at the very least exaggerated claim.
[3] Donelha tells how a Fula (Fulbe) army, allegedly *c*.1480, at one point blocked the river in order to pass over.

go on pilgrimage to Mecca House, and make their way over all the hinterland of Ethiopia.

...[1] And the reason that this river draws more water higher up - as those who go there say, 'the further up the river, the greater the flow of water' - is that up-stream all the waters (of the various named rivers) are joined (in a single stream). (Moreover), from Bottom Struck down-stream, apart from the islands in the middle of the river which I have spoken about, there are others, much bigger, surrounded by creeks, especially on the south side; and also /f.31/ there are many *bulanhas* (water-terraces) and pools, and the river spreads over low-lying and marshy land. This may not be the case up-stream from Bottom Struck. Also, it could be that they (the *tangomaos*) go up past the point where the Gambia River and the Rio Grande de Bonabo separate. It is certain that there is more water there, since the separation is of two rivers as full-flowing as those I have named. This is how it appears to me, lacking a better explanation.

FRANCISCO DE LEMOS COELHO (1669/1684)

Francisco de Lemos Coelho, known only through his two versions of an account of western Guinea, was apparently born in the Cape Verde Islands, from an early age lived on the mainland in various Portuguese settlements or trading stations, traded in the major trading districts between Senegal and Sierra Leone, latterly as a ship's captain and owner of a ship, and eventually retired to the islands, where he wrote his account. He seems to have been active as a trader between the later 1640s and the early 1660s, during which period he lived three 'or nearly four' years at Kasang in River Gambia, probably in the later 1650s, as well as visiting the river at other times, making 'great profit'. The section in his account on River Gambia is extensive but concentrates on navigational and commercial detail. It was written in, or at least completed in, 1669. Fifteen years later, in 1684, it was rewritten, but although in a few places updated (especially in relation to the increased English activity in the lower river), this was not done consistently, so that the later version is mainly an extension of the earlier version, based (it seems) on further recollections of the author, just possibly suggested to him by discussions with contemporaries among the Cape Verde Islands trading community. The conditions of navigation and commerce he describes appear to relate to the period of his personal knowledge, that is, mainly the 1650s and early 1660s, one generation later than Jobson's experience of the river. Coelho's account, which was only published in full in 1953, is relevant to Jobson's account in two respects. First, its navigational detail fills out Jobson's briefer account of the course of the lower and middle river. Secondly, it demonstrates that Portuguese (or perhaps, more correctly, Afro-Portuguese) trade in the river remained vigorous at least up to the mid seventeenth century, and that what Jobson stated and implied about Portuguese decline was premature and exaggerated.

The enlarged 1684 version is given, in translation, in full below. Where the two

---

[1] Omission of a long and fanciful account of the interior waterways, thought to be conjoined, based partly on information and misinformation supplied by a *bixirin*.

versions differ minimally, material appearing only in the later version is indicated in double round brackets (( )), while additional material in the earlier version is presented in italics. But where there is a substantial difference between the two, extending to whole sentences or paragraphs, the earlier version is given in a footnote, in italics. Editorial notes are in square brackets.[1]

CHAPTER 2: Description from Rio de Borçallo to River Gambia and of the whole of the latter.

1. After leaving the port of Borçalo you come to the Rio de Gambia, which does not have a difficult entrance, although it has shallows on the north and south sides. Hence, seek out the shallows to the north, and when you reach them, run along them until you find you have entered the river. If /f. 7ʸ/ it is necessary to tack, do not go beyond a depth of four fathoms in (the direction of) the southern shallows and into deeper water, because the deeper you go, the nearer you approach these shallows. This is the way to enter the river. Two leagues before you reach the port of the king of the Bar (Barra), you will find a river called Rio de Felam, whose people are subject to the king of the Bar, and their trade is fishing. After drying it in the sun they take the fish, which is mullet, and trade it up the whole Rio de Gambia, carrying it as far as Caçao, 60 leagues up-river. At this point there are so many canoes that it is said that they certainly number over one thousand. Rio de Felam is very difficult, having many sandbars, and it is impossible to take a ship there without a local pilot.
2. In this river lie two islands called Ilhas de Felam, and on these islands salt is produced in great quantity. It is obtained in almost the same way as in Portugal, although, since the salt pans are of clay, having been made by the blacks, the salt is darker, but in every other respect there is no difference. To this place all the settlers on the river owning ships come annually in order to load salt to carry up-river, the salt being very cheap. In the three years I was here, I sent my ship thrice, for a load, and I did good business with the salt. The caravans of the Jagancazes and Conjuros also come here for salt, travelling overland in search of it more than 300 leagues. This distance is no exaggeration, for they depart from their lands in November and

---

[1] [For the Portuguese texts of the two versions, see Damião Peres, ed. *Duas descrições seiscentistas da Guiné*, Lisbon, 1953, pp. 12–28, 110–14, 117–38. In the manuscript the texts are called, in translation, 'Description of the Coast of Guinea from Cape Verde to Sierra Leone with all the rivers sailed by the whites' (1669), 'Description of the Coast of Guinea and the situation of all the ports and rivers therein, and a Nautical Guide to enable all the rivers to be sailed'. The English translation given is that in P. E. H. Hair, ed. and trans., *Francisco de Lemos Coelho: Description of the Coast of Guinea (1684)*, Department of History, University of Liverpool, 1985, ch. 2, very occasionally revised; the introduction discusses the career of Francisco de Lemos Coelho. This translation is of a new transcription made by and for the late Avelino Teixeira da Mota and is occasionally different from the printed text. The translation of the 1664 version, as given in the printed text, is by PEHH (note that the accentuation shown in the printed text and here followed is erratic and inconsistent). The paragraph numbering of the 1684 version is that given in the Liverpool translation; the paragraph numbering of the 1669 version is based on the unnumbered paragraphing of the printed text, but (following a new transcription by Texeira da Mota) paras 6–7 represent para. 6 in print, paras 8–9 represent para. 7, and paras 16–17 represent para. 14, the later paragraphs being renumbered accordingly.]

return in July, spending all this time on the journey there and back, without stopping on the way.¹

3. After you leave this river, two leagues further on lies the port of the Bar (Barra), which is very easily recognized. The king's house lies along the waterside. The king is appointed by the king of Borçalo and is his tributary. Since we have reached the mouth of Rio de Gambia, it will be useful to state what (commodities) are necessary for (trade on) it. Because there are English now on the river, the best commodities (to buy there) are hides, wax and ivory; these can be sold to the foreigner in exchange for goods in demand on the river, iron, brandy, small black and white beads, red cloth, number 22 crystal (glass) /f.8/, and a large quantity of paper, for in one day twenty reams of paper can be sold. If you are coming from Cacheo, bring up to 50 barrels of cola, and some of the fine cloth (*roupa alta*) of Santiago Island; also some *pataca* coins, since at the very least they will give an eighth (⅛ oz.) of gold for one of these (?), and this will be the poorest return you can have for it. This trade is with the heathen. For the whites, (bring) all kinds of dress and food, and what they need for their ships, for not all of them visit the English lodge and they welcome having these goods brought to where they live, even though they are more expensive. Above all, the best commodity for the river from Caçam upwards is salt, and still more salt, for there is always too little available.

4. There is no trade in the port of the Bar, nor does anyone anchor there unless waiting for the tide in order to proceed to the port of Julufré. This is in the same kingdom and lies opposite the English fortress.

5. Facing the port of the Bar is the island of Bamgú, which has a very good harbour, although it contains a number of shoals. Some ships of ours once went there when they needed water, and it seemed deserted, and not finding (running) water, they made do with some waterholes (*casombus*). It should be noted that people do live on the island, and that they are (the dependants) of a Mandinga man to whom the king of Combo, the owner, gave the island. He lives less than a quarter of a league inland. There is very good water to be found not far from the sea, and much rice if one wanted to buy some, but no other trade.²

6. From the port of the Bar to the port of Julufré is five leagues, and just before the latter is the port of Aldabar. You can anchor in both ports, but the blacks are ill-disposed to whites, and there is no trade. In the port of Julufré there used to be a

---

¹ *Description of River Gambia.*
 1. *River Gambia is the most navigable river on the whole coast of Guinea. The English, Flemings, French, Courlanders, and Castilians have drawn from it, and still do draw from it, greater profits than from the whole of the rest of the Guinea coast. Their intermediaries are the Portuguese who live along the river and in the neighbouring districts, and through whose hands pass all the river provides, that is, ivory, wax, hides, blacks and a little gold, the chief goods for this trade being country cloths and salt. These goods are found in the river itself and do not need to be brought there, so that it is only necessary to import bars of Swedish iron, brandy, basins, glass, beads (avelloria de massa and larger ones), crystal glass (no. 30 to no. 18), paper, red cloth, tankards, Rouen cloth and all Indian cloths, muskets, gunpowder and machetes - all this for the blacks. And for the whites all that is required to clothe and feed them, and also to build ships. This is all that is necessary for this river, whose description follows.*

² 2. *On entering River Gambia, on the north side the port of the King of the Bar* (da Barra) *is immediately met. Here there is no trade because the ships go further to another port of the same country. On the south side lies the island of Banjú, which is populated, and provides a watering place and much husked rice, the island being fertile. Having passed the island, ...*

village of many whites with suitable houses, but today there are only locally-born Christians who live on crumbs from the foreigner.¹

7. Opposite the port of Julufré, in the middle of the river, which at this point must be more than a league wide, is an islet which the English keep well fortified [*sc.* James Island]. Although it is small, it is easily defended. It has a fortress of stone and mortar, with rooms inside for the general, stores for goods, and more than twenty pieces of artillery. Around the islet, between its headlands, the English have built, at water level, three platforms with four pieces of artillery on each; (the guns are) so level with the water that, when there is a wave, it washes their mouths. Ships from England come here /f.8ᵛ/ to unload the iron and other goods with which the fortress is normally well provided, and to load up with hides made ready in the river – cow-hides and hides from other animals such as *antas, sinsins, tancões* and *gimguisangas*, which are better and more valuable than cow-hides – and also with ivory and wax. In an average year, the English buy on the river and take away 50,000 hides and 1,500 quintals of wax and ivory. They also buy many blacks whom they ship to Barbados. To carry out this trade and the defence of the fortress they normally have 80–100 men there, soldiers, sailors, and traders who travel around in launches. They usually have two launches to convey the commodities purchased by the factors who are stationed in various ports of this river. They also buy some gold, even though the Portuguese do not do this for them - the Portuguese being the intermediaries through whose hands the greater part of the trade passes - because the Portuguese prefer to keep the coin with which the gold has to be bought and use it for other purchases. Thus, while the English buy in all about six pounds of gold each year, if it were bought for them it would be much more.²

8. On the other side of this islet, on the land to the south, and after passing Ilha do Banju or Bangu, lies the kingdom of Combo. The king is of the Falupo nation; and they say that the village in which he lives is the largest anywhere on the river. The land has much wax and rice and some blacks; the king and the people are heathen, without any kind of religion. On the same bank, further up and opposite the islet, is the port of Barefete *Beréfete*, which is under the control of the kingdom of Bintam. In this port can be found a village of whites, because the blacks are now Banhús, and although the Banhús are barbarians in religion, they are great friends of whites. They trade rice, many hides, many blacks and much wax.³

---

¹ *3. Opposite the Rio de Beréfete is the village of Juluferé belonging to the King of the Bar. It used be very great and many whites lived in it; today it has a few Christian 'sons of the land'. It trades in hides, blacks and some ivory. From Rio de Berefere to this port is a league and a half, the width of the river here.*

² *4. Midway between these two ports and in the middle of the river is an island, a small one (yet today extremely busy), the English having in it at present a fort with a trading house. On the island resides the governor or chief merchant with (a responsibility) for all the trading business that needs be undertaken in the river. To here come ships from England to unload, and to pick up the products of the river, the chief being cow hides, hides of wild animals (the best being those of* antas, sinsins, tancois *and* gimguicamgas*), ivory, and wax, which the blacks do not make use of themselves. Only a few [blacks] are bought, for Barbados. One year with another, 500 hides and 1,500 quintals of wax and ivory leave the river. The Portuguese are not concerned with gold, theirs being the hands through which pass all trading with the foreigner, because money means more to them since they have to use it to buy other goods. All told they buy annually 6 lb, but if interested they would have more.*

³ *2. ... Having passed the island [Banju], on the south side is Rio de Combo, a kingdom abounding in foodstuffs, many blacks [as slaves], hides, and wax. On the same south side, six leagues past this river, is Rio de Beréfete, a league up which is the village, which also has many foodstuffs of all kinds, and it belongs to the Kingdom of Fonhi. There is much trade in hides, wax, ivory, blacks and an abundance of meat. Whites live there.*

9. Five leagues further along the same bank lies Rio de Bintam. It has great sandbanks on the south side; so that to take a ship up it one must follow the central channel of the river, and edge more to the north than to the south bank. If you want to go into Rio de Bintam, line up the stern on a river called Rio de Joba, five leagues up from the port of Julufré, and direct the prow towards a cotton-tree (*poulão*) on the river called 'O Poulão de Solinto *Zolinto*' (Solinto's Cotton-Tree), and by using these landmarks you can make your way in. Once in, there is no difficulty in reaching the village, where there are white residents. Nowadays it is the best village on /f.9/ the river, having much trade in hides as well as in wax, ivory and blacks, and it is very well provided with all that is necessary for human life. ((For small purchases such as purchases of hens and foodstuffs, the commodity chiefly used is iron, cut up into little pieces. The majority of the people in the village are Mandingas, and they are Mahommedans, so that this evil seed begins at this point (on the river); but they are good friends of the whiteman. Anyone living there should, however, be on good terms with the king of the land, and should not seduce the black men's wives or slave-women, seduction of the latter being the worse offence, because they value them more highly; the lightest penalty will be loss of property. Use civil words to all the blacks, and do not be deceived into thinking that they know no Portuguese because they do not speak the language, since most of them can understand it adequately.))

10. If you wish to go further up Rio de Bintam, which is a branch of Rio de Gambia, six leagues from this port of Bintam is another kingdom of Banhús called the kingdom of Sangedegú, and in sight of the port is a village called the Village of the Heretics (*Aldea dos Herejes*). This used to be the village with most trade on Rio de Gambia; ((and even today the English have a commercial establishment there.)) Many hides, much wax and a number of blacks can be obtained, and some locally-born (*filhos da terra*) whites live there.[1] ((The heathen are savages, and the country is usually split between two warring factions, each wanting to have its own king; in consequence the roads across the country are not very safe. The kingdom contains, apart from the Banhús who would like to turn the land into two kingdoms, many Falupos who are (their) subjects. These Falupos are very wicked here, being highwaymen on the roads and great thieves, so that no-one is safe from them along the whole route. In spite of this, people travel over-land, and one can go from this village of the Heretics by land to Cacheo. I went this way three times because it is short. I made these journeys with files of blacks and plenty of merchandise, and nothing ever happened to me; except that one night, coming from Cacheo, these Falupos tried to rob me in the house in which I was lodging in the village of Boaguer, and actually stole from me 25 fine cloths (*panos altos*). Although it may be thought a digression, since I have mentioned this journey I shall describe how to

---

[1] 5. *On the north side, above Julúfere, is the Rio de Jobá, with about five leagues from one port to the other. Opposite this river, on the south side, is the Rio de Bintam. ... If you wish to go further up by this tributary of the river, six leagues from here is Porto dos Hereges, which belongs to the Kingdom of Sangue de Gú, a kingdom of the Banhús. This village used to have more trade than there was in the river, and even today it supplies large quantities of hides, wax and blacks. White 'sons of the country' live there. But with the number of them in decline, trade also has declined.*

carry it out, so that anyone who has to undertake it will be able to do so in the manner detailed below.))

[paragraphs 11–15 on a route from River Gambia to Cacheu omitted]

16. Returning to our Rio de Gambia and the port of Bintam and withdrawing into the main channel of the river, Rio de Joba is opposite (Rio de Bintam), on the north side. Six leagues further on is the port of Tagamdaba, on the same side; and another six leagues on is the port of Badibo, whose name the king of the land bears, calling himself the king of Badibo. All the land from Rio de Joba as far as here belongs to this king; and all these rivers can be entered in canoes to obtain salt, which is the chief trade of the land. The same is the case on the south side, which is the land of Rio de Bintam. Its first port is that of Tancoroale, where there is a fine village. The kingdom is called the kingdom of Quiam, and the king (the) Faram /f.10ᵛ/ of Quiam.

17. ((From here on, all the land on each side of the river belongs to the Mandingas, a nation which, coming from the land of Mandimança, and arriving as guests, has become naturalized here. Either that or the (original) inhabitants of the land, having adopted the religious rites of the Mandingas, have also taken their name, for now they are all Mohammedans, though with many corrupt beliefs, more even than the Jalofos. Among them are those learned in the (Islamic) Law, whom they call *bexerins*, and their doctors or 'bishops', whom they call *fodigés*. The *fodigés* can be distinguished from the rest, in that all *fodigés* wear capes over their shirts, and hats with strings: but this garb is often very ancient. These [Islamic clerics] are great beggars for alms, and they have a rule that, while they are training youths to become *bexerins*, they should be supported with alms. There is no *bexerim* who is not accompanied by ten or twelve youths whom he is teaching to read and write. They write on, and read from, wooden tablets only; and they are taught at night, by the light of fires, (the pupils) reciting in a very loud voice. By day they make intercessions to God, without excepting any individual and they take no offense even though someone may have spoken against them or done them ill. In the village of Cação, where I was then living, I saw a *fodigé* bless water in his fashion and sprinkle all the people in the village, which action they received as a holy deed. The *fodigé* performed this as he was entering the village. Like the men, the women are circumcised. This is done by cutting off a small piece of flesh which they have in the middle of their shameful parts; and what is cut off they put to dry in the sun, and afterwards string them together, the chaplet produced being held in great veneration. Women make their *salas* (prayers) too, just like men, making them in any open place near their houses. Some Mandingas observe the (Islamic) Law more than others, but not even for this default are they reprehended. In everything else both sexes are very corrupt and given to lasciviousness.))[1]

---

[1] *30. Other than the kingdom of Combo, inhabited by Falupos who keep the rituals of their own nation, the remainder of the river on each side is peopled by Mandingas, all Muslims although their beliefs stray far from that doctrine. Among them is a sect or [religious] order called bexerins. These are individuals literate in the Law. All read and write Arabic, although with many errors. They boast of being great diviners and sorcerers, and the blacks greatly fear them. Certain of them have a higher dignity, like doctors or bishops with us. These are called* fodigués, *and they consider it glorious to observe continence to the extent that they even have only three wives, and a female slave who is also their wife, whom they*

18. In this village of Tancoroale, which lies along the waterside, whites always used to live, but today there are none there other than 'sons of the land', and there is not much trade, other than that in salt and a few hides. In this land of Faram de Quiam, there are many ports, such as Jasabo, Tubabó-colom, and Sitato. All these ports have rivers that can be entered in canoes, in order to buy salt, just as on the northern side, for salt is only found in this district, and it is black like the salt on the other side.

19. When travelling by ship in these parts, you should always approach on the northern side, because the southern side has many shoals of mud and some sandbanks. /f.11/[1]

20. From Tagamdabá, which we also know by the name of Cabaçeira (Calabash-tree), because there is a large calabash-tree in the port, it is eight leagues to the port of Jagra, a port it is extremely difficult to recognize. On the southern side there is a very small river emerging among thick mangroves; this river takes the name of the kingdom. ((You go up it in a boat or canoe – since ships cannot enter it – for some two leagues to the port, and from there to the village is a short distance. Whites live there.)) It is very prosperous and well-supplied, and has plenty of trade, chiefly in the form of many blacks, husked rice, some wax, hides and ivory. ((The blacks are Mandingas, like all the other blacks on this river, and they observe the same (Islamic) rites.))[2]

21. From Jagra to Degumasamsam, a port of the same kingdom (of Jagra), is five leagues. Between the two ports is an islet called Elephant Isle, which has mangroves everywhere, most of it being a swamp. Three leagues long, it has little width, and elephants are found only in its name. Those going to Degumasamsam must pass it on the southern side, and then there is no danger whatever. ((After passing this island and skirting the land of Jagra, they will come to the mouth of a fairly large river which they should enter and proceed along for a league until they reach the village, which is situated in the port itself. The port has a very good anchorage and landing-place. Whites once lived in this village.)) The trade here is the same as in Jagra, since it is all the same land and belongs to the same king.[3]

---

call tala, and she is the most esteemed. These people are much venerated by all the heathen. This is how they can be recognized; they wear a hat with cord-like straps, and a cape, and they carry a stick, but no weapon.

31. Here as on the Jalofo coast, there is another caste of blacks called Judeos. Their occupation consists of playing their instruments, of which they have a great variety, dancing, singing, and acting as jesters. They lend assistance in battle by encouraging the soldiers with their songs. They only marry woman of their own caste and they are considered people without shame. When they die, they are not buried, for it is said that the earth would not accept Judeos, so they are put in the forks of trees. I mention this to show how Jews are little esteemed even among savages.
[The corresponding section on 'Judeos' appears in 1684 in an earlier chapter on the Jalofo, chapter 1, paras 18–19.]

[1] *6. Returning to emerge from Rio de Bintám, there follows, about six leagues on, on the south side, the village of Tancoroale, which belongs to the Kingdom of Quiám. In this land there is little trade, but Tancoroale has it all, chiefly in a large quantity of black salt, available only in this kingdom and the kingdom opposite, that of the King of Badibu. This is the salt used in the river.*
7. The ports in the Kingdom of Badibu on the north side are these. Two leagues beyond Rio de Jobá is Rio de Gerume, and six leagues further is the port of Tagamdabá. A further six leagues is the port of Badibú, the [name of] the kingdom. In all these ports salt is available, as I have said; and the same on the south side, whose ports are Jásabo, Tubabocolom, and Sitato, all having rivers by which you enter to obtain the salt. A ship travelling in these stretches ... .

[2] *8. From Tagamdabá ... and many blacks.*

[3] *9. From Jagra to Deguma Sansam ... a port of the same king.*

22. From Degumasamsam to the port of Manjaguar is eight leagues, the latter being situated on the north side. It is also very difficult to recognize, because its entrance too is by a very small stream. But there is no other stream in this land, and by this point the river does not have as many mangroves. You make your way in by boat, and the port is close by, as is the village too, but not in sight.

23. The kingdom of Manjagar is subject to the king of Borçalo, as are the kingdoms of Barra and Badibo. ((The king regularly comes here to trade with the ships when they do not go to his land, and he sells many blacks.)) This land produces many hides, the best on the whole river; also much husked rice, and some *many* blacks. In the same land, at a distance of one league, is the port of Caur, easily recognized because there are two large trees in the port, one a *sibe* and the other a *manepulo*. It is a good port; and the village lies within /f.ll^v/ sight and is the biggest village of blacks on the whole river, except for the village in the kingdom of Combo. Its inhabitants are *bixirim* traders. Much trade is done here. Many cloths and some gold can be bought, also very good hides like those at Manjagar. You can very successfully sell cola, which these Mandingas are very fond of. A ship anchoring here or in the previous port should be particularly careful with its anchors, because the mud attaches itself so tightly that they cannot in due course be raised; hence it is necessary to hoist them every tide.[1]

24. It is eight leagues from Manjagar to Nanhijaga. Nanhijaga can be recognized by a *little* river on the same side which has a *sibe* tree at the mouth. A ship can enter this river and travel up it six leagues to the village, ((the river lacking any obstacles or dangers)). Very good *many* hides are available here and a large quantity of the local staple foodstuff, which is *milho* (millet). ((This obliges (us) to go there often, for although the river abounds (with *milho*), here it is both plentiful and very cheap.)) Some ivory is also bought here. This port is the first one in the kingdom of Nhani.

25. Opposite the mouth of this river, on the south side, is the port of Nhamena, indicated by a small river. This river can be clearly seen because the country is by now open. Nhamena has hides and plenty of foodstuffs. You can go further inland (by this river) and exit at Degumasamsam. Ships do this when they come down-river in the rainy season in order to avoid the current in the main stream.[2]

26. From Rio de Nhamena to Rio de Cudan is four leagues. Before reaching the latter you see two islets so close together that they seem one. Leave them to the north, and take care at the entrance (to the river) as it has a shallow at its point. Between them and the land there is no danger. The river is small, as all of these are, and it cannot be entered by a ship. But by boat it can, and you travel a league to the village which has foodstuffs, hides and ivory. After passing these islets and proceeding along the main channel of the river, you come to the port of Nhanhimargo, a journey of four leagues. The village, which is the same as the port, is on the waterside on the north bank. ((It is very attractive,)) has very friendly inhabitants, and

---

[1] *10. From Deguma Sánsam to Macugár ... a manipoleiro ... bexerins, that is to say, men lettered in Islamic law (ley de Mafoma), and all are traders, so there is much trade and business. ... many blacks ... each tide.*

[2] *11. From Manjagar to Nanhijagá ... This port has more provisions than all the [other] rivers. ... the first port of the Kingdom of Nhanimançá. ... come out at Deguma Sánsam, which is what we do when we go up-river in the rainy season.*

offers a large quantity of foodstuffs, as well as hides, cloths, and blacks, also some ivory. ((Cola can be profitably sold here.))[1]

27. From Nanhimargo to the port of Cação is four leagues. There are many mudbanks on each side, so the course to take is midstream. ((From here upwards the countryside is very pleasant, there being no longer any mangroves; its ports are very attractive, the villages lying on the waterside, so that all is fine.)) An islet reached midway /f.12/ is called Ilheo das Caboupas, because of its many *caboupeira* trees. From these trees excellent planking for ships is made, this being the sort of timber with which ships are normally boarded up. When I was living in Cação I had many such planks made, and made with little trouble; and you pay nothing for (the timber of) these planks since the islet is uninhabited. Returning to the course up-river, anyone going to Cação must leave this islet to the north, and take care at the approach to the island and when leaving it, because there are sand banks at these points.

28. Before reaching the port of Cação there is a bare red rock called *Rocha do Ouro* (Gold Rock). The blacks say that it is enchanted, and they cut no wood there. They also say that English engineers went there to open mines, but that they all died. ((It may be God's will that this paradise should be reserved for the Portuguese nation, as the first-born daughter of this river, and that the key made of gold which the kings of Portugal sent to the king of Barra – and which they still have today – was to be a prophetic symbol of the great quantities of gold which will be obtained from the river.))

29. The port of Cação is very attractive, the houses of the village being visible from the river. Whites have always lived there, but in a village separate from the heathen, which is not the case in any other port. In this village the whites live in greater freedom than elsewhere. Only from this place can they remove themselves at any time they wish to do so, without the king impeding their removal or doing them any harm; and this freedom does not exist in the other ports of this river. In these, as in the ports on the coast, whites wishing to leave must act deviously, because if the blacks found out, they would rob them. Only in the port of Cação is this custom not followed. The blacks say that the whiteman is like a fish - the water brings both, and the water takes both away again. Nevertheless, if a white dies in the land, the king is his heir and takes all. I resided at Cação for three years, and actually lived there all the time I was not away on a voyage; and if I were to return to the river today, I would live nowhere else but here. For the blacks, the bravest and most spirited on the whole river, are very affable and homely towards the whiteman, and his very good friends. The black females are the most beautiful women on the whole river; and are unsurpassed by any on the Coast of Guinea. Even if a white gets into trouble with one of these women - and may God deliver him from this - not even then will the blacks rob him. The land has much trade /f.12v/ in hides, as well as in blacks, cloths and some ivory. Cola sells there extremely well. The blacks have many foodstuffs such as hens, husked rice (all high-quality and cheap), plenty of milk, and

---

[1] 12. From Nanhijagā [sic] to the Rio de Cudam ... an islet, [one of those] called the Islets of Cudam. ... The mouth is narrow like that of all these rivers. After emerging past the islet, and another which closely follows it, sail ... This is the first port of the Kingdom of Nhani. The people are good ... The country abounds in everything and is very pleasant to live in.

excellent fat (*manteiga*, 'butter'), both unprepared and prepared (*cosida*). This is because the whole kingdom of Nhani is full of villages of Fulos, who have these foodstuffs in abundance. A cow costs only a *pataca* or its equivalent. There are many sheep, which are unlike those of Europe in that they have no wool, but their meat is most tasty, the best being from the female animals; also many goats, which are equally excellent, and a large quantity of game, both animals and birds, and of many species. Thus everything necessary for human existence is found in this land in great plenty and sumptuousness. There are very good facilities for repairing ships, because of the timber available in bulk at every point along the bank of the river, so that even if large numbers of ships were built here the timber would not run out for many years; and the islet, which is in sight of the port, can be used for the planking, the supply of timber again being immense. But above all, this is a very healthy land, well cleansed by the wind, and since it is sixty leagues from the mouth of the river seldom sultry. In all the time I lived there I never knew illness or fever, and in such remote parts this is not the least (blessing) one might desire. Cação also has the convenience that from there one can contact Farim or Cacheo quickly, as witness what once happened to me. I had a ship which was careened, in order to be freshly planked, when all the caulkers, carpenters and deck-hands (*grumetes*) cleared out, leaving me without a single helper. Finding myself in this plight I sent a black to the land of Cacheo to find caulkers and deckhands, and within twelve days he had travelled there and returned with them.[1]

30. ((A white living in this land must guard against killing or wounding anyone - spilling blood on the earth, as they say - because if he does this he will die. And if any slave of his does it, he will be robbed. I have had experience of this practice, for when a black of mine who was held in chains killed his companion who was in the same chains, it led them to rob me. Only the fact that I was well liked enabled me to get my property returned; but I still lost a good amount. The king, with the nobles and other people of the land, led away the killer, to wash off the blood, as they put it. It cost me more than three blacks, so you can see it is the greatest crime in the land.))

31. ((Upstream from here the river is more tortuous, winding about like the rivers in Portugal, and it has many narrow stretches. Do not be surprised, therefore, if my description is fuller and more extended, since it will not only supply information (about the land) /f.13/ but will also serve as sailing directions for anyone who cares to make this journey. I believe that the journey should not be undertaken without a

---

[1] 13. *From Nanhimargo to Cação ... At two leagues from Nanhimargo is an islet .... Before reaching ... May God reserve this advantage for us Portuguese! ... The port of Cação is most pleasant, and the houses of the town can be seen from the river. Whites live here, in a village separate from the blacks, which happens in no other port. This is the port where whites live more freely than in any other. The writer lived here for three years and can supply much information on the country. The people are Mandinga, as everywhere on the river, being the most open-handed non-Christians, the men and women the most beautiful on the river, all good friends of the Whiteman. The land has much trade in whatever there is on the river, except wax. It abounds in foodstuffs and meats of all sorts, very cheap ones, as anyone who goes there will find. If I return [to the river] this is where I will live. It is very convenient for the building of ships since there is much timber along the river, and on Cabopa Island which is covered with trees from which can be made excellent planks, in such quantities that it would not be exhausted in many years. The port is very agreeable, and much bathed by winds, and is so healthy that in all the time I was there I never knew what it was to have a fever or any other sickness, which in faraway lands is not the least thing one can wish for.*

pilot experienced in this river, yet it will be no small comfort to have insight into its straits and dangers, so that you can inquire concerning points about which you have already a certain amount of knowledge. And if the ship follows these sailing directions, the journey will be undertaken without incident.))

32. From this port of Cação to the port of Conicomco is six leagues. At one league (from Cação) is an islet, which will lie to the north of you, and you should take great care as you approach it because there are shoals. But having approached it, go close to it, because the southern side (of the river) is all shoals. When you have made your way along this islet, you will see another, which you must leave to the south, and you will come out between the two. As you emerge, watch out for the shoals which extend from the northern islet. Thereafter travel in the middle of the river, for on each side there are mud-banks *shallows*, ((and you will miss the tide if you run aground, though there are no other dangers)). The port of Conicomco is situated on the south bank, at the foot of a rock, which is its landmark. Wax, hides and ivory are available in this port.[1]

33. From this port to that of Jurume is four leagues. When setting out, we keep well in towards the land on the north side. The other side - where you will see a little islet and a stream in the land to the south, these being called the port and islet of Sapugo - is full of shoals. So following the land to the north, as I have told you, you will proceed until you see another islet ahead. This one (also) must stay to the south of you. Then you can go into mid-stream, where there is no danger, until you have passed another port further ahead, called the port of Malor, which lies on the north bank, half-way along. It can be known by some *jala* trees, which resemble cork-oaks (*serveiros*) in Portugal.[2]

34. This port of Malor has foodstuffs and hides, and in exchange for these, and for cloths, ((which are also plentiful)), much salt can be sold. From here to Jurume you skirt the south bank, out of fear of a reef of rocks which runs out from the north side. You proceed this way until close to Jurume, then you anchor in the port on the north side. ((It can be recognized by some palmtrees in the port and the houses of the village near the water.)) There is much trade here, and much salt can be sold. The village is the grandest in the kingdom of Nhani.[3]

35. From this port to that of Lamé is eight leagues. As soon as you set out, you will see two islets. Go between them and steer close to the southern islet. When you are nearly past, pay /f.13ᵛ/ attention to the northern islet because it has a shallow. Once past them, the port of Lamé can be found at the foot of a rock on the north side. Much salt can be sold here in exchange for cloths, hides, foodstuffs and some ivory.[4]

36. Once past the port of Lamé, skirt the land to the south. If you are taking a large ship, turn the yards ((and proceed by oar)), running close in to the land to the south, but taking care that the trees on land which hang over the river do not get

---

[1] 14. From Cação to Coniconco ... hides, wax and ivory.
[2] 15. From Coniconco to Jerumê ... Rio de Sapugo ... . In this river there is also trade in hides and ivory. ... sail in the middle of the river. If tacking, you can go from one side to the other, since there are no shallows until you have passed the port of Malor ... Portugal.
[3] 15. This port has many provisions ... There is as much trade in cloths as in selling salt. ...
[4] 16. From Jagra to Lamé ... ivory.

entangled with the masts, for this is *Os Passos dos Fulos* (Fula's Passage). Make your way with lead in hand, and you will know you have gone through when at extreme low tide you have one fathom and immediately after the depth increases. When you gain three fathoms you will be able to go along the middle of the channel, that is, along the middle of the river, for you have now passed (the obstacle). This is one of the most dangerous passages on the river ((because the bottom consists of rocks and the current is very strong, the water always running like a mountain torrent. The blacks say that an army of the Fulos travelling in these parts wished to cross this river and could not do so. But there were so many of them that when their general ordered each one to throw a stone into the river, so many stones were thrown that they (blocked the river and) formed a 'pass', which enabled them to ford it. Later the currents of water opened the small channel through which one can sail. I always took this to be a myth. But let anyone believe it if he likes, for this is what the 'sons of the land' affirm.))[1]

37. When you have passed this narrow stretch, you can proceed at will, without (further) danger, to the port of Bunhacú, which lies on the south side and is the first port of the kingdom of Frincabó *Farim Cabo*, ((all the ports on this side mentioned earlier belonging to the kingdom of Jagra. The best trade on the river is up-stream from this port;)) and salt can be sold in all the ports there, salt being the best commodity we can carry. Bunhacú can be recognized by two *sibe* trees. Before you reach there, head to the north side, to avoid some rocks stretching out ((from a little river in the port)). But when you leave, go to the south side, because on the north side is a large sand-bank. Proceed along this (side) for about one league, and on the same side you will see a village at the water's edge. This is the port of Peripho [margin: Perifo], ((a very attractive port. I never went along this river without alighting here because of its cheerful and fresh appearance.))[2]

38. From this port to the port of Maresamsam is seven leagues, and there are no difficulties on the way there. The port lies in front of a rock, after you pass some palm trees, all these being on the north side. It belongs to the kingdom of Nhani, and has the same trade as the other ports. This is the first port where the owners of ships making the voyage to Cantor are in the habit of leaving a trader ashore – they call them /f.14/ *tavernas* – to buy hides, so that when the ships come down-river they can be loaded up. However I myself always left the first one in the land of the kingdom of Manjagar, in the port of Caur, because it has especially good hides, and because I always found there upwards of 500 - so I recommend this to anyone sailing this river.[3]

---

[1] *17. ... keep to the south bank and you will pass a river which you will at once find on the south side. ... the most dangerous points on the river. You can then ...*

[2] *17. ... You can then reach by land the port of Bunhocó on the south side. It is the first port of the Kingdom of Farim Cabu, which is the country of Cantor. It is 8 leagues from Lame. Here the trade is in hides and wax ... rocks lying off the south side.*
18. When leaving Bunhocó ... Perifô. The houses are on the water's edge and within the port is a village of whites. The trade is the same as at Bunhocó. ... only one league.

[3] *19. From Perifô to Maresánán ... In this port the trade is as I have stated. It is the first of the ports in which the owners of ships making the journey to Cantor, which they make annually in February, have the custom of landing an agent to buy hides, with which, on the return in June, they will load their ships. They do the same in the ports higher up, some in certain ones, others in others, according to the friends on land. But in all these ports an agent can be left, for in each there are many hides and a little ivory. Some owners ...*

39. Some ship owners are in the habit of going (up-river) at the beginning of the rainy season to winter in one or other of these ports, but they never go beyond this one, which costs a lot of effort to reach. Here they buy hides, some ivory and many blacks. ((It is useful in this case to carry liquor, because although the blacks are Mohammedans, they are great drinkers of wine and brandy, and they supply a black in exchange for seven or eight flasks of brandy, a third of which is river water.))[1]

40. From this port to Rio de Aleá is eight leagues. When you set out, follow the land to the north until you lose sight of the port, for on the other side are shoals of sand, and on this part of the route there is no more than a depth of ten palm-spans of water at low tide. Once you lose sight of the port, you can sail out into the river as far as you like.

41. On the way to Rio de Aleá is a rock on the south side. The largest and highest rock on the whole river, it is known as the Rock of Nicolão Saquedo, ((which in Portuguese sounds the same as '*Nicolão esta quieto!* – Nicholas, be quiet!')). The blacks say that this rock is enchanted, and every person who passes this way for the first time must dance naked in front of it, otherwise before completing his journey he will die. However I passed by it, and I neither danced nor allowed anyone else to perform the ceremony, yet ((thanks to the Lord)), I am still alive, ((and not one of the others who failed to dance in fact died.)) So I give no credence to all this (nonsense) ((and the other tales they tell about this river. But the whites and the blacks who live here take it all as Gospel truth, and the devil frequently does things which give the appearance of confirming it, so that they accept it without question.))[2]

42. You enter Rio de Aleá in boats or canoes, and travel inland along it for 8–10 leagues. In its many villages hides and ivory can be bought, and you can sell much; ((but this trade is more useful for the 'sons of the land', who travel about in canoes than for the owners of ships, apart from those whose ships are going to winter in the port of Marisamsam. This river lies on the north side and is easily recognized because there is no other river here.))

43. From the mouth of this river to the port of Ponor, which is on the south side, is six leagues. Its landmark is a steep cliff with a large thorn-tree on it. Before you reach the port, and as soon as you catch sight of it, head /f.14ᵛ/ close to the north side, because on the south side at this point a reef of rocks takes up more than three quarters of the river. But when you draw level with the large thorn-tree in the port, *point the prow to the opposite bank*, then head to the south side because the north side has a sand bank which takes up more than half the river. The port has the same trade as the others.[3]

44. From the port of Ponor to the port of Jalacuna is six leagues. You can, on the south side, sail without any anxiety on the first stage, up to the point where you come upon some large *jala* trees, which I have already explained are like cork-oaks. As soon as you see them, go up close to, indeed very close to – ((take good heed of this warning!)) – the north side, so that until you pass the trees, you advance with your oars (almost) touching the bank, because the river further out is rocks the whole way. This passage is hazardous because of the current. On the north side,

---

[1] 19. ... Some owners ... but they do not go past this port because of the currents in the river ... .
[2] 20. ... But I have passed by without dancing and I who write am still alive to put down these cock-and-bull stories.
[3] 20. ... Once you are level with the tree, point the prow to the opposite bank ...

opposite the port, is another village *port* with the same name. ((In each I have had good hides cheaply, since one *alqueire* of salt purchased five or six hides. All these ports are abundantly supplied with good things, and Epicureans will be able to get by very comfortably here. For the Fulo women – who are here in great numbers and are extremely beautiful and very neat, but not at all chaste – bring to the ship everything that could be desired.))[1]

45. From the port of Jalacuna to the port of Nhamenhancunda is eight leagues. The route is extremely difficult, ((so it is necessary to take very great care and travel only by day)). When you leave Jalacuna, advance until you catch sight of a small rock on the north side, and on the rock a *sibe* tree, which is a kind of palm. When you see this, flee away from the north side, for there are rocks there, also a small cliff facing them which enables you to recognize them more easily, *and approach on the south side*. Once you have passed them, you can proceed as you wish until you reach the second turn in the river, then you go closer to the north side, because the open water in the curve to the south also has treacherous places. Once past this turn, you can advance without anxiety as far as the port, which lies on the south side, and the village, near the water, ((which has a very good landing place. I always left a trader here, both because the port is convenient and because the local people are friendly; and I always found some ivory to buy.))

46. From the port of Nhamenhacunda to the port of Sumacunda is eight leagues. One league further, on the south side, is the port of Nhacobé *Canhobé*. And one league short of Sumacunda is the port of Perifo, *also on the south side*. When you have passed Perifo, go to the north side, because the south has shallow water and a sandbank. The landmark of the port of Perifo is a number of *poulão* trees, (cotton-trees), large ones standing in the village; and the landmark of the port of Sumacunda /f.15/ is a bare bank with a tamarind tree on top *and a little stream along it*.

47. From the port of Sumacunda [I suppose that the text is saying it is from here][2] to the port of Nhacoi is four leagues. This port is on the north side, and there are no dangers. Its landmark is a *sibe* tree at the foot of a rock. In the village there is much trade in hides, as in the other ports.[3]

48. From the port of Nhacoi to the port of Findifeto, which lies on the south side, is six leagues. The people of the land are evil, as all of them are upwards from here on the south side. ((Hence on this side up-stream we never left *tavernas*, because the blacks are wicked, being thieves and robbers. But the people on the north side are good, and are friends of the whites, and in all their ports we trade and leave traders, as in the other ports further back.)) Findifeto's landmark is a *sibe* tree. Before reaching it you will see a bare cliff on the north side, which you will head for and skirt, because on the opposite side is a sandbank. As soon as you have passed the cliff, head straight for the south side, where you will see another cliff, and skirt this,

---

[1] 20. ... *The passage here is dangerous and care is called for. It is called the Rocks of Jalacúna, from the name of the port on the south side once this is passed.* ...

[2] This insertion, and later insertions of this kind, were made by a copyist of the manuscript c.1700.

[3] 23. ... *at the foot of a rock and before reaching there, on the same side, twice, two sibes. In these two places a port exists but neither is the chief port.* ...

because on the north side there is also a sandbank. Then when you draw level with a low cliff on the north side, you are free of all the difficulties and can advance as you wish to the port of Fatatenda. This is four leagues up from the port of Nhacoi and lies on the north side, its landmark being a steep bank at the foot of a rock with some large trees. ((The port is very pleasant and attractive.)) In earlier times the ships which carried out the Cantor trade went no further than this, ((and the caravans came here,)) but desire for gain drove the ships further up-river. This is the first port of the kingdom of Oli *Ulimança*, ((for the kingdom of Nhani reaches as far as here. All these kingdoms also bear the name '*Mansa*' [*sc.* Mandinka for 'king'], such as the ones just mentioned, Nhanimansa and Olimansa.))

49. From the port of Fatatenda to the port of Pirai, which lies on the south side, is four leagues. Pirai's landmark is a *poulão* (cotton-tree) in the port *in the centre of the anchorage*. The route is this: once the port of Fatatenda is out of sight, head to the south side, and continue thus as far as the port (of Pirai). ((We trade in these ports on board the ship and even then with great caution.))

50. From the port of Pirai to the port of Cantor, which is on the north side, and is called Sume or Same [I could not make out from the text which it is, but I am inclined to think that Same is intended], is four leagues. Ships used to go here after they had passed Fatatenda. The land and the port /f. 15ᵛ/ on the south side are called Cantor, and belong to the Farim-Cabo, but the people are very wicked. Hence we do not leave *tavernas* there. Nor do we buy there other than at the water's edge, although Cantor has much trade especially in the skins and hides of wild animals, of which the hides of *antas* are the most prized by foreigners.

51. Ships going further up-river make a payment to the Farim do Cabo at this point, this being a kind of tribute. His officials arrive to receive it, and they are never satisfied, because it is not a fixed sum, which would be better, and because those who arrive to receive it are men of the worst sort.

52. Similarly you make a payment, either here or at Baracunda, to the king of Oli or Uli. But this is not a cause of grumbling because Oli is a land where whites and their slaves live and work without any aggravation occurring, and instead the local people on this side of the river treat the white man with notable affection. Moreover, the urge of the whites to draw greater profit from the river drove them even further up it in their ships, to another port called Baracunda, which is situated in the same land on the north side.[1]

53. From this port to the port of Baracunda is twelve leagues. It is a very hazardous route, full of many sand-banks, and no vessel can go there which requires more than seven palm-spans of water. A tale is told, and regarded as true, to the effect that a caravel from Portugal, fleeing before a French sloop, went up this river for twelve

---

[1] 26. *From Pirai to the port of Cantor ... four leagues. This is where the ships go. ... Ships going there give a present as a form of tribute. The north side belongs to the Kingdom of Ulimança, the people are friendly, and whites land without anxiety. Ship owners make these people presents too. It is to here that the caravans of blacks come, bringing ivory, cloths, blacks and gold, all in great quantities, apart from gold, for what the whites buy is not more than 6–7 lb, although the blacks of the country buy much. On each side (of the river) many hides and much ivory can be bought. Here too desire for gain has driven the whites to proceed 12 leagues up-river, to the port of Baracunda. The port of Samé can be easily recognized. When going from Pirai, it must be approached on the north side which has greater depth.*

leagues beyond the port of Baracunda, to a port which, in memory (of the event), the blacks call Tubabo Sita – in Portuguese 'Porto da Branco' ('Port of the Whiteman'). The French ship pursued the caravel as far as the port of Baracunda, but there by chance the French ship caught fire and hence the Portuguese escaped. The crew of the caravel suspected the blacks of giving the enemy information about their ship, so as they made their way down-stream, with two pieces of artillery they carried, they did much damage to the villages they passed, these being all on the waterside, and thus they wreaked vengeance.[1]

54. In this port of Baracunda, whites are as confident (in their security) as if they were living in any port in Portugal. To this port come the caravans of black merchants, the most impressive being the caravan of a race of blacks called Jagancases or Jagamcases [I am inclined to think that it is Jagancases]. This caravan comprises more than 3,000 persons and more than 2,000 donkeys. They leave their land in November and arrive here in July, as I already have stated.

55. It is reported that three caravans leave these (interior) parts at the same time, /f.16/ and that one comes to this river, another goes to Rio de Sanaga, and a third to Rio de Deponga. Their chief business is to obtain salt. Those who come to this river do not buy it from the whiteman but instead make their way to the mouth of the river, to the Ilhas de Felam, where they purchase it in exchange for cloth. To the owners of ships they sell many blacks, and much ivory and gold, this caravan being the one that brings the greatest quantity of the latter commodity. I understand that they bring the best part of an *arroba* of gold, since the local people also buy it in larger quantities than the whiteman, who does not care too much for buying gold, preferring to lay out his resources on other commodities, any one of which brings more profit.

56. What these blacks sell in the greatest quantity is cloth, the whites buying as much as they care to and their time permits. The resources with which it is bought are paper and small glass beads, black and white. The way the business is conducted is that as soon as the caravan arrives at this port, its leaders, called *solatiguis*, come to speak with the owners of ships, and they settle the price to be paid, either in paper or in beads, for a length of cloth. Once this is done, they all then arrive, and begin to buy and sell at a rate unsurpassed even in the meat-market at Lisbon on Easter eve. The buying and selling continues day and night for twenty-four hours, giving the whites not a moment to eat or sleep. No stealing whatsoever occurs.[2]

57. The Jagancases are a very sincere people, and they take no revenge for any ill done to them, leaving it to God. They say that when they leave their land, their leaders report to the king, who tells them to remember that they are going to alien

---

[1] 27. *From Samé or Cantor to Baracúnda, the last port to which whites go, ... Today this is where the caravans come. The most important ...*

29. Seventy or eighty years ago, according to a tradition and history accounted true, a caravel from Portugal arrived just as French pirates were making their way along the coast of Guinea. Having met one of the French ships, the caravel entered the river and ascended, the pirates in hot pursuit a tide behind. Having passed by 8 leagues the port of Baracunda, the caravel reached a place the blacks call Tubabocitá, which in the local language means 'Whiteman's Figtree', because of a figtree growing in this port. This is the furthest point reached by whites that is known about.

[2] 27. *... caravans come. The most important is that of a caste* (casta) *of blacks called Jagamcazes. The caravan comprises more than 3,000 persons and more than 1,000 beasts. What they chiefly come to obtain is salt, and the salt they carry away they buy from blacks who have travelled to obtain it in the islands called Islands of Feilám at the*

lands and must instruct their people to keep their mouths shut and their trouser-slits sewn up; and that if anyone does to them any harm they should leave it to God (to act). This is not the pronouncement of a barbarian but of a very great philosopher – if the Portuguese, when they left their country and came to these parts, had kept this saying in mind and followed it, many of the disasters which occurred would not have happened. But it is certain that when a disaster does occur, in whatever land it may be, it cannot be without the Portuguese giving some cause for it and deserving to some extent what the blacks do to them. Another caravan comes here too, not such a big one, yet one with more than a thousand persons, these belonging to another race of blacks called Conjuros. In this case, when their leaders take leave of their king, he commands them as follows. Where-ever they pass through to, or go, they must speak the truth and must not take what belongs to others. /f.16ᵛ/ But if people try to take away from them what is theirs, they carry weapons with which to defend themselves. If these weapons are insufficient, then they should avail themselves of the means they know. For all the Conjuros are great magicians (*feiticeiros*) and therefore very much feared by all the heathen, hence their property is sacred amongst all the blacks, and no one lays a finger on it against their will. But note that all the heathen of this river, especially those of the north side, take nothing from anyone and our goods can 'sleep in the street'.[1]

58. Many smaller caravans with 40–100 men, come from here from all the neighbouring kingdoms, bringing ivory and blacks; and they wish to be paid chiefly in salt. The way in which the whiteman buys ivory is most curious, and because it is not used anywhere else, I shall set it down here.

59. These blacks arrive with their ivory, and when they reach the water's edge they pile it all together. Then the owners of ships go ashore, and if there are six of them the blacks put the ivory into six piles, without the owners handling it at all; and so that there can be no cheating or complaints of 'my share was smaller', the owners draw lots to decide which pile goes to which owner. As soon as the ivory is shared out, each owner orders what falls to him to be loaded on to his own ship, as if he had already bought it, without the owners (of the ivory), who watch the transaction, saying anything, but each one noting where his tusks are being taken.

60. After this has been done, the blacks call up the boats, and each black goes aboard the ship where his ivory went, accompanied by his head-man, who is the one who makes the sale - but not because he gets a better price, because the price is already known. Then, when the whiteman buys the ivory, each tusk is brought to him, and a stick is inserted into the hollow (end), so that the length which is hollow can be judged from the stick, and a scratch is made on the outside (to mark its length). For the (hollow) part (of the tusk) nothing is paid, and only the length between the scratch and the tip, measured in palm-spans, counts towards the price, each palm-span costing the value of two cloths. This price is paid without any further argument.

---

mouth of the bar of River Gambia. They buy it with cloths. To whites in the ships they sell ivory, blacks, gold and also many cloths, in return for paper, beads, avelorio, and other trade goods such as crystal, red cloth, daggers, and red and yellow woollen thread.

[1] 28. Another caste of blacks called Conjuros arrive to seek the same things as the Jagamcazes. A great number of caravans of blacks called Bachares also arrive, each comprising 80–100 men, and bringing only ivory and blacks. What they chiefly seek is salt, which they buy from the whites. Here is how trading in ivory is carried out. ...

61. The largest and best tusk has at the most five palm-spans of solid ivory and costs ten cloths. This value is paid in salt, whose price is also fixed. For as soon as the ships reach the port, and presents have been given to the kings, then the royal officials set the price and measure for salt. They do this with a bowl they carry, a little bigger than one of our soup bowls; and no one asks more or is given less than the price they set. They similarly set the price for the remaining goods, especially paper and beads. The most salt given for one cloth is an *alqueire*; so for the tusk which I said costs ten cloths, they give /f.17/ ten *alqueires* of salt. I have bought a tusk weighing two *arrobas* or three *alqueires* of salt. This is the reason why gold is not bought, because they sell it in such a way that normally an *oitava* (⅛ of an ounce) of gold works out at two cloths, which are either two *alqueires* of salt, or four hands of paper, or 80 strings of glass beads, or a *ramal* of no. 24 crystal, these all being worth two cloths. It is more profitable to expend our resources on ivory or blacks rather than on gold; therefore very little gold is bought. This is not a unique state of affairs in the trades we have discovered, for the same thing has happened at the city of Malaca (in Asia). When there was plenty of gold coming from Samatra and Lequio, no merchant would buy it, because it was more profitable doing business in other commodities. The blacks whom these caravans bring to sell are infinite in number. But most of them are from a race called Bachares, who prove sickly, many dying. We attribute this to the quantity of salt they eat. Because the ships are full of salt, we cannot stop them; yet they have no salt in their own land and do not eat salt there, because it is expensive. Those whom we buy of this race, the most numerous of the blacks on offer, cost very little money, despite their being a most handsome people.[1]

62. This river also has much gum arabic, which comes from trees called *simbrão*, or *zimbrão*, which bear edible fruits resembling the damsons of Portugal.[2]

63. The river also has a tree-resin called *fumadoura* ('smoking place') which is like incense, and there is so much of it that quintals of it can be bought.

64. There is also a type of wool called *lã de poulão* (cotton-tree wool, *sc.* kapok) which the foreigners value highly, it being very fine, finer than silk; and which grows in large *foles* (containers). It is said that it is exported for the manufacture of expensive hats, and there is such a quantity of it that the mattresses used by the inhabitants of the river are all made of this wool. This land has so much husked rice that ships could be loaded with it. There is so much cloth in the river that large quantities can be purchased there, not only for use on the river but to trade with all over Guinea.[3]

---

[1] 28. *The whites select the largest tusks, weighing one arrôba or more each, and a single tusk can weigh 5–6 arrôbas. Each tusk is then measured this way. A stick is inserted into the hollow part and the length of the hollow is noted on the outside. The distance between the end of the stick and the point [of the tusk] is measured in palm-spans, and for each palm-span is given two cloths or their value in salt. This method is so regularly used that it raises no dispute. Salt counting as one cloth weighs one alqueire. The largest tusk has at most five palm-spans of solid ivory, so that it costs 10 cloths or 10 alqueires of salt. I have bought a task weighing two arrôbas for three alqueires of salt. ... The second reason is that it uses up to such an extent the other goods that the ships trading up-river are always short of resources. ...*

[2] 32. *... trees called zimbarão which produce fruits like very red and very sweet damsons, from which the blacks make wine which intoxicates like wine in Portugal.*

[3] 32. *... hats, which, it is said, is why the foreigners carry it away. ... There is so much husked rice that ships could be loaded with it. ... throughout Guinea.*

65. Throughout the land there are many animals providing hides such as *gingigangas*, which have as much flesh as two European bulls, ((hence their hides are very large and worth twice as much as any others)); *antas*, which resemble cows; *sinsins*, which are like horses, and the males being different only in that they have antlers ((like roe deer or stags)); /f.17ᵛ/ and *tancões* which are like donkeys, the males also having antlers. The land is well-stocked with game for hunting, such as wild boar of several kinds; gazelles, which are like roe deer *wild goats*, ((also of many kinds)); *many pigs of many kinds*; *herds of* buffalos, which are like wild cows, and have such a thick hide that it is no use for tanning; ((a vast number of elephants)); and many game-birds such as wildfowl – we call them forest hens ((or African hens, they being larger than the domestic variety –)), different kinds of ducks, ((some of them very beautiful, very large, wild peacocks, and many others)) – *innumerable sorts of animals and birds to give delight to the sight and pleasure to the palate*.

66. The river has a large quantity of fish. Between its mouth and Cação fish is abundant; and at Cantor an infinite quantity is obtainable. ((The fish comes from the rivers which join the Gambia inland. It is of excellent quality, and there is so much of it that some is dried and conveyed (for sale) downstream.)) The river also has many hippopotamuses (*peixes-cavalos*, 'fish horses'), which have as much flesh as an elephant, ((and the blacks and certain of the whites consider the meat very good to eat)).[1]

67. There are also many dangerous animals, ((as in all Africa)); lions, tigers, leopards (*onças*) and wolves (*sc.* hyenas). And also venomous ones, such as snakes of many kinds. Snakes of one kind, very large and very thick ((and called *irans*, lack poison)) and the blacks eat them. All the way up the river there are very large crocodiles *which in the Indies are called caymans* ((which can swallow a person whole, and I can testify to this since one of these crocodiles swallowed a fairly large black in the port of Antula, in front of my very eyes)).

REFLECTION

68. From all I have said, it will be obvious that great profits can be gained from this river, for it can be truly claimed that if Guinea is an egg, then this (district) is the yolk. It is a pity that the foreigner profits from it, since it was we (Portuguese) who discovered it. But the foreigner replies that, if we were the discoverers, we did it for them, and this is only too true. All this can in the future be remedied, without the foreigners being offended, only when His Highness, whom God protects, establishes a trading post (*feitoria*) on some part of the river. It would offer the trade goods the foreigner brings and those bought at first hand, so that these can be supplied with the same ease as the foreigner supplies. Similarly, when the Portuguese have these trade-goods, the settlers in the river will sell what they possess to their own people rather than to the foreigners. Once there was an (intermediate) trade

---

[1] 32. *The whole river as far as Cantor has much fish, and both in the river and in the lagoons which we regularly drain dry. ... elephant.*

conducted by the Portuguese, this trade being the means by which the foreigners profit, then the river would be cleared (of them). But if His Highness did not wish to involve crown property, /f.18/ he could assign the business to traders who would organise a *bolsa* (regulated trading system, chartered company). ((But it should be done here, and not at Cacheo, where no profit can be made, and where its only effect would be to ruin both this island and the whole of Guinea, as experience would prove.))

69. With the *bolsa* set up on this river, as is done in all parts of the north [*sc.* northern Europe], the vassals would have recourse to it and benefit from its advantages while His Highness would profit from its taxes. If the traders who would have to join the *bolsa* were to tell me that the hides bought on the river are small and not tanned, (I would reply that) they should realize that with four penny-worth of goods in Portugal (*ventens de emprego de Portugal*) the foreigner buys a hide from the whites, and not a few for less; and that if the foreigner did not make so much profit from hides, he would not make the efforts he does to obtain them, for he values them higher than ivory or gold.[1]

70. I have stated what the river possesses and indicated its ports, courses and difficult passages.[2] And now, in concluding, I also wish to mention a point which always shocked me when I lived there. It is this. Along the whole course of the river, which has so many towns where whites live, I never saw, or heard, that there was, or ever had been, a church anywhere there, a state of affairs not the case in any other part of the Coast of Guinea. And yet the priests who go from Cacheo to administer the sacraments (along the river) go there at wickedly long intervals. Hence, since they (the Portuguese) lack church buildings and teachers, and the people of the river with whom they live are heretics and Mohammedans, and there is no one to admonish the Portuguese, their behaviour is dissolute in the extreme. In general they keep the commandments of Holy Mother Church badly, eating meat without scruple on all those days when it is prohibited. May Our Lord God help them (to amend).

71. As I have stated, I lived on this river for close on four years, and all that time I lived there without the (regular) consolation (of religion), since, in order to make my confessions, it was sometimes necessary for me to go overland to Cacheo. I was very fortunate that while I was living there, it was God's will that the general of the (English) islet was a secret Catholic, and so had in his company a priest, and together with the general, in his chamber, at times I heard mass and made my confessions. This deprivation made me leave the river more quickly, abandoning the great profits it had promised me.

---

[1] 33. *May it please Your Majesty, there could be drawn from this river, for yourself and your vassals, the profit that foreigners draw from it, and this without clashing with them, simply by having established a factory at one point in this river and providing it with the necessary merchandise. ... ivory or gold.*

[2] 33. *Such is the information that I possess on this river, where I lived four years during which I observed those matters about which here I write, in order to obey the one who can command me. I am now going to give information on the rest of the coast and state what I observed there during the 23 years that I have been up and down it, sailing in most of the rivers and trading in most of the ports that I am going to describe. This description will extend as far as Sierra Leone, for this is as far as I have sailed, and such is moreover the instruction I received. Should these lines serve nothing else, they will at least inform those who wish to pass into these regions, how to set about sailing on many rivers and trading with many peoples of savage humours and diverse customs, and assist them many times in the frequent perils from which they will be delivered more by their labours than their riches.*

72. The water in the river is all fresh, and from Tancoroale upwards it can be drunk. May it be Our Lord God's will that just as the river runs sweet water, so sweet Christian instruction should also reach there, in order to succour its Portuguese residents and fructify its pagan peoples. For, if there were teachers, /f.18ᵛ/ it would not be difficult to draw the pagans away from their errors and bring them to the true Catholic religion.

PLATES

Plate I. Unloading salt carried from River Saalum to River Gambia (1950)

Plate II. Landing place at Kau-ur (intervening swamp with causeway) (1947)

Plate III. Swamp fishing (Wolof, Panchang, 1947; cf. *GT* 26)

Plate IV. Boy studying (Fatoto, 1953; cf. *GT* 67)

Plate V. Teaching House (Kumbija, 1953; cf. *GT* 67)

Plate VI.   Mandinka wife, recently married (Kerewaan, 1947; cf. *GT* 56)

Plate VII.   Shed for newly-circumcised boys (Kundam, 1953; c.f. *GT* 111)

Plate VIII.   Newly-circumcised boys wearing traditional dress (Bunting, 1951; c.f. *GT* 113)

Plate IX.   Fula cattle with herdsman's hut (Saare Mansajang, 1953; cf. *GT* 34)

Plate X.   Blacksmith at work (Njau, 1947; cf. *GT* 120–21)

Plate XI. Man playing *bolombato* (1953; cf. *GT* 105–6)

Plate XII. Fiddler (1953; cf. *GT* 105)

Plate XIII. Underarm drum (Njau, 1953; cf. *GT* 105)

BIBLIOGRAPHY

PUBLISHED SOURCES

Richard Jobson

(a) The Golden Trade: / OR, / A discovery of the River *Gambra*, and / the Golden Trade of the *Aethiopians*. / Also, / *The Commerce with a great blacke Mer- / chant, called* Buckor Sano, *and his report of* the / houses covered with Gold, and other strange / *observations for the good of our / owne countrey;* / Set downe as they were collected in travelling, part / of the yeares, 1620, and 1621. / By *Richard Jobson*, Gentleman. // LONDON, / Printed by *Nicolas Okes*, and are to be sold by / Nicholas Bourne, *dwelling at the entrance* / of the Royall Exchange, 1623.

[[4] + 166 pp]

Reprinted: Speight and Walpole, Teignmouth, 1904 (ed. Charles G. Kingsley, modernized characters, list of printing errors, xvii + 210 pp.); Penguin Press, London, 1932 (introduction by D. B. Thomas, misprints silently corrected, 214 pp.); Dawson, London, 1968 (introduction by Walter Rodney, misprints corrected in a separate list, with an index and a short glossary, mainly of place names, 204 pp.); facsimile, Theatrum Orbis Terrarum, Amsterdam / Da Capo Press, New York, 1968 (the facsimile from the copy in the Bodleian Library, Oxford; no apparatus).

Select summary versions, in English or translation: [Thomas Astley], *A New General Collection of Voyages and Travels*, 4 vols, London, 1745–7, 2, pp. 174–89; hence Antoine François Prévost, ed. and trans., *Histoire Générale des Voyages ou Nouvelle Collection de toutes les relations de voyages par mer et par terre …*, 80 vols, Paris, 1747–89, 3, pp. 26–46; and Johann Joachim Schwabe, ed. and trans., *Allgemeine Historie der Reisen zu Wasser und Lande …*, 21 vols, Leipzig, 1747–74, 3, pp. 27–52; Pieter van de Aa, ed., *Gottfried's Reysen*, Leiden, 1727, VI, item 19.

(b) 'Larger Observations of Master Richard Jobson, touching the River Gambra, with the People, Merchandise, and Creatures of those parts, then in his Journall is contayned, gathered out of his larger Notes', in Samuel Purchas, *Purchas His Pilgrimes…*, 4 vols, London, 1625, 1/9/13, 1567–76; reprinted, 20 vols, Glasgow, 1905, with original book/chapter/page shown.

(c) 'A true Relation of Master Richard Jobsons Voyage, employed by Sir William Saint John, Knight, and others; for the Discoverie of Gambra, in the Sion, a ship of two hundred tuns, Admirall; and the Saint John fiftie, Vice-Admirall. In which they passed nine hundred and sixtie miles up the River into the Continent. Extracted out of his large Journall', in Purchas, *Pilgrimes*, 1/7/1/, 921–6.

Other (books and articles)

Africa Pilot, 1 (Cabo Espartel–Calabar River), 12th edn, Hydrographer of the Navy, London, 1967

Almada, André Álvares de, *Tratado breve dos rios de Guiné (c. 1594)*, ed. Luís Silveira, Lisbon, 1946 [cited version]; another version, *Tratado breve dos rios de Guiné do Cabo Verde*, ed. António Brásio, Lisbon 1964; English version, trans. and ed. P. E. H. Hair, with additional notes by Jean Boulègue, *Brief Treatise on the Rivers of Guinea (c. 1594)*, 2 vols, Department of History, University of Liverpool, 1984

Álvares, Manuel, [S.J.], *Ethiopia Minor and a Geographical Account of the Province of Sierra Leone (c.1615)*, trans. P. E. H. Hair, Department of History, University of Liverpool, 1990 [of a transcript by the late A. Teixeira da Mota and Luís de Matos of an unpublished text of the Biblioteca da Sociedade de geografia de Lisboa, by permission of the Centro de estudos de história e cartografia antiga, Lisbon, and the Sociedade de geografia]

Ames, D. W., 'The Use of a Transitional Cloth-money Token among the Wolof', *American Anthropologist*, 57, 1955, pp. 1016–24

Andersons, Edgars, *Tur Plīvoja Kurzemes Karogi*, [New York], 1970

Anguiano, Mateo de, *Misiones Capuchinas en África*, ed. Buenaventura de Carrocera, Madrid, 1957

[Anon.], 'La Navigabilité de la Gambie en amont de la frontière anglaise', *L'Afrique française*, 19, 1909, pp. 330–42

Arber, E., ed., *A Transcript of the Register of the Company of Stationers of London 1554–1640*, 5 vols, London, 1875–94

[Astley, Thomas], *A New General Collection of Voyages ...*, 4 vols, London, 1745–7, reprinted 1968

Atkinson, J. E., *A Commentary on Q. Curtius Rufus' Historiae ...*, Amsterdam, 1980

Ba, Cheikh, *Les Peuls du Sénégal: étude géographique*, Dakar, 1986

Balandier, G., 'Notes sur l'exploitation de sel par les vieilles femmes de Bargny (environs de Rufisque)', *Notes africaines*, 32, October 1946, p. 22

Barbot, John, *A Description of the Coast of North and South Guinea*, London, 1732 – see Hair, Jones and Law, *Barbot on Guinea*, below

Barros, João de, *Ásia de João de Barros* [1552–1613], eds H. Cidade and M. Murias, 4 vols, Lisbon, 1945; English version of a Guinea section in Crone, *Voyages of Cadamosto*, below

Bensaude, Joaquim, and Baião, António, eds, *O manuscrito «Valentim Fernandes»*, Academia Portuguesa da História, Lisbon, 1940; Th. Monod, A. Teixeira da Mota, and R. Mauny, eds, *Description de la côte occidentale d'Afrique (Sénégal au Cape de Monte, Archipels) par Valentim Fernandes (1506–1510)*, Bissau, 1951

Bérenger-Féraud, [-], 'Le Mariage chez les nègres Sénégambiens', *Revue d'anthropologie*, 12, 1883, pp. 284–97

Blake, John W., *European Beginnings in West Africa 1545–1578*, London, 1937; revised edn *West Africa: the Quest for God and Gold 1454–1578*, London, 1977

——, 'English Trade with the Portuguese Empire in West Africa, 1581–1629', *Quarto Congresso do Mundo Português*, VI/1, Lisbon, 1940, pp. 314–41

——, ed., *Europeans in West Africa, 1450–1560*, Hakluyt Society, 2 vols, 2nd ser., 86–8, London, 1942

——, 'The English Guinea Company, 1618–1660', *Proceedings of the Belfast Natural History and Philosophical Society*, III/1, 1945/6, pp. 14–27

——, 'The Farm of the Guinea Trade, 1631', in H. A. Cronne, T. W. Moody and D. B. Quinn, eds, *Essays in British and Irish History in Honour of James Eadie Todd*, London, 1949, pp. 85–105

Boilat, P. D., *Esquisses sénégalaises*, Paris, 1853

Borel, E. L., 'Voyage à la Gambie: description des rives de ce fleuve et des populations qui les habitent', *Le Globe: Mémoires de la Société de géographie de Genève*, 5, 1866, pp. 5–31

Boulègue, Jean, 'Relation de Francisco d'Andrade sur les îles du Cap-Vert et la côte occidentale d'Afrique (1582)', *Bulletin de l'Institut fondamental d'Afrique noire*, sér. B, 29, 1967, pp. 67–87

——, *Les Luso-Africains de Sénégambie, XVIe–XIXe siècle*, Dakar, 1972

——, *Les Anciens Royaumes Wolof (Sénégal): Le Grand Jolof (XIIIe–XVIe siècle)*, Blois/Paris, 1987

Bouttiaux, Anne-Marie, *Senegal behind Glass: Images of Religious and Daily Life*, Munich/New York, 1994

Bradshaw, A. T. von S., 'Vestiges of Portuguese in the Languages of Sierra Leone', *Sierra Leone Language Review*, 4, 1965, pp. 5–37

Brásio, António, ed., *Monumenta missionaria africana. África ocidental*, 20 vols, Lisbon, 1952–88

Brett-Smith, Sarah C., *The Making of Bamana Sculpture: Creativity and Gender*, Cambridge, 1994

Brooks, George E., *Kola Trade and State Building: Upper Guinea Coast and Senegambia, 15th–17th centuries*, Working Paper 38, African Studies Center, Boston University, 1980

——, *Landlords and Strangers: Ecology, Society, and Trade in Western Africa, 1000–1630*, Boulder, 1992

Bühnen, Stephan, 'Place Names as an Historical Source: an Introduction with Examples from Southern Senegambia and Germany', *History in Africa*, 19, 1992, pp. 45–101

C., R., R. C., *A True Historical Discourse of Muley Hamet's rising to the three Kingdoms of Moruecos, Fes and Sus*, London, 1609

Cadamosto – see Leporace, below

Camara, Sory, *Gens de la Parole: essai sur la condition et le rôle des griots dans la société malinke*, The Hague, 1976

Carr, C. T., ed., *Select Charters of Trading Companies A.D. 1530–1707*, London, 1913

Carty, James, *Ireland from the Flight of the Earls to Grattan's Parliament (1607–1782)*, Dublin, 1949

Church, R. J. Harrison, *West Africa*, London, 1957

Cissé, Youssouf Tata, *La Confrérie des chasseurs Malinké et Bambara*, Paris, 1994

Cissoko Mody, and Kaoussou, Sambou, *Receuil des traditions orales des Mandingues de Gambie*, Niamey: Centre régional de documentation pour la tradition orale, 1974

Cissoko, Sékéné-Mody, 'La Royauté (*mansaya*) chez les Mandingues occidentaux, d'après leurs traditions orales', *Bulletin de l'Institut français de l'Afrique noire*, sér. B, 21, 1969, pp. 325–38

Coelho, Francisco de Lemos, *Duas descrições seiscentistas da Guiné* [1669, enlarged version 1684], ed. D. Peres (Lisbon, 1953; English version, *Description of the Coast of Guinea (1684)*, trans. and ed. P. E. H. Hair, Department of History, University of Liverpool, 1985

Conrad, David C., and Frank, Barbara E., *Status and Identity in West Africa: nyamakalaw of Mande*, Bloomington, 1995

Crone, G. R., ed., *The Voyages of Cadamosto and other Documents on Western Africa in the Second Half of the Fifteenth Century*, Hakluyt Society, 2nd ser., 80, London, 1937

Cultru, Prosper, ed., *Premier voyage de Sieur de la Courbe fait á la coste d'Afrique en 1685*, Paris, 1913

Curtin, Philip D., 'Pre-Colonial Trading Networks and Traders: The Diakhanke', in Claude Meillassoux, ed., *The Development of Indigenous Trade and Markets in West Africa*, London, 1971, pp. 228–39

Danvers, F. C., and Foster, William, eds, *Letters Received by the East India Company*, 6 vols, London, 1896–1902

Dapper, Olfert, *Naukeurige Beschijvinge der Afrikaensche Gewesten*, Amsterdam, 1660, 1676

Davies, K. G., *The Royal African Company*, London, 1957

Diallo, Ousmane, 'Le "Tuppal"', *Notes africaines*, 21, January 1944, p. 19

Diallo, Ousmane, 'Encres et teintures au Fouta Djallon', *Notes africaines*, 28, October 1945, p. 17

Donelha, André, *Descrição da Serra Leoa e dos Rios de Guiné do Cabo Verde (1625) / An Account of Sierra Leone and the Rivers of Guinea of Cape Verde (1625)*, ed. A. Teixeira da Mota and P. E. H. Hair, Lisbon, 1977

Dunlop, R., 'Sixteenth-Century Maps of Ireland', *English Historical Review*, 20, 1905, pp. 309–37

Dupuy, André R., ed., *Le Niokolo-Koba, premier grand Parc National de la République du Sénégal*, Dakar, 1971

English/British State Papers

Acts of the Privy Council [APC] 1588, ed. J. R. Dasent, London, 1897; *1613–1614*, London, 1921; *July 1623–March 1625*, London, 1933; *1626 June–December*, London, 1938; *1627 Jan.–Aug.*, London, 1938; *1628 July–1629 April*, London, 1958

Calendar of State Papers [CSP] Domestic 1598–1601, ed. M. A. E. Green, London, 1869; *1603–1610*, ed. M. A. E. Green, London, 1857; *1611–1618*, ed. M. A. E. Green, London, 1858; *1619–1623*, ed. M. A. E. Green, London, 1858; *1625–1626*, ed. J. Bruce, London, 1858; *1628–1629*, ed. J. Bruce, London, 1859; *1629–1631*, ed. J. Bruce, London, 1860

Calendar of State Papers [CSP] Colonial 1574–1660, ed. W. N. Sainsbury, London, 1860

Calendar of State Papers [CSP] Ireland, 1598 Jan.–1599 Mar., ed. E. G. Atkinson, London, 1895; *1606–8*, eds. C. W. Russell and J. P. Prendergast, London, 1874

Calendar of State Papers [CSP] Spanish 1587–1603, ed. M. A. S. Hume, London, 1899

Calendar of State Papers [CSP] Venetian 1619–1621, ed. A. B. Hinde, London, 1910

Correspondence relating to the recent expedition to the Upper Gambia under Administrator V. S. Gouldsbury [C.3065], London, 1881

Reports exhibiting the past and present State of Her Majesty's Colonial Possessions for 1848, London, 1849 [includes report on the Gambia Protectorate from Governor Mac-Donnell, 16.6.1849]

Farias, P. F. de Moraes, 'Silent Trade: Myth and Historical Evidence', *History in Africa*, 1, 1974, pp. 9–24

Fernandes, Valentim – see Bensaude and Brásio, above

Figueiredo, Manuel de, *Hydrografia … Com os Roteiros do Brasil, …Guiné, …*, Lisbon, 1614

Fitch, M., ed., *Index to Testamentary Records in the Commissary Court of London 1571–1625*, III, London, 1985

Francis-Boeuf, Claude, 'L'Industrie autochtone du fer en Afrique Occidentale Française', *Bulletin du Comité d'études historiques et scientifiques de l'A.O.F.*, 20, 1937, pp. 403–64

Froger, F., *Relation du voyage de Mr. de Gennes …*, Paris, 1698

Gamble, David P., *The Wolof of Senegambia*, London, 1967

——, with Sperling, Louise, *A General Bibliography of The Gambia*, Boston, Mass., 1979

——, *Terms found in Old Writings about Senegambia*, Gambian Studies 28, Brisbane, California, 1993

Gomes, Diogo – see Monod, Mauny and Duval, below

Gray, J. M., *History of the Gambia*, London, 1940

Gray, William, and Dochard, [-], *Travels in Western Africa …*, London, 1825

Guerreiro, Fernão, ed., *Relaçam anual das cousas que fizeram os padres da Companhia de Jesus nas partes da India Oriental, & em algũas outras …*, Evora/Lisbon, 5 parts, 1603–11, reprinted Coimbra, ed. A. Viegas, 1930–42; English version of Guinea sections, in Hair, *Jesuit Documents*, below

Hair, P. E. H., 'The Use of African Languages in Afro-European Contacts in Guinea: 1440–1560', *Sierra Leone Language Review*, 5, 1966, pp. 5–26; reprinted in Hair, *Africa Encountered*, below

——, 'The Text of Valentim Fernandes's Account of Upper Guinea', *Bulletin de de l'Institut fondamental d'Afrique noire*, sér. B, 31, 1969, pp. 1030–8

——, 'Guinea', in D. B. Quinn, ed., *The Hakluyt Handbook*, Hakluyt Society, 2nd ser., 144–5, London, 1974, pp. 197–207

——, 'The Falls of Félou: a Bibliographical Exploration', *History in Africa*, 11, 1984, pp. 113–30

——, 'Material on Africa (other than the Mediterranean and Red Sea lands) and on the Atlantic Islands, in the Publications of Samuel Purchas 1613–1626', *History in Africa*, 13, 1986, p. 126; also, reprinted, slightly abbreviated, in Pennington, *Purchas Handbook*, below, pp. 190–218

——, ed. and trans., *Jesuit Documents on the Guinea of Cape Verde and the Cape Verde Islands 1585–1617*, Department of History, University of Liverpool, 1989

———, 'Dutch Voyage Accounts in English Translation 1580–1625: a Checklist', *Itinerario*, 14, 1990, pp. 95–106
———, *Africa Encountered: European Contacts and Evidence 1450–1700*, Aldershot, 1997
———, 'Heretics, Slaves, and Witches – as seen by Guinea Jesuits c.1610', *Journal of Religion in Africa*, 28, 1998, pp. 131–44
———, and Alsop, J. D., eds, *English Seamen and Traders in Guinea 1553–1565: the New Evidence of their Wills*, Lewiston/Lampeter, 1992
———, Jones, Adam, and Law, Robin, eds, *Barbot on Guinea: The Writings of Jean Barbot on West Africa 1678–1712*, 2 vols, Hakluyt Society, 2nd ser., 175–6, London, 1992
Hakluyt, Richard, *Principall Navigations ...*, London, 1589; reprinted, eds D. B. Quinn and R. A. Skelton, Hakluyt Society, Cambridge, 1965; extended version, *Principal Navigations ...*, 3 vols, London, 1598–1600; reprinted 12 vols, Glasgow, 1903–5, original pagination shown
Hallett, Robin, *The Penetration of Africa*, London, 1966
Hardinge, Rex, *Gambia and Beyond*, London, 1934
Hill, R. H. Ernest, ed., *Index of Wills Proved in the Prerogative Court of Canterbury, vol. VI, 1620–9*, London, 1912
Historical Manuscripts Commission, *Twelfth Report, Appendix 1*, London, 1890
Holden, M., and Reed, W., *West African Freshwater Fish*, London, 1972
Holderer, P., 'Notes sur la coutume mandingue du Ouli (Cercle de Tombacounda)', in *Coutumes juridiques de l'Afrique Occidentale Française: Sénégal*, Paris, 1939, pp. 323–48
[**House of Commons**], *Journal of the House of Commons I*, London [n.d.]
Howard, Allen M., 'Trade and Islam in Sierra Leone, 18th–20th centuries', in Alusine Jalloh and David E. Skinner, *Islam and Trade in Sierra Leone*, Trenton/Asmara, 1997, pp. 21–63
Jallow, Cherno A., 'Hippo Lands and Islands', *The Gambia News Bulletin*, 18.6.1982, 2
Jammeh, Momodu, with Sheehan, Nancy Ann, *My Own Beef*, Land Tenure Centre, University of Wisconsin-Madison, 1993
Jannequin, Claude, *Le Voyage de Lybie au Royaume de Senega ...*, Paris, 1643
Jessup, Lynne, *The Mandinka Balafon: an Introduction with Notation for Teaching*, La Mesa, California, 1983
Joucla, E., *Bibliographie de l'Afrique Occidentale Française*, Paris, 1937
Keulen, Jan van, *De nieuwe groote lightende zee-fakkel*, pt. 5, Amsterdam, 1683
Labat, Jean-Baptiste, *Nouvelle relation de l'Afrique Occidentale ...*, 5 vols, Paris, 1728
Lasnet, Dr. [-], *Les Races du Sénégal: Sénégambie et Casamance*, Paris, 1900, pp. 173–8
Lemos Coelho – see Coelho above
Leporace, Tulia Gasparrini, ed., *Le navigazioni atlantiche del veneziano Alvise da Mosto*, Venice, 1966; Portuguese version of Cadamosto text, Damião Peres, ed., *Viagens de Luís de Cadamosto e de Pedro de Sintra*, Academia Portuguesa da história, Lisbon, 1948; English version, in Crone, *Voyages of Cadamosto*, above; French version, Frédérique Verrier, trans. and ed., *Voyages en Afrique noire d'Alvise Ca' da Mosto (1455 et 1456)*, Paris, 1994

Levtzion, Nehemia, 'Merchants vs. Scholars and Clerics in West Africa: Differential and Complementary Roles', in N. Levtzion and Humphrey J. Fisher, eds, *Rural and Urban Islam in West Africa*, Boulder, 1987, pp. 21–37

Lockyer, Roger, *Buckingham*, London, 1981

Loiser, Chris M., and Barber, Anthony D., 'The Crocodile Pools of the Western Division, The Gambia', *British Herpetological Bulletin*, 47 (1994), pp. 16–22

Lorimer, Joyce, ed., *English and Irish Settlement on the River Amazon 1550–1646*, Hakluyt Society, 2nd ser., 171, London, 1989

Madeira Santos, Maria Emília, ed., *História geral de Cabo Verde*, II, Lisbon/Praia, 1995

[**Marees**, Pieter de], *Beschryvinge ende Historische verhael vant Gout Koninckrijck van Gunea ...* (Amsterdam, 1602); English version, Pieter de Marees, *Description and Historical Account of the Gold Kingdom of Guinea (1602)*, trans. and ed. Albert van Dantzig and Adam Jones, Oxford, 1987

Mark, Peter, *The Wild Bull and the Sacred Forest: Form, Meaning, and Change in Senegambian Initiation Masks*, Cambridge, 1992

Masseville, [-], *Histoire sommaire de Normandie* (Paris, 1668)

Mattiesen, Otto Heinz, *Die Kolonial- und Überseepolitik der kurländischen Herzöge im 17. und 18. Jahrhundert*, 2 vols, Stuttgart, 1940 ['Journael van den Crocodel', 1651, I, pp. 45–170]

Mauny, Raymond, 'Notes d'histoire sur Rufisque d'après quelques textes anciens', *Notes africaines*, 46, April, 1950, pp. 47–9

Mitchinson, A. W., *The Expiring Continent: a Narrative of Travel in Senegambia*, London, 1881

Monod, T., Mauny, R., and Duval, G., trans. and eds, *De la première découverte de la Guinée récit par Diogo Gomes (fin XVe siècle)*, Bissau, 1959

Monod, T. – see also Bensaude above

Moore, Francis, *Travels into the Inland Parts of Africa*, London, 1738

Moraes, Nize Isabel de, 'Le commerce des peaux à la Petite Côte au XVIIe siècle (Sénégal)', *Notes africaines*, 134, 1972, pp. 37–45

——, ed., *À la découverte de la Petite Côte au XVIIe siècle (Sénégal et Gambie)*, 4 vols, Dakar, 1993–5

Munter, Robert, and Grose, Clyde L., *Englishmen Abroad, being an Account of their Travels in the Seventeenth Century*, Lewiston, 1986

Neveu-Lemaire, Dr M., *Deux Voyages cynégetiques et scientifiques en Afrique Occidentale Française, 1911–1914*, Paris, 1920

Nunes Costa, Mario Alberto, 'D. António e o trato Inglês da Guiné', *Boletim cultural da Guiné Portuguesa*, 8, 1953, pp. 683–797

Oliver, Roland, ed., *The Cambridge History of Africa*, III, Cambridge, 1977

Pacheco Pereira, Duarte, *Esmeraldo de Situ Orbis: Côte occidentale d'Afrique du Sud Marocain au Gabon* [Portuguese/French], ed. R. Mauny, Bissau, 1956; English version, in G. H. T. Kimble, ed, *Esmeraldo de Situ Orbis*, Hakluyt Society, 2nd ser., 79, London, 1937

Park, Mungo, *Travels in the Interior Districts of Africa ...*, London, 1799

Pelissier, Paul, *Les Paysans du Sénégal*, Saint-Yrieix, Haute-Vienne, 1966

Pennington, L. E., ed., *The Purchas Handbook*, 2 vols, Hakluyt Society, 2nd ser., 185–6, London, 1997

Peres, Damião – see Coelho above

Pina, Ruy de, *Crónica de el-Rei D. João II*, ed. Alberto Martins de Carvalho, Coimbra, 1950

Poole, Thomas Eyre, *Life, Scenery and Customs in Sierra Leone and the Gambia*, 2 vols, London, 1850

Porter, R., 'The Crispe Family and the African Trade in the Seventeenth Century', *Journal of African History*, 9, 1968, pp. 57–77

Portères, Roland, 'Encres et tablettes à écrire de fabrication et d'utilisation locales à Dalaba (Fouta-Djalon, République de Guinée)', *Notes africaines*, 101, 1964, pp. 28–9

Portugaliae Monumenta Africana, I, eds Luís de Albuquerque, Maria Emília Madeira Santos, *et al.*, Lisbon, 1993

Pory, John, *A geographical historie of Africa*, London, 1600; reprinted in Robert Brown, ed., *The History and Description of Africa ...*, 3 vols, Hakluyt Society, 1st ser., 92–4, London, 1896

Purchas, Samuel, *Purchas his Pilgrimage ...*, London, 1613, 1614, 1617, 1626

——, *Purchas his Pilgrimes ...*, 4 vols, London, 1625; reprinted 20 vols, Glasgow, 1905–7, original pagination shown

Quinn, D. B., *The Elizabethans and the Irish*, Ithaca, 1966

Rabb, T. K., *Enterprise & Empire: Merchant and Gentry Investment in the Expansion of England*, Cambridge (MA), 1967

Raffenel, Anne, *Voyages dans l'Afrique occidentale... et de la Gambie, depuis Baracounda jusqu'à l'Océan en 1841 et 1844*, Paris, 1846

Rançon, André, *Dans la Haute-Gambie: Voyage d'exploration scientifique, 1891–1892*, Paris, 1894

Ratelband, K., ed., *Reizen naar West-Afrika van Pieter van den Broecke 1605–1614*, 's Gravenhage, 1950

Reeve, Henry Fenwick, *The Gambia: its History, Ancient, Mediaeval and Modern*, London, 1912

Rose, J. Holland, Newton, A. P., and Benians, E. A., eds, *Cambridge History of the British Empire*, I, Cambridge, 1929

Rougé, Jean-Louis, *Petit dictionnaire étymologique du Kriol de Guiné-Bissau et Casamance*, Bissau, 1988

Roure, G., *La Haute Gambie et le Parc national du Niokolo Koba*, Dakar, 1956

Ruiters, Dierick, *Toortse der Zeevaerten*, Vlissinghen, 1623; reprinted, ed. S. P. L. Naber, Linschoten-Vereeniging, 's Gravenhage, 1913

S Lo, Alexis de, *Relation du Voyage du Cap-Verd*, Paris, 1637

Saad, Elias N., *Social History of Timbuktu: the Role of Muslim Scholars and Notables 1400–1900*, Cambridge, 1983

Sanneh, Lamin O., 'Field-work among the Jakhanke of Senegambia', *Présence africaine*, 93, 1975, pp. 92–112

——, 'Slavery, Islam, and the Jakhanke People of West Africa', *Africa*, 46, 1976, pp. 80–97

——, *The Jakhanke*, London, 1979
——, *The Jakhanke Muslim Clerics: a Religious and Historical Study of Islam in Senegambia*, Lanham, 1989
——, 'A Childhood Muslim education: *Barakah*, Identity, and the Roots of Change', in *The Crown and the Turban: Muslims and West African Pluralism*, New York, 1997, pp. 121–46
Santarém, Visconde de, *Quadro elementar das relações políticas e diplomáticas de Portugal*, III, Paris, 1854
Scantamburlo, Luigi, *Gramática e dicionário da língua Criol da Guiné-Bissau (GCr)*, Bologna, 1981
Scott, W. R., *The Constitution and Finances of English, Scottish and Irish Joint-Stock Companies to 1720*, 3 vols, Cambridge, 1910–12
Shoberl, Frederic, ed., *The World in Miniature* [translation of Geoffroy de Villeneuve, *L'Afrique, ou histoire, moeurs, usages et coutumes des Africains*, Paris, 1814], 4 vols, London, 1821–5
Skelton, R. A., *Geographical Magazine*, 25, 1956, p. 153
——, *Explorers' Maps*, London, 1970
So, Sire, 'Le Tribunal des singes de Monkey Kote', *Notes africaines*, 40, October 1948, pp. 4–5
Squibb, C. D., *Wiltshire Visitation Pedigrees 1623*, Harleian Society, London, 1954
Stibbs, Bartholomew, 'A Voyage from James Fort up the River Gambia by Command of the Royal African Comp. of England', in Moore, *Travels*, above, pp. 245–97
Stokes, E., ed., *Index of Wills proved in the Prerogative Court of Canterbury, 1605–19*, V, London, 1912
Stone, Thora G., 'The Journey of Cornelius Hodges in Senegambia 1689–90', *English Historical Review*, 39, 1924, pp. 89–95
Tarawale, Ba, *Mandinka-English Dictionary*, Banjul, 1980
Teixeira da Mota, A., 'Un document nouveau pour l'histoire des Peuls au Sénégal pendant les XVème et XVIème siècles', *Boletim cultural da Guiné Portuguesa*, 24, 1969, pp. 781–860, also in série separatas 56, Agrupamento de estudos de cartografia antiga, Lisbon, 1969, repaginated 1–86
——, *Dois escritores quinhentistas de Cabo Verde: Ándre Álvares de Almada e André Dornelas*, série separatas 61, Agrupamento de estudos de cartografia antiga, Lisbon, 1971, reprinted from *Liga dos Amigos de Cabo Verde – Boletim cultural, suplemento*, November 1970, pp. 10–44
——, *Mar, além Mar: estudos e ensaios de história e geografia*, Lisbon, 1972
Thévenot, Melchisedek, *Relation de divers voyages*, 4 vols, Paris, 1661–4
Thilmans, Guy, 'La Relation de François de Paris (1682–1683)', *Bulletin de l'Institut fondamental d'Afrique noire*, sér. B, 38, 1976, pp. 1–51
United Nations Food and Agricultural Organization, *Report to the Government of Gambia and Senegal: Integrated Agricultural Development in the Gambia River Basin*, Rome, 1964
Van-Chi-Bonnardel, Régine, 'Exemple de migrations multiformes intégrées: les migrations des Niominka (Îles du bas Saloum, Sénégal)', *Bulletin de l'Institut fondamental d'Afrique noire*, sér. B, 19, 1977, pp. 836–9

Vinche, P. P., Singleton, M., and Diouf, P. S. D., 'Techniques et instruments de chasse chez les Séreer du Sine (Sénégal)', *Notes africaines*, 138, October 1985, pp. 113–18

Viterbo, Sousa, 'Noticia de alguns arabistas e interpretes de linguas africanas e orientaes', *O Instituto*, 52, 1905; 53, 1906 [many instalments]

Warburton, Eliot, *Memoirs of Prince Rupert*, 3 vols, London, 1849

Weil, Peter M., 'Slavery, Groundnuts, and European Capitalism in the Wuli Kingdom of Senegambia, 1820–1930', *Research in Economic Anthropology*, 6, 1984, pp. 77–119

Westermann, Dietrich, and Bryan, M. A., *Languages of West Africa*, Oxford, 1952

Willan, T. S., *Studies in Elizabethan Foreign Trade*, Manchester, 1959

Wilson, W. A. A., *The Crioulo of Guiné*, Johannesburg, 1962

Wright, Donald R., 'Darbo Jula: The Role of a Mandinka Jula Clan in the Long-distance Trade of the Gambia River and its Hinterland', *African Economic History*, 3, 1977, pp. 33–45

——, *The World and a Very Small Place in Africa*, New York, 1997

Zimmerman, Robert, *Enchantment of the World: The Gambia*, Chicago, 1994

UNPUBLISHED SOURCES

Baum, Robert M., 'Incomplete Assimilation: Koonjaen and Diola in Pre-Colonial Senegambia', paper, American Historical Association annual meeting, 1983

Bühnen, Stephan, 'Geschichte der Bainunk und Kasanga', Ph.D. thesis, Justus-Liebig-Universität Giessen, 1994

Collins, Edward D., 'The Royal African Company', Ph.D. thesis, Yale, 1899

Galloway, Winifred, 'A History of Wuli from the Thirteenth to the Nineteenth Century', Ph.D. thesis, Indiana University, 1975

Gamble, David P., field notes

Hunter, Thomas Charles, 'The Development of an Islamic Tradition of Learning among the Jahanke of West Africa', Ph.D. thesis, University of Chicago, 1977

Maloney, Father, 'Notes, compiled by Father Maloney and John Balde' [1946], Roman Catholic Mission, Basse

Pryce, H. Lloyd, 'The Laws and Customs of the Mandingos of the North Bank Territory of the Gambia Protectorate' [1907], Gambian National Archives

Sanneh, L. O., 'The Muslim Education of an African Child: Stresses and Tensions', paper, Manding Studies Conference, School of Oriental and African Studies, University of London, 1972

Stanley, W. B., 'Notes on the Physical Distribution of the Country, and Political Organisation of the Fullahs of the Gambia, their Customs, Laws, etc. [1907]', Gambian National Archives

Traouré, Amadou, 'African Games', paper, Manding Studies Conference, School of Oriental and African Studies, University of London, 1972

Weil, Peter M., 'Land Use, Labour and Intensification among the Mandinka of Eastern Gambia', paper, African Studies Association annual meeting, 1980

MAPS OF RIVER GAMBIA, 1468–1980

For reproductions of the earliest maps noting River Gambia, see Teixeira da Mota, 1972, plates 1, Benincasa 1468, 3, Reinel *c.* 1485, 7, Cantino 1502, 10, 'Paris' *c.*1468, etc. Sixteenth-century maps showed only a notional course of River Gambia and the toponym 'Cantor'.

Luís Teixeira, 'Effigies ... Guineae', printed map, Amsterdam, 1602; reproduced and discussed in Marees, 1987, pp. xxiii–xxv; section on Senegambia reproduced in Donelha, 1977, fig. 3; notional course of river but earliest map to show several up-stream toponyms as far as 'Cassan' and 'Cantozy'.

Gerbier manuscript map, 'La Rivière Gambia et La Cité de Cassan en Afrique', British Library, two copies, King/Topographical CXVII, 98, and Add. MS 16371.h (coloured, decorative illustrations redrawn, with attached and apparently related text, see Appendix C), a large map (120 x 31 cm), the text signed B. Gerbier Douuily, no date, ? 1610s or 1620s. The latter copy reproduced in Donelha, 1977, fig 4. Shows the course of the river in some detail, the earliest extant map to do so, with many toponyms, up to just beyond Kasang. Largely copied in *De Lichtende Zeefakkel*, Amsterdam, ?1683, 2: map 45.

Huntington manuscript map, Henry E. Huntington Library, San Marino, California, HM 2098, nautical chart of west coast of Europe and Africa, anon., English (? Thames School), ? 1620–50, section includes crude course of river to beyond 'sette-coe', few toponyms.

Courlander manuscript maps, (a) the river to just beyond Tendabaa, 1651, reproduced in Andersons, 1970, (b) the river as far as Kasang, 1652, German, reproduced in Mattiesen, 1940, Karte 1, the configuration and toponymy only partly correspond, the many toponyms and other inscriptions are difficult to read.

Vermuyden manuscript map, 'The Mapp of the River of Gambia', British Library, King/Topographical, CXVII, 96, apparently recording a 1661 up-river voyage of John Vermuyden, the river to far beyond Baarakunda, the lower portion of the map reproduced in Skelton, 1956, p. 153; and 1970, fig. 183, the upper portion representing the earliest map of the upper river, this portion with many toponyms, but mainly invented ones

Fitzhugh manuscript map, 'Guinea', 1684, Bibliothèque Nationale, Paris, Depôt, III.1.11, the river section reproduced in Donelha, 1977, fig. 6, the river shown to just beyond 'Maresansan', many toponyms

Leach printed map, 1732, 'A Draught of the river Gambia with the Soundings; also all the Kingdoms, most of the principall Towns, and the Royal African Companies Factories ... in a course of 500 miles', in Moore, 1738, detailed map of the river to Baarakunda (215 x 678 mm), many toponyms and political boundaries, often reproduced, e.g. in Skelton, 1970, fig. 184, and, with modern equivalents added, in Douglas Grant, *The Fortunate Slave*, London, 1968

'Map of the Upper Gambia for 105 miles above the Falls of Barraconda ...', 1848, in the Despatch of Governor **MacDonnell** no. 41, in *Reports ... 1848*, 1849

'Map of the Upper Gambia from Yarbutenda to Bady Wharf' (1: 145,800), 'Expedition to the Upper Gambia, 1881, Route Survey' (1: 729,000), maps by Lieutenant H. N. **Dumbleton**, War Office, Intelligence Department, nos. 65–66, in *Correspondence* ... (Colonial Office Report C. 3065), 1881, reprinted in *Irish University Series of British Parliamentary Papers* (Colonies, Africa, 56); see p. 193 above

'Gambia', 1906, War Office, Topographical Section, map no. 1958 (2 sheets, 1: 250,000), based on the work of the 1904–1905 Anglo-French Boundary Commission, revised and reissued 1931

'French West Africa', 1941, War Office, Geographical Section, no. 4149, upper river section D. 28 (4 sheets, 1: 200,000), based on 1925 map, Service Géographique de l'Afrique Occidentale, Dakar

Chart of River Gambia, 1942, Admiralty charts, nos 608–609 (1: 75,000), later revisions

Maps of Gambia Colony, 1948/1956, Directorate of Colonial Surveys, series 15 (later 415), 30 sheets (1: 50,000)

Maps of Gambia Colony, 1956, Directorate of Colonial Surveys, series 502 (3 sheets, 1: 125,000)

Maps of The Gambia, 1980, Directorate of Overseas Surveys for the Government of The Gambia (2 sheets, 1: 250,000)

LIST OF MARGINAL NOTES (SIDENOTES)

THE GOLDEN TRADE

The Invitement to this golden Trade, shewing the cause of the first undertaking it and orderly proceeding therein.

The trade of the Moores in in [sic] Barbary for their gold.	78
The Kings Majesties Letters Patents. The first voyage.	79
The ship taken by the vagrant Portingall, and the men slaine. The second voyage.	80
Captaine *Thompson* slain. The third voyage.	81
The whole way from England to the River, runne in 20. dayes. The particulers handled in this booke.	82

The description of the River

These are all more largly written of, where the tillage of the ground is handled. We were 10 of our owne company, that went up in a shallop, and 4 Blacke[s] that I hired to carry up a Canoe.	86
These places are more largly written of when I set downe the manner of our trade, at the highest we went in the River.	87
The Country people have no boates or Canoos above the ebbing and flowing. We laboured to get up the River onely 7. houres in 24. No townes neare the River side after wee past the ebbing and flowing.	89
The description of the Crocodile, whom the Country people call Bumbo. How the people do feare him. The manner they passe their cattle over the River, for feare of Bumbo. The Blackes would not goe into the River.	90
Their answere when they went in. The strong sent of the Crocodile, changing the tast of the River water, & the fish were taken in it. The description of the sea-horse.	91
The sea-horse feedes in the night upon the shore. A daungerous blow, by a sea-horse. A neglect to be hereafter, carefully provided for.	92
The Country people esteeme the sea-horse, for excellent meate.	93
A strange operation of a fish. The running fish.	94
The nature of the river foule. The manner of the peoples fishing.	95

The severall Inhabitants, &c.

The vagrant Portingall.	97
An especiall Caveat.	98
The curteous usage of the natural inhabitants.	99
The reward of treachery.	100

The wandering Fulbie.

The misery of the Fulbie.	101
The cleanelines of the Fulby women.	103

The Maudingo *or* Ethiopian, *being the naturall Inhabitants distinguished by the name of the* Maudingos.

The time and manner of the peoples feeding.	104
A digression, by the Writer, for the better preserving of mens lives and healths.	106
The Writers opinion concerning dyet. The Caveat must be lookt carefully to, in the setting forth. The manner of their building.	107
Strange Ante-hils. The towne of *Cassan* with the manner of fortification.	108
The armes or weapons the people have in use.	109
The inhabitan[t]s custome in the night.	110
The great King of *Cantore*. The great King of *Bursall*. The great King of *Wolley*.	111
These great Kings are likewise tributaries to one great King far ahove [*sic:* above] in the land, as is reported to us. The reverence of the people to the petty Kings. The Religious ceremony of these people.	112
The manner of their apparrel. The description of their Gregories, which are charmes they receive from their Mary-buckes.	113
The number of their wives. Allowance of other women for necessitie sake. The reason of that necessity.	114
Strict punishment for unchastity. The men buy their wives.	115
The widdowes buy their husbands. The subjection of the women. No outward daliance seene amongst them. A strange report.	116
The womens clothing. The manner of taking away their wives, which in some sort is used in *Ireland* at this day. The modesty of a new married woman	118
These people stand much upon their dignity. A dangerous quarrell betwixt them. The certaine knowledge of their Kings & Governours and their successors.	119
Their titles of honor. Wherein their Riches consists. Great Beggers. The temporall people great drinkers of Aquavitæ. The life of their Kings truly described.	120
The deposing of Kings. The ceremony used betwixt the King and us, when first we meet.	121
He gives his chiefe Gregory drinke first. Their women not allowed to drinke in publike, although they love it well.	122

The discourse of their Maribuckes or religious men.

They observe the leviticall Law. And have great knowledge of the old Testament. They marry in their own tribe and breede up their children in their owne sects.	122
The Mary-bucks have the same allowance of women the Kings or temporall people have. This *Fodee Kareere* was my Alchade and bought & sold for me.	123
Fodee Bram was the chiefe Mary-bucke of all the Country. Both Priest & people weare one manner of apparrell. The chiefe Mary-bucke, daungerously sicke.	124
His manner of entertaining me. The valuation of the present I gave him, which was so highly esteemed. The description of the town called Setico.	125

LIST OF MARGINAL NOTES

The Chiefe Mary-bucke wonderfull desirous to confer with me about our Religion. They worship the true God above, whom they call Alle. They have no manner of Image. They have no Churches. They observe not their Sabboth. The manner of teaching their male children to write, and reade.	126
The manner of their character. Their law is not written in the publicke tongue. We suppose they performe their religious ceremonies, under the shady trees.	127
They called our dwelling the white mens towne. In any occasion of falling out betweene the people and us, this old man would come with his Assegy presently to ayde us. Their manner of devotion. The death of the chiefe Mary-bucke.	128
The great resort to his buriall. They do bury the body, with all sweete savors and perfumes they can get. The manner of buriall. Verses and Orations in commendations of the deceased. A Relique of great esteeme.	129
The investing of the eldest sonne in the fathers place. A Ramme for sacrifice. They cal Christ by the name of Nale. The opinion they hold concerning him.	130
They have bookes of great volumes all manuscripts. The wonderfull sobernesse of these Mary-buckes.	131
They abstaine from all sweete things. A strange example of abstinence. The true way to know a mary-bucke. They will tell no lies.	132
The marybucks manner of travaile. They will beg of us without deniall. The Marybucke free to travaile in all places. Their report of gold.	133
Old Mahome his ceremony at my going up. The subtilty of his neighbour Hammet. The trade and travaile of the mary-bucks of Setico. Wherewith he maintaines his greatnesse.	134
An ill opinion of the Marybuckes, to bury their golde. A good commoditie. The reason of looking after Buckor Sano.	135
Note.	136

Our travell up the River

The Marybuckes name was Selyman, the other Tombo. Samgulley a blacke boy.	136
Eleven dayes travell against the streame, wherein wee wrought eighty eight houres. Some of our men grew fearfull.	137
As bigge in body as a great Stagge, and had wreathed hornes. The returne of one of our messengers. The comming of Buckor Sano. Provision the people brought. His going aboord the boat and report of our powder. A Stalker.	138
The saying of Buckor Sano aboord the boate. Hee was but once overtaken with our strong drinkes. The great esteeme of salt. He makes a proclamation.	139
He offers women to sell unto us. Their commodities. A markethouse made a shore. Warning not to take notice of their gold.	140
Buckor Sanos report of gold and of the houses above covered therewith. He seem'd wonderous willing of our companies.	141
Buckor Sano his subtill speech. His declaration of the Moores of *Barbary*. Pleasing intelligence, being the maine businesse wee ayme at.	142
Their course of trading. An oath they observe carefully. An unhappy accident. A people that never saw white men before.	143

Strange breeches the common people did weare. An encouragement to search further up the River. Those people had another language.	144
These people expect our returne. The King comes unto us. Buckor Sano did always eate with us in the boate. These exercises did commonly hold three houres in the night.	145
Buckor Sano made the white mens Alchade. The acknowledgment of his new title. His mediation to the King in our behalfes. The Kings answere. Buckor Sanos gratification. The Kings acceptance and faire reply. He gives us the Country.	146
A strange recemoney [*sic:* ceremony] in takeing of possession. The possession given unto me. A great protestation of defence. A people markt in the face. Observe this Mary-bucke.	147
Ferambra was Lord of his Country, and when the Portingals had got the King of Nany to send horsmen to kill Thompson and his small company, hee did preserve them, and put himselfe & Country in armes for their defence. The marybucks first tale. His second discourse.	148
Our opinion cercerning [*sic:* concerning] Trombutto [*sic:* Tombutto] and Gago. More incouragment to go further up the River. The manner of merchandising without speech or sight one of the other. The report of the people with the great lippe.	149
The people who bought our salt had no use of it, but for sale. This ought speedily to be considered of. The place called St. *John* Marte. Our curteous parting with Buckor Sano.	150
The fashion of the Irish Rimer. Upon this instrument only they play with their fingers. A strange consortship. Their chiefest instrument.	151
The manner of this instrument. Their manner of daunsing. Their Fidlers rich.	152
They are basely esteemed of, and being dead are not buried. The affection of Samgulley our blacke boy unto us.	153
This Bo John was brother to Ferambra. The feast of their Circumcision. Samgulley taken from us to be circumsised. The great resort to this solemnity.	154
They that were cut, kept all together. The curtesie and mirth that past betwixt us. women looke upon the circumcision.	155
Our boy circumcisised [*sic:* circumcised], and the manner thereof. We were not suffered to go amongst the new circumcised. No use of medicines to cure them.	156
The discourse of their divell Ho-re.	157
He is a monstrous eater. Sometimes 8 or 9 atonce, are carried away, and sayd to be in his belly. Comming forth they speake not for certaine daies. Our opinion concerning Ho-re.	158
How he was partly discovered. An example of the divells converse with the Fidlers.	159
The Divell could not tell the Portingall where [whether] we were friends or foes. The trades or occupations they have in use, their painfull season of thunder and lightning, also what fruites & plants the Countrey yeeldes, and are growing there amongst them. The Smith. An excellent charcole to worke their Iron.	160
The Sepatero they of this trade are most ingenious. The Potter & tobacco pipe maker.	161

They have in the highest of the River, excellent mattes. A market kept every monday. No mony or coyne amongst them. All labour to till the earth and sow their graine.	162
They understand not to make their cattle worke. The manner of their painefull labours. Their corne, or graine. The manner of their Rice.	163
The planting of cotton. The misery of the people. The times of their raines & the fearefulnes thereof.	164
A faire intreaty to men of judgment. The great aboundance of poyson. The nature of the first raines. An observation to be kept.	165
A note of experience. An observation of the tempestuous times. They heare & speak of Christ but will not beleeve. Gods mercy to us.	166
A comfort to the traveller. Plantans. Limes. Orenges. Good wine forth of a tree.	167
Severall sorts thereof. Palmeta apples. A made drinke, called Dullo. Gowrdes. Locusts.	168
Wild hony. Munkies meat. A stony apple. This fruite is of great esteeme.	169
They are not growing within the limit we saw. Great store brought us, when we were above. This is like our water Lilly.	170
The sensible tree.	171

The discourse of the wild beasts.

All ravenous beasts in the day time keepe their dennes. The Lyon. His small servant. His manner of hunting. The causes of our knowledge.	171
Ounces, and Leopards. The Ounce dangerous. A true tale of a Child. Civit Cat. Porcupine. The Elephant.	172
The nature of the great ones. He browses like a deare. A false opinion. This was that *Ferambra* I noted before. Elephants flesh good to feede upon.	173
The manner how he killed them. I brought two of these tayles away with me. The Elephant a fearefull beast. The peoples admiration, we durst set upon them. Buffelos. Blew boores	174
Antelops. Deare of all sorts. Munkies. Babownes a strange story.	175
A government amongst them. The people of the Country eate them. The Spaniards opinion of them. The pleoples [*sic:* peoples] report of a Unicorne.	176
The Stalker. The wake. Ginney hens. Patridges.	177
Quailes. Pigeons. Parats. Paraquetos. Variety of smal birds. A small bird without legges. A bird with foure wings, about the bignes of a turtle Dove.	178
How the birds preserve their young from the Babouns, and Munkyes. The subtilty the Babowne. Another meanes of preservation.	179
Hawkes that will kill a Vallow deare. Bastard Eagles. How the people finde the dead beasts. The Inhabitants want knowledge to take them. The Kings manner of Hawking.	180

The Conclusion.

The vagrant Portingall.	182

JOURNALL

A true Relation of Master RICHARD JOBSONS Voyage ... Extracted out of his large Journall

The Katherine betrayed. Gambra Portugals which trade.	185
The generall winds. Tankorovalle. Tindobauge. The Voyage up the River. Pudding Iland. Maugegar.	186
Wolley, Wolley. Cassan. Portugals perfidie. Pometon. Jeraconde. English at Oranto.	187
Oranto. Ferambas faith. Batto.	188
Sea-horses, high-wayes. Marybuckes, sacred persons, by the superstition of those parts, and are their Priests and Merchants. Sea-horses abounding in the fresh water, both in the water and on shoare. They are like a horse, but with clawes on their feet, and short legs, tuskes, manes, etc. Monkies and Baboones.	189
Crocodiles thirty foot long. Elephants. Muskie water distastefull. Tinda. Muskie fish.	190
Antelope. Gun-thunder. Salt, chiefe trade.	191
Bajay Dinko usko was the chiefe man, & called by the name of his Countrey, under the great King of Cantor. Juddies or Fidlers. He bought and sold etc. for us. Nuts of precious esteeme. He seemeth to be the Cola.	192
The Countrey given to the English. In this manner the Kings take possession of the lands they came [? come] to. Iron preferred before Gold.	194
Much Gold. Combaconda. Tombuto. Barraconde. Circumcision. Daunces.	195
Hore. Setico. Blow by a Sea-horse. Marybucks Funerall.	196
Devils oracles.	197

INDEX

PERSONS AND PEOPLES

The following terms, occurring frequently, are not listed: African/s, English.

Adam/Adama, 122, 126, 163, 203
Admedi, 260
Aethiopians, 1 (*see also* Ethiopian/s)
Affonso, Gonçalo, 261
Affonso, King, 264
Afro-Portuguese, 5, 9, 18, 20, 33, 34, 41, 50, 56, 68–71, 243, 288, 291 (*see also* Portingale/s)
Aholah, 115
Aholibah, 115
Alexander [the Great], 77
Almada, André Álvares de, 239, 242, 274; *text* 274–84
Andrade, Francisco de, 239, 242, 273; *text* 273–4
Andrea, 250
Anhadalen, 278
Anriques, Diego, 290
António, Dom, 10
Antoniotto, 267 (*see also* Usodimare)
Apsley, Sir Allen, 39, 75
Arabecke/s, 142–3, 148, 194–5
Arabs, 270

Bacan Tombo, 136
Bacay Tombo, 136, 189
Bachares, 308 (*see also* Basari)
Baiage Dinggo/Bajay Dinko, 144, 192
Bainunka, 67
Bambara, 58, 68
Banhús/Banhuns, 280–81, 283, 294–5
Barbacijs/Barbacins/Barbaçins, 264, 271, 274, 276, 283
Barros, João de, 239, 242, 265, 271; *text* 271–2
Basari, 68 (*see also* Bachares)
Batimansa/Batimaussa, 251–2, 256, 261
Beafares/Beafar, 281–2
Blithe, John, 25, 187
Bo John/Boo John, 119–20, 154–5, 159, 188, 195, 224, 228–9

Bormelli, 261 (*see also* Mormelli)
Brewer, [-], 23–5, 188
Bridges, Henry, 25, 187
Broad, Matthew, 25, 36, 187–8, 192
Brown, Mr, 234
Bucker, 259
Bucker, 194 (*see also* Buckor Sano)
Buckingham, Duke of, 232
Buckor Sano, 1, 22–4, 28, 30, 32, 34–7, 44–5, 68, 81, 83, 135–43, 145–8, 150, 190–92, 194, 218–19 (*see also* Bucker)
  misses contact with Thompson, 137; meets Jobson in Tinda, 138, 191; drunk, 139; directs trade, 139; informs re gold trade, 140–42; made intermediary by Jobson, 145; introduces local king, 145–7, 192–4; departs, 150
Buramos, 280
Burgundy, Duchess of, 255
Button, Sir Thomas, 39, 75.

Cadamosto [Alvise Ca' da Mosto], 40, 239, 241–2, 244, 256, 258, 267 (*see also* Mosto); *text* 244–56
Cafres, 278
Caiado, Christavão, 289
Canaan, 114
Cassangas, 280–81
Charles [I], King, 2, 15, 39, 199
Child, Lieutenant, 234
Christian/s, 41, 42, 57, 69, 70, 98, 130, 248–9, 252, 256–60, 262–4, 267, 269, 276, 294, 311
Columbus, 204
Conjuros, 292, 307
Consalvos, Jasper, 159
Correa da Silva, Francisco, 290
Corseen, Emanuel, 186
Coxe, Thomas, 21

Cramp, M., 185
Cruz, Izidro da, 289
Curtius, Quintus, 40, 77

Davies, John, 13, 28
Davis, Humphrey, 25, 187
Delgado, João, 264
Dowry, 155
Dutch, 9, 69

Egyptians, 100
Elving, Robert, 26
Ethiopian/s, 82, 104, 132, 200, 203 (*see also* Aethiopians)
Eve/Evahaha, 122, 126, 203
Ezechiel, 115

Falupos, 295
Farisangul/Farosangoli/Forisongul, 252, 258, 261
Ferambra, 20, 44, 60, 120, 147–8, 154, 158, 173–4, 188, 213, 224, 226, 228–9
Fernandes, Valentim, 239, 241–2, 267; *text* 267–71
Ferran, 120
Fodee Bram/Fodea Brani, 124, 196
Fodee Careere, 28, 34–5, 63, 123, 215
  hired, 123, 189; travels up-river with Jobson, 124, 136; nearly drowned, 132, 190; as interpreter, 34–5, 143; accompanies Jobson to circumcision, 154; accompanies to Setico, 124; introduces to senior marabout, 124
Forbicher, 232
Francisco, 185
Frangazick, 258
French, 5, 9, 10, 69, 281, 283, 289, 305–6
Fula/Fulo/s/, 32, 44, 52, 56, 64–6, 228, 276–9, 281–3, 300, 302, 304 (*see also* Fulbe, Pholey)
Fulbe/Fulbie/s, 43–4, 56, 61, 64–7, 82, 100–101, 103–4, 156, 158, 224, 228 (*see also* Fula, Pholey)

Galloway, Winifred, 59–60
Geloffa/Gyloffos, 262, 267 (*see also* Jalofo, Jolof, Wolof)
Gerbier/Gerbier Douilly, 50, 231–2
German, 247, 258, 267
Gnumimenssa, 254
Gomes, Diogo, 5, 239, 241–2, 258; *text* 258–65
Grand Fulo, 273, 277–8, 283
Grant, John, 235

Hakluyt, Richard, 6, 11, 12, 40
Ham, 42, 114
Hammet, 128, 134, 227
Hawkins, John, 31
Henrique, Dom, [the Infante], 255, 258 (*see also* Henry, Infante)
Henry, [the Infante], 263 (*see also* Henrique, Infante)
Henry [VII], King, 204
Hobab, 132
Hodges, Cornelius, 47–8, 52, 55

Infante, the, 255, 258, 260, 262–4, 267 (*see also* Henry, Henrique)
Irish, 39, 103, 129, 151, 192

Jacob, 263
Jagancases/Jagancazes, 292, 306
Jalofo/s, 266, 276, 281, 283, 285, 289–90, 296 (*see also* Geloffa, Jolof, Wolof)
James [I and VI], King, 15, 49, 209, 232, 294
Jaroale, 282, 284–5
Jatta, 68
Jaxanke, 51, 58, 61–4
Jeremy, 132
Jesus, 42
João [II], King, 272
Jobson, Richard
  in River Gambia, 24–38; itinerary, 308 ff; career and character, 38–46; employed by Guinea Company, 82; in lower river, 84–5; informed of Thompson's murder, 186; leads shallop up-river, 186; receives gift from a Portuguese, 186; at middle river base, gifts from Ferambra, eats elephant meat, 173–4, 188; meets various kings, 111, 120; intervenes in quarrel, 119; shown child carried off by animal, 122; inquires re Hore, 158–9; helped by aged marabout, 128; in boat party up-river, 86–7, 136–8, 188 ff; fed by forewarned Portuguese host, 159–60, 197; health on journey, 106; sees elephants, 190; contacts Tinda, 87, 138; trade at Tinda, 140, 144, 191; meets Buckor Sano, 138 ff; 191; inquires re gold trade, 140–41, 143, 148–9, 194–5; discussion hindered, 143, 195; sends to interior, 143; visited by local king, 145–7, 192; honours Buckor Sano, 145, 192; awarded land rights, 146–7, 194; leaves Tinda, 150; 195; attends circumcision, 153–7, 195–6; visits Setico, 123, 196;

INDEX

attends senior marabout, 124–5; attends burial, 128–30, 196; buys monkey, 175; observes mimosa, 170–71; spends month at Kasang, 197; advocates more voyages, 180–84; petitions crown, 199 ff.
Jolof/s, 61, 262, 266, 269, 271–2 (*see also* Geloffa, Jalofo, Wolof)
Jonadab, 132
Josephus, 204
Jugurtha/Jugurth, 40, 78

Lemos Coelho, Francisco de, 239, 242; *text* 291–311
Lowe, Henry, 24–5, 27, 29, 36–7, 186–7, 212

MacDonnell, Governor, 234
Mahome, 128, 134, 227–8
Mahomet, 126, 266 (*see also* Mohammed, Muhammed)
Mandimansa/Mandimança, 267–8, 279, 290, 296
Mandinga/Mandingo/Maudingo/s, 56, 82, 96–7, 100–101, 104, 236, 266, 271–2, 274–5, 277, 279–85, 288, 290, 293, 295–8 (*see also* Mandinka)
Mandinka/s, 34–5, 43–4, 46, 50–51, 56–9, 61–2, 64–9, 227, 229, 265–6, 268–71, 305 (*see also* Mandinga)
Marko, 159
Mohammed/Mafoma, 41, 268, 288–90 (*see also* Mahomet, Muhammed)
Mohammedans, 296, 303, 310
Molatos/Molatoes, 17, 70, 80, 97
Molhughes, Richard (?), 24
Moor/s/Moore/s/Moorish, 21, 23, 44, 78–80, 88, 94, 137, 142, 148–9, 181, 190, 195, 201, 226, 232, 252–3, 264, 269, 272, 279, 282
Mormelli, 261 (*see also* Bormelli)
Moses, 122, 126, 132, 203
Mosto, Luys de, 266 (*see also* Cadamosto)
Muhammad, 253, 263 (*see also* Mahomet, Mohammed)
Musgrove, Hugh, 24

Nicholas, 25, 187
Noah, 122, 203
Nomimans/Nomymans, 263–4
Numez, Hector, 185
Nuno, 263
Nyoominka, 68

Pacheco Pereira, Duarte, 239, 241–2, 265; *text* 265–6
Pangloss, 42
Park, Mungo, 51, 60, 63, 235
Pholey, 102 (*and see* Fula, Fulbe)
Portingale/Portingal/Portingall/s, 17, 19, 20, 34, 69–70, 79–80, 82, 94, 97–100, 110, 115, 135, 147, 159, 160, 170, 182 (*see also* Portugals, Portuguese)
Portugals/Portugalls, 11, 12, 70–71, 159, 185–8, 196–7, 213 (*see also* Portingales, Portuguese)
Portuguese, 4–10, 12, 18, 23, 32–35, 37, 40–41, 43, 45, 47, 49–50, 52–6, 65, 67, 69–72, 207, 213, 220, 229, 231, 241–2, 244, 258, 266, 268, 270–71, 273, 284, 291, 294–5, 299, 303, 306–7, 309–11 (*see also* Portingales, Portugals)
Prester John, 257
Ptolemy, 271
Purchas/Purchus, Samuel, 1–4, 20, 27, 35, 77, 185, 207, 224–5, 229

Rechabites, 132
Roderigo, Bastian, 186
Romans, 78

Sambegeny, 260
Samgully/Samgulley, 68, 136, 153–5, 215 (*see also* Sanguli)
  joined Thompson's party, 136; hired by Jobson, 136, 188–9; interprets at Tinda, 34–5, 143, 195; circumcised, 154–6, 195–6
Sandie, 94
Sanguli/Sangully, 28, 30, 33–6, 68, 188 (*see also* Samgully)
Sanneh, L. O., 63
Saquedo, Nicolão, 303
Saracen, 211, 220
Selyman, 136
Semanagu, 260
Serahuli, 61, 68
Serer/Serreos, 67–8, 264
Solinto/Zolinto, 294
Solomon, King, 3, 73, 199, 201–3
Soninke, 57–9, 61, 64
Soto de Cassa, Abbot of, 264
Spaniard/s/Spanish, 11, 43, 97, 110, 119, 176, 200
St John, Sir Oliver 39
St/Saint John, Sir William, 39, 185
Stibbs, Bartolomew, 47–55

333

Sula Moros, 61
Sumaway/Summaway, 120, 188, 215, 229
Summa Tumba, 188

Thompson/Tompson/Tomson, George, 17–25, 27–9, 33, 35–6, 45, 50, 54, 70, 79–81, 136–7, 147–8, 153, 166, 186, 188–9, 195, 223, 227–9
    commands 1618 voyage, 79–80; defended at middle river base by Ferambra, 147; travels up-river to Tinda, 80–81, 132; tries to contact Buckor Sano, 137; sends marabout to inquire about the gold trade, 148; returning, is murdered 81

Tristão, Nuno, 241
Tucuraes, 271

Usodimare, Antoniotto, 239, 241, 256 (*see also* Antoniotto); *text* 256–8

Vaz, Gaspar, 288
Vermuyden, John, 48, 55
Vivaldi, 257

Wali, 58–61, 68
Wolof, 51, 56, 61, 67–8, 267 (*see also* Geloffa, Jalofo, Jolof)

PLACES

Note that titles ending in -mansa are at times used to denote localities. The following toponym, occurring frequently, is not listed: Gambia.

Africa/Affrica, 1, 6, 7, 9, 13, 24, 43, 65, 77–8, 177, 200, 202, 231, 264–5, 267, 271, 309
Albafur, 260
Alcacer dalquivi [Al Ksar el Kebir], 264
Alcuzet, 262
Aldabar, 293
Aleá, Rio de 303
Alentejo, 268
America, 43, 200, 202, 232
Animaijs [Niani], 259 (*see also* Nany, Naoy, Niani)
Arguim, 272
Arse Hill, 52

Baarakunda/Baraconda Baraconde/Baracunda, 28, 34, 46, 48–9, 52–5, 67, 71, 85, 89, 136, 138, 189, 195, 207, 215–16, 219, 224–5, 229, 305–6
Badibo/Badibu, 56, 296–8
Balangar, 280
Bambaro, 283
Bambuk, 62
Bamgú/Bangu/Banjú, 293
Benanko, 53, 189, 215
Baniserile [Bani Israel], 63
Bantunding, 226
Baraconda Falls/Barrakunda Falls, 53, 60
Barbados, 294
Barbaciis, Rio dos, 265

Barbary/Barbaries, 17, 21, 23, 44, 78–80, 88, 137, 142, 148–9, 181, 201
Barra, 50, 285, 292–3, 298–9
Basse, iv, 52
Bathurst, 234–5
Batimansa, 256, 261
Batto, 28, 188, 207, 214, 219, 224, 229
Bereck, 188, 215, 224–5, 229 (*see also* Perai)
Berefete, 285–6, 294
Beretenda, 229
Bintam/Bintang, 67, 294–7
Boaguer, 295
Bohemia, 267
Bojangkunda, 229
Bondou/Bundu, 62, 234–6
Borçalo/Broçalo, 274, 292–3, 298 (*see also* Bursall, Saalum)
Borsall, River of, 185, 208 (*see also* Saalum River)
Bottom Struck, 284, 289–91 (*see also* Findefeto, Fundo Feito)
Bunhacú, 302
Bunting, Plate VIII
Bursall, 55–6, 67–8, 111–12, 121, 134, 187, 197 (*see also* Borçalo, Saalum)

Cabaçeira, Porto da, 297
Cabo, 56, 189, 282, 284, 302, 305 (*see also* Guabuu, Kaabu)

INDEX

Cabopa Island/Ilheo das Caboupas, 283, 288, 299
Caçan/Cação, 275, 296, 299–301, 309 (*see also* Casan, Kasang)
Cacheo/Cacho, 170, 293, 295, 300, 310
Cairo, 260, 272
Camboya, 231
Cameroons, 65
Canary Island, 82, 208
Canhobé, 304
Cantor/Cantore, 5, 11, 50, 52, 55–6, 69, 111, 120, 188–9, 192, 215, 217, 225, 227, 229, 241, 258–62, 265, 267, 270–74, 284–5, 289–90, 302, 305, 309 (*see also* Kantora)
Cape St Mary, 270, 284
Cape Verde/de Verde, 4, 5, 8, 19–20, 49, 69, 99, 207, 241–2, 249, 258, 268, 271, 273–4, 284, 291
Cape Verde Islands, 5, 8, 49, 69, 242, 268, 273–74, 284, 291
Carthage, 260
Casan/Cassan, 11, 17, 44, 51, 93, 108, 120, 187, 197, 207, 211–12, 220, 231, 283–4, 288–90 (*see also* Caçan, Kasang)
Caur, 287, 298, 302 (*see also* Kau-ur)
Cereculle, 261
Codichar, 289
Çofala, 273 (*see also* Sofala)
Color, 229 (*see also* Kulari)
Combo/Cumbo, 55–6, 197, 221, 285, 293–4, 298 (*see also* Kombo)
Combomansa, 284
Commuberta. 261
Conicomco, 301
Cudan, Rio de, 298
Culenho, 287

Dartmouth, 24, 82, 106, 185, 200, 208
Degumasamsam, 297–8
Deponga, Rio de, 306
Devil's Turn, 50, 286
Dia/Ja, 62
Diakha/Jakha, 62
Dobancoo, 265
Dover, 14, 200, 234–5, 237

Elephant Island/Isle, 50, 211, 272, 286, 297
Emcalhor [Kayor], 266
Emin, [River], 260
England, 2, 6, 10, 22–5, 27, 30, 37, 40, 44, 82, 101, 154, 163, 166, 170, 177, 179, 199–200, 207, 221, 232, 294

Ethiopia, 77–8, 291

Fancassa, 131
Fatatenda/Fattatenda, 52, 226, 229–30, 234, 30
Fatoto, Plate IV
Fay, 149 (*see also* Jaye)
Faye, 159, 228
Felam, Rio de/Ilha de, 292
Fez, 260, 272
Findefeto/Findifeto, 226, 304 (*see also* Bottom Struck, Fundo Feito)
Fula Kunda, 228
Fundo Feito, 284 (*see also* Bottom Struck, Findefeto)
Fuuta Jalon, 67 (*see also* Gelu, Mount Gelu, Serra Geley)

Gago [Gao], 87–8, 149, 195, 278
Galalho, 278
Ganjal, 274
Gelu, 262 (*see also* Fuuta Jalon, Mount Gelu, Serra Geley)
Germany, 267
Gravesend, 185, 208
Grey River, 55, 235
Guian, 286
Guinea, 4–6, 8, 13, 16, 18, 28, 30–31, 39–41, 57, 69, 231, 237, 241–2, 244, 256, 258, 260, 264, 267–77, 279, 282–4, 287, 290–91, 299, 308–10
Guabuu, 266 (*see also* Cabo, Kaabu)

Herejes, Aldea dos, 295

India/Indian, 262, 277–8, 293
Ireland, 38–40, 118, 199

Jagra, 286, 297, 302
Jagrançura, 277
Jalacuna, 303–4
Jalancoo, 265
Jalofo Coast, 289
Jambor, 286
James Island, 49, 209, 294
Jamnamsura, 265
Janguemangue, 282
Jasabo, 297
Jaye, 22, 87, 148, 195 (*see also* Fay)
Jelicot, 44, 56, 145, 192, 219
Jeraconde, 56, 187, 213, 224–6
Joala, 12

335

Joba, Rio de, 295–6
Jubander, 274
Julacote, 286
Julufré, 293–5
Jurume, 301

Kaabu, 56, 58, 60, 266 (*see also* Cabo, Guabuu)
Kantora, 55, 59, 227–8 (*see also* Cantor)
Karantabaa, 52, 224–5
Kasang, 17, 19–21, 24–5, 27–8, 34, 50–52, 54, 56, 61, 67–8, 70–71, 207, 211–12, 220–21, 223–4, 227, 241, 284, 291 (*see also* Cação, Casan)
Kau-ur, 51, Plate II (*see also* Caur)
Kiam/Kiang, xvi, 56 (*see also* Quiam)
Kiriwani/Kerewaan, 63, Plate VI
Koina, 53
Kombo, 50, 221 (*see also* Combo)
Kukia, 260, 262 (*see also* Quioquum)
Kulari, 228–9 (*see also* Color)
Kumbija, Plate V
Kundam, PlateVII
Kuntaur, 51
Kwinela, xvi

Lagoa, 285
Lagos [Portugal], 258
Lagos, Rio de, 274
Lame/Lamé, 281, 289, 301
Launcerot, 82, 208
Lequio [Japan], 308
Libya, 271–2
Limbambulu, 226, 228
Little Coast [of Senegal], 5, 6, 9, 11, 20, 69, 207–8
London, 6, 14, 16, 38, 75, 78, 92, 142, 182, 189

MacCarthy's Island, 234, 236
Madina, 60
Madrid, 273
Malagueta [Coast], 231
Mali, 57, 58, 263 (*see also* Melli)
Malor, 277, 282, 301
Mamayungebi [Hill], 52
Mangegar/Manjagar/Manjaguar/Maugegar, 25, 33, 36, 51, 56, 70, 162, 186, 197, 207–8, 211, 221, 298, 302
Maresamsam, 302
Mecca, 291
Medina, 63, 226
Melli, 251, 257 (*see also* Mali)

Mina, 5, 269, 272, 279
Mondego, 272
Monkey Court, 51
Morocco, 6, 78, 267, 272
Mount Gelu, 260 (*see also* Fuuta Jalon, Gelu, Serra Geley)
Mumbar/Mumbarre, 87, 142
Munster, 38–9
Muscovie/Muscovy, 184

Nanhijaga, 298
Nanhimargo/Niani Maru, 51, 298–9
Nany [Niani], 20, 55, 147 (*see also* Animaijs, Naoy, Niani)
Naoy [Niani], 188 (*see also* Animaijs, Nany, Niani)
Nhacobé, 304
Nhacoi, 304–5
Nhamena, 298
Nhamenhacunda, 304
Niani/Nhani, 20, 51, 55–6, 67–8, 226, 228, 298, 300–302, 305 (*see also* Animaijs, Nany, Naoy)
Niérico/Nieriko/Nylarico, River, 55, 235
Niger, River, 57, 62
Niguer, 271
Niokolo Koba, River, 47(n.1), 48(n.1)
Niumi, 50, 56, 67
Njau, Plates X, XIII
Nunez, River, 235

Ola Ola, 287 (*see also* Oulaoula, Wolley Wolley)
Oli/Olimansa/Ollimansa/Uli, 261, 282, 305 (*see also* Ulimaijs, Wolley, Wuli)
Ophir/Ophyr, 3, 202
Oran, 260, 272
Oranto, 19–20, 25, 32, 56, 187–8, 207, 213–14, 223–30
Oulaoula, 283 (*see also* Ola Ola, Wolley Wolley,)

Perai/Pereck/Pirai, 56, 189, 215, 224–5, 229, 305
Perifo/Peripho, 302, 304
Peru, 290
Pinto's Island, 287–8
Pisania, 51
Pompetan/Pompetane/Pompeton, 33, 52, 71, 159, 187, 197, 207, 213, 220, 224–7
Ponor, 303
Port of Palm-trees, 287
Porto d'Ally [Portudal], 12
Portugal, 69–71, 159, 185–7, 196–7, 213, 248, 252, 255–7, 264, 266, 272, 292, 299–301, 305–6, 308, 310

INDEX

Pudding Island, 50, 211

Quiam, 296–7 (*see also* Kiam)
Quioquum, 260 (*see also* Kukia)

Red Hill, 288
Red Range, 287
Rio Grande, 276, 280–82, 290–91
Rio Grande de Bonabo, 291
Rocha do Ouro, 299
Rock of Nicolão Saquedo, 303
Rouen, 293
Rufisque, 208, 221 (*see also* Travisco)

Saalum, 5, 49, 55, 61, 67–8, 70, 208 (*see also* Borçalo, Bursall)
Saalum, River, 5, 49, 67–8, 70, 208, Plate I (*see also* Borsall, River of)
Saare Mansajang, Plate IX
Sahara, 6
Saint John's Mart, 150, 195
Samatra, 308
Same Tenda, 52
Sanaga [Senegal], Rio de, 306
Sangedegú/Sangue de Gú, 295
Sangedugu Creek, 52
São Domingos, Rio de, 280, 282
Sapugo, 301
Senegal, 5–6, 9–13, 16, 20, 40, 53, 55, 65, 67, 69–70, 207–8, 235, 241, 271–3, 291
Senegambia, 9–10, 12, 41, 70, 258, 271
Serra Geley, 260 (*see also* Fuuta Jalon, Gelu, Mount Gelu)
Serra Leoa/Lyoa, 260, 277, 282 (*see also* Sierra Leone)
Setico, 53, 61, 64, 71, 97, 99, 123–5, 134–6, 148, 196, 207, 220, 227–8 (*see also* Sutuko)
Sheba, 202
Sherbro, 16
Sierra Leone, 13, 16, 235, 260, 267, 270, 272, 274, 290–91 (*see also* Serra Leoa)
Sincu Faye, 228
Sitato, 297
Sofala, 278 (*see also* Çofala)
Somanda, 261
Spain/Spaine, 200, 249, 255
St Andrew's Island, 285
Sumacunda, 304
Sume, 305
Sutuko/Sutucoo/Sutukobaa, 24, 26–7, 51–2, 61, 64, 71, 220, 223–9, 265–6 (*see also* Setico)

Tagamdaba/Tagamdabá, 296–7 (*see also* Tendabaa, Tindobauge)
Tagus, [River], 272
Tambakunda, 60
Tambucutu/Tumbocutum/Tunbuqutum, 260, 262, 273, 278–9 (*see also* Timbuktu, Tombutto)
Tancoroale/Tankular/Taukorovalle, 50, 70, 196, 210, 296–7, 311
Tendabaa, 23–5, 33, 50–51, 56, 70, 207, 210 (*see also* Tagamdaba, Tindobauge)
Termezen, 260
Thames, [River], 184, 200
Timbuktu, 260, 262, 272 (*see also* Tambucutu, Tombutto)
Tinda/Tenda, 22–4, 36, 44, 50, 54–5, 68, 81, 136–7, 188, 190–92, 195–6, 207, 214, 217–20, 228, 235
Tinda River, 55, 191, 207
Tindobauge, 50, 186, 207, 210 (*see also* Tagamdaba, Tendabaa)
Tlemcen, 272
Tobabo Conda/Condo, 128, 147, 227
Tombutto, 149 (*see also* Tambucutu, Timbuktu)
Trading-place for Gold, 271–2
Travisco, 185, 197, 208, 221 (*see also* Rufisque)
Tuba Kuta Creek, 52, 225–6, 230
Tubabo Sita, 306
Tubabó-colom, 297
Tunis, 260, 272

Ulimaijs/Ulimança, 259, 305 (*see also* Oli, Wolley, Wuli)
Ulster, 38

Venice/Venetian, 244, 267, 277

West Africa, 1, 6, 9, 16, 65, 241, 265, 271
West India/Indies, 97, 166, 234
Wolley [Wuli], 55–6, 67, 70, 111, 187, 189, 192, 197, 212, 216, 219, 221 (*see also* Oli, Ulimaijs, Wuli)
Wolley, Wolley/Wolly Wolly, 51, 68, 70, 187, 197, 212, 221 (*see also* Ola Ola, Oulaoula)
Wolof Panchang, Plate III
Wuli, 55, 58–61, 67–8, 224, 226, 282 (*see also* Wolley, Ulimaijs)

Yabbatenda, 234
Yambor, 273

Zaza, 265

SELECT SUBJECTS

The modern form of spelling is given, except in instances where the earlier variant spellings are not easy to relate to the modern form. The following terms, occurring frequently, are not listed: boat, king, house, ship, trade. *Note the following general listings:* Animals, Birds, Dress, Fish, Occupations, Plants, Religion, Shipping and Navigation, Trade Commodities, Weapons.

Alcade/Alcaid/Alcaide/Alchade, 28, 59, 123–4, 129, 132, 136, 145–6, 154, 186–7, 192, 212, 219

ale, 168

ANIMALS

antelope, 175, 270; ant, 108; ape, 201–2, 252; ass, 64, 115, 125, 134–5, 137, 141, 196, 265, 269; baboon, 33, 51, 169, 189, 202, 216, 252; bat, 256; beafe/beefe/beeves/biefe, 29, 66, 90, 96, 100–101, 119, 129, 136, 138, 156, 189, 191–2, 196; bee, 169–70, 275, 280; boar, 174; buffalo, 174, 270, 275, 309; camel, 78, 142, 260; cattle, 61, 64–7, 90, 95, 100–101, 103–4, 125, 144, 156, 163, 172, 174, 272, 275–6, Plate IX; civet/civet cat, 172, 203, 253, 257, 270; cow, 256, 266, 269–70, 276–7, 280, 294, 300, 309; crocodile, 33, 45, 87, 90–91, 93–5, 101, 145, 180, 189–91, 217, 256, 270, 272, 276, 287, 309; deer, 95, 108, 138, 144, 161, 173, 175–6, 179–80, 190–91, 237, 275, 309; dog, 31, 172, 254, 270; donkey, 280, 306, 309; elephant, 2, 29, 33, 45, 68, 172–4, 180, 188, 190, 195, 202, 214, 217, 235, 254–7, 262, 270, 272, 275, 279, 286, 297, 309; fish-horse [hippopotamus], 270; gazelle, 270, 309; goat, 29, 96, 100, 129, 138, 161, 191, 196, 269, 270, 300, 309; hare, 270, 281; hippopotamus, 30, 45, 53, 87, 215–16, 220, 270, 272, 286, 309 (*see also* fish-horse, river-horse, sea horse, water horse); horse, 19, 61, 68, 91–3, 99, 101, 108, 110, 112, 114–15, 159, 189, 227, 255–6, 261, 264, 266, 268–70, 272, 276, 278, 281, 309; hyena, 309; jackall, 33; leopard, 70, 172, 262, 270, 272, 309; lion, 42, 101, 110, 171–2, 175, 309; lizard, 266, 272; monkey, 30, 51, 169, 175, 177, 179, 189, 202, 216, 268, 281; ounce, 101, 172, 186; 275; pig, 255, 264, 270, 272, 309; porcupine, 172; rabbit, 270; ram, 130, 196, 264, 290; rhinoceros, 232; river-horse [hippopotamus], 237; sea horse/sea-horse [hippopotamus], 87, 91–3, 189, 191, 196, 272; serpent, 165; sheep, 130, 264, 269, 300; snake, 91, 98, 165, 270, 272, 309; stag, 138, 173–4, 191; swine, 169, 254; tiger, 309; unicorn, 42, 176, 257; water horse/water-horse [hippopotamus], 266, 281; wolf, 309; worm, 165, 170

anthill, 33, 108

baize, 277

ballard, 151, 155, 195 (*see also* xylophone)

bandora, 151

beast, 42, 64, 83, 90–92, 101, 104, 108, 110, 125, 133, 135, 138, 141, 144, 146, 156, 161, 163, 171–6, 180, 190, 192, 196, 201–2, 266, 276

bed, 31, 124, 139, 162, 172

bedmate, 32

bee-hive, 270

beer, 29

betel, 282

BIRDS

African hen, 309; bird of Arabia, 178; bird with four wings, 178; bird without legs, 178; buzzard, 180; curlew, 95; crane, 30, 275; dove, 178, 275; duck, 95, 266, 275, 309; eagle, 180, 191; falcon, 264; flamingo, 275; forest hen, 308; fowl, 29–30, 138, 145, 160, 176–80, 190–91, 201, 232, 237, 264, 309; Ginney/Guinea fowl/hen, 30, 104, 177, 327; goose, 95, 264; hawk, 179–80; hen, 29, 96, 104–5, 125, 129, 138, 156, 159, 177, 180, 191, 288, 290, 295, 299, 309; herne, 95; heron, 275; jerfauchon, 179; kite, 180; mallard, 95; parakeet, 178; parrot, 179, 256–7, 262, 268; partridge, 29, 104, 177, 180; peacock, 30, 201–2, 309; pigeon, 178; plover, 95; poultry, 156, 174; quail, 178; stalker, 138, 177, 191; stork, 95; teal, 275; wake, 177; weaverbird, 33

bolombato, 151 (n.1), Plate XI

book, 1, 2, 4–7, 16, 25, 30–31, 34, 36, 39–41, 43, 45, 66, 70, 78, 82, 126–7, 131, 133, 185, 201, 203, 207, 227, 229, 277, Plate IV

338

INDEX

bowl, 152, 279, 289, 308
brass, 142, 266, 268, 278
butter, 29, 103, 288, 300
candle, 30, 92–3, 146
caravan, 22, 60, 63–4, 69, 71, 88, 137, 187, 236, 260, 272, 277–9, 292, 305–8
child, 31, 35, 41, 63, 97–8, 114, 121–2, 126–8, 132, 155, 172, 178, 203, 257
commerce, 5, 67, 69–70, 83, 99, 131, 145, 154, 158, 192, 199–202, 236–7, 291
compass, 30, 38, 110, 194, 235
copper, 273, 277–8
crafts, 161, 269
cup, 29, 105, 122, 131, 139, 145, 150, 192
customs, 38, 186, 210, 267
discovery, 1, 4, 6, 8, 14, 21–2, 28, 35, 39, 41, 73, 75, 78, 80–81, 90, 95, 142, 150, 172, 182–3, 185, 199, 204, 235

DRESS
breeches, 30, 113, 144, 155, 192, 268–9; cap, 30, 155, 247, 266, 279, 283; cape, 277, 296; clothes/clothing, 30, 44, 118, 131, 138, 140, 147, 155, 165, 168, 194, 251, 276–7, 279, 287; gown, 30, 139; handkerchief, 254; hat, 266, 277, 296, 308; mantle, 259, 268; necklace, 259; pantaloons, 266; shirt, 30, 113, 146, 155, 194, 296; shoe, 27, 113, 287; surplice, 113; smock, 247, 251, 259, 266; surcoat, 252; tippet, 277; trousers, 279

drum/drumming, 110, 151, 154, 160, 195–6
dry season, 46, 66, 234–5
earth, 21, 77, 107, 129–30, 146–7, 150, 160–66, 171, 194, 196, 202–3, 261, 264, 268, 276, 300
emulation, 23, 35–6, 81, 188, 195
farm/farming, 59, 62, 65–6, 228
female, 32, 42, 93, 158, 175, 196, 219, 262, 268, 299–300
fish/fishing, 25, 29, 68–9, 91, 93–5, 104–5, 145, 190, 232, 256, 270, 272, 276, 292, 299, 309, Plate II

FISH
eel, 33; mullet, 93, 292; plaice, 276

fruit, 96, 132, 160, 166, 169, 204, 253, 269–70, 274, 277, 282–3, 288, 308
genitalia, 43–4
globe, 30, 38, 194
gold mine, 232, 261
golden, 1, 3, 16, 73, 75–9, 82, 95, 131, 170, 181, 188, 200, 202
grain, 66, 83, 96, 104–5, 116, 125, 129, 162–3
Guinea Company, 16, 31, 39

headache, 282
honey, 169–70, 270, 275
hook, 29, 91, 145
hunting, 33, 68, 93, 171, 227, 235, 254, 264, 270, 309
irons, 151, 163
kiss, 43, 116
knife, 103, 156, 196, 232, 275, 279
leather, 44, 58, 110, 113, 161, 195, 247, 277–8, 289
lute, 30, 153
manuscript, 1–2, 131, 133, 203, 242, 267
market, 66, 140–41, 162, 168, 197, 219, 221, 235, 272, 288, 306
milk, 29, 66–7, 103, 250, 271, 283, 288, 299
mine, 43, 97, 232, 260–61, 299
music, 31, 33, 110, 138, 145–6, 150–51, 153–5, 192, 195
noble, 61, 261, 263, 300
nut, 30, 134, 146, 169–70, 181, 192, 271, 282

OCCUPATIONS
bishop, 263, 277, 296; blacksmith/smith, 58, 61–2, 65, 160–61, Plate X; carpenter, 26, 197, 221, 299; chirurgeon, 25, 106, 154, 156; cowman, 33; craftsman, 44, 289; drummer, 65, Plate XIII; factor, 16, 24–5, 27, 36, 39, 60, 106, 120–21, 221, 236, 290, 294; farmer, 58; fiddler, 83, 110, 145–6, 150, 152, 159, 192, Plate XII; fisherman, 276; goldsmith, 232; herdsman, Plate IX; Joulah, 236–7; Juddy, 83, 150, 153, 159, 192; leatherworker, 61–2; marabout, 22, 27–30, 34–6, 40, 43–5, 53, 57–9, 62–4, 72, 82, 214–16, 220, 226, 228 (see also Marybucke); marchant, 39, 199, 201–2 (see also merchant); Marybucke, 44–5, 58, 62–3, 82, 90, 94, 112–13, 122–4, 126–9, 130–36, 146–8, 154, 158–9, 162, 188–9, 194–6, 203, 227–8 (see also marabout); merchant, 5, 6, 10, 13, 15, 22, 24, 28, 31, 36–9, 44–5, 60, 68, 218, 220, 234–7, 256, 260, 266–7, 270, 272, 277–9, 306, 308 (see also marchant); musician, 30, 58, 61; potter, 161; priest, 42, 90, 112, 122–5, 130, 136, 162, 189, 196, 264, 269, 277, 289–90, 310; sailor, 25, 32, 37, 248–50, 253, 282, 286, 294; seaman, 18, 21, 25–7, 37–8, 106, 165, 182–3, 202; silversmith, 280; trader, 8, 24–6, 32, 36, 49–51, 55, 60–64, 69–72, 207, 236, 242, 244, 248, 274, 277, 284, 290–91, 294, 298, 302, 304, 310

odour, 202, 203
oil, 269–271

palm/palmeta wine, 29, 166, 173, 232, 270
pipe, 161, 162
PLANTS
  apple, 168–9; barbery, 170; banana, 29, 96; *cabopa*, 282–3, 288; cola/*gola*, 30, 134, 169, 192, 274, 277, 282, 288–9, 293, 298–9; corn, 38, 59, 92, 178, 196; cotton, 30, 44, 57, 62, 110, 113, 118, 140, 164, 173, 191, 247, 251–2, 254, 266, 268–9, 273, 275, 282–3, 287, 289, 295, 304–5, 308; cotton-tree, 295, 304–5, 308; date, 21, 48, 226, 253, 271, 284; dyewood, 13, 15; garlic, 170; gourd, 103, 116, 122, 128, 151–2, 162, 166, 168, 173, 270, 286; lemon, 166, 262; lime, 166; mangrove, 46, 50–51, 274, 278, 286, 297–9; mimosa, 33; mustard, 163; onion, 170; orange, 29; orras, 29, 129; palm/palmtree/palm-tree, 262, 271, 287, 289, 301; plantain, 96, 166; pumpion, 168; reed, 105, 107–9, 122, 139–40, 143, 146, 154, 169, 173–4, 179–80, 262, 279; rice, 11, 29, 58–9, 92, 105, 116, 129, 163, 177, 196, 253, 269–71, 273, 275, 281, 283–5, 288, 293–4, 297–9, 308; samphire, 170; senna, 283; sugar, 132, 279; tamarind, 283, 304; water-reed, 262, 311; yam, 268, 270
poison/poisoned, 21, 42, 109–10, 160, 165, 174, 246–7, 254, 257, 262, 275, 309
pole star, 257
port, 19–20, 47, 51–4, 67, 69, 71–2, 108, 124, 129, 134, 153, 213, 224–9, 265, 273, 275, 277, 283–90, 292–306, 308–10
provisions, 32, 107, 136, 138, 158, 207, 209, 237, 257
rainy season, 21, 45, 52, 66, 234–5, 298, 303 (*see also* wet season, winter)
RELIGION
  ablutions, 33, 43; amulet, 41, 61–3, 290 (*see also* gregory); Bible, 40, 122; Christianity, 40, 64; circumcision, 24, 28, 33, 35, 42, 59, 62, 83, 126, 153–5, 157–8, 195–6, 207, 215, 219, Plates VII, VIII; Devil/devil, 41, 50, 83, 150, 153, 155, 157, 159–60, 197, 277, 286, 303; gregory, 113–14, 121–2, 133, 161 (*see also* amulet); Hore/Ho-re, 41–2, 45, 83, 143, 150, 153, 157–9, 196; idolatry/idolaters, 253, 269; Islam/Islamic, 6, 24, 28, 31, 33, 40–43, 57–9, 61–2, 65, 68, 72, 207, 220, 296; Leviticall law, 122, 203; Old Testament, 40, 42–3, 64, 122; Protestant, 40; psalm, 31, 40, 194; Roman Catholic, 38, 40–41; teaching house, Plate V
ring, 30, 135, 151, 194, 201, 232, 252, 277, 280

rosa solis, 29, 145
sack, 29, 106, 122
sexual service, 32
SHIPPING AND NAVIGATION
  canoe, 29, 46, 49, 54, 68–9, 86, 94, 129, 144–5, 189, 214, 216, 245–8, 250–51, 253, 258, 260, 263, 281–3, 286–7, 289, 292, 296–7, 303; caravel, 245–6, 248, 251, 254, 256, 258–9, 261–5, 267, 305, 306; Catherine, the, 17–18, 79; crew, 10, 25–6, 50, 221, 241, 247–8, 275, 286, 306; oar, 14, 22, 54, 80, 137, 189, 246–7, 251, 253, 301, 303; pinnace, 21, 24–5, 48, 53, 80–81, 106, 121, 208, 220; sail, 244, 247, 251, 259, 285, 287; Saint John, the, 21, 24–7, 34, 37, 80–81, 150, 165, 185–6, 196–7, 207–8, 211–13, 220–21, 227; shallop, 14, 17–18, 25, 27, 29–30, 34, 37, 48–9, 53, 79, 86, 106, 186, 188, 197, 200, 209, 211–14, 221, 228–9; Sion, the, 24–6, 34, 36–7, 185–6, 197, 207–8, 210, 220, 221; tide, 30, 46, 50, 53, 93, 153, 186, 188–9, 214, 224, 229, 246, 258, 265, 273, 280, 283, 286–7, 289, 293, 298, 301–3; wind, 46, 49–52, 86, 108, 151, 164, 186, 195, 245–6, 249, 258, 280, 284, 287, 289, 300
silk, 252, 254, 266, 278, 308
straw, 127, 172, 227, 262, 269–70, 275
sun, 30, 49, 105, 107–8, 140–41, 149, 153, 155, 164–6, 171, 176, 189, 232, 250, 266, 269, 292, 296
swamp, 46, 52, 58–9, 66, 285, 287, 297, Plate II
tablet, 296
thunder, 83, 86, 138, 144, 160, 164, 166, 191, 197, 201, 248, 250
TRADE COMMODITIES
  ambergris, 203; basin, 277–8; bead, 30, 268, 277, 293, 306, 308; blew stone, 135, 152; bracelet, 142, 149, 266, 268, 273, 277–8, 280; brandy, 29, 106, 293, 303; cloth, 15, 30, 57, 60, 62, 109, 118–19, 155–6, 172, 174, 191, 194, 232, 252, 257, 259, 266–9, 271, 273, 275, 277–9, 282–3, 287, 289, 293, 295, 298–9, 301, 306–8; coin, 162, 255, 277, 279, 293–4; coral, 30, 145–6, 192, 194; cowry, 277; crystal, 30, 145, 192, 194; elephant's tooth, 11, 125, 140, 161, 191, 202 (*see also* ivory, teeth, tusk); gold, 1, 5, 6, 11, 16, 22, 24, 30, 35–6, 46, 64, 69, 71, 78, 88, 97, 128, 133–5, 140–41, 143–4, 148–9, 170, 191–2, 194–6, 201–3, 207, 219–20, 231–2, 235, 245, 252, 257–61, 266–8, 270–74, 277–80, 282–4, 289, 293–4, 298–9, 306, 308, 310; gum, 9, 15, 232, 308; hides, 6, 9, 11, 22,

30, 44, 96, 139–40, 144, 161, 175, 191–2, 268, 289, 293–5, 297–9, 301–5, 309–10; iron, 29, 105, 109, 139, 151–2, 160–61, 163, 174, 194–5, 201, 270, 275, 280, 293–5; ivory, 9, 30, 44, 201–2, 232, 273–5, 283–4, 289, 293–5, 297–9, 301, 303–8, 310 (*see also* elephant's tooth, teeth, tusk); liquor, 29, 37, 107, 122, 131–2, 147, 168, 188, 303; mat, 112, 116, 121, 124, 141, 145–6, 155, 162, 172, 192, 194, 268, 289; musk/musky, 91, 190, 203, 253, 257, 268; paper, 23, 29–30, 81, 126–7, 133, 143, 161, 204, 277–8, 290, 293, 306, 308; salt, 22, 29, 36, 46, 49, 51, 57, 60, 67–9, 88, 103, 134, 137, 139–40, 142, 144, 148–50, 187–8, 191–2, 195–6, 219, 232, 263, 273, 279, 282–3, 285, 292–3, 296–7, 301–2, 304, 306–8, Plate I; silver, 30, 145, 192, 201, 232, 278, 280; slave, 31, 43–4, 57–8, 61–3, 68–9, 71, 97, 120, 122, 128, 134, 140, 187, 191, 196, 219, 232, 252, 257, 268, 273–7, 282–3, 295, 300, 305; spice, 29, 245; teeth, 11, 49, 92, 140, 161, 169, 173–4, 191–2, 202, 262, 281 (*see also* elephant's tooth, ivory, tusk); timber, 13, 196, 283, 286, 288, 299–300; tobacco, 144, 161–2, 164, 169; tusk, 173–4, 189, 255–7, 266, 307–8 (*see also* elephant's tooth. Ivory, teeth); wax, 9, 11, 232, 270, 273–5, 283–4, 287, 289, 293–5, 297, 301; wine, 29, 31, 107, 132, 166–7, 169, 173, 232, 262, 264, 269–71, 277–8, 286, 303

vegetables, 250

war, 54, 58–9, 61, 78, 139, 226, 260, 262, 265, 285

WEAPONS

arrow, 90, 108–9, 131, 138, 145, 147, 160, 165, 192, 194, 246–8, 254, 257, 262, 270, 275–6, 279–80; artillery, 294, 306; assigie/azagaie/azagay, 160, 254, 262; bombard, 245, 247–8; bow, 30, 90, 101, 109–10, 113–14, 131, 138, 145, 147, 154, 178, 192, 194, 246–8, 254, 257, 268, 279; cannon, 30, 286–7, 289; dart, 108–9, 160, 275; fortress, 272, 276, 293–4; gun, 30, 61, 119, 138, 141, 144–5, 150, 158, 176, 191, 201, 278; musket, 30, 192, 197, 201, 232, 282, 286–7, 289; peece [gun], 30, 92, 95, 99, 124, 173, 190, 195, 201–2; pelletbow, 30, 178; powder, 30, 61, 92, 95, 138, 145, 277, 289; shield, 78, 247, 262, 279, 289; spear, 30, 202; stonebow, 30; sword, 30, 109, 131, 139, 141–2, 152, 160, 196, 262, 264, 278–9, 289

wet season, 46 (*see also* rainy season, winter)

white man/whiteman 31–2, 91, 128, 131, 138, 140, 143, 145, 149, 151, 159, 188, 192, 194, 199, 246, 251, 258, 305

white people, 131, 153, 181, 192, 200–201, 204

whites, 45, 54, 60, 214, 260, 268, 275, 283, 293–5, 297, 299, 303–6, 309–10

wife, 30, 32, 43, 98, 110, 114–16, 118, 121–4, 131–2, 142, 145–7, 152, 155, 170, 186, 188–9, 191–2, 203, 228, 263, 266, 268, 270, 277, 295, Plate VI

winter, 85, 250, 271, 275, 283, 285, 287–8, 303 (*see also* rainy season, wet season)

woman, 30–33, 41, 43–5, 59, 66–7, 70, 100–101, 103–5, 114–16, 118, 120–23, 125, 129–31, 135, 138, 140–42, 144, 147, 152, 155, 159, 162, 192, 194–5, 197, 203, 219, 253–4, 261, 263, 268–70, 288, 295–6, 299, 304

xylophone, 152 (*see also* bolombato)